COMPREHENSIVE CHROMATOGRAPHY IN COMBINATION WITH MASS SPECTROMETRY

WILEY-INTERSCIENCE SERIES IN MASS SPECTROMETRY

Series Editors

A complete list of the titles in this series appears at the end of this volume.

COMPREHENSIVE CHROMATOGRAPHY IN COMBINATION WITH MASS SPECTROMETRY

Edited by

LUIGI MONDELLO

A JOHN WILEY & SONS, INC., PUBLICATION

Published by John Wiley & Sons, Inc., Hoboken, New Jersey.
Published simultaneously in Canada.

Library of Congress Cataloging-in-Publication Data:

Comprehensive chromatography in combination with mass spectrometry / edited by Luigi Mondello
p. cm.
Includes Index.
ISBN 978-0-470-43407-9 (cloth)
1. Chromatographic analysis. 2. Multidimensional chromatography. I. Mondello, Luigi.
QD79.C4C66 2011
543′.8–dc22

2010036838

Printed in Singapore

oBook ISBN: 9781118003466
ePDF ISBN: 9781118003442
ePub ISBN: 9781118003459

10 9 8 7 6 5 4 3 2 1

CONTENTS

CONTRIBUTORS

Keith D. Bartle, University of Leeds, Leeds, UK

Leonid Blumberg, Fast GC Consulting, Hockessin, Delaware

Francesco Cacciola, Chromaleont s.r.l., A spin-off of the University of Messina, Messina, Italy and University of Messina, Messina, Italy

Paola Donato, University Campus Bio-Medico, Rome, Italy and University of Messina, Messina, Italy

Paola Dugo, University of Messina, Messina, Italy

Isabelle François, University of Gent, Gent, Belgium; currently at Waters NV/SA, Zellik, Belgium, Division of Waters Corporation, Milford, Massachusetts

Tadeusz Górecki, University of Waterloo, Waterloo, Ontario, Canada

Elizabeth M. Humston, University of Washington, Seattle, Washington

Pavel Jandera, University of Pardubice, Pardubice, Czech Republic

Philip J. Marriott, Monash University, Clayton, Victoria, Australia

Luigi Mondello, University of Messina, Messina, Italy

Ahmed Mostafa, University of Waterloo, Waterloo, Ontario, Canada

Samuel D. H. Poynter, University of Tasmania, Hobart, Tasmania, Australia

Koen Sandra, Metablys, Research Institute for Chromatography, Kortrijk, Belgium

Pat Sandra, University of Gent, Gent, Belgium

Danilo Sciarrone, University of Messina, Messina, Italy

John V. Seeley, Oakland University, Rochester, Michigan

Robert A. Shellie, University of Tasmania, Hobart, Tasmania, Australia

Robert E. Synovec, University of Washington, Seattle, Washington

Peter Q. Tranchida, University of Messina, Messina, Italy

PREFACE

Over the last half-century, single-column chromatography processes have been widely exploited for untangling constituents forming real-world samples. Many separation scientists are still acquainted with a single chromatography view, that is, the alignment of a series of peaks along a single, rather restricted separation axis. In many cases, one-dimensional separation spaces are enough for the isolation and detection of all the compounds of interest; however, in others, analysts must surrender themselves to an overwhelming antagonist: sample complexity.

In recent years, the great advances made in the field of instrumental analytical chemistry have made it increasingly apparent that natural or synthetic samples, characterized by hundreds, thousands, or even tens of thousands of constituents, are a common occurrence. In one-dimensional chromatography applications, the presence of tangled analytes at the column outlet is a frequent and undesired phenomenon. The most effective way to circumvent such an obstacle is to expand the separation space by using multiple analytical dimensions of a chromatographic and mass spectrometric (MS) nature.

The great analytical benefits provided by comprehensive chromatographic (CC) techniques have been exploited and emphasized by a constantly increasing part of the separation-science community during the last two decades. The term *well-known* has been stripped from a multitude of real-world samples, the true composition of which has been revealed through CC methodologies. The amount of separation space generated by current-day CC processes is unprecedented, making theses methods best suited for the unraveling of highly complex samples. The addition of a third mass spectrometric dimension to a comprehensive chromatography system generates a very powerful analytical tool: two selectively distinct chromatographic dimensions and a third mass-differentiating dimension.

A series of factors stimulated me to edit the present contribution, devoted to comprehensive two-dimensional chromatography in combination with mass spectrometry: first, and foremost, my personal excitement and passion for CC–MS technology, my main field of research; second, recent instrumental advances and the expanding popularity of CC–MS methods; and finally, and simply, the fact that there is still an immense wealth of information to be revealed on the composition of samples in all scientific fields of research.

Finally, I hope that this book will contribute to the promotion and development of CC–MS methods, which are still far from well established. Although I have been operating in the chromatographic world for quite some time, it is still very exciting for me to "play" with a CC–MS system, run a sample, and reveal its unsuspected complexity. In a way, CC–MS methodologies give us the pleasure to discover things for the first time.

Acknowledgments

As editor of the book, I would like to thank the many people who provided support; read, wrote, offered comments, and gave precious suggestions; and assisted in the editing and proof reading.

I am grateful to the authors for the considerable amount of work devoted to the preparation of the chapters, covering a variety of CC–MS aspects, ranging from historical aspects, to theoretical and optimization considerations, to pure applications, and on to hardware and software evolution.

Special thanks to Dr. Paola Donato and Prof. Peter Quinto Tranchida, for helping in the process of selection and editing, and to Prof. Giovanni Dugo, my father-in-law and mentor, who initiated me into this wonderful world of separation sciences.

Above all, I want to thank my wife, Paola, and my daughters, Alice and Viola, who supported and encouraged me in spite of all the time I was away from them.

LUIGI MONDELLO

1

INTRODUCTION

LUIGI MONDELLO AND PETER Q. TRANCHIDA
University of Messina, Messina, Italy
KEITH D. BARTLE
University of Leeds, Leeds, UK

The world surrounding us is characterized by an enormous number of heterogeneous samples, in terms of both complexity and chemical composition. Some mixtures, such as natural fats and oils (e.g., butter, olive oil) are composed of a relatively small number of constituents, whereas others, such as roasted coffee aroma, motor fuels, or protein hydrosylates, are highly complex. Chromatography is, without doubt, the technique of choice for the separation of a real-world sample either into a series of low-complexity subsamples or (ideally) into its individual constituents. To count the number of real-world samples that can be analyzed by using a chromatography technique is akin to counting the number of particles of sand on a beach.

The first use of a chromatographic method, as well as the employment of the word *chromatography*, were reported over a century ago (Tswett, 1906a,b). Prior to the invention of chromatography, target analyte separation from the rest of the matrix was achieved mainly through distillation, liquid extraction, and crystallization. Mikhail Semenovich Tswett, the inventor of chromatography, described the separation of plant pigments as follows: "Like light rays in the spectrum, the different components of a pigment mixture, obeying a law, are resolved on the calcium carbonate column and then can be qualitatively and quantitatively determined. I call such a preparation a chromatogram and the corresponding method

Comprehensive Chromatography in Combination with Mass Spectrometry, First Edition.
Edited by Luigi Mondello.

the chromatographic method." A variety of aspects related to liquid–solid chromatography were discussed in the two initial fundamental papers, suggesting the possibility of achieving *two-dimensional chromatography* by developing the column with another eluent after the primary separation (Ettre, 2003; Tswett, 1906a,b). As for many revolutionary inventions, the scientific community showed scepticism toward a technique that would have changed the world of analytical chemistry. It was not until the 1930 s that the potential of chromatography was to be fully appreciated, unfortunately long after the death of its first promoter.

1.1 TWO-DIMENSIONAL CHROMATOGRAPHY–MASS SPECTROMETRY: A 50-YEAR-OLD COMBINATION

Up until about half a century after the invention of chromatography, the structural elucidation of unknown analytes eluting from a chromatography column was a rather tedious task. Peak assignment was commonly achieved through two approaches: (1) by comparing the retention times of unknown analytes with those of known compounds, using two or more chromatographic columns with different selectivities; or (2) by collecting each chromatographic band eluting from the column and subjecting each fraction to an instrumental identification procedure (e.g., infrared spectroscopy, mass spectrometry). The unreliability of the first option is evident, due mainly to the limited peak capacity of the chromatography systems used in that historical period. Consider the difficulty, or better the impossibility, of qualitatively analyzing a 100-plus component sample on a packed column using only retention times! The second approach, and certainly more preferable, was commonly achieved by using a cold trap, but was characterized by the disadvantages related to excessive sample handling.

It was not until the end of the 1950s that the first online chromatography–mass spectrometry (MS) experiments were reported. In particular, Gohlke (1959) described a two-dimensional gas chromatography (GC)–time-of-flight (TOF) mass spectrometry (MS) system characterized by four parallel packed columns plus a thermal conductivity detector. The TOF MS employed enabled mass unit resolution up to a mass of 200 and generated 2000 spectra per second. A GC–TOF MS chromatogram for a five-compound mixture separated on a 10-ft-long packed column is illustrated in Figure 1.1. The outstanding results reported are certainly not diminished by the fact that probably not more than 50 compounds could have been identified reliably using the two-dimensional method. In fact, a rough visualization of the chromatogram suggests that not more than 15 peaks can be stacked side by side in a 20-min one-dimensional separation space. Furthermore, Gohlke affirmed that "single chromatographic peaks containing two or three components can usually be successfully resolved by a careful examination of several mass spectra obtained at various times during the development of the chromatographic peak." This was a very interesting statement; clearly, Gohlke fully comprehended the complementary nature of the two analytical dimensions.

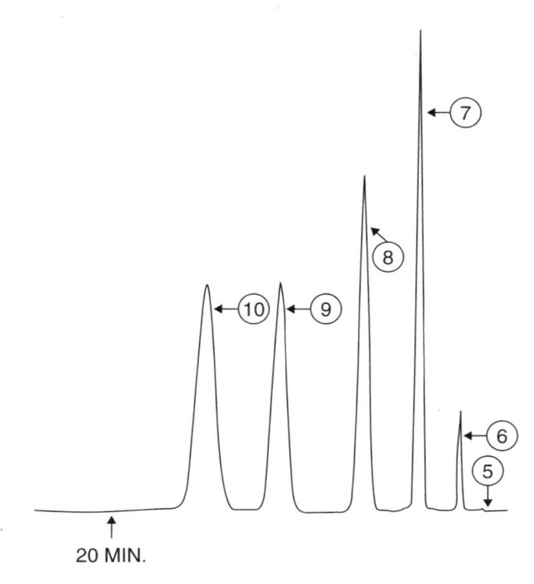

Figure 1.1 GC–TOF MS chromatogram of acetone (6), benzene (7), toluene (8), ethylbenzene (9), and styrene (10). [From Gohlke (1959), with permission. Copyright © 1959 by The American Chemical Society.]

In the field of separation science today, analysts still tend to fall within one of two well-defined groups:

1. Chromatography experts, who tend to have great faith in their capability to optimize the chromatographic process and are inclined to treat the mass spectrometer as little more than a detector. Such an outlook is appropriate as long as the ion source receives analytes resolved entirely (then identified, for example, by using MS libraries). However, problems can arise when extensive peak overlap occurs and a thorough exploitation of the MS dimension becomes necessary.
2. MS experts, who tend to have great faith in their capability to untangle a multicomponent band that enters the ion source, because the mass analyzer can then resolve a group of ions on a mass basis. It is true that mass spectrometry can be very useful for the unraveling of overlapping analytes. However, the reliability of peak identification is inversely proportional to the extent of compound coelution. In truth, chromatography and MS processes are equally important and complementary, and should be pushed to their full potential.

1.2 SHORTCOMINGS OF ONE-DIMENSIONAL CHROMATOGRAPHY

At present, one-dimensional chromatography is the method most commonly employed for the separation of real-world samples. However, in the past three decades it has become increasingly clear that the baseline separation of all the constituents of a sample or of specific target analytes from the rest of the matrix is often an unreasonable challenge when using a single chromatography column. The two fundamental aspects that govern all chromatography processes are peak capacity (n_c) and stationary-phase selectivity. The former parameter is related to the column characteristics (i.e., length, internal diameter, particle diameter, stationary-phase thickness, intensity of analyte–stationary phase interactions, etc.) and to the experimental conditions (i.e., mobile-phase flow and type, temperature, outlet pressure, etc.). The other feature is related to the chemical composition of the stationary phase, and hence to the specific type of analyte–stationary phase interactions (i.e., dispersion, dipole–dipole, hydrogen bonding, electrostatic, size exclusion, etc.). Selectivity is also dependent on analyte solubility in the mobile phase, whenever this type of interaction occurs. Ideally, a chromatographic analysis will be achieved in the minimum time for a given sample if the column is characterized by the most appropriate separation phase and generates the minimum peak capacity required. Nothing more than this goal is sought by all chromatographers.

An experienced chromatographer with a knowledge of basic theory will easily get the best out of a column or, in other words, will maximize the number of peaks that can be stacked side by side (with a specific resolution value) in a one-dimensional space. However, it has been emphasized that such an analytical capability will fall far short of the peak capacity requirements for many applications. In inspirational work, Giddings demonstrated from a theoretical viewpoint that "no more than 37% of the peak capacity can be used to generate peak resolution" and that "many of the peaks observed under these circumstances represent the grouping of two or more close-lying components," concluding that "*s* (the number of single component peaks) can never exceed 18% of n_c (Giddings, 1990). Although such a value does not take stationary-phase selectivity into account, it provides an excellent indication of the separation power of a one-dimensional chromatography system.

The well-known master equation for the calculation of resolution between two compounds with retention factors equal to k_1 and k_2, is

$$R_s = \frac{\sqrt{N}}{4}\frac{\alpha - 1}{\alpha}\frac{k_2}{k_2 + 1} \tag{1.1}$$

The different degrees of influence of N, α, and k on R_s can be observed in the excellent example shown in Figure 1.2, where the separation of two analytes ($k_1 = 4.8$; $k_2 = 5.0$; $\alpha = 1.05$) on a GC column ($N = 20{,}000$) under fixed conditions is considered. If we direct our attention to the three variables contained in Eq. (1.1) and to the effects of their variation on resolution (visualized in Figure 1.2), we can draw the following conclusions:

k If the column phase ratio is reduced (or a lower temperature is used), leading to an increase in the retention factors, the benefits gained are very limited in terms of resolution. An increase in k has a substantial effect on R_s only for analytes with low k values (≤ 3).

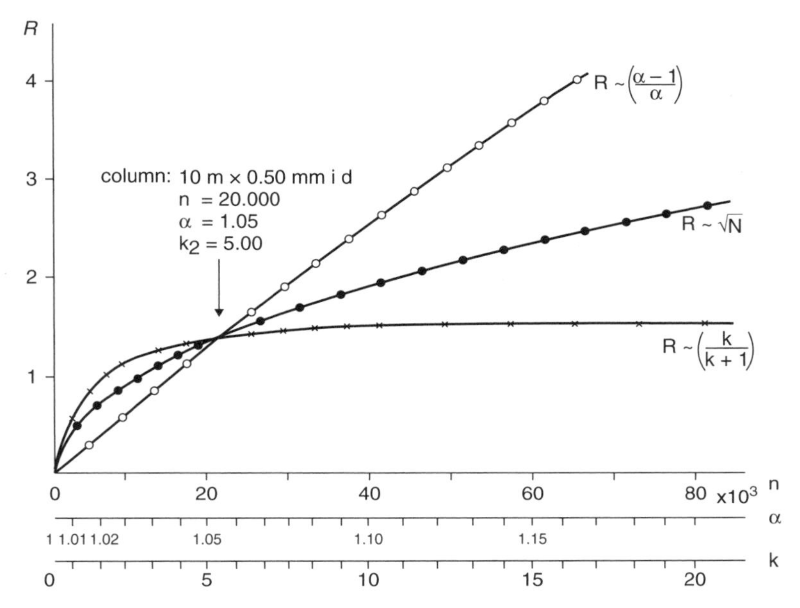

Figure 1.2 Resolution: influence of N, α, and k. [From Sandra and David (1990), with permission. Copyright © 1990 by Taylor & Francis Group LLC.]

α If a more selective stationary phase is employed, thus increasing the separation factor, resolution will benefit greatly. From Eq. (1.1) it can be concluded that at lower values, an increase in α will lead to a considerable improvement in resolution, up to an α value of about 3. At higher separation factor values, the function tends to level off. Of the three variables, selectivity has the greatest effect on resolution, and thus it is fundamental to select the most suitable stationary phase for a given separation. However, it must also be noted that Eq. (1.1) is valid only for a single pair of analytes and not for a complex mixture of compounds; in the latter case, a stationary-phase change will often lead to an improvement in resolution for some analytes and a poorer result for others. The choice of the most selective stationary phase has the best effects only when a low-complexity sample is subjected to separation.

N If the column length is extended fourfold, leading to an increase in N by the same factor, resolution of the two analytes is only doubled. It follows that an evident improvement in peak resolution can only be achieved by extending the column length considerably. Such a modification is usually not desirable and certainly is not a practical solution in view of the greatly increased analysis time. However, enhancing the plate number is without doubt the best choice whenever a highly complex mixture is subjected to chromatography. In fact, an increase in N will lead to the same percentage increase in resolution for all the constituents of a sample.

1.3 BENEFITS OF TWO-DIMENSIONAL CHROMATOGRAPHY

The most effective way to enhance the peak capacity (and the selectivity) of a chromatography system, with equivalent detection conditions, is by using a multidimensional chromatographic (MDC) system. An online MDC instrument is generally characterized by the combination of two columns of different selectivity, with a transfer device located between the first and second dimensions. MDC processes can be divided into two major categories: the heart-cutting and the comprehensive methodologies. Classical heart-cutting MDC enables the transfer of selected bands of overlapping compounds from a primary to a secondary column, generally of the same or similar efficiency. The number of samples that can be reinjected onto the second dimension is limited because excessive (or continuous) heart cutting would cause the loss of a substantial fraction of the primary column resolution (Mondello et al., 2002). In terms of MDC peak capacity, this equals the sum of the resolution of the first and second dimensions, the latter multiplied by the number (x) of heart cuts $[n_{c1} + (n_{c2} \times x)]$. The benefits of combining two independent separation processes—namely, two-dimensional chromatography—were recognized very early within the chromatography community. For example, heart-cutting two-dimensional gas chromatography was introduced in 1958 by Simmons and Snyder, who described a first-dimension boiling-point separation of C_5 to C_8 hydrocarbons and a polarity-based separation of each of the four classes in the second dimension, in four distinct analyses (Figure 1.3). Transfer of chromatographic bands between the two dimensions was achieved using a valve-based interface. The results obtained were quite remarkable, and it was stated that "separations can be obtained with this column arrangement which are not normally possible with previously described arrangements of single columns and multiple columns connected in series." Although GC column technology has evolved considerably during the last half-century, this statement is still fully valid.

If the entire initial sample requires analysis in two different dimensions, a different analytical route must be taken: a multidimensional comprehensive chromatography (MDCC) technique. In an ideal MDCC system, the total peak capacity becomes that of the first dimension multiplied by that of the second dimension ($n_{c1} \times n_{c2}$). The first example of comprehensive multidimensional chromatography dates back over 60 years. In 1944, chromatography pioneers described a two-dimensional procedure for the analysis of amino acids on cellulose as follows (Consden et al., 1944): "A considerable number of solvents has been tried. The relative positions of the amino-acids in the developed chromatogram depend upon the solvent used. Hence by development first in one direction with one solvent followed by development in a direction at right angles with another solvent, amino-acids (e.g., a drop of protein hydrolysate) placed near the corner of a sheet of paper become distributed in a pattern across the sheet to give a two-dimensional chromatogram characteristic of the pair of solvents used." The combination of solvents to be employed in that two-dimensional analysis was chosen on the basis of R_F values, a parameter (movement of band/movement

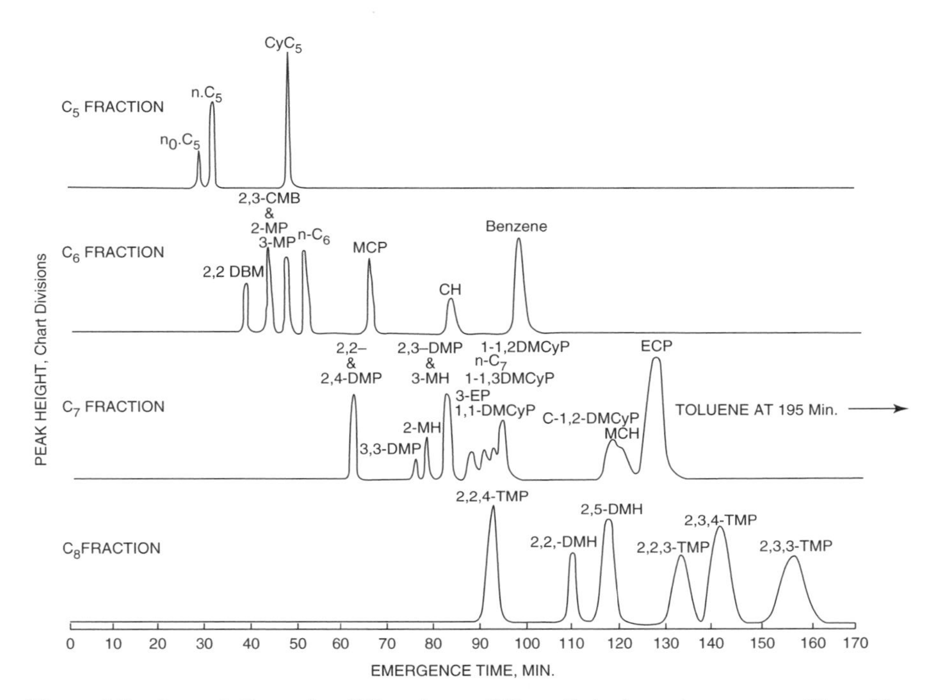

Figure 1.3 Second-dimension GC analyses of C_5 to C_8 hydrocarbon groups. [From Simmons and Snyder (1958), with permission. Copyright © 1959 by The American Chemical Society.]

of solvent front) introduced in the Consden et al. paper. In fact, the proper combination of solvents would enable a more extensive occupation of the two-dimensional space. The amino acid R_F values for a series of solvent combinations were used both to predict and to construct what today we would define as *dot plots*. A predicted two-dimensional chromatogram using collidine to develop the first dimension and a phenol–ammonia mixture to develop the second dimension is shown in Figure 1.4. The excellent agreement between the results predicted and the experimental results can be seen in Figure 1.5.

It is outside the scope of this chapter to elaborate the details regarding MDCC instrumental methodologies introduced over the last 30 years. It will be seen, however, that MDCC techniques generate very high peak capacities, making these approaches a prime choice to meet the challenge of the separation of highly complex mixtures.

1.4 BOOK CONTENT

The title and content of this book are focused mainly on the combination of a third mass spectrometric dimension to a comprehensive chromatographic (CC–MS) system. It will be seen that three-dimensional CC–MS methods are the most

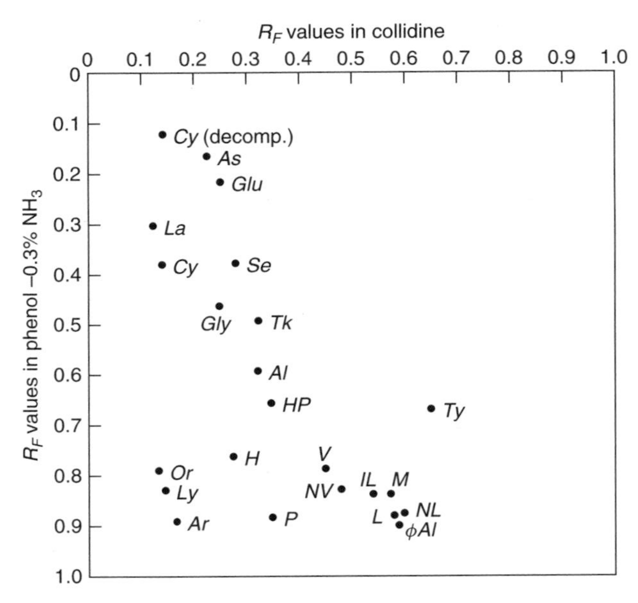

Figure 1.4 Expected positions of a series of amino acids on a two-dimensional paper chromatogram, developed using collidine in the first dimension and a phenol–ammonia mixture in the second dimension. Al, alanine; Ar, arginine; As, aspartic acid; Cy, cystine; Glu, glutamic acid; Gly, glycine; H, histidine; HP, hydroxyproline; IL, isoleucine; L, leucine; La, lanthionine; Ly, lysine; M, methionine; NL, norleucine; NV, norvaline; Or, ornithine; øAl, phenylalanine; P, proline; Se, serine; Th, threonine; Tr, tryptophan; Ty, tyrosine; V, valine. [From Consden et al. (1944), with permission. Copyright © 1944 by The Biochemical Society.]

powerful analytical tools currently available for the purposes of analyte separation and identification. They involve unprecedented selectivity (three different separation dimensions), high sensitivity (mainly comprehensive gas chromatographic methods), enhanced separation power, speed (the number of resolved peaks per unit of time), and structured chromatograms (the formation of group-type patterns makes peak identification easier and more reliable).

The importance and necessity of using comprehensive multidimensional GC (GC × GC) and LC (LC × LC) chromatography systems whenever a sample consists of several constituents is demonstrated in Chapters 2 and 3 from a theoretical standpoint. It is shown that a single column does not generally possess sufficient peak capacity for the baseline separation of all the sample components, and that the extent of peak overlap is enhanced as the number of compounds increases. The reason that ideal and practical peak capacities are still rather far apart is addressed particularly in Chapter 2.

In Chapter 4, a detailed description is provided of GC × GC since its introduction. Particular attention is devoted to hardware developments in this specific field. The present-day weak points and future prospects of GC × GC are also

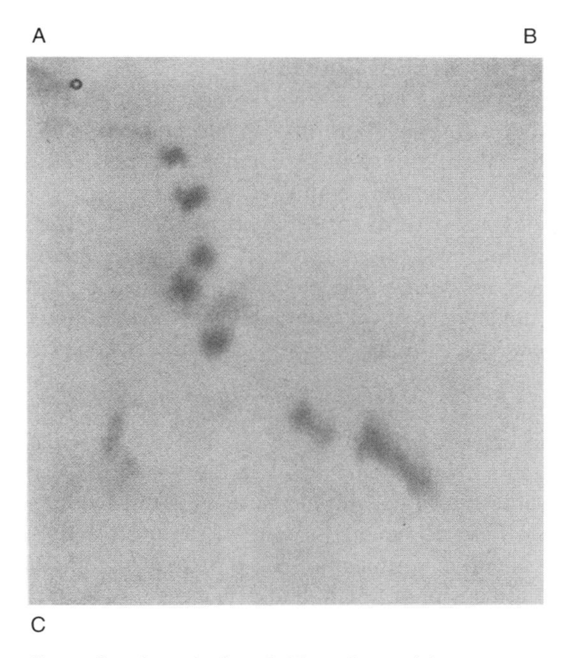

Figure 1.5 Two-dimensional analysis of 22 amino acids on paper, developed using collidine in the first dimension and a phenol–ammonia mixture in the second dimension. [From Consden et al. (1944), with permission. Copyright © 1944 by The Biochemical Society.]

considered and discussed. Chapter 4 is also devoted to one of the most difficult tasks related to GC × GC analysis: namely, method optimization. Apart from the modulation conditions (modulation period, trapping, and reinjection temperatures), the most important parameters that must be considered are stationary-phase selectivities, temperature gradient(s), column dimensions, gas linear velocities, outlet pressure, and detection settings (including MS). It is clear that the scenario is much more complex than that of a single-column system, while experience acquired in the field of conventional, classical multidimensional, and very fast GC is of great help.

Cryogenic GC × GC modulation is usually a rather costly issue, due to the high requirements for cooling gases for an analysis. It follows that the concept of cryogenic-free modulation is a very attractive option. Consequently, pneumatically modulated GC × GC is certainly a highly interesting and desirable approach. Although flow modulation technology has evolved considerably during the past decade, the method is still far from replacing cryogenic modulation.

Chapter 5 is devoted to the history, evolution, and optimization aspects related to flow-modulated GC × GC. Current advantages, disadvantages, and future prospects are also discussed.

Chapter 6 is focused on the rather brief but not insubstantial history of three-dimensional GC × GC–MS, which is the most powerful approach for the separation and identification of volatile molecules. The first GC × GC–MS experiments were carried out at the end of the 1990s, and since then the methodology has been studied and applied increasingly. Almost all applications have been carried out using either a time-of-flight or a quadrupole mass analyzer. Significant experiments relative to a variety of research fields, as well as advantages and disadvantages of the MS systems employed, are discussed.

Chapter 7 is devoted to GC × GC applications (i.e., food and fragrances, petrochemical and tobacco, pharmaceutical, environmental pollutants, biological samples, etc.), considering all detectors other than the mass spectrometer. The considerable advantages of the two-dimensional technique over conventional GC [separation power, sensitivity, selectivity, speed (resolved peaks/min), spatial order] are illustrated.

Chapter 8 is focused on a detailed description of LC × LC since its introduction at the end of 1970. As in the GC × GC chapter (Chapter 4), particular attention is directed to instrumental hardware evolution; the present-day shortcomings and possible future developments of LC × LC are also considered and discussed. Chapter 8 is also devoted to one of the most delicate tasks related to LC × LC analysis: method optimization. Two columns with a different selectivity and with different dimensions are employed, and distinct flow rates are applied in each dimension. Moreover, with respect to GC, the number of LC methods with distinct separation mechanisms is greater, and therefore in theory the number of orthogonal combinations is higher.

Chapter 9 is focused on the history of three-dimensional LC × LC–MS: the most powerful approach for the separation and identification of nonvolatile analytes. In GC × GC–MS, fragmentation is generally carried out by using electron ionization, and the spectral fingerprint aspect overshadows the m/z separation potential of mass spectrometry. In contrast, in LC × LC–MS it is emphasized that "softer" ionization MS techniques are generally employed. The low degree of fragmentation and the formation of intense molecular ion peaks highlight the third-dimensional separation capabilities of mass spectrometry. Applications in a variety of research areas, as well as advantages and disadvantages of the MS systems employed, are discussed.

Chapter 10 is devoted to LC × LC applications (e.g., polymers, food antioxidants, pharmaceutical compounds, proteins, peptides, etc.), invoving all detectors except the mass spectrometer.

Chapter 11 is focused on less common comprehensive chromatographic methods recently introduced. The principles, history, and evolution of each technique are described, as well as their use in combination with mass spectrometry. The methods described are comprehensive two-dimensional liquid–gas (LC × GC), supercritical fluid (SFC × SFC), and liquid–supercritical fluid (LC × SFC) chromatography.

Chapter 12 covers the fundamental differences in the data structure acquired with comprehensive chromatography methodologies compared to data from one-dimensional instrumentation. The requirements and opportunities for data analysis associated with multidimensional data (in particular, GC × GC) are highlighted.

Furthermore, the benefits of fully utilizing the added information in the data structure with chemometrics are demonstrated. Although the examples presented in Chapter 12 are for GC $\times$ GC, the basic principles of the data analysis are readily applicable to other forms of comprehensive chromatography analysis.

1.5 FINAL CONSIDERATIONS

Whenever a new sample is subjected to CC–MS analysis, one feels like a child opening a Christmas present. The time necessary to open the CC–MS data file using dedicated comprehensive chromatography software is the unwrapping process; two-dimensional visualization is like discovering the surprise and is often extremely rewarding. If further hardware and software improvements occur, CC–MS methodologies will undergo a gradual and constant expansion in future years, enabling a deeper insight into the world surrounding us and what is within ourselves. A final note is devoted to the previously described original GC–TOF MS, MDC, and MDCC experiments, and to the work of many other scientists, who have made advances and revolutionary discoveries in the field of separation science. The wealth of intuition contained in many old papers can still be exploited to bring progress in this wonderful field of research.

REFERENCES

Consden R, Gordon AH, Martin AJP. *Biochem. J.* 1944; 38:224–232.

Ettre LS. *LC-GC N. Am.* 2003; 21:458–467.

Giddings JC. In: Cortes HJ, Ed., *Multidimensional Chromatography*. New York: Marcel Dekker, 1990, pp. 1–27.

Gohlke RS. *Anal. Chem.* 1959; 31:535–541.

Mondello L, Lewis AC, Bartle KD, Eds. *Multidimensional Chromatography*. Chichester, UK: Wiley, 2002.

Sandra P, David F. In: Cortes HJ, Ed., *Multidimensional Chromatography: Techniques and Applications*. New York: Marcel Dekker, 1990, pp. 145–189.

Simmons MC, Snyder LR. *Anal. Chem.* 1958; 30:32–35.

Tswett M. *Ber. Dtsch. Bot. Ges.* 1906a; 24:316–323.

Tswett M. *Ber. Dtsch. Bot. Ges.* 1906b; 24:384–393.

2

MULTIDIMENSIONAL GAS CHROMATOGRAPHY: THEORETICAL CONSIDERATIONS

LEONID M. BLUMBERG
Fast GC Consulting, Hockessin, Delaware

Invented by Phillips in the early 1990s (Liu and Phillips, 1991; Phillips and Liu, 1992), *comprehensive two-dimensional* (2D) *gas chromatography* (GC), or GC × GC, is probably the most promising innovation in GC since the discovery of capillary columns (Golay, 1958a). The main driving force behind the interest in *multidimensional* (MD) separations (Bushey and Jorgenson, 1990; Consden et al., 1944; Erni and Frei, 1978; Giddings, 1984, 1987, 1990, 1995; Liu and Phillips, 1991; Zakaria et al., 1983), including GC × GC, was the unrelenting need to *resolve* (identifiably and quantifiably separate) more components in complex mixtures.

The number of peaks that a chromatographic system can resolve is a trade-off between that number, the analysis time, and detection of the smallest peaks (Beens et al., 2005; Blumberg, 2003; Blumberg et al., 2008; Klee, 1995; Klee and Blumberg, 2002; Steenackers and Sandra, 1995). The number of peaks that a chromatographic analysis can resolve can be expressed via the system's *peak capacity* (Giddings, 1967) (n_c). Its ability to detect and quantify small peaks can be expressed as the *minimal detectable concentration* (MDC) (Noij, 1988). As shown below, using longer columns makes it possible to increase the peak capacity of one-dimensional (1D) GC analysis without changing its MDC (Blumberg, 2003; Blumberg et al., 2008). Unfortunately, this will increase the analysis time in proportion to the third power of peak capacity. As a result, the peak capacity

Comprehensive Chromatography in Combination with Mass Spectrometry, First Edition.
Edited by Luigi Mondello.

of 1D GC can be increased only at the expense of prohibitively long analysis time. Thus, a twofold increase in n_c requires an eightfold-longer analysis (a 1-h analysis becomes an 8-h analysis).

Theoretically (Blumberg, 2003; Blumberg et al., 2008), the peak capacity of GC × GC can be more than an order of magnitude greater than the peak capacity of 1D GC with an equal analysis time and MDC. However, due to insufficiently sharp sample reinjection into the secondary columns in currently used routine GC × GC analyses (Blumberg et al., 2008), their peak capacities are about the same as the peak capacities of 1D analyses with similar analysis times and detection limits (Blumberg, 2002, 2003; Blumberg et al., 2008; Mydlová-Memersheimerová et al., 2009).

There is no doubt that the design problems of GC × GC will be solved and that the full theoretical potential of GC × GC will eventually be realized. This could revolutionize GC instrumentation and applications.

The potential ability to provide a substantially larger peak capacity is not the only advantage of GC × GC over 1D GC. Commercially available GC × GC systems offer several features that cannot be matched by 1D GC. Thus, GC × GC can expose the *structural composition* of a sample in a way that 1D GC cannot (Adahchour et al., 2008; Dimandja, 2004; Focant et al., 2004b; Marriott and Morrison, 2006; Mondello et al., 2003; Ryan et al., 2004; Tranchida et al., 2009). GC × GC also has a unique *class selectivity* feature that makes it possible to remove a majority of peaks of no interest from the 2D separation space in favor of resolving more peaks of interest than 1D GC can resolve with equal peak capacity (Adahchour et al., 2008; Beens et al., 2000; Dallüge et al., 2002; Phillips and Xu, 1995; Tranchida et al., 2009).

The main topic of this chapter is the evaluation of the number of peaks that can be resolved by GC × GC and the factors affecting that number. Of the three advantages of GC × GC over 1D GC—potentially larger peak capacity, class selectivity, and analysis of structural composition of the sample—only the first two are discussed. To study the potential advantages of GC × GC, its optimal operation is considered. The differences between theoretically required performance and actual underperformance of the critical component of GC × GC and its impact are also discussed. Only **capillary** (*open-tubular*) *columns* with **uniform** (the same everywhere along the column) **circular** cross section and with **liquid** stationary-phase film are considered. Other limitations (in **boldface** type) are introduced in the text.

Theoretical analysis of potential performance of GC × GC is based on the theory of temperature-programmed 1D GC. In view of that, about one-third of the chapter is dedicated to a brief review of theoretical results concerning 1D GC that are necessary for the study of GC × GC. Additional details can be found elsewhere (Blumberg, 2010). It is assumed, however, that the reader is familiar with the basic concepts of 1D GC and with the basic structures of 2D GC, such as *heart cutting* (Bertsch, 1990) (not considered in this chapter) and comprehensive techniques. Specifically, it is assumed that the reader is familiar with such concepts as linear *velocity*, u, of a carrier gas at an arbitrary location along the column as well

TABLE 2.1 Column Dimensions in the Examples

Type	Dimensions (Length × Diameter)
High efficiency	80 m × 0.1 mm
Conventional	30 m × 0.25 mm

as *outlet* and *average velocity*, u_o and $\overline{u} = L/t_M$, where t_M is the *hold-up time* and L is a column *length*. A *high-efficiency column* and a *conventional column* are used in several examples below. Their *dimensions—internal diameter* (briefly, *i.d.* or *diameter*) and *length*—are specified in Table 2.1. Experimental conditions are described in the text. Experimental results for 1D analyses under those conditions were described elsewhere (Blumberg et al., 2008). It is assumed throughout the chapter that the *high-efficiency analyses* (1D or 2D) utilize the high-efficiency column and *conventional analyses* utilize the conventional column.

2.1 SYMBOLS

2.1.1 Common Subscripts

Subscript	Description
end	at the end of analysis
init	initial (at the beginning of analysis)
max	maximum
min	minimum
opt	parameter corresponding to maximal efficiency
Opt	parameter corresponding to minimal analysis time at a given efficiency
R	at retention time
ref	reference parameter

2.1.2 Common Symbols

Symbol	Description
d_c	column internal diameter
d_f	stationary-phase film thickness
E	column efficiency, Eq. (2.6)
F	flow rate, Eq. (2.2)
f	specific flow rate, Eq. (2.3)
G_n	peak capacity gain, Eq. (2.61)
G_s	separation capacity gain, Eq. (2.59)
H	plate height
h	dimensionless plate height, Eq. (2.6)
k	retention factor
${}^2k_{\mathrm{last}}$	k of the second-dimension peak eluting at ${}^2t_M$ of the next cycle
L	column length
m	number of components in a sample

Symbol	Description
n_c	peak capacity, Eq. (2.19)
n_{co}	peak capacity unaffected by resampling
$n_{c,2D}$	peak capacity of GC × GC, Eqs. (2.48) and (2.49)
p	pressure
Δp	pressure drop
p_i	inlet pressure
p_o	outlet pressure
p_{sing}	number of resolved single-component peaks
p_{st}	1 atm—standard pressure
R_T	column heating rate
S	peak separation, separation capacity of interval, Eq. (2.16)
$S_{\min}$	required separation
s_c	separation capacity of analysis, Eq. (2.18)
s_{co}	separation capacity unaffected by resampling
$s_{c,2D}$	separation capacity of GC × GC, Eqs. (2.46) and (2.47)
T	column temperature
T_{init}	initial temperature of a heating ramp
T_{norm}	298.15 K (25°C)—normal temperature
T_{st}	273.15 K (0°C)—standard temperature
t	time
t_{anal}	analysis time
t_M	hold-up time
$t_{M,\text{init}}$	hold-up time at T_{init}
Δt_s	sampling period of resampling
U_s	utilization due to resampling, Eqs. (2.51) and (2.52)
u	gas velocity
u_o	gas outlet velocity
$\bar{u}$	gas average velocity
α	saturation, Eq. (2.20)
γ_{gas}	gas parameter, Table 2.2
η	gas viscosity
μ	solute mobility, Eq. (2.7)
ρ_s	sampling density, Eq. (2.63)
ρ_n	noise spectral density
σ	peak width (standard deviation)
σ_{filt}	width of filter's impulse response
${}^1\sigma_o$	peak width at the outlet of the primary column
${}^2\sigma_i$	width of reinjection pulse into the secondary column
${}^2\sigma_M$	width of unretained peak in the secondary column
${}^2\sigma_{M0}$	width of unretained peak in the secondary column with sharp reinjection pulses
τ_{anal}	dimensionless analysis time, Eq. (2.30)
$\Delta\tau_s$	dimensionless sampling period, Eq. (2.63)
φ	relative film thickness, Eq. (2.39)

2.2 ONE-DIMENSIONAL GC

2.2.1 Operational Parameters

The carrier gas in GC is a *compressible* fluid. **Ideally**, its *density* is proportional to *pressure*, p. To maintain gas flow, a negative pressure gradient is required along the column; the *inlet pressure*, p_i, is higher than the *outlet pressure*, p_o. Pressure reduction in the direction from column *inlet* to *outlet* causes gas *decompression* in that direction. In the case of **mass conserving flow**, the decompression causes an increase in the gas *velocity*, u, so that the product, pu, does not change along the column. The change in gas velocity along the column is an important factor of column operation in GC.

Gas decompression and the subsequent change in gas velocity along the column can be either *strong* or *weak*, depending on the *pressure drop*, $\Delta p = p_i - p_o$. The

$$\text{decompression is} \begin{cases} \textit{weak} & \text{when } \Delta p \ll p_o \\ \textit{strong} & \text{when } \Delta p \gg p_o \end{cases} \tag{2.1}$$

Strong decompression occurs in GC–MS, where the column outlet is at *vacuum* ($p_o = 0$), and in the analyses of complex mixtures typically using relatively long columns with relatively small diameters. Weak decompression is typical for analyses using relatively short columns having relatively large diameters and atmospheric pressure at the column outlet.

The gas *flow rate*,

$$F = \frac{p_o}{p_{st}} \frac{T_{norm}}{T} \frac{\pi d_c^2 u_o}{4} \tag{2.2}$$

in a column with *diameter* d_c is typically measured at **standard pressure,** $p_{st} = 1$ atm, and **normal temperature**, $T_{norm} = 298.15$ K (25°C), regardless of actual column temperature (Blumberg, 1999; Klee and Blumberg, 2002).

In studies of column performance, it is convenient to deal with the *specific flow rate* (Blumberg, 2010),

$$f = \frac{T}{T_{norm}} \frac{F}{d_c} \tag{2.3}$$

which is the flow rate per unit of column diameter. Normalization by a factor T/T_{norm} removes the dependence of f on specifics of flow measurement in GC (sometimes, F is measured at 0°C, at ambient temperature, etc.). Due to Eq. (2.2), f can be expressed as

$$f = \frac{p_o}{p_{st}} \frac{\pi d_c u_o}{4} \tag{2.4}$$

Unlike optimal F, which is proportional to column diameter (Blumberg, 1999), optimal f is independent of column dimensions (see below). From that perspective, f is similar to Giddings' *dimensionless* (*reduced*) *gas velocity*, v (Giddings, 1963a, 1965, 1991). Evaluation of a GC parameter at a fixed f or v means that one evaluates that parameter at a fixed ratio of F to its optimal value. On the other hand, due to its simple relation to gas flow rate (F), the quantity f is more practically oriented than is the dimensionless gas velocity. When f is known, F can be found as

$$F = \frac{T_{\text{norm}}}{T} d_c f \tag{2.5}$$

2.2.2 Peak Width

Peak width is one of the most basic metrics of a chromatogram. Throughout this chapter, **peak width** is identified by its *standard deviation*, σ (the square root of the peak's *variance*) (Blumberg, 2010; Grushka et al., 1969; Jönsson, 1987; Korn and Korn, 1968; Kučera, 1965; Sternberg, 1966). Among widely known *peak width metrics* (Ettre, 1993), standard deviation is the only one that can be found for non-*Gaussian* peaks (Ettre, 1993) (like those resulting from resampling in GC × GC) (Blumberg, 2008a) from experimental conditions.

Let L, H, N, t_M, and k be column *length, plate height, plate number, hold-up time*, and solute *retention factor*, respectively. The quantities

$$E = \sqrt{N} = \sqrt{\frac{L}{H}} = \sqrt{\frac{L}{hd_c}}, \qquad h = \frac{H}{d_c} \tag{2.6}$$

$$\mu = \frac{1}{1+k} \tag{2.7}$$

are, respectively, the column *efficiency* (Blumberg et al., 2008), *dimensionless* (*reduced*) (Giddings, 1963a, 1965, 1991) *plate height*, and the solute *mobility factor* (Blumberg, 2010; Blumberg and Klee, 2001b). The latter is the fraction of a solute in the mobile phase relative to the solute total amount in the column.

In *isothermal* and *temperature-programmed analysis*, peak widths (σ) can be found as (Blumberg, 2010; Giddings, 1962b; Habgood and Harris, 1960; Harris and Habgood, 1966)

$$\sigma = \frac{t_{M,R}}{E\mu_R} \tag{2.8}$$

In temperature-programmed analysis, the quantities μ_R and $t_{M,R}$ are, respectively, the solute *elution mobility* and the *temperature-dependent hold-up time* measured in isothermal analysis at the temperature equal to the solute *elution temperature* in actual temperature-programmed analysis.

2.2.3 Plate Height

To a significant degree, the balance between a column's separation performance and analysis time depends on plate height (H). Dependence of H on column parameters and operational conditions is known from Golay (1958b). Among the factors affecting H is the *stationary-phase film thickness*. A *thin-film* column is one where the film thickness has a negligible effect on H. *Thick-film* columns are typically used when it is necessary to inject a larger sample amount without *overloading* a column. The same benefit can be obtained by a proportional increase in column length, diameter, and film thickness without creating the negative impact of the thick film on the plate height. Only **thin-film** columns are considered in this chapter.

Gas compressibility can increase H by up to 12.5% (Blumberg, 2010; Giddings et al., 1960; Stewart et al., 1959), reducing the peak separation by up to 6%. This negligible effect is **ignored** in this chapter. While having a negligible effect on H, gas decompression can significantly complicate some formulas for H and for optimal conditions corresponding to the lowest H. The Golay formula for H (Golay, 1958b) does not account for the effect of gas compressibility on H. According to Golay (1958b), H can be expressed as a function, $H(u_o)$, of a carrier gas outlet velocity (u_o), which is not easy to measure when gas decompression is strong. It is much easier to measure carrier gas *average velocity*, $\overline{u}$. Unfortunately, due to gas decompression, $H(\overline{u})$ is a complex function of $\overline{u}$ (Blumberg, 2010; Blumberg, 1997a,b)—much more complex than the widely used expression $H = B/\overline{u} + C\overline{u}$, which has been attributed incorrectly to Golay (1958b) and van Deemter et al. (1956). Also complex is the dependence of *optimal average velocity*, $\overline{u}_{\text{opt}}$, on column dimensions and outlet pressure.[†]

A better alternative comes from expressing the Golay formula in this *symmetric* form:

$$h = \frac{H}{d_c} = \frac{h_{\min}}{2}\left(\frac{f_{\text{opt}}}{f} + \frac{f}{f_{\text{opt}}}\right), \qquad h_{\min} = \frac{\sqrt{1 + 6k + 11k^2}}{2\sqrt{3}(1 + k)} \tag{2.9}$$

$$f_{\text{opt}} = \frac{\pi D_{\text{pst}}}{h_{\min}} \tag{2.10}$$

The quantities $h_{\min}$, f_{opt}, and D_{pst} are, respectively, the *lowest dimensionless plate height*, the carrier gas *optimal specific flow rate*, and the solute *diffusivity* (Blumberg, 2010; Fuller et al., 1966) in the gas at standard pressure. By direct substitution of Eqs. (2.4) and (2.10) in Eq. (2.9), one can verify that Eq. (2.9) is indeed a *Golay formula* (Golay, 1958b) for H.

[†]Values frequently recommended for $\overline{u}_{\text{opt}}$ (35 cm/s for helium and 50 cm/s for hydrogen) are suitable only for conventional columns. For a 1m $\times$ 0.1mm column in GC–MS (a popular choice of secondary column in GC $\times$ GC), $\overline{u}_{\text{opt}} \approx 150$ cm/s for helium and $\overline{u}_{\text{opt}} \approx 250$ cm/s for hydrogen (Blumberg, 2010).

2.2.4 Speed-Optimized Conditions

At $f = f_{opt}$ (and therefore at $F = F_{opt}$), plate height is the lowest and column efficiency E [Eq. (2.6)] is the highest. Obtaining the highest efficiency is not the only possible *column optimization* goal. Another *optimization goal* could be *speed optimization*, i.e., obtaining the shortest analysis time at a predetermined efficiency (Blumberg, 1997c, 1999; Klee and Blumberg, 2002). The optimal flow rate in speed optimization is denoted here as F_{Opt}. The quantities F_{opt} and F_{Opt} are also known, respectively, as *efficiency-optimized flow rate* (EOF) and *speed-optimized flow rate* (SOF). These quantities, as well as the corresponding *optimal specific flow rates*, are related as (Blumberg, 1999)

$$F_{Opt} = \sqrt{2}\ F_{opt}, \qquad f_{Opt} = \sqrt{2} f_{opt} \tag{2.11}$$

The quantities f_{Opt} for several gases at 25°C are listed in Table 2.2. For this temperature, F_{Opt} can be found from Eq. (2.5) as (Blumberg, 1999; Klee and Blumberg, 2002)

$$F_{Opt} = d_c f_{Opt} \qquad (\text{at } T = 25^\circ\text{C}) \tag{2.12}$$

The quantity F_{Opt} drops in approximate proportion with $T^{-0.6}$ (Blumberg, 2010; Blumberg et al., 1999); that is,

$$F_{Opt} \sim T^{-0.6} \tag{2.13}$$

This dependence is relatively weak and can be ignored, suggesting that F_{Opt} found from Eq. (2.12) is satisfactory for all practical GC temperatures.

The two different optimization approaches (efficiency and speed optimization) do not lead to substantially different column efficiencies and analysis times (Blumberg, 1997c). However, only the *speed optimization* approach leads to solutions for the problem of *optimal heating rate*, $R_{T,Opt}$, in a temperature-programmed analysis. Thus, for a **heating ramp** covering a **wide temperature range** at **constant pressure** (Blumberg and Klee, 1998, 2000a,b;Klee and Blumberg, 2002),

$$R_{T,Opt} \approx 10^\circ\text{C}/t_{M,\text{init}} \tag{2.14}$$

TABLE 2.2 Quantities $f_{Opt}{}^{a}$ and γ_{gas} in Eqs. (2.12) and (2.36) for Several Carrier Gases

	Hydrogen	Helium	Nitrogen	Argon
f_{Opt} (mL/min/mm, at 25°C)	10	8	2.5	2.2
γ_{gas} (ms/m, at 150°C[b])	4.4	7.3	12.4	15.0

[a]Reported by Blumberg (1999) and Klee and Blumberg (2002).
[b]Estimates reported previously (Blumberg, 2003; Blumberg et al., 2008) were for 50°C.

where $t_{M,\mathrm{init}}$ is the *initial hold-up time* (the hold-up time at the beginning of the heating ramp). In all further discussions of optimal performance in this chapter, **speed optimization** is assumed.

Being a function of a solute retention and diffusivity [Eqs. (2.9) and (2.10)], plate height (H) and therefore column efficiency (E) can be different for different solutes. In a temperature-programmed analysis, E can also change with *column temperature*, T. However, all these dependencies are weak and the changes in E are practically insignificant (Golay, 1958b; Habgood and Harris, 1960; Harris and Habgood, 1966). It is assumed hereafter that E **does not change** during the analysis. At optimal flow,

$$H \approx d_c, \qquad h \approx 1, \qquad E \approx \sqrt{\frac{L}{d_c}} \qquad (\text{at } f = f_{\mathrm{Opt}}) \tag{2.15}$$

2.2.5 Separation Capacity and Peak Capacity

There are several metrics of separation in chromatography. Peak *resolution* (Ambrose et al., 1960; Ettre, 1993; Jones and Kieselbach, 1958; Martin et al., 1958), R_s; *separation number* (*Trennzahl*) (Ettre, 1993; Kaiser, 1962; Kaiser and Rieder, 1975), SN; and *peak capacity* (Giddings, 1967), n_c, are some of them. The metrics R_s and SN have significant shortcomings (Blumberg and Klee, 2001c), making them unsuitable for systematic studies of column performance. Many of the currently accepted definitions of peak capacity (Blumberg, 2003; Blumberg and Klee, 2001c; Davis and Giddings, 1983, 1985; Giddings, 1967, 1991; Grushka, 1970; Krupčík et al., 1984; Lan and Jorgenson, 1999; Martin et al., 1986; Medina et al., 2001; Shen and Lee, 1998) are incompatible with each other (Blumberg and Klee, 2001c) and with Giddings' original definition (Giddings, 1967).

The system of mutually compatible metrics of peak separation (Blumberg and Klee, 2001c) adopted in this chapter has already been used for comparison and optimization of different GC techniques (Blumberg, 2003, 2008a,b; Blumberg and Klee, 2001d; Blumberg et al., 2008). The basis of the system is the *separation measure* (Blumberg and Klee, 2001c), S, which is simply the number of σ-*slots* (σ-wide segments) between the retention times of two peaks or within any Δt-long time segment in a chromatogram. If all peaks in a chromatogram have the same width (a reasonable approximation for temperature-programmed GC), the separation measure of a Δt-long time segment is $S = \Delta t/\sigma$. Extension of this formula to an arbitrary time interval (t_1, t_2) and *varying* peak width is

$$S = S(t_1, t_2) = \int_{t_1}^{t_2} \frac{dt}{\sigma} \tag{2.16}$$

If t_1 and t_2 are retention times of two peaks, S in Eq. (2.16) is the peak *separation*.[†] If, on the other hand, t_1 and t_2 are arbitrary time markers on the time axis of a chromatogram, S in Eq. (2.16) is the *separation capacity* of the time interval (t_1, t_2).

Important for this chapter is a *running separation capacity*,

$$s = s(t) = S(t_M, t) = \int_{t_M}^{t} \frac{dt}{\sigma} \tag{2.17}$$

of analysis up to an arbitrary time, t. Equation (2.16) can be expressed as $S = s(t_2) - s(t_1)$. The *separation capacity* of the *entire analysis* up to retention time t_{anal} is the number

$$s_c = s(t_{\text{anal}}) = \int_{t_{\text{M}}}^{t_{\text{anal}}} \frac{dt}{\sigma} \tag{2.18}$$

of σ-slots in its chromatogram.

The quantity s_c describes a column's separation performance. However, it does not describe the number of peaks that a GC system as a whole can *resolve* (quantifiably and identifiably separate). That number depends not only on the column, but also on the ability of data analysis to quantify and identify two *closely spaced neighboring* peaks. That ability can be expressed via the *required separation*, S_{min}, the lowest S that the data analysis process requires for resolving two peaks. For example, in some cases, $S_{\text{min}} = 6$ (peaks should be separated by at least six σ-slots). For more sophisticated data analyses based on peak *deconvolution*, S_{min} can be lower than 1, meaning that even peaks that are less than one σ-slot apart from each other can be resolved (Gong et al., 2003; LECO Corporation, 2007; Prazen et al., 1999; Shao et al., 2004; Shen et al., 2005). Generally, S_{min} depends on the relative heights of neighboring peaks and on other factors. However, these dependencies are not considered in this chapter, and S_{min} is assumed to be the **same for all peak pairs** in a given analysis.

A $(S_{\text{min}}\sigma)$-wide interval [briefly, $(S_{\text{min}}\sigma)$-*slot*] occupied by a resolved peak can be called a *resolution slot*. The *peak capacity*, n_c, of a GC *analysis* is the number

$$n_c = \frac{s_c}{S_{\text{min}}} \tag{2.19}$$

of the resolution slots in its chromatogram. This definition (expressed differently) was used by Davis and Giddings in their studies of peak *overlap statistics* (Davis, 1994; Davis and Giddings, 1983; Giddings, 1991).

Two parameters in Eq. (2.19) *identify two factors affecting the peak capacity of GC analysis. The separation capacity* (s_c) *represents the column separation*

[†]If (a) two peaks are Gaussian and (b) their widths are different by less than a factor of 2, their resolution (R_s) can be approximated with less than 4% error as (Blumberg and Klee, 2001c) $R_s \approx S/4$.

performance. The required separation (S_{min}) represents the capability of data analysis to resolve poorly separated peaks.

2.2.6 Expected Number of Resolved Peaks

Equal separation of all peaks in a real chromatogram is highly unlikely. Therefore, the peak capacity (n_c) of a given GC analysis is not the number of peaks that the analysis can realistically resolve, but a *benchmark* that can serve as a basis for estimating that number. The number depends on statistics of peak distribution in a chromatogram. If that distribution is *statistically uniform* (Blumberg, 2008b; Blumberg and Klee, 2010; Davis, 1994), the *number of resolved single-component peaks*, p_{sing}, expected in the analysis of an m-component mixture can be found as (Davis and Giddings, 1983; Giddings, 1991) (Figure 2.1)

$$p_{sing} = me^{-2\alpha} = n_c\alpha e^{-2\alpha}, \qquad \alpha = \frac{m}{n_c} \tag{2.20}$$

where α is the *saturation* of a chromatogram.

The fraction $p_{sing}/m = e^{-2\alpha}$ of resolved peaks is an exponentially declining function of α (Figure 2.1). To raise the fraction from a small minority to a large majority of all components in a given mixture, a substantial reduction in α is required. According to the definition of α in Eq. (2.20), this requires a substantial increase in peak capacity.

Example 2.1 According to Eq. (2.20), raising p_{sing} from 10% to 90% of all components in a mixture requires a 22-fold increase in peak capacity.

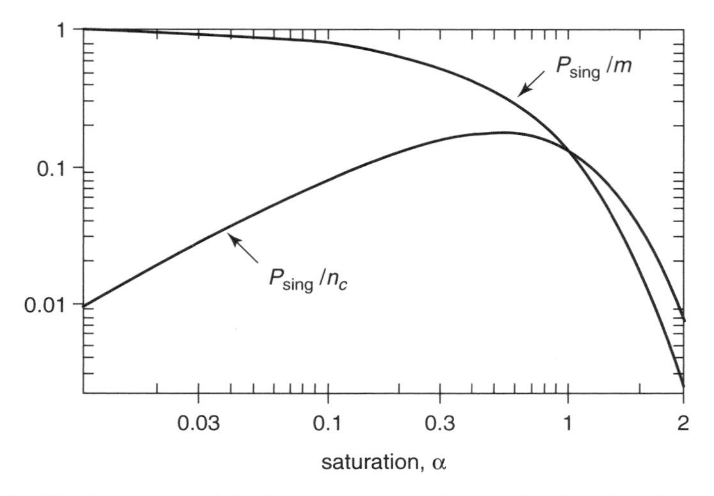

Figure 2.1 Number, p_{sing}, of single-component peaks as a fraction of peak capacity, n_c, and of a number, m, of components in a sample for different saturations, α.

The ratio $p_{\rm sing}/n_c = \alpha e^{-2\alpha}$ describing $p_{\rm sing}$ as a fraction of peak capacity has a maximum,

$$\left(\frac{p_{\rm sing}}{n_c}\right)_{\rm max} = \frac{1}{2e} \approx 0.184 \tag{2.21}$$

at $\alpha = 0.5$ (Figure 2.1). The number of single-component peaks that a system can resolve can be up to 18.4% of the system peak capacity (n_c), but only when the number, m, of components in a sample is one-half of n_c. In this case, 18.4% of peak capacity represents 37% of the total number (m) of components in a sample.

Example 2.2 In a system with $n_c = 1000$, $p_{\rm sing,max} \approx 184$. That many peaks is expected to be resolved when $\alpha = 0.5$ and therefore when $m = 500$ and $p_{\rm sing,max}/m \approx 37\%$. For a different number (m) of components in a test mixture, $p_{\rm sing}$ would always be smaller than 184. Thus, for $m = 250$, Eq. (2.20) yields $\alpha = 0.25$, $p_{\rm sing} \approx 152$, and $p_{\rm sing,max}/m \approx 61\%$. For $m = 1000$, it yields $\alpha = 1$, $p_{\rm sing} \approx 135$, and $p_{\rm sing,max}/m \approx 13.5\%$.

2.2.7 Separation Capacity and Column Parameters

Temperature-Programmed Analysis The theory of a column optimal performance in temperature-programmed GC was outlined by Giddings (1962a,c). Additionally, Harris and Habgood (1966) described several aspects of temperature programming. Other details of the subject are scattered throughout the literature. The following brief review summarizes the details that are essential for this chapter.

The separation capacity (s_c) of a GC analysis is a function [Eq. (2.18)] of peak width (σ) and analysis time. Of the three factors (E, $t_{M,R}$, and μ_R) affecting σ in Eq. (2.8), column efficiency (E) in a given analysis is essentially a fixed quantity. Furthermore, all solutes eluting during a *linear heating ramp* covering a wide temperature range elute with roughly the same retention factor (k_R) and mobility (μ_R) (Blumberg, 2010; Blumberg and Klee, 2000a, 2001b). Temperature-dependent hold-up time ($t_{M,R}$), being proportional to gas *viscosity* (Blumberg, 2010; Blumberg, 1995; Blumberg and Klee, 1998; Guiochon, 1966), increases in approximate proportion with $T^{0.7}$ (Blumberg, 2010; Hinshaw and Ettre, 1997), a relatively weak function of T. It is sufficient for the estimates in this chapter to treat $t_{M,R}$ as a **fixed quantity** equal to its value somewhere in the middle of a typical temperature range (say, at **150°C**). With this, peak width (σ) in Eq. (2.8) becomes a fixed quantity. Equation (2.18) yields

$$s_c = \frac{\Delta t_{\rm anal}}{\sigma} = E\mu_R\frac{\Delta t_{\rm anal}}{t_{M,R}}, \qquad \Delta t_{\rm anal} = t_{\rm anal} - t_{M,\rm init} \approx t_{\rm anal} \tag{2.22}$$

At optimal heating rate, quantities k_R and μ_R can be estimated as (Blumberg and Klee, 2000a,b, 2001b,d; Blumberg et al., 2008)

$$k_R \approx 1.75, \qquad \mu_R \approx \frac{1}{1+k_R} = \frac{1}{2.75} \qquad (\text{at } R_T = R_{T,\text{Opt}}) \tag{2.23}$$

Equation (2.22) can be further simplified as[†] (Blumberg and Klee, 1998, 2000a,b, 2001a,b,d; Blumberg et al., 2008; Klee and Blumberg, 2002)

$$s_c \approx E\frac{\Delta T_{\text{anal}}}{37^\circ\text{C}}, \qquad \Delta T_{\text{anal}} = R_{T,\text{Opt}} t_{\text{anal}} \tag{2.24}$$

Separation capacity (s_c) at an optimal heating ramp is approximately equal to column efficiency (E) per each 37°C span of the ramp.

The temperature range (ΔT_{anal}) in Eq. (2.24) depends strongly on the sample. For consistent comparison of different techniques, it is convenient to always use the same temperature range: for example, 100°C. It follows from Eq. (2.24) that at optimal heating rate,

$$s_c \approx 2.7E \qquad (\text{per } 100^\circ\text{C}) \tag{2.25}$$

Example 2.3 An optimized high-efficiency column (Table 2.1) has $E = 890$ [Eq. (2.15)]. For a 100°C-wide temperature range, Eqs. (2.24), (2.19), and (2.21) yield $s_c \cong 2400$. Therefore, at $S_{\text{min}} = 1, n_c \approx 2400$ and $p_{\text{sing,max}} \approx 440$. At $S_{\text{min}} = 6, n_c \approx 400$ and $p_{\text{sing,max}} \approx 75$. Similar parameters for a conventional column are $E \approx 350, s_c \approx 940, n_c \approx 940$, and $p_{\text{sing,max}} \approx 175$ at $S_{\text{min}} = 1, n_c \approx 155$, and $p_{\text{sing,max}} \approx 30$ at $S_{\text{min}} = 6$.

Isothermal Analysis During isothermal analysis, the carrier gas hold-up time and mobility of any solute remain fixed. Therefore, there is no need for the subscripts R in $t_{M,R}$ and μ_R in Eq. (2.8). A solute having mobility μ elutes at time $t = t_M/\mu$. Equation (2.8) becomes

$$\sigma = \frac{t}{E} \tag{2.26}$$

In isothermal analysis, peak width increases proportionally with time.

There are specifics that affect Eq. (2.18) in the case of GC × GC. *The separation space* of a 1D analysis occupies the time interval (t_M, t_{anal}) because there are no peaks in the interval (0, t_M). In isothermal analysis in the secondary column of GC × GC, the interval $(0, {}^2t_M)$ is available for the peaks that *wrap around* (Focant et al., 2004b; Ryan et al., 2005; Schoenmakers et al., 2003) from the previous sampling period and elute before ${}^2t_M$ in the next period. [This useful

[†]The parameter 37°C in the denominator of Eq. (2.24) relates directly to the *characteristic thermal constant* (Blumberg and Klee, 2000a, 2001a) of a solute [the inverse of the slope of its function $-\ln k(T)$ at $k = 1$] eluting at 150°C.

wraparound not interfering with other peaks is not always viewed as a wraparound (Adahchour et al., 2008).] As a result, the separation space of each analysis in the secondary column can stretch from ${}^2t_M$ to ${}^2t_{\text{anal}} + {}^2t_M$. Changing the integration limits in Eq. (2.18) accordingly and accounting for Eq. (2.26) leads to the following expression for the separation capacity, ${}^2s_{\text{co}}$, of the secondary column with *ideal* (*infinitely sharp*) sample introduction:

$${}^2s_{\text{co}} = {}^2E \int_{{}^2t_M}^{{}^2t_{\text{anal}}+{}^2t_M} \frac{\mathrm{d}t}{t} = {}^2E \ln \frac{{}^2t_{\text{anal}} + {}^2t_M}{{}^2t_M} = {}^2E \ln(2 + {}^2k_{\text{end}}) \qquad (2.27)$$

where

$${}^2k_{\text{end}} = \frac{{}^2t_{\text{anal}}}{{}^2t_M} - 1 \qquad (2.28)$$

is the retention factor at the time ${}^2t_{\text{anal}}$ of the *latest non-wrapped-around* second-dimension peak. Retention factor ${}^2k_{\text{last}}$ of the last peak eluting at the hold-up time in the next sampling period can be found as

$${}^2k_{\text{last}} = \frac{{}^2t_{\text{anal}} + {}^2t_M}{{}^2t_M} - 1 = {}^2k_{\text{end}} + 1 \qquad (2.29)$$

2.2.8 Analysis Time and Hold-up Time

The *analysis time* (t_{anal}) of isothermal and temperature-programmed analysis at constant pressure is proportional to the hold-up time (t_M) and can be expressed as (Blumberg and Klee, 1998)

$$t_{\text{anal}} = t_{M,\text{init}} \tau_{\text{anal}} \qquad (2.30)$$

where τ_{anal} is the *dimensionless analysis time* and $t_{M,\text{init}}$ is the *initial hold-up time* (t_M at $t = 0$). Due to Eq. (2.28), τ_{anal} in isothermal analysis can be expressed as

$$\tau_{\text{anal}} = 1 + k_{\text{end}} \qquad \text{(isothermal analysis)} \qquad (2.31)$$

The quantity t_M can be found as (Blumberg, 2010; Blumberg and Klee, 1998; Maurer et al., 1990)

$$t_M = \frac{32L^2\eta}{j_H d_c^2 \Delta p} = \frac{32E^4h^2\eta}{j_H \Delta p}, \qquad j_{\text{H}} = \frac{3(p_o + p_i)^2}{4(p_o^2 + p_o p_i + p_i^2)} \qquad (2.32)$$

Columns of different dimensions typically operate at different pressures but at the same specific flow rate (f) (e.g., at $f = f_{\text{Opt}}$). Therefore, there is a need to express t_M as a function of f rather than of Δp as in Eq. (2.32). Combining

Eq. (2.32) with known formulas (Blumberg, 1995) for p_i as a function of u_o and with Eq. (2.4), one can express t_M as functions of L and E (Blumberg, 2010):

$$t_M = t_M(L) = \frac{\pi^2 d_c^4 p_o^3}{1536\eta f^2 p_{\text{st}}^2}\left[\left(1+\frac{256L\eta p_{\text{st}} f}{\pi d_c^3 p_o^2}\right)^{3/2} - 1\right]$$
$$= \begin{cases} \dfrac{\pi d_c L}{4f}\dfrac{p_o}{p_{\text{st}}}, & \Delta p \ll p_o \\[2ex] \dfrac{8}{3}\sqrt{\dfrac{\pi L^3 \eta}{d_c f p_{\text{st}}}}, & \Delta p \gg p_o \end{cases} \tag{2.33a}$$

$$t_M = t_M(E) = \frac{\pi^2 d_c^4 p_o^3}{1536 f^2 p_{\text{st}}^2 \eta}\left[\left(1+\frac{256E^2 p_{\text{st}}\eta}{\pi d_c^2 p_o^2} fh\right)^{3/2} - 1\right]$$
$$= \begin{cases} \dfrac{\pi d_c^2 E^2}{4}\dfrac{h}{f}\dfrac{p_o}{p_{\text{st}}}, & \Delta p \ll p_o \\[2ex] \dfrac{8 d_c E^3}{3}\sqrt{\dfrac{\pi\eta}{p_{\text{st}}}\dfrac{h^3}{f}}, & \Delta p \gg p_o \end{cases} \tag{2.33b}$$

Solution of Eq. (2.33a) for L is

$$L = \frac{\pi d_c^3 p_o^2}{256 f p_{\text{st}}\eta}\left[\left(1+\frac{1536 f^2 p_{\text{st}}^2 t_M \eta}{\pi^2 d_c^4 p_o^3}\right)^{2/3} - 1\right]$$
$$= \begin{cases} \dfrac{4 f p_{\text{st}} t_M}{\pi d_c p_o}, & \Delta p \ll p_o \\[2ex] \left(\dfrac{9 d_c f p_{\text{st}} t_M^2}{64\pi\eta}\right)^{1/3}, & \Delta p \gg p_o \end{cases} \tag{2.34}$$

Substitution of Eq. (2.33b) in Eq. (2.8) and accounting for Eq. (2.6) yields

$$\sigma = \frac{\pi^2 d_c^4 p_o^3}{1536 E\mu_R \eta p_{\text{st}}^2 f^2}\left[\left(1+\frac{256E^2 p_{\text{st}}\eta}{\pi d_c^2 p_o^2} fh\right)^{3/2} - 1\right]$$
$$= \begin{cases} \dfrac{\pi d_c^2 E}{4\mu_R}\dfrac{h}{f}\dfrac{p_o}{p_{\text{st}}}, & \Delta p \ll p_o \\[2ex] \dfrac{8L}{3\mu_R}\sqrt{\dfrac{\pi\eta}{p_{\text{st}}}\dfrac{h}{f}}, & \Delta p \gg p_o \end{cases} \tag{2.35}$$

The last formula shows that as has been found elsewhere (Blumberg, 2010; Blumberg and Berger, 1993),

in a thin-film column, σ at strong gas decompression ($\Delta p \gg p_o$) is independent of column diameter and proportional to column length.

Due to Eq. (2.9), a combination of parameters f and h in Eqs. (2.33b) and (2.35) is a function of a single variable f. At optimal flow and heating rates ($f = f_{\rm Opt} = \sqrt{2} f_{\rm opt}$, $R_T = R_{T,\rm Opt}$), Eqs. (2.23), (2.10), and (2.9) yield $h \approx 1$, transforming Eq. (2.35) at $\Delta p \gg p_o$ and 150°C into

$$\sigma = \frac{4.73L}{\mu_R}\sqrt{\frac{\eta}{f_{\rm Opt} p_{\rm st}}} = \frac{\gamma_{\rm gas} L}{\mu_R} \qquad (\text{at } \Delta p \gg p_o,\ f = f_{\rm Opt},\ R_T \approx R_{T,\rm Opt}) \tag{2.36}$$

where the parameter $\gamma_{\rm gas}$ is listed for several gases in Table 2.2.

2.2.9 Detection Limits

The *detection limit* (*DL*) is a measure of the ability of a chromatographic system to detect and to quantify low-level solutes in the presence of *baseline noise* (Coleman and Vanatta, 2009; Crummett et al., 1980; Dal Nogare and Juvet, 1962; Ettre and Hinshaw, 1993; Guiochon and Guillemin, 1988; Lee et al., 1984; Littlewood, 1970; Noij, 1988; Novák, 1988). The noise can come from carrier gas impurities, column bleeding, detector electronics, and other sources. Some noise components can be reduced by using high-purity gases, low-bleed columns, avoiding too high temperatures, and so on. However, more fundamental *random noise* such as the *electronic noise* cannot be removed.

Important characteristics of random noise are its *spectral density*, ρ_n, and *intensity* (Motchenbacher and Connely, 1993; Papoulis, 1965). The latter can be expressed via the *root mean square* (r.m.s.) of the noise, and can be controlled by *noise filtering*. The filtering characteristics of a *noise filter* can be expressed via the width, $\sigma_{\rm filt}$, of the filter's *impulse response* (Papoulis, 1965), i.e., the filter's output resulting from an extremely sharp input spike having unity area. The r.m.s. of the filtered **white noise** (a typical noise type and the only type considered below) can be found as

$$\text{r.m.s.} = \sqrt{\frac{\rho_n}{2\pi\sigma_{\rm filt}}} \tag{2.37}$$

indicating that the wider the filter's impulse response, the lower the noise. Unfortunately, too wide an impulse response can unacceptably broaden the peaks and cause substantial reduction in a system's peak capacity. Therefore, it is necessary to maintain a certain balance between the widths (σ) of chromatographic peaks and the width ($\sigma_{\rm filt}$) of the noise filter's impulse response. It is assumed in this chapter that when the effects on DL of different operational factors are compared, the **ratio $\sigma_{\rm filt}/\sigma$ does not change**, and therefore the noise intensity is proportional to $\sqrt{1/\sigma}$.

Two types of DL—*minimal detectable amount* (MDA) and *minimal detectable concentration* (MDC)—can be considered (Noij, 1988).

A column parameter that might affect DL is the column *loadability*, i.e., the largest *nonoverloading* sample amount, $A_{\rm max}$, that can be introduced into the column. The quantity $A_{\max}$ is proportional to the product $d_c d_f \sqrt{HL}$ (Blumberg, 2003), which, due to Eqs. (2.6) and (2.39), can be expressed as

$$A_{\max} \sim d_c d_f \sqrt{HL} = \varphi d_c^{5/2} \sqrt{hL} \tag{2.38}$$

where

$$\varphi = \frac{d_f}{d_c} \tag{2.39}$$

is a stationary-phase *relative film thickness*.

MDC is inversely proportional to the *linear dynamic range*, Λ, the ratio of the largest height of a nonoverloaded peak to the noise r.m.s. As shown elsewhere (Blumberg, 2003), Λ at strong gas decompression is proportional to $d_c^{3/2} d_f/(1+k)^{1/2}$. Applying the same logic to other cases and accounting for Eq. (2.6), one can find the following relations of proportionality ($\sim$) for MDA and MDC in isothermal and temperature-programmed analyses:

$$\text{MDA} \sim \begin{cases} \sqrt{\dfrac{p_o \rho_n \sqrt{d_c^3 Lh}}{f \mu_R}}, & \Delta p \ll p_o \\ \sqrt{\dfrac{L \rho_n}{\mu_R} \sqrt{\dfrac{h\eta}{f}}}, & \Delta p \gg p_o \end{cases} \sim \begin{cases} d_c \sqrt{\dfrac{Ehp_o \rho_n}{f \mu_R}}, & \Delta p \ll p_o \\ E\sqrt{\dfrac{d_c h^3 \eta \rho_n}{f \mu_R}}, & \Delta p \gg p_o \end{cases} \tag{2.40}$$

$$\text{MDC} \sim \frac{1}{\varphi} \begin{cases} \sqrt{\dfrac{p_o \rho_n}{f \mu_R \sqrt{d_c^7 hL}}}, & \Delta p \ll p_o \\ \sqrt{\dfrac{\rho_n \sqrt{\eta}}{d_c^5 \mu_R \sqrt{fh}}}, & \Delta p \gg p_o \end{cases} \sim \frac{1}{\varphi} \begin{cases} \dfrac{1}{d_c^2} \sqrt{\dfrac{p_o \rho_n}{Efh\mu_R}}, & \Delta p \ll p_o \\ \sqrt{\dfrac{\rho_n \sqrt{\eta}}{d_c^5 \mu_R \sqrt{fh}}}, & \Delta p \gg p_o \end{cases} \tag{2.41}$$

Equation (2.41) shows that MDC at strong gas decompression is independent of a column length (L). This has a simple explanation. Higher loadability of a longer column [Eq. (2.38)] improves MDC. However, this is more than compensated by the peak broadening proportional to L [Eq. (2.36)], a factor that reduces the peak heights and worsens MDC. Independence of MDC from L comes as the net result of these factors combined with the noise reduction due to peak broadening [Eq. (2.37)].

2.2.10 Relations Among Separation Capacity, Analysis Time, and Detection Limit

The main purpose of chromatography is the separation. However, it is not the only aspect of performance of GC analysis. Analysis time and DL are other important metrics of GC analysis. With no concern for the analysis time and/or DL, any separation capacity would be possible. In reality, however, the separation capacity is a result of trade-off with analysis time and DL.

According to Eqs. (2.33b), (2.40), and (2.6), reducing column diameter and length while keeping their ratio fixed reduces t_M and MDA without changing E. This means that

there is no fundamental conflict among separation, analysis time, and detection of a fixed amount of a trace-level solute.

This has a simple explanation. Making a column smaller without changing its efficiency (E) reduces the analysis time (t_{anal}) and makes peaks taller and sharper, thus increasing the signal-to-noise ratio and reducing (improving) MDA. The reduced MDA can be traded for larger efficiency and/or a shorter analysis time.

In analyses of complex mixtures, one needs to worry not only about detection of small peaks, but also about prevention of column overloading by large peaks. The sample amount (including the amount of the trace-level components) becomes a function of column dimensions. MDA becomes unsuitable as a metric of DL for columns of different dimensions. On the other hand, nonoverloading sample concentration is independent of column dimensions and can be treated as a fixed quantity, making MDC suitable as a metric of DL. Unless the contrary is stated explicitly, only **MDC** is considered hereafter to be the metric of DL.

Exclusion of parameter d_c from Eqs. (2.33b) and (2.41) leads to the following relations among E, t_M, and MDC:

$$E^3 \sim \frac{\varphi^2 \mu_R}{p_o^3 \rho_n} \frac{f^3}{h} t_M^2 \cdot \text{MDC}^2, \qquad \Delta p \ll p_o \tag{2.42}$$

$$E^{15} \sim \frac{\varphi^2 \mu_R}{\eta^3 \rho_n} \frac{f^3}{h^7} t_M^5 \cdot \text{MDC}^2, \qquad \Delta p \gg p_o \tag{2.43}$$

Unlike the case for MDA,

there is a conflict among E, t_M, and MDC: An improvement in any two of these metrics worsens the third.

The trade-offs among column efficiency, analysis time, and MDC are sometimes illustrated by a *triangle of compromise* (Klee, 1995; Klee and Blumberg, 2002). Equations (2.42) and (2.43) are the mathematical expressions of that "triangle." Furthermore, an increase in the number of resolved peaks is inevitably associated with strong gas decompression. This makes Eq. (2.43) the *ultimate* expression of the "triangle of compromise." It follows from Eqs. (2.19), (2.22), (2.27), and (2.30) that $n_c \sim s_c \sim E$ and $t_{\text{anal}} \sim t_M$. For optimal flow and heating rates, Eq. (2.43) can be rephrased as

$$n_c^{15} \sim s_c^{15} \sim E^{15} \sim \frac{\varphi^2}{\gamma_{\text{gas}}^6 \sqrt{\rho_n}} t_{\text{anal}}^5 \cdot \text{MDC}^2, \qquad \Delta p \gg p_o \tag{2.44}$$

Raising the column efficiency from E_1 to E_2 makes the analysis time $(E_2/E_1)^3$ times larger if MDC remains fixed; it makes MDC $(E_2/E_1)^{7.5}$ times larger if the analysis time remains fixed. The carrier gas effect on the trade-off among n_c, t_{anal}, and MDC depends on the parameter γ_{gas} (Table 2.2).

Additional details regarding the pressure and the column dimensions at $\Delta p \gg p_o$ can be found from Eqs. (2.33) and (2.41). Some of them are:

Equation (2.41): *Column length does not affect MDC.*

Equations (2.32) and (2.33b): *Raising E from E_1 to E_2 without increasing analysis time requires $(E_2/E_1)^4$ times higher pressure and $(E_2/E_1)^3$ smaller column diameter.*

Example 2.4 Under optimal conditions, a high-efficiency column (Table 2.1) with hydrogen operates at 1 mL/min flow rate (Table 2.2), 1000 kPa (1 MPa) pressure, 4 min hold-up time, and 2.5°C/min heating rate. At this heating rate, it takes 40 min to cover each 100°C temperature span, which corresponds to about the longest practically acceptable analysis time. The efficiency of this column is 890 (Example 2.3). To double the efficiency without raising the analysis time requires, according to Eqs. (2.33b) and (2.32), an eightfold smaller column diameter (i.e., 12.5 μm) and a 16-fold higher pressure (i.e., 16 MPa).

The last example allows one to conclude that the high-efficiency column of Table 2.1 provides about the highest efficiency that can be obtained on commercial instruments within a reasonable time.

2.3 COMPREHENSIVE GC x GC

2.3.1 Overview

The structure of a GC × GC chromatograph is shown in Figure 2.2. Both columns could be in the same oven or in different ovens with different temperature programs (Tranchida et al., 2009). The heating of the secondary column is much slower than the optimal heating rate of $10°C/^2t_M$ [Eq. (2.14)]. As a result, each second-dimension cycle behaves as an **isothermal** analysis. *Resampling* (Blumberg, 2008a), *modulation* (Blumberg, 2003; Blumberg et al., 2008; Bueno and Seeley, 2004; Davis et al., 2008; Focant et al., 2004a; Harju et al., 2003; Horie et al., 2007; Khummueng et al., 2006; LaClair et al., 2004; Liu and Phillips, 1991; Schoenmakers et al., 2003; Seeley, 2002), and *sampling* (Bushey and

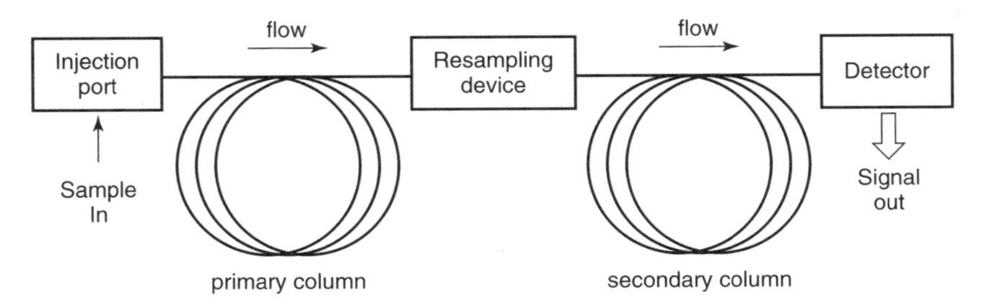

Figure 2.2 Structure of a GC × GC chromatograph.

Jorgenson, 1990; Davis et al., 2008; Horie et al., 2007; Khummueng et al., 2006; Murphy et al., 1998; Seeley, 2002) are some of the names for the repetitive process of slicing the *primary* column *eluite* (Guiochon and Guillemin, 1988), refocusing each slice, and reinjecting them or their *representative fractions* as a stream of sharp *reinjection pulses* into the secondary column.

Comprehensive 2D separation techniques were known (Consden et al., 1944; Erni and Frei, 1978; Zakaria et al., 1983) long before the invention of GC × GC. A conceptual description of 2D and MD separations is known primarily from Giddings (1984, 1987, 1990), who defined MD separation as one where the analytes are "subjected to two or more largely independent separative displacements" and that it is "structured such that whenever two components are adequately resolved in any one displacement step, they will generally remain resolved throughout the process." The purpose of the *second* constraint was to "rule out purely tandem arrangements of two or more columns" (Giddings, 1963b; Kaiser and Rieder, 1979; Repka et al., 1990) where "the resolution gained in one column can be partially or entirely nullified."

Conventional definitions of MD separations treat Giddings' second constraint as an unconditional requirement that the separation obtained in any displacement step should *never* be lost in subsequent steps. This disqualifies many GC × GC implementations from being considered 2D techniques (Blumberg and Klee, 2010) and can lead to the suboptimal design of instruments or method conditions.

The following definition (Blumberg and Klee, 2010) is based on the properties of the outcome of a separation process rather than on its internal characteristics: A *ν-dimensional (νD) analysis of multicomponent mixture is one that generates ν-dimensional displacement information regarding the mixture*. According to this definition, GC analysis is two-dimensional if the output, y, of its detector is a function, $y({}^1t, {}^2t)$, of two *independent (each carrying information) time coordinates (displacement coordinates)*, ${}^1t$ and ${}^2t$, as in Figure 2.3.

Functions such as $y({}^1t, {}^2t)$ are 3D objects. However, the *dimensionality* of analysis is defined not by the dimensionality of its output, but by the dimensionality of its displacement space. The definition speaks of *analysis* and applies only to *analytical separations* that generate *information* regarding a sample. In the remainder of this chapter, only **analytical separations** are considered.

An MD displacement space can also be called an MD *separation space*. In that context, the 2D time space of GC × GC analysis is its 2D *separation space*, as in Figure 2.3. In *comprehensive* (Bushey and Jorgenson, 1990) νD analysis, the entire sample is subjected to νD separation, whereas in a *partial* νD analysis such as *heart-cut* 2D GC (Bertsch, 1990), only a part of the sample is subjected to νD separation. The notation GC × GC is an adaptation of the pattern *chromatography × chromatography* proposed by Giddings (1984). Here, GC × GC always means *comprehensive* 2D GC. Only **comprehensive** 2D GC is considered in the rest of this chapter.

A component of a test mixture occupies a certain *spot* (Davis, 1991, 1993) (Figure 2.3) in the separation space of a given MD analysis. Spot distribution in MD space is *statistically uniform*, as in Figure 2.3, if the likelihood of a spot

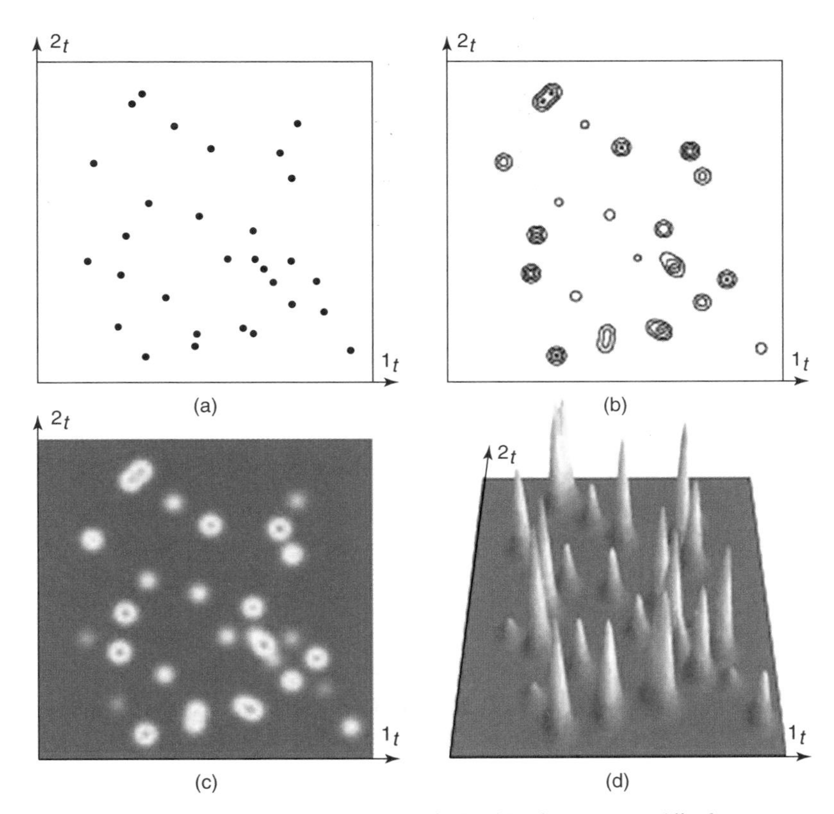

Figure 2.3 Results of the same 2D analysis in 2D time–space (displacement space, separation space) with time axes ^1t and ^2t: (a) peak distribution map (peaks are shown as circular spots); (b) contour plot; (c) density plot; (d) 3D plot. Small peaks are invisible in (b)–(d).

in a subspace of that space does not depend on the location of that subspace within the space. Otherwise, the distribution is *nonuniform*, as, for example, in Figure 2.4. A certain degree of *mutual independence* of the displacement mechanisms is necessary for statistically uniform spot distribution, although total independence can result in highly nonuniform distribution (Blumberg and Klee, 2010). Additional details regarding the properties of MD separations may be found elsewhere (Blumberg and Klee, 2010).

2.3.2 Performance Metrics and Parameters

Peak Capacity: Overview Let ^1t and ^2t be the pair of *time axes* of 2D separation space. Generally, the cross section of a 3D peak in GC × GC analysis can have any shape. Ideally, however, a cross section parallel to the $(^1t, {}^2t)$-plane is an **elliptic spot** with its axes parallel to the time axes, as in Figure 2.5a. Only these types of spots are considered below.

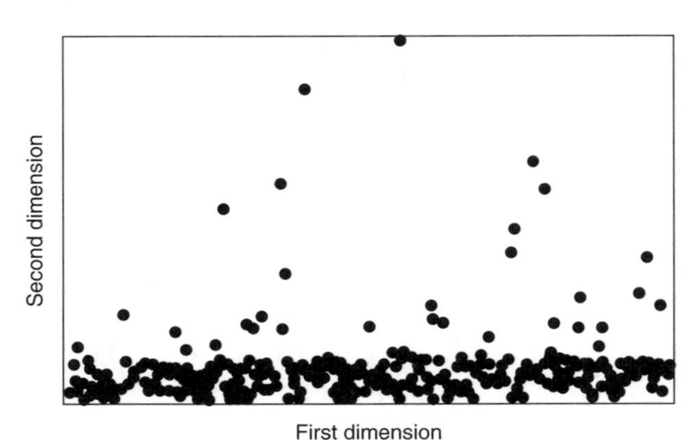

Figure 2.4 Nonuniform spot distribution in 2D space. The spots tend to be at the bottom of the space.

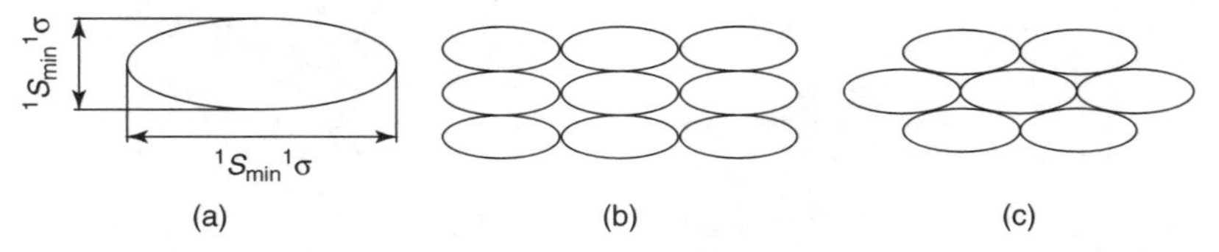

Figure 2.5 Elliptic resolution spots in 2D separation space. (a) A spot and its diameters, ${}^1S_{\min} \cdot {}^1\sigma$ and ${}^2S_{\min} \cdot {}^2\sigma$. Orthogonal (b) and flower (c) packing of the spots.

Let $({}^1\sigma, {}^2\sigma)$ and $({}^1S_{\min}, {}^2S_{\min})$ be, respectively, a pair of peak widths of an elliptic spot and a pair of respective required separations. The 2D extension of the concept of a *resolution slot* in 1D space is an elliptic *resolution spot* having diameter pair $({}^1S_{\min} \cdot {}^1\sigma, {}^2S_{\min} \cdot {}^2\sigma)$, as illustrated in Figure 2.5a. It is assumed in this chapter that, as in the case of 1D GC, ${}^1S_{\min}$ in a given 2D analysis is a **fixed** quantity (the same for all spots). The same is true for quantity ${}^2S_{\min}$.

Recall that the peak capacity of a 1D analysis is the number of resolution slots in the time interval (t_M, t_{anal}). Things are not as simple in 2D separation space, even if it has a simple topology where all resolution spots have the same size. As shown in Figure 2.5, there is more than one option for packing the spots in 2D space. The number of spots that can be packed in a given space also depends on the interference of the packing with the space boundaries. However, the *boundary conditions* are **ignored** in this chapter.

The variety of packing of elliptic spots in 2D space gives rise to a variety of definitions of *peak capacity* of GC × GC. Consider the following three:

$$n_{c,\text{2D,orth}} = {}^1n_c \cdot {}^2n_c, \qquad n_{c,\text{2D,flower}} = \frac{\sqrt{3}}{2} \cdot {}^1n_c \cdot {}^2n_c,$$

$$n_{c,\text{2D,0}} = \frac{4}{\pi} \cdot {}^1n_c \cdot {}^2n_c \tag{2.45}$$

where ${}^1n_c$ and ${}^2n_c$ are *axial peak capacities*. The quantities $n_{c,\text{2D,orth}}$ and $n_{c,\text{2D,flower}}$ reflect *orthogonal* (Giddings, 1987) and *flower packing* in Figure 2.5, respectively. The quantity $n_{c,\text{2D},0}$ represents the definition of peak capacity as the ratio of the total area of the separation space to the area of the resolution spot (Davis, 1991).

The peak capacities in Eq. (2.45) represent differently scaled products ${}^1n_c \cdot {}^2n_c$. The existence of several differently scaled peak capacities of the *same* 2D analysis suggests that the peak capacity is not a fundamental property of 2D analysis, but (as mentioned earlier) a *benchmark* for estimations of peak overlap statistics. The statistics themselves do not depend on the peak capacity scaling (Davis, 1991). Thus, the number of resolved single-component peaks expected in a given analysis is unique for that analysis and does not depend on the definition of peak capacity. For the peak capacity to uniquely represent the number of resolved peaks expected in an analysis of *any dimensionality*, the peak capacity should be scaled such that as long as the number of resolved peaks expected in two analyses is the same, the peak capacities of these analyses should also be the same, even if the analyses have different dimensionalities. All definitions in Eq. (2.45) do not satisfy this criterion. Thus, a 2D analysis with a peak capacity of, say, 1000 found from any formula in Eq. (2.45) is expected to resolve a different number of single-component peaks (Blumberg, 2008b) than would a 1D analysis with peak capacity 1000.

One-dimensional separation space can be viewed as *linear*. This gave rise to the terms *linear separation capacity* and *linear peak capacity*, $n_{c,\text{2D}}$, of 2D analysis (Blumberg, 2008b), where $n_{c,\text{2D}}$ is scaled such that if $n_{c,\text{2D}}$ is equal to the peak capacity of a 1D analysis, then in any analysis of the same mixture yielding statistically uniform peak distribution in both cases, both are expected to resolve an equal number of single-component peaks. Here, the same concepts are simply called, respectively, the separation capacity and the peak cpacity of 2D analyses. They can be described as follows (Blumberg, 2008b).

Generally, each parameter, ${}^1\sigma$ and ${}^2\sigma$, of a spot can be an arbitrary function of both time coordinates, ${}^1t$ and ${}^2t$, of the separation space. However, it is assumed in this chapter that ${}^1\sigma$ **depends only** on ${}^1t$, while ${}^2\sigma$ **depends only** on ${}^2t$. In this case, each time axis can be transformed such that ${}^1\sigma$ and ${}^2\sigma$ become fixed quantities.

As in 1D analysis, it is convenient to base the definition of the peak capacity of 2D analysis on the definition of the separation capacity of that analysis. The *separation capacity*, $s_{c,\text{2D}}$, of a 2D space is the ratio (Blumberg, 2008b)

$$s_{c,\text{2D}} = \frac{A}{2A_0} = \frac{2A}{\pi \cdot {}^1\sigma \cdot {}^2\sigma} \tag{2.46}$$

where A is the *area* of the space and $A_0 = (\pi/4) \cdot {}^1\sigma \cdot {}^2\sigma$ is the *area* of a $({}^1\sigma, {}^2\sigma)$-spot. For a *rectangular* separation space with $A = ({}^1s_c \cdot {}^1\sigma)({}^2s_c \cdot {}^2\sigma)$, where ${}^1s_c$ and ${}^2s_c$ are the *separation capacities* of the *primary* and *secondary* columns,

Eq. (2.46) yields (Blumberg, 2008b)

$$s_{c,2D} = \frac{2}{\pi} \cdot {}^1s_c \cdot {}^2s_c \approx 0.637 \cdot {}^1s_c \cdot {}^2s_c \tag{2.47}$$

The *peak capacity*,

$$n_{c,2D} = \frac{s_{c,2D}}{{}^1S_{min} \cdot {}^2S_{min}} \tag{2.48}$$

of this space can be expressed as (Blumberg, 2008b)

$$n_{c,2D} = \frac{2}{\pi} \cdot {}^1n_c \cdot {}^2n_c \approx 0.637 \cdot {}^1n_c \cdot {}^2n_c \tag{2.49}$$

where, in line with the definition of 1D peak capacity in Eq. (2.19),

$${}^1n_c = \frac{{}^1s_c}{{}^1S_{min}}, \qquad {}^2n_c = \frac{{}^2s_c}{{}^2S_{min}} \tag{2.50}$$

With these definitions, Eq. (2.20) for the expected number (p_{sing}) of single-component peaks is valid (Blumberg, 2008b) for 1D and 2D analyses [provided that in the latter case, n_c in Eq. (2.20) is $n_{c,2D}$]. In a similar way, Eq. (2.20) can be extended (Blumberg, 2008b) to analysis of a larger dimensionality.

Equations (2.47) and (2.49) show that *the peak capacity* ($n_{c,2D}$) *of a 2D analysis is 36% lower than the product*, ${}^1n_c \cdot {}^2n_c$, *of peak capacities of its 1D components. The same is true of the separation capacity* ($s_{c,2D}$) *and the product* ${}^1s_c \cdot {}^2s_c$.

An important factor in the separation performance of GC × GC is resampling (Figure 2.2). It can significantly broaden the peaks in both dimensions (Blumberg, 2003, 2008a; Blumberg et al., 2008; Horie et al., 2007; Khummueng et al., 2006; Murphy et al., 1998; Seeley, 2002) and reduce the separation capacities ${}^1s_c$ and ${}^2s_c$ compared to their counterparts ${}^1s_{co}$ and ${}^2s_{co}$, undisturbed by the resampling. This effect can be expressed as

$${}^1s_c = {}^1U_s \cdot {}^1s_{co} \tag{2.51}$$

$${}^2s_c = {}^2U_s \cdot {}^2s_{co} \tag{2.52}$$

where ${}^1U_s$ and ${}^2U_s$ are *utilizations* of the primary and secondary columns due to the resampling. Equations (2.47) and (2.49) become

$$s_{c,2D} = \frac{2}{\pi} \cdot {}^1s_c \cdot {}^2s_c = \frac{2}{\pi}({}^1U_s \cdot {}^1s_{co})({}^2U_s \cdot {}^2s_{co}) \tag{2.53}$$

$$n_{c,2D} = \frac{2}{\pi}({}^1U_s \cdot {}^1n_{co})({}^2U_s \cdot {}^2n_{co}) \tag{2.54}$$

where

$$ {}^1n_{\mathrm{co}} = \frac{{}^1s_{\mathrm{co}}}{{}^1S_{\min}}, \qquad {}^2n_{\mathrm{co}} = \frac{{}^2s_{\mathrm{co}}}{{}^2S_{\min}} \tag{2.55} $$

Peak Capacity Gain A possible advantage of GC × GC over 1D GC in peak capacity can be expressed via the *peak capacity gain* (Blumberg, 2003, 2008b; Blumberg et al., 2008),

$$ G_n = \frac{n_{c,\mathrm{2D}}}{n_{c,\mathrm{ref}}} \tag{2.56} $$

where $n_{c,\mathrm{2D}}$ is the peak capacity of GC × GC in question and

$$ n_{c,\mathrm{ref}} = \frac{s_{c,\mathrm{ref}}}{S_{\min,\mathrm{ref}}} \tag{2.57} $$

is the peak capacity of some *reference* 1D *analysis* having separation capacity $s_{c,\mathrm{ref}}$ and requiring separation $S_{\min,\mathrm{ref}}$ for adequate peak resolution. Substitution of Eqs. (2.48) and (2.57) in Eq. (2.56) yields

$$ G_n = \frac{s_{c,\mathrm{2D}}}{s_{c,\mathrm{ref}}} \frac{S_{\min,\mathrm{ref}}}{{}^1S_{\min} \cdot {}^2S_{\min}} = G_s \frac{S_{\min,\mathrm{ref}}}{{}^1S_{\min} \cdot {}^2S_{\min}} \tag{2.58} $$

where

$$ G_s = \frac{s_{c,\mathrm{2D}}}{s_{c,\mathrm{ref}}} \tag{2.59} $$

is the *separation capacity gain*.

At the current state of the art, data analysis systems capable of using peak deconvolution to resolve poorly separated peaks can be used in 1D GC and in the second dimension of GC × GC. For that reason, when the potential advantage of GC × GC over 1D GC is considered, it is proper to **assume** that in Eq. (2.57),

$$ S_{\min,\mathrm{ref}} = {}^2S_{\min} \tag{2.60} $$

This transforms Eq. (2.58) into (Blumberg et al., 2008)

$$ G_n = \frac{G_s}{{}^1S_{\min}} \tag{2.61} $$

This formula highlights the two ways of increasing the peak capacity of GC × GC over that of 1D GC (Blumberg, 2003):

1. *Increasing the separation capacity gain* (G_s) *(i.e., making the separation capacity of GC × GC larger than that of 1D GC)*

2. *Improving data analysis along the first dimension of $GC \times GC$ in order to reduce its required separation* ($^1S_{\min}$)

The effect of the peak capacity gain on the expected number (p_{sing}) of single-component peaks resolvable by GC × GC can be found from substitution of Eq. (2.56) in Eq. (2.20). One has (Figure 2.6)

$$p_{\text{sing}} = me^{-2\alpha_{1D}/G_n} = n_{c,1D}\alpha_{1D}e^{-2\alpha_{1D}/G_n}, \qquad \alpha_{1D} = \frac{m}{n_{c,1D}} \tag{2.62}$$

2.3.3 Components of a *GC* x *GC* System

Resampling Following is a brief review of the effects of resampling (Figure 2.2) on the GC × GC performance. Additional details may be found elsewhere (Blumberg, 2008a).

Let $^1\sigma_o$ be the *width* of a peak at the outlet of primary column, and Δt_s be the *sampling period* of the resampling. The number, $r_s = 1/\Delta t_s$, of samples per unit time is the *sampling rate* (or *frequency*); the number, $\rho_s = {}^1\sigma_o/\Delta t_s$, of samples per $^1\sigma_o$ is the *sampling density*, and its inverse,

$$\Delta\tau_s = \frac{1}{\rho_s} = \frac{\Delta t_s}{^1\sigma_o} \tag{2.63}$$

(the number of the $^1\sigma_o$-long intervals per sampling period), is the *dimensionless sampling period* (Seeley, 2002). Resampling and subsequent *reconstruction* of a 2D chromatogram increases the first-dimension width, $^1\sigma$, of each peak and reduces the first-dimension separation capacity, ^1s_c, compared to their levels, $^1\sigma_o$ and $^1s_{co}$, at the outlet of the primary column.

In the case of data reconstruction by *linear interpolation* (connection of neighboring data points by straight lines), *utilization* [^1U_s, Eq. (2.51)] of the first

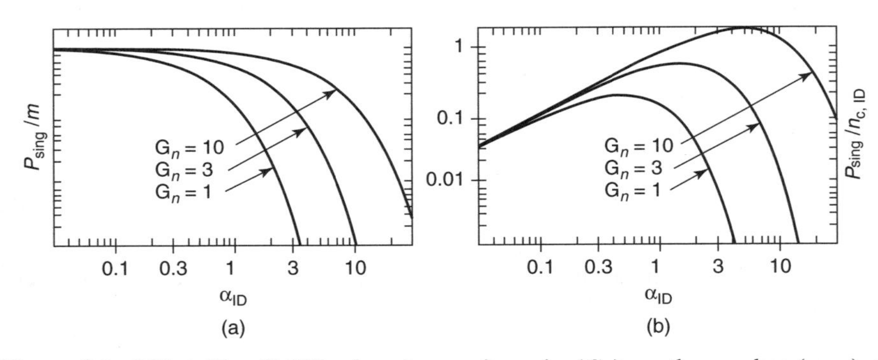

Figure 2.6 Effect [Eq. (2.62)] of peak capacity gain (G_n) on the number (p_{sing}) of single-component peaks resolvable by GC × GC.

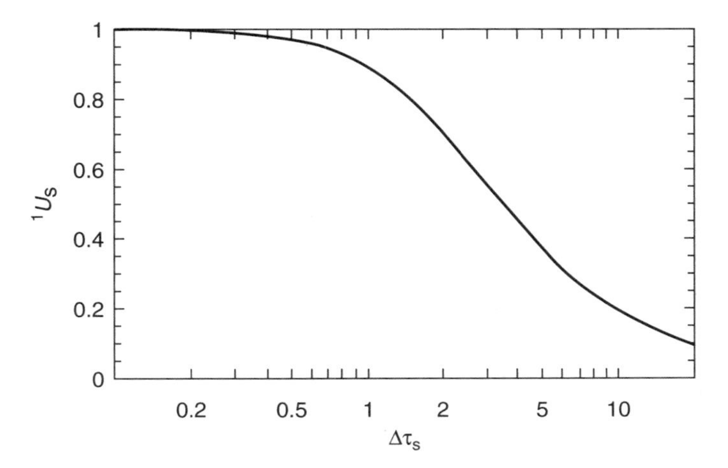

Figure 2.7 Utilization (^1U_s) [Eq. (2.64)] of the first dimension separation capacity as a function of dimensionless sampling period ($\Delta\tau_s$) [Eq. (2.63)].

dimension separation capacity can be found for any peak shape at the outlet of the primary column (Blumberg, 2008a), as illustrated in Figure 2.7, as

$$^1U_s = \frac{^1s_c}{^1s_{co}} = \frac{^1\sigma_o}{^1\sigma} = \frac{\rho_s}{\sqrt{0.25+\rho_s^2}} = \frac{1}{\sqrt{1+0.25\Delta\tau_s^2}} \approx \begin{cases} 1, & \Delta\tau_s < 1 \\ 2/\Delta\tau_s, & \Delta\tau_s > 4 \end{cases} \tag{2.64}$$

This together with Eq. (2.63) allows one to approximate the first-dimension widths ($^1\sigma$) of the reconstructed peaks as

$$^1\sigma = {}^1\sigma_o\sqrt{1+0.25\Delta\tau_s^2} \approx \begin{cases} ^1\sigma_o, & \Delta\tau_s < 1 \\ 0.5\Delta t_s, & \Delta\tau_s > 4 \end{cases} \tag{2.65}$$

Davis at al. (2008) confirmed Eq. (2.65) by Monte Carlo testing for Gaussian peaks.

The condition $\Delta\tau_s < 1$ (more than one sample per $^1\sigma_o$) in Eqs. (2.64) and (2.65) can be qualified as *oversampling*. Conversely, condition $\Delta\tau_s > 4$ (the sampling period is wider than $4 \cdot {}^1\sigma_o$) is *undersampling*. Significant undersampling can substantially broaden the first-dimension peaks.

Consider an *ideal resampling*, where reinjection pulses are *sufficiently sharp* so that

$$^2\sigma_i \ll {}^2\sigma_M \tag{2.66}$$

where $^2\sigma_i$ and $^2\sigma_M$ are, respectively, the *width of the reinjection pulse* into the secondary column and the *width of the unretained peak* in that column. As a result, ${}^2s_c = {}^2s_{co}$ and, according to Eq. (2.52),

$$ {}^2U_s = 1 \tag{2.67} $$

(Nonideal resampling is considered later.)

Typically, the second-dimension analysis time (${}^2t_{\text{anal}}$) in GC × GC is equal to the sampling period (Δt_s) (i.e., ${}^2t_{\text{anal}} = \Delta t_s$). It follows from Eqs. (2.30) and (2.31) that a change in ${}^2t_{\text{anal}}$ can be obtained at the expense of a change in hold-up time (${}^2t_M$) of the secondary column, a change in retention factors (${}^2k_{\text{end}}$) at the end of each second-dimension analysis, or a combination of both. Suppose that the approach to designing GC × GC is such that a change in ${}^2t_{\text{anal}}$ should not affect ${}^2k_{\text{end}}$ (see the discussion below). It follows from Eqs. (2.30) and (2.31) that a change in ${}^2t_{\text{anal}}$ can be obtained only at the expense of a proportional change in ${}^2t_M$. According to Eqs. (2.32) and (2.27) and condition ${}^2t_{\text{anal}} = \Delta t_s$, this implies that an increase in Δt_s allows one to increase the secondary column efficiency (${}^2E$) and separation capacity (${}^2s_c$). Therefore, although it can be harmful for separation in the first dimension, undersampling (too large Δt_s) is beneficial for separation in the second dimension. These observations suggest that there can be an optimal Δt_s (Blumberg, 2003, 2008a; Blumberg et al., 2008) corresponding to the highest separation capacity ($s_{c,\text{2D}}$) of GC × GC overall. Let's find that optimum.

Solving Eqs. (2.27), (2.33b), (2.30), and (2.31) and condition ${}^2t_{\text{anal}} = \Delta t_s$ together, one has

$$ {}^2s_{co} = {}^2P_k \begin{cases} \dfrac{1}{{}^2d_c}\sqrt{\dfrac{8 \cdot {}^2fp_{\text{st}}\Delta t_s}{\pi^2 \cdot {}^2h \cdot {}^2p_o}}, & {}^2\Delta p \ll {}^2p_o \\ \left(\dfrac{\Delta t_s}{{}^2d_c}\sqrt{\dfrac{9 \cdot {}^2fp_{\text{st}}}{32\pi\eta \cdot {}^2h^3}}\right)^{1/3}, & {}^2\Delta p \gg {}^2p_o \end{cases} \tag{2.68} $$

where

$$ {}^2P_k = \begin{cases} \dfrac{\sqrt{\pi}\ln(2 + {}^2k_{\text{end}})}{\sqrt{2(1 + {}^2k_{\text{end}})}}, & {}^2\Delta p \ll {}^2p_o \\ \dfrac{\ln(2 + {}^2k_{\text{end}})}{(\sqrt{2}(1 + {}^2k_{\text{end}}))^{1/3}}, & {}^2\Delta p \gg {}^2p_o \end{cases} \tag{2.69} $$

Combining Eqs. (2.67), (2.51), (2.47), (2.63), (2.64), and (2.68), one has

$$ s_{c,\text{2D}} = {}^1s_{co}X_s \cdot {}^2P_k \begin{cases} \dfrac{4\sqrt{2}}{\pi^2 \cdot {}^2d_c}\sqrt{\dfrac{{}^1\sigma_o \cdot {}^2fp_{\text{st}}}{{}^2h \cdot {}^2p_o}}, & {}^2\Delta p \ll {}^2p_o \\ \left(\dfrac{{}^1\sigma_o}{{}^2d_c}\sqrt{\dfrac{32 \cdot {}^2fp_{\text{st}}}{3\pi^7\eta \cdot {}^2h^3}}\right)^{1/3}, & {}^2\Delta p \gg {}^2p_o \end{cases} \tag{2.70} $$

where the parameter

$$X_s = {}^1U_s \begin{cases} \sqrt{\Delta\tau_s}, & {}^2\Delta p \ll {}^2p_o \\ \sqrt{3}\left(\dfrac{\Delta\tau_s}{4}\right)^{1/3}, & {}^2\Delta p \gg {}^2p_o \end{cases} = \begin{cases} \sqrt{\dfrac{4\Delta\tau_s}{4+\Delta\tau_s^2}}, & {}^2\Delta p \ll {}^2p_o \\ \dfrac{\sqrt{3}(2\Delta\tau_s)^{1/3}}{\sqrt{4+\Delta\tau_s^2}}, & {}^2\Delta p \gg {}^2p_o \end{cases} \tag{2.71}$$

represents the dependence (Figure 2.8) of $s_{c,2\mathrm{D}}$ on a dimensionless sampling period ($\Delta\tau_s$).

The maximum in X_s and therefore in $s_{c,2\mathrm{D}}$ takes place at the *optimal dimensionless sampling period*, $\Delta\tau_{s,\mathrm{Opt}}$, or, equivalently, at the *optimal sampling density*, $\rho_{s,\mathrm{Opt}}$, which can be found from Eq. (2.71) as (Blumberg, 2008a)

$$\Delta\tau_{s,\mathrm{Opt}} = \begin{cases} 2, & {}^2\Delta p \ll {}^2p_o \\ \sqrt{2}, & {}^2\Delta p \gg {}^2p_o \end{cases}, \qquad \rho_{s,\mathrm{Opt}} = \frac{1}{\Delta\tau_{s,\mathrm{Opt}}} = \begin{cases} 0.5, & {}^2\Delta p \ll {}^2p_o \\ \sqrt{0.5}, & {}^2\Delta p \gg {}^2p_o \end{cases} \tag{2.72}$$

Optimal utilization [^1U_s, Eq. (2.64)] of the first dimension and peak capacity ($n_{c,2\mathrm{D}}$) of GC × GC becomes (Blumberg, 2008a)

$${}^1U_s = \begin{cases} 0.71, & {}^2\Delta p \ll {}^2p_o \\ 0.82, & {}^2\Delta p \gg {}^2p_o \end{cases} \qquad (\text{at } \Delta\tau_s = \Delta\tau_{s,\mathrm{Opt}}) \tag{2.73}$$

$$n_{c,2\mathrm{D}} = {}^1n_{co} \cdot {}^2n_{co} \cdot {}^2U_s \cdot \begin{cases} 0.45, & {}^2\Delta p \ll {}^2p_o \\ 0.52, & {}^2\Delta p \gg {}^2p_o \end{cases} \qquad (\text{at } \Delta\tau_s = \Delta\tau_{s,\mathrm{Opt}}) \tag{2.74}$$

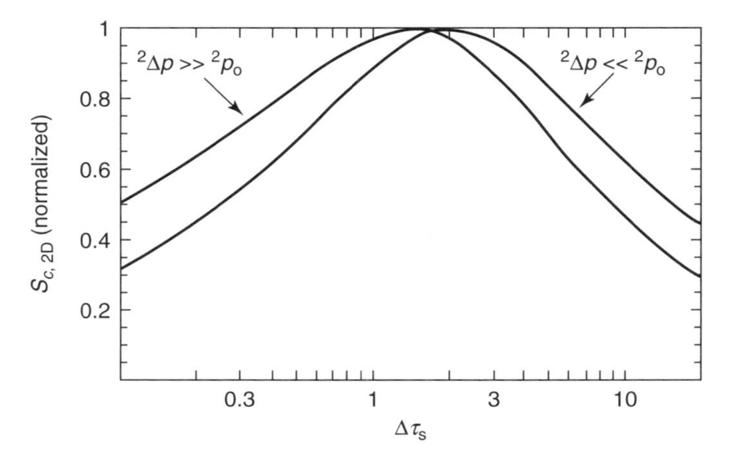

Figure 2.8 Normalized dependence (X_s [Eq. (2.71)] of separation capacity ($s_{c,2\mathrm{D}}$ [Eq. (2.70)] of GC × GC on a dimensionless sampling period ($\Delta\tau_s$).

The last formula shows that (Blumberg, 2008a)

even under the most favorable conditions of ideal resampling ($^2U_s = 1$) with the optimal rate, the peak capacity ($^2n_{co}$) of the second dimension should be at least 2 in order for the peak capacity ($n_{c,2D}$) of GC × GC to equal the peak capacity ($^1n_{co}$) of its first dimension unaffected by resampling.

Equation (2.71) is illustrated in Figure 2.8, showing the effect of the sampling period (Δt_s) on the separation capacity of GC × GC when a change in Δt_s is accompanied by the change in the secondary column length (^2L) required for maintaining a fixed $\Delta t_s/^2t_M$ ratio. When ^2L is not changed to accommodate a change in Δt_s, the resulting harm to the first dimension (Figure 2.7) uncompensated by the improvement in the second dimension can significantly reduce the separation capacity of GC × GC overall. It can also cause harmful wraparound (Focant et al., 2004b; Ryan et al., 2005; Shellie and Marriott, 2002). It is assumed in this chapter that unless the contrary is stated explicitly, any change in Δt_s is accompanied by a corresponding change in ^2L, so that the change in Δt_s has **no effect on retention factors** in the secondary column. The quantities $\Delta\tau_{s,\mathrm{Opt}}$ and $\rho_{s,\mathrm{Opt}}$ found in several sources (Table 2.3) are close to each other, especially if compared from the perspective of Figure 2.8, showing that $s_{c,2D}$ has a reasonable tolerance of departures of Δt_s from their optimal values.

Film Thickness and Temperature of the Secondary Column Through parameter ^2P_k [Eq. (2.69)], the separation capacities ($^2s_{co}$ and $s_{c,2D}$) of the secondary column and GC × GC as a whole depend [Eqs. (2.68) and (2.70)] on the retention factor (k_{end}) at the end of analysis in the secondary column. As mentioned earlier, the second-dimension analysis time ($^2t_{anal}$) can be allocated differently between the hold-up time (^2t_M) and $^2k_{end}$ [Eqs. (2.30) and (2.31)]. When $^2t_{anal}$ is fixed, ^2t_M can be increased only at the expense of lowering $^2k_{end}$, and vice versa. What is the optimal choice of parameters ^2t_M and $^2k_{end}$ at a given $^2t_{anal}$?

In a column with a given diameter and flow rate, ^2t_M can be raised by raising the column length. This increases the column efficiency (^2E). As a result, the trade-off between $^2k_{end}$ and ^2t_M becomes the trade-off between the impacts [Eq. (2.27)] of parameters $^2k_{end}$ and ^2E on $^2s_{co}$. In Eq. (2.69), this impact is

TABLE 2.3 Optimal Dimensionless Sampling Periods ($\Delta\tau_{s,\mathrm{Opt}}$) and Optimal Sampling Densities ($\rho_{s,\mathrm{Opt}}$)[a]

	Source[b]							
	[1]	[2]	[3]	[4]	[5]	[6]	[7]	[7]
Conditions	—	—	LI, SD	—	—	LI, SD	LI, WD	LI, SD
$\Delta\tau_{s,\mathrm{Opt}}$	2	1.5	1.7	1.3	4	1.4	2	1.4
$\rho_{s,\mathrm{Opt}}$	0.5	0.7	0.6	0.75	0.25	0.7	0.5	0.7

[a]From several sources (in chronological order). LI, linear interpolation; SD, strong decompression in secondary column ($^2\Delta p \gg {^2p_o}$); WD, weak decompression in secondary column ($^2\Delta p \ll {^2p_o}$).
[b][1] Murphy et al., 1998; [2] Seeley, 2002; [3] Blumberg, 2003; [4] Khummueng et al., 2006; [5] Horie et al., 2007; [6] Blumberg et al., 2008; [7] Blumberg, 2008a.

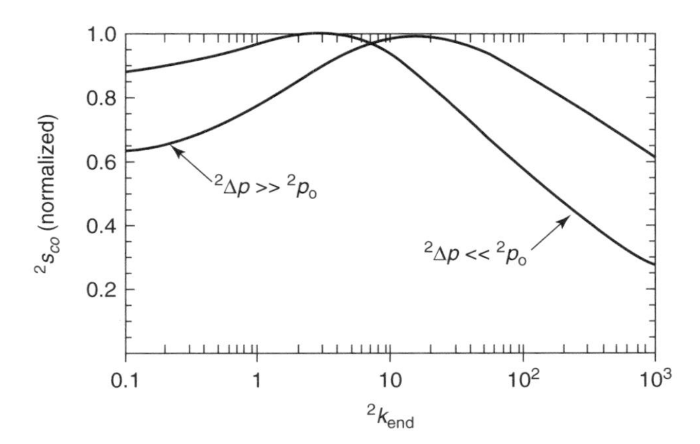

Figure 2.9 Normalized dependence (^2P_k) [Eq. (2.69)] of separation capacity ($^2s_{co}$) [Eq. (2.68)] of a secondary column on the retention factor ($^2k_{end}$) at the end of the sampling period.

reduced to the effect of a single parameter, $^2k_{\text{end}}$, on ^2P_k, which has a maximum (Figure 2.9) at†

$$^2k_{\text{end,Opt}} = \begin{cases} 3, & ^2\Delta p \ll {}^2p_o \\ 15, & ^2\Delta p \gg {}^2p_o \end{cases} \tag{2.75}$$

$$^2t_{M,\text{Opt}} = \begin{cases} 0.25\Delta t_s, & ^2\Delta p \ll {}^2p_o \\ 0.063\Delta t_s, & ^2\Delta p \gg {}^2p_o \end{cases} \qquad (\text{at } {}^2k_{\text{end}} = {}^2k_{\text{end,Opt}}) \tag{2.76}$$

The curves in Figure 2.9 suggest that the quantity $^2s_{\text{co}}$ is tolerant of departures of $^2k_{\text{end}}$ from $^2k_{\text{end,Opt}}$. Thus, at strong gas decompression in the secondary column, $^2s_{co}$ remains above 90% of its maximum when $3 < {}^2k_{\text{end}} < 80$. The weak dependence of $^2s_{co}$ on $^2k_{\text{end}}$ reduces the effect of the gas decompression on $^2k_{\text{end,Opt}}$. Thus, when $3 < {}^2k_{\text{end}} < 15$, $^2s_{co}$ remains above 90% of its maximum at any gas decompression.

Retention factors of components of a given sample depend on the stationary-phase film thickness and on column temperature. Each of these parameters can be chosen to assure that no solute resides in the secondary column longer than $^2t_{\text{anal}} + {}^2t_M$. Generally, it is better to use the film thickness for lowering MDC by having the thickest possible film that does not significantly affect the column plate height. In this case, optimal mapping of second-dimension retention factors can be obtained by using a separate oven for independent control of the secondary column temperature.

†It was assumed previously (Blumberg, 2003; Blumberg et al., 2008) that $k_{\text{end}} \approx 4$. The difference between this assumption and the data in Eq. (2.75) is the source of some differences between numerical data in previous reports and forthcoming numerical data in this chapter.

Column Diameters Column diameters are an important factor of GC × GC operation. Typically (Beens et al., 2005; Dallüge et al., 2002; David et al., 2008; Focant et al., 2004a; Marriott et al., 2003; Tranchida et al., 2007, 2008, 2009), a secondary column is narrower than the primary column (${}^2d_c < {}^1d_c$), although there is no theoretical or experimental evidence that this offers the best trade-off among peak capacity, analysis time, and MDC. To the contrary, because the column diameter significantly affects MDC [Eq. (2.41)], the second-dimension MDC in the case of ${}^2d_c < {}^1d_c$ becomes the limiting factor for MDC of GC × GC as a whole (Harynuk et al., 2005; Marriott et al., 2000; Shellie and Marriott, 2002). An opposite choice of ${}^2d_c > {}^1d_c$ appears to offer a better trade-off among separation performance, analysis time, and MDC. That being said, for the sake of simplicity, it is assumed from now on that unless the contrary is stated explicitly, both columns in GC × GC have **equal diameters** and **flow rates**. This implies equality of specific flow rates: that is,

$$ {}^2d_c = {}^1d_c = d_c, \qquad {}^2f = {}^1f = f \tag{2.77} $$

Equality of specific flow rates implies approximate **equality of dimensionless plate heights**: that is,

$$ {}^2h \approx {}^1h \approx h \tag{2.78} $$

Finally, the presence of a secondary column at the outlet of much longer primary column having the same diameter has a **negligible** effect on operation of the primary column.

Equations (2.77) and (2.78) together with Eqs. (2.22) and (2.35) transform Eq. (2.70) into

$$ s_{c,\mathrm{2D}} = {}^2P_k X_s \left(\frac{{}^1\Delta T_{\mathrm{anal}}}{{}^1R_{TM}} - 1 \right) \left\{ \begin{array}{l} \left. \begin{array}{ll} {}^1E^{3/2}\sqrt{\dfrac{8\cdot {}^1p_o \cdot {}^1\mu_R}{\pi^3 \cdot {}^2p_o}}, & {}^1\Delta p \ll {}^1p_o \\ {}^1E^2\sqrt{\dfrac{256\cdot {}^1\mu_R \sqrt{fhp_{\mathrm{st}}\eta}}{3\pi^{7/2} d_c \cdot {}^2p_o}}, & {}^1\Delta p \gg {}^1p_o \end{array} \right\}, \quad {}^2\Delta p \ll {}^2p_o \\ \left. \begin{array}{ll} {}^1E^{4/3}\left(\dfrac{\sqrt{2}d_c \cdot {}^1p_o \cdot {}^1\mu_R^2}{\sqrt{3\pi^5 fhp_{\mathrm{st}}\eta}} \right)^{\frac{1}{3}}, & {}^1\Delta p \ll {}^1p_o \\ {}^1E^{5/3}\cdot \dfrac{2^{11/6}}{\sqrt{3\pi}} \cdot {}^1\mu_R^{2/3}, & {}^1\Delta p \gg {}^1p_o \end{array} \right\}, \quad {}^2\Delta p \gg {}^2p_o \end{array} \right. \tag{2.79} $$

where

$$^{1}\Delta T_{\text{anal}} = {}^{1}T_{\text{end}} - {}^{1}T_{\text{init}}, \qquad {}^{1}R_{TM} = {}^{1}R_{T} \cdot {}^{1}t_{M,R}$$

and, as stated earlier, the quantities d_c and f are as defined in Eq. (2.77). As also stated earlier, the quantities $^{1}t_{M,R}$ and η are assumed to be measured at one temperature (such as at 150°C) in the middle of the heating ramp.

General Design Considerations Separation capacity ($s_{c,\text{2D}}$) is an important metric of performance of GC × GC. Let's review its parameters in Eq. (2.79). Except for a constant $p_{st} = 1$ atm, all parameters in Eq. (2.79) can be selected by a method developer. Thus, d_c is a column i.d., f represents a carrier gas flow rate, parameters $^{1}p_o$ and $^{2}p_o$ depend on a choice of detector, η is a property of a carrier gas, the gas and parameter f define the dimensionless plate height, h [Eq. (2.9)].

An important column parameter is its length (L), which for a given d_c and h sets the column efficiency (E). It is frequently convenient to treat E as an independent parameter determined by the needs of a column separation performance. In that case, Eq. (2.6) can be used to find L for known d_c, E, and h. This approach is taken in Eq. (2.79), where ^{1}E is treated as an independent parameter, assuming that ^{1}L can be found from Eq. (2.6).

Other parameters of choice in Eq. (2.79) are *temperature range*, $^{1}\Delta T_{\text{anal}}$, and *normalized heating rate*, $^{1}R_{TM}$—the heating rate expressed in units of temperature per hold-up time (Blumberg and Klee, 1998, 2000b). Once $^{1}R_{TM}$ is chosen, the absolute heating rate $^{1}R_{T}$ (in units of temperature per time, such as °C/min) can be found as $^{1}R_{T} = {}^{1}R_{TM}/{}^{1}t_{M,R}$, where $^{1}t_{M,R}$ can be found from Eq. (2.33b). Three remaining parameters in Eq. (2.79) are $^{2}P_k$ [Eq. (2.69)], X_s [Eq. (2.71)], and $^{1}\mu_R$. The former two are functions of retention factor ($^{2}k_{\text{end}}$) at the end of each second dimension cycle and dimensionless sampling period ($\Delta\tau_s$), respectively. The elution mobility (μ_R) of all solutes eluting during a linear heating ramp covering a wide temperature range can be found as (Blumberg, 2010; Blumberg and Klee, 2001b) $^{1}\mu_R = 1 - \exp(-{}^{1}r)$, where ^{1}r is the primary column's *dimensionless heating rate* (Blumberg and Klee, 2000a), which in the context of this chapter can be estimated as $^{1}r = {}^{1}R_T t_{M,\text{init}}/22°\text{C}$. For example, at an optimal heating rate of $10°\text{C}/t_{M,\text{init}}$ [Eq. (2.14)], $^{1}r \approx 0.45$, yielding $^{1}\mu_R \approx 0.365$, posted in Eq. (2.23). The parameters $^{2}k_{\text{end}}$, $\Delta\tau_s$, and $^{1}R_T$ are subjects of a method developer's choice, and therefore are parameters $^{2}P_k$, X_s, and $^{1}\mu_R$.

This confirms that, indeed, all parameters of Eq. (2.79) can be controlled directly by a method developer. This also suggests that Eq. (2.79) reveals the most basic independent factors affecting $s_{c,\text{2D}}$.

Equation (2.79) does not address all specifics of designing a GC × GC system. Thus, Eq. (2.79) is silent about sampling period (Δt_s), secondary column length (^{2}L), the largest acceptable widths ($^{2}\sigma_i$ and $^{2}\sigma_{\text{filt}}$) of reinjection pulses and noise filter impulse response, and so on. These parameters may be found from the formulas provided earlier and from practical considerations.

If a dimensionless sampling period ($\Delta\tau_s$) is chosen, Δt_S can be found from Eqs. (2.63) and (2.35). If ${}^2k_{\text{end}}$ is also known, ${}^2t_M$ can be found from Eq. (2.28), where ${}^2t_{\text{anal}} = \Delta t_s$. Once ${}^2t_M$ is known, ${}^2L$ can be found from Eq. (2.34). This can be followed by finding ${}^2E$ from Eqs. (2.6) and (2.77), and that, in turn, can be followed by finding the width (${}^2\sigma_{M0}$) of unretained second-dimension peak from Eqs. (2.35) at $\mu_R = 1$. The value of ${}^2\sigma_{M0}$ affects requirements to ${}^2\sigma_i$ and ${}^2\sigma_{\text{filt}}$. As stated earlier, it is desirable that ${}^2\sigma_i < {}^2\sigma_{M0}$ and ${}^2\sigma_{\text{filt}} < {}^2\sigma_{M0}$.

Optimal values of all key parameters of a GC × GC system were described earlier in the chapter. It does not mean, however, that designing a well-performing system is automatic and requires no judgment. Take, for example, selection of parameter ${}^2k_{\text{end}}$. Optimal values (${}^2k_{\text{end,Opt}}$) of ${}^2k_{\text{end}}$ are listed in Eq. (2.75). Proper combination of secondary-column stationary-phase type, thickness, and temperature might allow a method developer to set ${}^2k_{\text{end}}$ at ${}^2k_{\text{end,Opt}}$. However, it might not be worth it to spend time on fine tuning of ${}^2k_{\text{end}}$. As noted in the discussion of Eq. (2.75) and Figure 2.9 a significant departure of ${}^2k_{\text{end}}$ from ${}^2k_{\text{end,Opt}}$ can cause only a minor reduction in $s_{c,2D}$—a minor reduction, that is, when other system components could be adjusted accordingly. Unfortunately, this is not always possible. An increase in ${}^2k_{\text{end}}$ requires almost proportional reduction in ${}^2t_M$ [Eq. (2.28)]. That, in turn, calls for a reduction in ${}^2L$ [Eq. (2.34)], which reduces ${}^2\sigma_{M0}$ [Eq. (2.35)] and therefore requires proportionally narrower reinjection pulses. This shifts the burden of increasing ${}^2k_{\text{end}}$ on a resampling device, which typically is the weakest component of GC × GC hardware. Due to their inability to generate sufficiently sharp reinjection pulses, currently available reinjection devices (modulators) act as the bottleneck of GC × GC instrumentation (Blumberg et al., 2008). To reduce the burden on resampling devices, the smallest possible ${}^2k_{\text{end}}$ should be used whenever possible. For that, the thinner film and higher temperature of the secondary column, along with a smaller difference in the polarities of the primary and secondary columns, should be used whenever possible.

An example of the effect of ${}^2k_{\text{end}}$ on the required width of reinjection pulses is just one of many practical factors that a GC × GC method developer should consider. More detailed study of this and similar factors falls outside the main topic of this chapter—the study of the separation performance of GC × GC theoretically possible and the factors preventing its current realization.

The rest of the chapter is dedicated to an analysis of the most promising case of GC × GC, the one that has the largest potential peak capacity and peak capacity gain: the case of GC × GC–MS with vacuum at the outlet of the secondary column (${}^2\Delta p \gg {}^2p_o$), causing a strong gas decompression in both columns under optimal conditions: ${}^2\Delta p \gg {}^2p_o$ and ${}^1\Delta p \gg {}^1p_o$.

2.3.4 Potential Separation Performance

To evaluate the best potential peak capacity gain obtainable from GC × GC, the best-performing GC × GC should be compared with the best-performing 1D GC having equal analysis time and equal MDC. This way of comparing GC ×

GC and 1D GC is important because it is possible, for example, to raise the peak capacity of substantially suboptimal 1D GC analysis by adding to it the secondary column and making it a GC × GC analysis. However, it would be much easier and much less expensive just to optimize the 1D GC analysis. In all forthcoming comparisons of separation performance of GC × GC and 1D GC, **the same analysis time and MDC** is always assumed in both cases. Furthermore, the potential peak capacity gain of GC × GC over 1D GC is evaluated in **this section** under the following conditions.

1. Resampling is ideal. (Reinjection pulses are sharp [Eq. (2.66)], so that ${}^2U_s = 1$ [Eq. (2.67)].)
2. Only the GC × GC–MS (${}^1\Delta p > {}^1p_o, {}^2\Delta p > {}^2p_o$) that has the largest potential peak capacity gain [Eq. (2.79)] is considered.
3. The carrier gas of choice is hydrogen, which provides the best trade-offs among peak capacity, analysis time, and MDC [Eq. (2.44), Table 2.2].
4. Column pressure does not exceed 1000 kPa and analysis time does not exceed 1 h per 100°C span of a heating ramp. This implies (Example 2.4) that 0.1 mm is the smallest diameter of the primary column that can be used for the comparison.

From Eqs. (2.79) and (2.22), one has

$$s_{c,2\mathrm{D}} = \frac{2^{11/6} X_s \cdot {}^2P_k}{\sqrt{3}\pi} \frac{{}^1E^{2/3}}{{}^1\mu_R^{1/3}} \cdot {}^1s_{\mathrm{co}} \tag{2.80}$$

which, due to Eqs. (2.71), (2.51), and (2.47), implies that

$${}^2s_{\mathrm{co}} = 2^{1/6} \cdot \frac{{}^2P_k \Delta\tau_s^{1/3}}{{}^1\mu_R^{1/3}} \cdot {}^1E^{2/3} \tag{2.81}$$

The *separation capacity* (${}^2s_{\mathrm{co}}$) *of the secondary column in GC × GC–MS is proportional to the 2/3-power of the primary column efficiency* (${}^1E$).

Some parameters of optimized GC × GC-MS are listed in Table 2.4. To focus further on the *potential* separation performance of GC × GC, three of the four **optimal conditions** summarized in Table 2.4, excluding $f = f_{Opt}$ (flow does not affect forthcoming results in this section), are assumed from now on unless the contrary is stated explicitly. Equations (2.80), (2.81), and (2.24) then yield

$$s_{c,2\mathrm{D}} = 0.92 \cdot {}^1E^{2/3} \cdot {}^1s_{\mathrm{co}} \approx \frac{\Delta T_{\mathrm{anal}}}{40^\circ\mathrm{C}} \cdot {}^1E^{5/3} \tag{2.82}$$

$${}^2s_{\mathrm{co}} = 1.8 \cdot {}^1E^{2/3} \tag{2.83}$$

TABLE 2.4 Optimal Conditions and Derived Parameters of GC x GC–MS with Ideal Resampling

	f_{Opt}	$R_{T,\text{Opt}}$	$\Delta\tau_{s,\text{Opt}}$	${}^2k_{\text{end,Opt}}$	${}^2U_s$
Value or source	Table 2.2	$10°\text{C}/t_{M,\text{init}}$	1.4	15	1
Parameter	$h = 1$, ${}^1E = ({}^1L/d_c)^{1/2}$	${}^1\mu_R = 1/2.75$	$X_s = 1$, ${}^1U_s = 0.82$	${}^2P_k = 1$	1

The potential separation capacity gain of GC × GC over its primary column undisturbed by resampling follows from Eq. (2.59) at

$$s_{c,\text{ref}} = {}^1s_{co} \tag{2.84}$$

Equations (2.59) and (2.82) yield

$$G_s = 0.92 \cdot {}^1E^{2/3} \tag{2.85}$$

Example 2.5 For primary columns of Example 2.3, $G_s \approx 85$ for ${}^1E = 890$ and ≈ 45 for ${}^1E = 350$.

With the current state of GC instrumentation ($E < 1000$), *the potential separation capacity gain of GC × GC over 1D GC is limited to 100*.

Due to Eqs. (2.61) and (2.85), the peak capacity gain (G_n) of GC × GC over its first dimension undisturbed by resampling becomes (Figure 2.10)

$$G_n = \frac{0.92}{{}^1S_{\min}} \cdot {}^1E^{2/3} \tag{2.86}$$

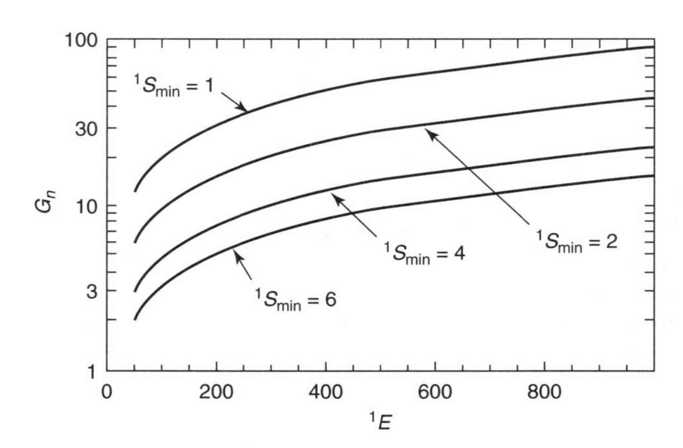

Figure 2.10 Peak capacity gain (G_n) of optimized GC × GC as a function of the efficiency (${}^1E$) of the primary column, and the separation (${}^1S_{\min}$) required for resolving the peaks along the first dimension of GC × GC.

The potential peak capacity gain (G_n) of GC × GC over its first dimension undisturbed by resampling is proportional to $^1E^{2/3}$. At the current state of the art ($^1E < 1000$, $^1S_{min} \approx 6$), $G_n < 15$.

Additional details of optimal high-efficiency implementations providing nearly the largest theoretically possible peak capacities of contemporary 1D GC and GC × GC are listed in Tables 2.5 and 2.6. Parameters of optimal implementations with conventional column are listed for comparison.

Practical implementation of the theoretically predicted performance of 1D GC is possible (Blumberg et al., 2008; Mydlová-Memersheimerová et al., 2009). What about GC × GC?

If achieved in practice, the theoretically possible 15-fold peak capacity gain in GC × GC-MS would be an enormous improvement. The same peak capacity increase in 1D GC would require a prohibitive 3400-fold-longer analysis time. To better bracket the significance of this gain, it should be mentioned that as significant as it would be, it would not solve all problems of separation of complex mixtures. Thus, the gain would not meet the challenge of Example 2.1—raising the fraction of unresolved peaks from 10% to 90% of all components of a complex test mixture. According to Eq. (2.86), additional peak capacity gain can come from a reduction in the quantity S_{min} (improvement in the data analysis along the first dimension). However, this is not the most immediate challenge of contemporary GC × GC.

TABLE 2.5 Parameters of Optimized Temperature-Programmed Analyses with Hydrogen as a Carrier Gas

	L (m)	d_c (mm)	R_T (°C/min)	σ (s)	E	s_c^a	t_{anal} (min)
High efficiency	80	0.1	2.5	0.96	890	2400	108
Conventional	30	0.25	17.2	0.36	350	950	14.7

[a] Per 100°C span of heating ramp.

TABLE 2.6 Parameters of Optimized GC x GC−MS with Primary Columns Operating with Parameters in Table 2.4

	^2L (m)	$\Delta t_{s,Opt}$ (s)	^2t_M (ms)	$^2\sigma_{M0}$ (ms)	$^2s_{co}$	$G_n{}^a$	$s_{c,2D}{}^b$
Formula	$0.4 (d_c \cdot {}^1L^2)^{1/3}$	$1.4 \cdot {}^1\sigma_o$	$0.89 \cdot {}^1\sigma_o$	$\gamma_{gas} \cdot {}^2L$	Eq. (2.83)	Eq. (2.86)	Eq. (2.53)
High efficiency	0.34	1.36	85	1.5	165	14	325×10^3
Conventional	0.24	0.51	32	1.0	90	7.5	70×10^3

[a] At $^1S_{min} = 6$.
[b] Per 100°C span of heating ramp.

Although theoretically possible and unquestionably beneficial, an order-of-magnitude peak capacity gain is numerically not overwhelming. Insufficient optimization here and there can reduce the gain substantially or even wipe it out completely. In reality, one operation in common implementations of contemporary GC × GC—too slow reinjection into the secondary column—does just that (Blumberg, 2002, 2003; Blumberg et al., 2008).

2.3.5 Slow Reinjection: The Bottleneck of Contemporary GC × GC

Slow Reinjection in GC × GC Optimized for Ideal Resampling So far, ideal resampling (sharp reinjection pulses) has been assumed in all evaluations of peak capacity of GC × GC. Again let $^2\sigma_i$ be the width of *reinjection pulses* into the secondary column and $^2\sigma_{M0}$ be the width of unretained peaks in the secondary column with sharp injection pulses. Reinjection is *slow* (*insufficiently sharp*) when

$$^2\sigma_i > {}^2\sigma_{M0} \tag{2.87}$$

Its effect on utilization (^2U_s) of the secondary column can be found as (Figure 2.11)

$$^2U_s = \frac{^2s_c}{^2s_{co}} = \frac{1}{\ln X} \ln \frac{X + \sqrt{X^2 + x^2}}{1 + \sqrt{1 + x^2}}, \qquad X = 2 +^2 k_{\mathrm{end}}, \qquad x = \frac{^2\sigma_i}{^2\sigma_{M0}} \tag{2.88}$$

In a favorable testing environment, currently available *thermal* (Beens et al., 2001; Kinghorn and Marriott, 1999; Ledford et al., 2003) and *flow-switching* (Bueno and Seeley, 2004; LaClair et al., 2004) resampling devices can generate reinjection pulses with $^2\sigma_i$ ranging between 10 and 30 ms (Beens et al., 2001; Bueno and Seeley, 2004; Gaines and Frysinger, 2004; Junge et al., 2007; LaClair et al., 2004). Even these pulses are not sufficiently sharp. Indeed, in high-efficiency GC × GC, where $^2\sigma_{M0} = 1.5$ ms (Table 2.6), $^2\sigma_i = 10$ ms reduces ^2U_s

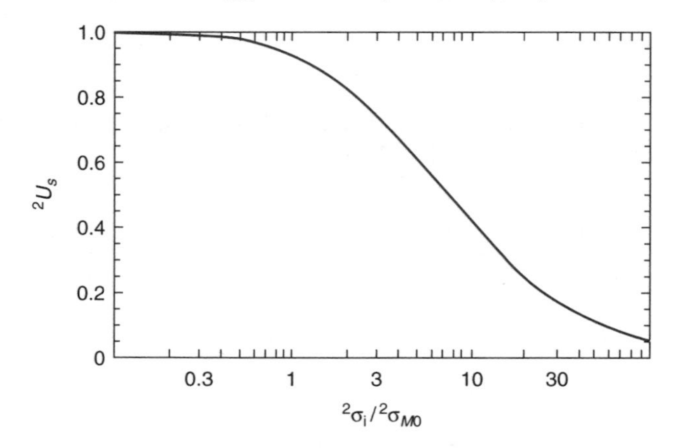

Figure 2.11 Utilization (^2U_s) [Eq. (2.87)] of the secondary column versus the relative width, $^2\sigma_i/^2\sigma_{M0}$, of reinjection pulses.

to 50%, slashing the peak capacity gain by half. Reinjection pulses 30 ms wide reduce the gain to nearly 20% of its theoretical maximum. The damage is even greater with conventional GC × GC, where ${}^2\sigma_{M0} = 1$ ms (Table 2.6).

In routine implementations of thermal resamplers, where σ_i is close to 50 ms (about 120 ms at half-height) or larger (Adahchour et al., 2006; Beens et al., 2000, 2005; Blumberg et al., 2008; Bueno and Seeley, 2004; Dimandja, 2004; Focant et al., 2004a,b; Harju et al., 2003; Harynuk and Górecki, 2007; LaClair et al., 2004; Ong and Marriott, 2002; Tranchida et al., 2007, 2009), all of the potential peak capacity gain can disappear. To be more specific, let's assume that ${}^2\sigma_i = 50$ ms. In GC × GC–MS with a conventional primary column (${}^2\sigma_{M0} = 1$ ms, Table 2.6), this results in ${}^2U_s \approx 0.11$ and a reduction in the peak capacity gain (G_n) from 7.5 to 0.8 (i.e., to the net *peak capacity loss*). This is a reflection of a general trend in the *peak generation rate* (Marriott and Morrison, 2006) (the number of resolved peaks per unit of time) that the

peak generation rate in GC × GC with existing resamplers cannot keep up with the rate obtainable in 1D GC.

Things are not as bad in high-efficiency GC × GC–MS (${}^2\sigma_{M0} = 1.5$ ms), where ${}^2U_s \approx 0.16$, reducing G_n from 14 to 2.2. The remaining gain of 2.2 is about the largest that can be obtained in GC × GC with *routinely used resamplers*. Unfortunately, it is not likely practically to realize even this modest gain. Wider reinjection pulses ($\sigma_i > 50$ ms), non-Gaussian peaks, only partial utilization of 2D separation space (Adahchour et al., 2008; Beens et al., 2000; Focant et al., 2004b; Mondello et al., 2003; Ryan et al., 2004), harmful wraparound, and other factors can reduce the gain further. Thus, less than 50% utilization of the separation space alone, observed frequently, reduces the gain by one-half. On the other hand, close to 100% utilization of the separation space is usually accompanied by massive wraparound (Dallüge et al., 2002). As a result (Blumberg, 2002; Blumberg et al., 2008),

with routinely used resamplers, one cannot expect much better than parity in the peak capacity of GC × GC with that obtainable in 1D GC.

The *parity* implies that

the largest number of single-component peaks that can be resolved by GC × GC with routinely used resamplers does not exceed the number obtainable in 1D GC.

If peak distribution is statistically uniform in both cases, then the number of single-component peaks resolvable in analysis of a given mixture by GC × GC with routinely used resamplers is about the same as the number resolvable by 1D GC.

At the time of this writing, there were no known reports demonstrating practical realization of definitively higher peak capacity of GC × GC compared to optimized 1D GC. On the other hand, Mydlová-Memersheimerová et al. (2009) have demonstrated that the number of peaks resolved in 1D GC analysis of 209 polychlorinated biphenyl (PCB) congeners can match or exceed the number of

PCBs resolved in GC × GC analyses (Focant et al., 2004b; Harju et al., 2003) with comparable or longer analysis times. The MDC was not measured in these studies. However, based on the columns used in the analyses, comparable MDC values should be expected. It is also important that only one apolar stationary phase and typical single-ramp temperature program was used in 1D GC, whereas several combinations of polar and apolar stationary phases along with fine tuning of temperature programs were used in GC × GC.

Slow Reinjection Compensated by Undersampling Not only does too-slow reinjection prevent realization of the theoretically possible peak capacity of GC × GC, but it can also destroy the ability of GC × GC to expose the structure of the sample, as illustrated in the following analysis. Due to Eqs. (2.84), (2.61), (2.59), (2.47), and (2.51), the separation capacity (${}^2s_c$) of the secondary column in GC × GC optimized for ideal resampling can be expressed as

$$ {}^2s_c = \frac{\pi}{2} \frac{G_n \cdot {}^1S_{\min}}{{}^1U_s} \tag{2.89} $$

Slow reinjection reduces the peak capacity gain (G_n). Suppose that $G_n = 1$. Substitution of $G_n = 1$, ${}^1S_{\min} = 6$, and ${}^1U_s = 0.82$ [Eq. (2.73)] in Eq. (2.89) yields ${}^2s_c \approx 11.5$ (11.5 σ-slots or barely two 6σ-spaced peaks). Together with other nonideal practical conditions, inadequate resampling can reduce the peak capacity of the second dimension to the point where it has hardly any room for more than one resolved peak along the second dimension. With almost no room for the peak separation along the second dimension, the ability of GC × GC to display structural information regarding a sample vanishes. GC × GC behaves like 1D GC.

The peak capacity (${}^2n_c$) of the second dimension can be raised by increasing its analysis time. This requires increasing the sampling period to the point of significant undersampling of *primary peaks* (the peaks at the outlet of a primary column). This does not recover the lost overall peak capacity of GC × GC, but it does help in raising ${}^2n_c$ at the expense of lowering ${}^1n_c$. Consider, for example, high-efficiency GC × GC with a sampling period $\Delta t_s = 6$s, which is 4.4 times larger that $\Delta t_{s,\text{Opt}} = 1.36$s (Table 2.6). The longer Δt_s allows for proportionally longer ${}^2t_M$ [Eq. (2.76)], which, in turn, requires a 2.7-fold longer column [Eq. (2.33a)] producing 2.7-fold wider [Eq. (2.36)] unretained peaks (${}^2\sigma_{M0} = 4$ms instead of ${}^2\sigma_{M0} = 1.5$ms in Table 2.6). This reduces the harm from too-wide reinjection pulses by raising the utilization of the secondary column from ${}^2U_s = 0.16$ to ${}^2U_s = 0.36$ (Figure 2.11). Furthermore, a 2.7-fold longer secondary column is 1.6 times more efficient [Eq. (2.33b)]. This, together with better utilization of the secondary column, raises ${}^2s_c$ from 11.5 to 41.5, making a room for almost seven (${}^2n_c \approx 7$) 6σ-spaced peaks. While raising ${}^2s_c$, the undersampling reduced the utilization (${}^1U_s$) of the primary column from an optimal 0.82 [Eq. (2.73)] to 0.3 [Eq. (2.64)], making ${}^1n_c$ 2.7-fold smaller. Nevertheless, there is a modest 30% recovery in the peak capacity gain (G_n) of GC × GC as a whole.

These observations are summarized in Table 2.7, where three methods are compared. Method 1 represents an optimal system with ideal resampling. Method

TABLE 2.7 Effect of the Sharpness of the Reinjection ($^2\sigma_i$) and Sampling Period (Δt_s) on Parameters of High-Efficiency (Table 2.5) GC x GC–MS

Method	$^2\sigma_i$ (s)	Δt_s (s)	$\Delta\tau_s$	^2L (cm)	^1U_s	^2U_s	^2s_c	G_n[a]
1	0	1.36	1.4	34	0.82	1	165	14
2	50	1.36	1.4	34	0.82	0.16	11.5	1[b]
3	50	6	6.25	94	0.3	0.36	41.5	1.3[b]

[a] $S_{\min} = 6$.
[b] Includes about a factor of 2 allowance for various nonideal conditions (see the text).

2 represents slow reinjections ($^2\sigma_i = 50$ ms) in a system optimized for ideal resampling. The slow reinjection alone reduces the peak capacity gain (G_n) to 2.2. It is assumed in Table 2.7 that the slow reinjection, combined with other earlier mentioned nonideal conditions, reduces G_n to 1. Method 3 represents slow reinjection compensated by increased oversampling ($\Delta\tau_s = 6.25$ instead $\Delta\tau_s = 1.4$ in methods 1 and 2). The fact that method 3 of Table 2.7 has a larger G_n than that of method 2 has an interesting implication.

Optimal dimensionless sampling period and sampling density [$\Delta\tau_{s,\text{Opt}}$ and $\rho_{s,\text{Opt}}$ in Eq. (2.72) and in Table 2.3] *are optimal only for ideal resampling* ($^2\sigma_i \ll {}^2\sigma_{M0}$). *For resampling with nonideal slow reinjection* ($^2\sigma_i > {}^2\sigma_{M0}$), *greater (suboptimal) undersampling of primary peaks* ($\Delta\tau_s \gg \Delta\tau_{s,\text{Opt}}$, $\rho x_s \ll \rho_{s,\text{Opt}}$) *can lead to better-performing* GC × GC *overall*.

It should be emphasized, however, that *when the reinjection is too slow, undersampling can recover only a small fraction of the G_n that is theoretically possible for ideal resampling*.

Thus, in method 1 of Table 2.7 (ideal resampling at an optimal sampling rate), $G_n = 14$, whereas in method 3 (slow reinjection compensated by undersampling), $G_n = 1.3$. Two factors prevent realization of theoretically possible large peak capacity gain ($G_n = 14$) in method 3: (1) substantial reduction in the peak capacity of the first dimension, and (2) peak capacity loss in the second dimension. Too-wide reinjection pulses [Eq. (2.87)] are the underlying cause of both factors.

Slow Reinjection and Column Dimensions Suboptimal undersampling, described in method 3 of Table 2.7, worked together with increasing the length of the secondary column from 0.34 m to 0.94 m. The latter column is close to a frequent choice (1 m × 0.1 mm) for the secondary column (Beens et al., 2005; Dallüge et al., 2002; David et al., 2008; Focant et al., 2004a; Harynuk and Górecki, 2007; Junge et al., 2007; Marriott et al., 2003; Mondello et al., 2003; Ryan et al., 2004; Tranchida et al., 2007, 2008, 2009). The undersampling imposed by too slow reinjection relaxes the requirement on the primary column performance. Indeed, the undersampling increases the first-dimension widths of resampled and reconstructed peaks. In that regard, the undersampling is similar to replacing a higher-efficiency primary column with a lower-efficiency column or to operating the primary column at suboptimal conditions, such as too slow a heating rate.

A typical choice for the primary column is a conventional 30 m × 0.25 mm or similar column (Dallüge et al., 2002; David et al., 2008; Harynuk and Górecki, 2007; Junge et al., 2007; Marriott et al., 2003; Mondello et al., 2003; Ryan et al., 2004; Tranchida et al., 2007, 2008, 2009). In a typical GC × GC–MS implementation reexamined elsewhere (Blumberg et al., 2008), the following experimental conditions were used (additional details may be found in the source reference). Dimensions of secondary columns: 1 m × 0.1 mm; carrier gas: helium at constant pressure, initial flow: 2.56 mL/min; column heating: from 50°C at 2.5°C/min; sampling period: 6 s. In forthcoming discussions, this implementation would be called *conventional suboptimal* GC × GC–MS. It is suboptimal because, to accommodate inadequate resampling, the column heating rate in it is only 2.5°C/min compared to the much faster rate of 17.2°C/min (Table 2.5) that is optimal for a conventional column. This and other suboptimal features of conventional suboptimal GC × GC–MS make its analysis time about seven times longer than the analysis time of optimal conventional GC × GC–MS and 1D GC–MS analyses (Blumberg et al., 2008).

The analysis time of the suboptimal conventional GC × GC–MS is about the same (Blumberg et al., 2008) as it is in all three high-efficiency GC × GC–MS analyses listed in Table 2.7. In addition to that, all three analyses of Table 2.7, as well as the suboptimal conventional GC × GC–MS analysis, have comparable MDC. Indeed, the MDC of a GC × GC–MS is independent of column length and is defined by a column of smaller diameter [Eq. (2.41) at $\Delta p \gg p_o$], which is a column with $d_c = 0.1$ mm in all four analyses being considered. These observations suggest that it is proper to compare the separation capacity of the suboptimal conventional GC × GC–MS analysis with that of any GC × GC–MS analysis listed in Table 2.7. Of the three methods listed in the table, only the *optimal* high-efficiency GC × GC–MS of method 1 (ideal resampling at optimal sampling period) and *suboptimal* high-efficiency GC × GC–MS of method 3 (slow reinjection with ${}^2\sigma_i = 50$ ms accompanied by suboptimal undersampling with a 6-s sampling period) are considered below.

The widths (${}^1\sigma_o$) of the primary peaks are, respectively, 0.96 s in high-efficiency GC × GC–MS (Table 2.5) and 2.1 s in suboptimal conventional GC × GC–MS (Blumberg et al., 2008). Because these analyses have about the same analysis times, the widths of their primary peaks suggest that the separation capacity (${}^1s_{co}$) of the primary column in the suboptimal conventional analysis is about two times lower than ${}^1s_{co}$ in high-efficiency analyses. The 6-s sampling period ($\Delta t_s = 6$ s) in suboptimal conventional and high-efficiency analyses represents suboptimal undersampling in both cases. However, the impact of the undersampling is greater in the latter, with its narrower peaks, than it is in the former. The undersampling substantially mitigates the twofold advantage in ${}^1s_{co}$ in high-efficiency analysis over ${}^1s_{co}$ in conventional analysis. One can find from Eqs. (2.63) and (2.65) that the first-dimension widths (${}^1\sigma$) of resampled and reconstructed peaks in the former are only about 15% narrower than they are in the latter. As a result, the separation capacity (${}^1s$) in undersampled high-efficiency

GC × GC–MS analysis is only about 15% larger than ${}^1s$ in undersampled conventional GC × GC–MS analysis.

This comparison teaches several interesting lessons. In contemporary GC × GC, resampling with about one sample per *three* ${}^1\sigma_o$, as it is in suboptimal conventional GC × GC–MS, is commonplace. Not only does it represent a sampling rate that is lower than all but one (Horie et al., 2007) optimal rate in Table 2.3, but it is also a source of the nearly two-fold reduction in ${}^1s_c$ compared ${}^1s_{co}$. This is a clear violation of the widely shared perception that for GC × GC to qualify as 2D separation, the resampling should not reduce the separation obtained in the primary column.

Moreover, in undersampled high-efficiency GC × GC–MS, there is one sample per more than *six* ${}^1\sigma_o$'s, well outside all the optimal sampling rates in Table 2.3. The undersampling reduces ${}^1s_c$ to 30% of ${}^1s_{co}$. And yet, ${}^1s_c$ in this analysis is 15% higher than it is in undersampled conventional analysis, where sampling density is twice as high. This suggests that even in the presence of substantial undersampling, using higher-efficiency primary column can improve ${}^1s_c$. *Slow reinjection in GC × GC does not justify the use of low-efficiency primary columns*.

The definition of a multidimensional analysis in this chapter imposes no requirement on the relationship between ${}^1s_c$ and ${}^1s_{co}$.

Does the suboptimal conventional GC × GC – MS provide a meaningful peak capacity gain (G_n) relative to the peak capacity of its primary column? The answer to this question is positive (Blumberg et al., 2008). The peak capacity of suboptimal conventional GC × GC–MS is significantly larger that that of its own primary column. However, this answer does not tell the entire story. The suboptimal conventional GC × GC–MS runs more than seven times longer and has 10 times higher (worse) MDC [Eq. (2.41) at $\Delta p \gg p_o$] compared to its *optimal* 1D counterpart. The peak capacity gain in suboptimal conventional GC × GC–MS comes at the expense of severely suboptimal operation of its primary column and is therefore not indicative of the ability of GC × GC to have a higher peak capacity than that of 1D GC that has the same analysis time and MDC.

Proper comparison of the peak capacity of suboptimal conventional GC × GC–MS with that of 1D GC is a comparison of that of the former with the peak capacity of optimal high-efficiency 1D GC–MS. Both analyses have the same analysis time (40 min per 100°C) and the same MDC (defined by 0.1-mm i.d. columns in both cases).

It has been shown elsewhere (Blumberg et al., 2008) that the peak capacity, $n_{c,\mathrm{2D,conv}}$, of suboptimal conventional GC × GC–MS is approximately equal to that of optimal high-efficiency 1D GC–MS. The same result follows from the earlier discussed comparison of $n_{c,\mathrm{2D,conv}}$ with the peak capacity of the high-efficiency GC × GC–MS of method 3 in Table 2.7, which is only about 30% higher than the peak capacity of high-efficiency 1D GC-MS. These observations support the previously stated conclusion that peak capacities generated in currently known routine GC × GC analyses are not significantly higher and may even be lower than the peak capacities obtainable by 1D GC analyses in a similar time frame and having a similar MDC.

Considerations for Designing Resampling Devices The previous discussion clearly indicates that the key to practical realization of potential separation performance of GC × GC is to improve significantly the sharpness of reinjection pulses. The following property of GC × GC is relevant to designing optimal resamplers. Equation (2.38) shows that the primary column has higher loadability than that of a much shorter secondary column. By increasing the amount of sample injected in the primary column and venting out a large portion of it during (e.g., flow switching) reinjection into the secondary column might help to sharpen the reinjection pulses (Blumberg, 2008a). Take, for example, optimized high-efficiency GC × GC–MS. The loadability of its 80-m-long primary column in optimized high-efficiency GC × GC is 15 times larger than loadability of its 0.34-m-long (Table 2.6) secondary column: therefore injecting 15 times more sample in the primary column and then using a 14 : 1 split at the reinjection would yield the same MDC while helping to sharpen the reinjection and substantially improve the peak capacity overall.

Furthermore, one might accept a degraded MDC in order to further sharpen the reinjection pulses by increasing the *split ratio* further. As stated earlier, trade-offs can be made among the MDC, analysis time, and/or peak capacity [Eq. (2.44)]. It is important, however, that each fraction reinjected be representative samplings of *all* analytes emerging from the primary column during the sampling period.

2.3.6 Class Selectivity and the Number of Peaks Resolved

It has been earlier suggested that, due to inadequate resampling, it is not currently possible for GC × GC to yield definitively higher peak capacity than that obtainable by 1D GC. That being said, are there cases wherein GC × GC might resolve more (single-component) peaks than 1D GC can in analysis of the same mixture? The answer to this question is positive, but only under certain conditions.

GC × GC has a much higher ability than 1D GC to separate *classes* of components in complex mixtures. One type of *class selectivity* readily available in GC × GC is *polarity-based selectivity*. In GC × GC with an *apolar* primary column and *polar* secondary column (Adahchour et al., 2008; Dallüge et al., 2002; Dimandja, 2004; Marriott et al., 2003; Tranchida et al., 2007, 2008), apolar peaks are retained much more strongly in the primary column than they are in the secondary column (Adahchour et al., 2008; Beens et al., 2000; Dallüge et al., 2002). As a result, apolar peaks tend to be grouped at the bottom of the 2D separation space (Phillips and Xu, 1995; Tranchida et al., 2009), as shown in Figure 2.4. An idealized illustration of this is shown in Figure 2.12d. The absence of apolar peaks in the bulk of 2D separation space reduces its saturation and can significantly increase the number of resolved polar peaks. In a similar manner, a reverse column configuration with polar primary and apolar secondary column (Marriott and Morrison, 2006; Mondello et al., 2003; Ryan et al., 2004, 2005) can improve the separation of the apolar peaks.

This approach works only for *oversaturated* analyses ($m > 0.5n_{c,\mathrm{2D}}$) and samples containing a chromatographically isolatable class of components of little or no analytical interest. The upper limit of $p_{\mathrm{sing}} = 0.5e^{-1}n_{c,\mathrm{2D}}$ [Eq. (2.20)] is

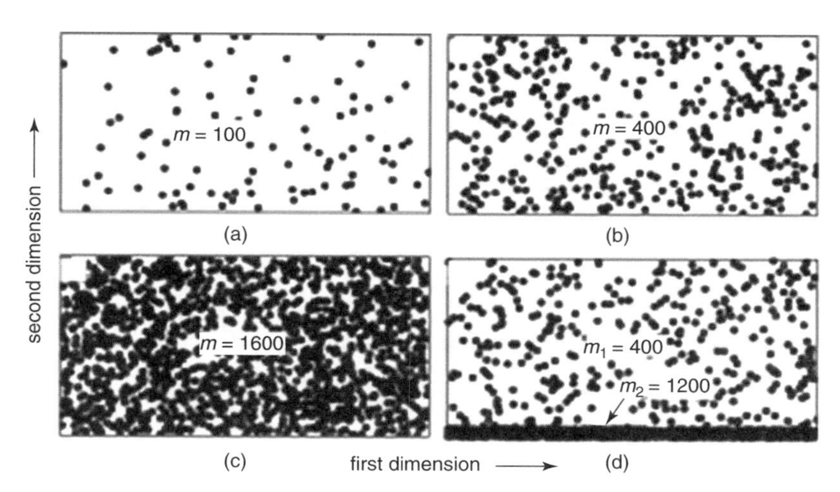

Figure 2.12 Statistically uniform (a–c) and nonuniform (d) distribution of circular spots in a 2D plane having a spot capacity $n_{c,2D} = 800$. The quantities m in (a–c) and $m_1 + m_2$ in (d) are the number of components in a sample. In (d), m_1 components have statistically uniform distribution in 2D space, while m_2 components are distributed along the bottom of the space. The expected numbers (p_{sing}) [Eq. (2.20)] of resolved (not overlapping) spots are (a) 88, (b) 147, (c) 29, (d) 147.

TABLE 2.8 Upper Limits of Expected Number (p_{sing}) of Resolved Single-Component Peaks per 100°C of a Heating Ramp in Multicolumn and GC × GC Analyses with Commercial Instrumentation[a]

	Number of Columns in Multicolumn Analysis				GC × GC	
	1	2	3	4	$G_n = 1$	$G_n = 10$
$p_{sing,max}$ at $S_{min} = 6$	75	125	160	190	75	750
$p_{sing,max}$ at $S_{min} = 1$	440	740	960	1150	440	4400

[a]Based on Table 2.5 data for a high-efficiency column.

reached when exactly $0.5n_{c,2D}$ peaks remain in 2D separation space, while other peaks of no analytical interest are lumped into a small fraction of the space (Table 2.8). In *undersaturated* analysis where $m \le 0.5n_{c,2D}$, a statistically uniform distribution of all peaks leads to the largest p_{sing}. The combined effect (Figure 2.13) for any m can be described as

$$p_{sing} = \begin{cases} me^{-2m/n_{c,2D}}, & m \le 0.5n_{c,2D} \\ \dfrac{n_{c,2D}}{2e}, & m > 0.5n_{c,2D} \end{cases}$$

$$= n_{c,2D} \begin{cases} \alpha e^{-2\alpha}, & \alpha \le 0.5 \\ \dfrac{1}{2e}, & \alpha > 0.5 \end{cases} = m \begin{cases} e^{-2\alpha}, & \alpha \le .5 \\ \dfrac{1}{2\alpha e}, & \alpha > 0.5 \end{cases}, \qquad \alpha = \frac{m}{n_{c,2D}} \tag{2.90}$$

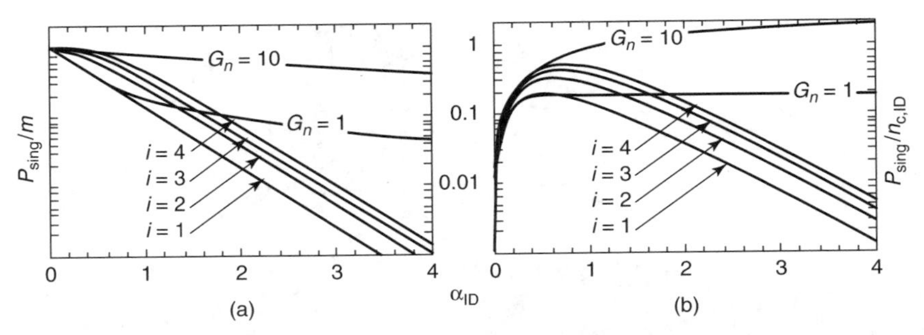

Figure 2.13 Expected number (p_{sing}) of single-component peaks resolved in different analyses. The curves at $G_n = 1$ and $G_n = 10$ represent GC × GC analyses having peak capacities $n_{c,1D}$ and $10n_{c,1D}$ and fully utilized class selectivity [Eq. (2.90)] whenever the total number (m) of solutes exceeds half of the peak capacity of the analysis. The curves marked $i = 1$ through $i = 4$ represent the number [Eq. (2.91)] of peaks resolved in at least one of i 1D analyses having the same peak capacity, $n_{c,1D}$, but sufficiently different retention mechanisms.

It is probably true that the ideal case of statistically uniform distribution of exactly $0.5n_{c,2D}$ peaks in 2D separation space in oversaturated GC × GC analysis is a practical impossibility. If so, Eq. (2.90) describes the upper limit rather than a realistic value of p_{sing} in oversaturated analysis. Nevertheless, Eq. (2.90) and Figure 2.13 explain and illustrate why and how GC × GC *can* resolve more peaks than 1D GC with equal peak capacity sometimes does. It also reveals the following limitations of class selectivity:

GC × GC can resolve significantly more peaks than can 1D GC with equal peak capacity but only when, due to oversaturation of its separation space, 1D GC resolves only a small fraction of all components in a sample. The number of peaks resolved in any analysis (1D or 2D with or without class selectivity) cannot exceed 18.4% of its peak capacity.

2.3.7 Other Ways of Resolving More Peaks

Since contemporary GC × GC has approximately the same peak capacity as that in 1D GC, other means of increasing the number of resolved peaks should be considered. In a *multicolumn* GC configuration (i × GC, where i is the number of columns), the sample is analyzed (sequentially or simultaneously) by i columns having *different retention mechanisms* (Chu and Hong, 2004; Davis and Blumberg, 2005). A peak is considered to be *resolved* in i × GC if it is resolved in at least one of i columns. If peak distribution in a single-column analysis is statistically uniform, the number (p_{sing}) of single-component peaks resolved in i × GC can be found as (Davis and Blumberg, 2005) (Figure 2.13)

$$p_{\text{sing}} = m[1 - (1 - e^{-2\alpha})^i] \tag{2.91}$$

Suppose that 1D GC (i.e., 1 × GC) and GC × GC configurations have equal peak capacities (i.e., $G_n = 1$). Figure 2.13 shows that with $i > 1$, the number

(p_{sing}) of peaks expected statistically that can be resolved by $i \times$ GC is larger than that in 1D GC or GC $\times$ GC without class selectivity. In the case of oversaturation, p_{sing} in $i \times$ GC is nearly proportional to i (Davis and Blumberg, 2005). Thus, a dual-column analysis can resolve almost twice as many peaks as can either an oversaturated single-column analysis or GC $\times$ GC analysis without class selectivity.

On the other hand, Figure 2.13 illustrates once more the power of class selectivity. When $\alpha > 1.3$, GC $\times$ GC with fully utilized class selectivity can resolve more peaks than can a dual-column analysis. At even higher saturations, a multicolumn approach quickly runs out of a reasonable number of columns necessary to resolve more peaks than a class-selective GC $\times$ GC can resolve.

The potential ability of GC $\times$ GC to provide superior separation performance is illustrated in Figure 2.13 and summarized in the following conclusions based on theoretical considerations discussed in this chapter.

GC $\times$ GC has the potential to provide at least an order-of-magnitude larger peak capacity than can 1D GC. Practical realization of this theoretical potential is currently limited by available resampling devices. When the hardware problems are solved, the separation performance of GC $\times$ GC will be out of reach for any known alternative GC technique.

Further significant increase in peak capacity and therefore in the overall separation performance of GC $\times$ GC can be obtained through the improved ability of a data analysis system to resolve poorly separated peaks along the first dimension.

The role of these two factors in raising the peak capacity gain (G_n) of GC $\times$ GC over 1D GC is described in Eq. (2.61). Solving the hardware problems can raise the separation capacity gain (G_s) of GC $\times$ GC over 1D GC up to theoretical limits described in this chapter. The goal of improving the data analysis system is to increase the peak capacity of GC $\times$ GC through reduction of the minimal separation (${}^1S_{min}$) required for resolving the peaks along the first dimension. The theoretical limit of reduction in ${}^1S_{min}$ is currently unknown.

REFERENCES

Adahchour M, Beens J, Vreuls RJJ, Brinkman UAT. *Trends Anal. Chem*. 2006; 25:821–840.

Adahchour M, Beens J, Brinkman UAT. *J. Chromatogr. A* 2008; 1186:67–108.

Ambrose D, James AT, Keulemans AIM, Kováts E, Röck H, Rouit F, Stross FH. *Pure Appl. Chem*. 1960; 1:177–186.

Beens J, Blomberg J, Schoenmakers PJ. *J. High Resolut. Chromatogr*. 2000; 23:182–188.

Beens J, Adahchour M, Vreuls RJJ, van Altena K, Brinkman UAT. *J. Chromatogr. A* 2001; 919:127–132.

Beens J, Janssen H-G, Adahchour M, Brinkman UAT. *J. Chromatogr. A* 2005; 1086: 141–150.

Bertsch W. In: Cortes HJ, Ed., *Multidimensional Chromatography: Techniques and Applications*. New York: Marcel Dekker, 1990, pp. 74–144.

Blumberg LM. *Chromatographia* 1995; 41:15–22.

Blumberg LM. *J. High Resolut. Chromatogr*. 1997a; 20: 704.

Blumberg LM. *J. High Resolut. Chromatogr*. 1997b; 20:597–604.

Blumberg LM. *J. High Resolut. Chromatogr*. 1997c; 20:679–687.

Blumberg LM. *J. High Resolut. Chromatogr*. 1999; 22:403–413.

Blumberg LM. *Anal. Chem*. 2002; 74: 503A.

Blumberg LM. *J. Chromatogr. A* 2003; 985:29–38.

Blumberg LM. *J. Sep. Sci*. 2008a; 31:3358–3365.

Blumberg LM. *J. Sep. Sci*. 2008b; 31:3352–3357.

Blumberg LM. *Temperature-Programmed Gas Chromatography*. Weinheim: Wiley-VCH, 2010.

Blumberg LM, Berger TA. *Anal. Chem*. 1993; 65:2686–2689.

Blumberg LM, Klee MS. *Anal. Chem*. 1998; 70:3828–3839.

Blumberg LM, Klee MS. *Anal. Chem*. 2000a; 72:4080–4089.

Blumberg LM, Klee MS. *J. Microcol. Sep*. 2000b; 12:508–514.

Blumberg LM, Klee MS. *Anal. Chem*. 2001a; 73:684–685.

Blumberg LM, Klee MS. *J. Chromatogr. A* 2001b; 918/(1):113–120.

Blumberg LM, Klee MS. *J. Chromatogr. A* 2001c; 933:1–11.

Blumberg LM, Klee MS. *J. Chromatogr. A* 2001d; 933:13–26.

Blumberg LM, Klee MS. *J. Chromatogr. A* 2010; 1217:99–103.

Blumberg LM, Wilson WH, Klee MS. *J. Chromatogr. A* 1999; 842(1–2): 15–28.

Blumberg LM, David F, Klee MS, Sandra P. *J. Chromatogr. A* 2008; 1188:2–16.

Bueno PA Jr, Seeley JV. *J. Chromatogr. A* 2004; 1027:3–10.

Bushey MM, Jorgenson JW. *Anal. Chem*. 1990; 62:161–167.

Chu S, Hong C-S. *Anal. Chem*. 2004; 76:5486–5497.

Coleman D, Vanatta L. *Am. Lab*. 2009; 41: 50, 52.

Consden R, Gordon AH, Martin AJP. *Biochem. J*. 1944; 38:224–232.

Crummett WB, Amore FJ, Freeman DH, Libby R, Laitinen HA, Phillips WF, Reddy MM, Taylor JK. *Anal. Chem*. 1980; 52:2242–2249.

Dal Nogare S, Juvet RS. *Gas–Liquid Chromatography: Theory and Practice*. New York: Wiley, 1962.

Dallüge J, van Stee LLP, Xu X, Williams J, Beens J, Vreuls RJJ, Brinkman UATh. *J. Chromatogr. A* 2002; 974:169–184.

David F, Tienpont B, Sandra P. *J. Sep. Sci*. 2008; 31:3395–3403.

Davis JM. *Anal. Chem*. 1991; 63:2141–2152.

Davis JM. *Anal. Chem*. 1993; 65:2014–2023.

Davis JM. *Anal. Chem*. 1994; 66:735–746.

Davis JM, Blumberg LM. *J. Chromatogr. A* 2005; 1096:28–39.

Davis JM, Giddings JC. *Anal. Chem*. 1983; 55:418–424.

Davis JM, Giddings JC. *Anal. Chem*. 1985; 57:2168–2177.

Davis JM, Stoll DR, Carr PW. *Anal. Chem*. 2008; 80:461–473.

Dimandja J-MD. *Anal. Chem*. 2004; 76: 167A–174A.

Erni F, Frei RW. *J. Chromatogr*. 1978; 149:561–569.

Ettre LS. *Pure Appl. Chem.* 1993; 65:819–872.

Ettre LS. Hinshaw JV. *Basic Relations of Gas Chromatography*. Cleveland, OH: Advanstar, 1993.

Focant J-F, Sjödin A, E. Turner W, Patterson DG. *Anal. Chem.* 2004a; 76:6313–6320.

Focant J-F, Sjödin A, Patterson DG. *J. Chromatogr. A* 2004b; 1040:227–238.

Fuller EN, Schettler PD, Giddings JC. *Ind. Eng. Chem.* 1966; 58:19–27.

Gaines RB, Frysinger GS. *J. Sep. Sci*. 2004; 27:380–388.

Giddings JC. *J. Chem. Educ*. 1962a; 39:569–573.

Giddings JC. *Anal. Chem.* 1962b; 34:722–725.

Giddings JC. In: Brenner N, Callen JE, Weiss MD, Eds., *Gas Chromatography*. New York: Academic Press, 1962c, pp. 57–77.

Giddings JC. *Anal. Chem.* 1963a; 35:1338–1341.

Giddings JC. *Anal. Chem.* 1963b; 35:353–356.

Giddings JC. *Dynamics of Chromatography*. New York: Marcel Dekker, 1965.

Giddings JC. *Anal. Chem.* 1967; 39:1027–1028.

Giddings JC. *Anal. Chem.* 1984; 56: 1258A–1270A.

Giddings JC. *J. High Resolut. Chromatogr*. 1987; 10:319–323.

Giddings JC. In: Cortes HJ, Ed., *Multidimensional Chromatography Techniques and Applications*. New York: Marcel Dekker, 1990, pp. 1–27.

Giddings JC. *Unified Separation Science*. New York: Wiley, 1991.

Giddings JC. *J. Chromatogr*. 1995; 703:3–15.

Giddings JC, Seager SL, Stucki LR, Stewart GH. *Anal. Chem.* 1960; 32:867–870.

Golay MJE. In: Coates VJ, Noebels HJ, Fagerson IS, Eds., *Gas Chromatography*. New York: Academic Press, 1958a, pp. 1–13.

Golay MJE. In: Desty DH, Ed., *Gas Chromatography*. New York: Academic Press, 1958b, pp. 36–55.

Gong F, Liang Y-Z, Chau F-T. *J. Sep. Sci*. 2003; 26:112–122.

Grushka E. *Anal. Chem.* 1970; 42:1142–1147.

Grushka E, Myers MN, Schettler PD, Giddings JC. *Anal. Chem.* 1969; 41: 889.

Guiochon G. In: Lederer M, Ed., *Chromatographic Review*. Amsterdam: Elsevier, 1966, pp. 1–47.

Guiochon G, Guillemin CL. *Quantitative Gas Chromatography for Laboratory Analysis and On-Line Control*. Amsterdam: Elsevier, 1988.

Habgood HW, Harris WE. *Anal. Chem.* 1960; 32:450–453.

Harju M, Danielsson C, Haglund P. *J. Chromatogr. A* 2003; 1019:111–126.

Harris WE, Habgood HW. *Programmed Temperature Gas Chromatography*. New York: Wiley, 1966.

Harynuk J, Górecki T. *Am. Lab. News* 2007; 39:36–39.

Harynuk J, Górecki T, de Zeeuw J. *J. Chromatogr. A* 2005; 1071:21–27.

Hinshaw JV, Ettre LS. *J. High Resolut. Chromatogr*. 1997; 20:471–481.

Horie K, Kimura H, Ikegami T, Iwatsuka A, Saad N, Fiehn O, Tanaka N. *Anal. Chem.* 2007; 79:3764–3770.

Jones WL, Kieselbach R. *Anal. Chem.* 1958; 30:1590–1592.

Jönsson JA. In: Jönsson JA, Ed., *Chromatographic Theory and Basic Principles*. New York: Marcel Dekker, 1987, pp. 27–102.

Junge M, Bieri S, Huegel H, Marriott PJ. *Anal. Chem*. 2007; 79:4448–4454.

Kaiser RE. *Z. Anal. Chem*. 1962; 189:1–14.

Kaiser RE, Rieder RI. *Chromatographia* 1975; 8:491–498.

Kaiser RE, Rieder RI. *J. High Resolut. Chromatogr*. 1979; 3:416–422.

Khummueng W, Harynuk J, Marriott PJ. *Anal. Chem*. 2006; 78:4578–4587.

Kinghorn RM, Marriott PJ. *J. High Resolut. Chromatogr*. 1999; 22:235–238.

Klee MS. In: Grob RL, Ed., *Modern Practice of Gas Chromatography*. New York: Wiley, 1995, pp. 225–264.

Klee MS, Blumberg LM. *J. Chromatogr. Sci*. 2002; 40:234–247.

Korn GA, Korn TM. *Mathematical Handbook for Scientists and Engineers*. New York: McGraw-Hill, 1968.

Krupcík J, Garaj J, Eellár P, Guiochon G. *J. Chromatogr*. 1984; 312:1–10.

Kucera E. *J. Chromatogr*. 1965; 19:237–248.

LaClair RW, Bueno PA Jr, Seeley JV. *J. Sep. Sci*. 2004; 27:389–396.

Lan K, Jorgenson JW. *Anal. Chem*. 1999; 71:709–714.

LECO Corporation (advertisement). *Am. Lab*. 2007; 39: 1.

Ledford EB, Billesbach CA, Termaat Jr. Transverse thermal modulation. U.S. Patent 6,547,852 B2, 2003.

Lee ML, Yang FJ, Bartle KD. *Open Tubular Gas Chromatography*. New York: Wiley, 1984.

Littlewood AB. *Gas Chromatography: Principles, Techniques, and Applications*, 2nd ed. New York: Academic Press, 1970.

Liu Z, Phillips JB. *J. Chromatogr. Sci*. 1991; 29:227–231.

Marriott PJ, Morrison PD. *LC-GC N. Am*. 2006; 24:1067–1076.

Marriott PJ, Kinghorn RM, Ong R, Morrison P, Haglund P, Harju M. *J. High Resolut. Chromatogr*. 2000; 23:253–258.

Marriott PJ, Dunn M, Shellie R, Morrison PD. *Anal. Chem*. 2003; 75:5532–5538.

Martin AJP, Ambrose D, Brandt WW, Keulemans AIM, Kieselbach R, Phillips CSG, Stross FH. In: Desty DH, Ed., *Gas Chromatography*. New York: Academic Press, 1958, p. xi.

Martin M, Herman DP, Guiochon G. *Anal. Chem*. 1986; 58:2200–2207.

Maurer T, Engewald W, Steinborn A. *J. Chromatogr*. 1990; 517:77–86.

Medina JC, Wu N, Lee ML. *Anal. Chem*. 2001; 73:1301–1306.

Mondello L, Casilli A, Tranchida PQ, Dugo P, Dugo G. *J. Chromatogr. A* 2003; 1019:187–196.

Motchenbacher CD, Connely JA. *Low Noise Electronic Design*. New York: Wiley, 1993.

Murphy RE, Schure MR, Foley JP. *Anal. Chem*. 1998; 70:1585–1594.

Mydlová-Memersheimerová J, Tienpont B, David F, Krupcík J, Sandra P. *J. Chromatogr. A* 2009; 1216:6043–6062.

Noij THM. Ph.D. dissertation, Technical University of Eindhoven, Eindhoven, The Netherlands, 1988.

Novák JP. *Quantitative Analysis by Gas Chromatography*. New York: Marcel Dekker, 1988.

Ong R, Marriott PJ. *J. Chromatogr. Sci*. 2002; 40:276–291.

Papoulis A. *Probability, Random Variables, and Stochastic Processes*. New York: McGraw-Hill, 1965.

Phillips JB, Liu Z. U.S. Patent 5,135,549, 1992.

Phillips JB, Xu J. *J. Chromatogr. A* 1995; 703:327–334.

Prazen BJ, Bruckner CA, Synovec RE, Kowalski BR. *J. Micro. Sep*. 1999; 11:97–107.

Repka D, Krupcík J, Benická E, Maurer T, Engewald W. *J. High Resolut. Chromatogr*. 1990; 13:333–337.

Ryan D, Shellie R, Tranchida P, Casilli A, Mondello L, Marriott P. *J. Chromatogr. A* 2004; 1054:57–65.

Ryan D, Morrison P, Marriott PJ. *J. Chromatogr. A* 2005; 1071:47–53.

Schoenmakers PJ, Marriott PJ, Beens J. *LC-GC Eur*. 2003; 16:335–339.

Seeley JV. *J. Chromatogr. A* 2002; 962:21–27.

Shao X, Wang G, Wang S, Su Q. *Anal. Chem*. 2004; 76:5143–5148.

Shellie R, Marriott PJ. *Anal. Chem*. 2002; 74:5426–5430.

Shen Y, Lee ML. *Anal. Chem*. 1998; 70:3853–3856.

Shen Y, Zhang R, Moore RJ, Kim J, Metz TO, Hixson KK, Zhao R, Livesay EA, Udseth HR, Smith RD. *Anal. Chem*. 2005; 77:3090–3100.

Steenackers D, Sandra P. *J. High Resolut. Chromatogr*. 1995; 18:77–82.

Sternberg JC. In: Giddings JC, Keller RA, Eds., *Advances in Chromatography*. New York: Marcel Dekker, 1966, pp. 205–270.

Stewart GH, Seager SL, Giddings JC. *Anal. Chem*. 1959; 31: 1738.

Tranchida PQ, Casilli A, Dugo P, Dugo G, Mondello L. *Anal. Chem*. 2007; 79:2266–2275.

Tranchida PQ, Costa R, Donato P, Sciarrone D, Ragonese C, Dugo P, Dugo G, Mondello L. *J. Sep. Sci*. 2008; 31:3347–3351.

Tranchida PQ, Purcaro G, Conte L, Dugo P, Dugo G, Mondello L. *Anal. Chem*. 2009; 81:8529–8537.

van Deemter JJJ, Zuiderweg FJ, Klinkenberg A. *Chem. Eng. Sci*. 1956; 5:271–289.

Zakaria M, Gonnord M-F, Guiochon G. *J. Chromatogr*. 1983; 271:127–192.

3

MULTIDIMENSIONAL LIQUID CHROMATOGRAPHY: THEORETICAL CONSIDERATIONS

PAVEL JANDERA
University of Pardubice, Pardubice, Czech Republic

Environmental analysis, food analysis, the analysis of biological samples (proteomics), natural products, and metabolites, and so on, pose ever-increasing demands for high-performance liquid chromatographic (HPLC) methods enabling identification and determination of large numbers of compounds in complex samples. Chromatographic resolution is essential for reliable identification and determination of sample compounds, especially if present at low concentration levels. Usually, samples containing no more than a few tens of compounds can be separated on a single HPLC column.

Two-dimensional (2D) LC methods dramatically improve possibilities for separations of a high number of sample components. Two-dimensional separations can be carried out in either the space or time domain (Guiochon et al., 2008). Planar bed techniques such as thin-layer chromatography or 2D polyacrylamide gel electrophoresis (PAGE) have long been used in 2D arrangements, especially in protein analysis. The first online direct real-time coupling of two LC columns was reported more than 30 years ago (Erni and Frei, 1978). Even though the development of necessary equipment has greatly advanced since (Bushey and Jorgenson, 1990), online 2D HPLC found practical applications no earlier than the past decade.

Even though the main focus of 2D HPLC techniques is on the increase in the number of compounds separated with respect to one-dimensional separations,

Comprehensive Chromatography in Combination with Mass Spectrometry, First Edition.
Edited by Luigi Mondello.

LC–LC offers another, equally important (even though often not fully appreciated) benefit. By careful selection of the separation systems in both the first and second dimensions, group separations of various samples based on molecular structure are often possible, so that various structurally related classes of compounds can be distinguished and located in different areas in 2D retention space, which helps in the interpretation of 2D data.

3.1 TWO-DIMENSIONAL LC TECHNIQUES

Multidimensional LC techniques have long been employed off-line in independent one-dimensional HPLC systems, as this approach is simple and easy to operate. However, off-line procedures are often time consuming, difficult to automate, and subject to sample loss or contamination. Pseudo-multidimensional separations can be achieved on a single column by using sequential applications of different selective mobile phases (Little et al., 1991) (e.g., combining solvent and pH gradients in reversed-phase separations of peptides or proteins) (François et al., 2009).

Online 2D HPLC techniques are technically more complex and less straightforward to optimize, but are more reproducible and are easy to automate. The columns in the first and second dimensions are connected via a modulator (sampling) interface. Heart-cutting techniques are commonly used for online sample transfer between two serially coupled columns, where only one or more selected fractions are collected from the effluent of the first column, to be reinjected onto the second column for analysis, providing improved resolution of compounds coeluting in the first dimension.

The separation power of the off-line and heart-cutting approaches is usually not sufficient for analyzing samples containing more than 100 to 200 relevant compounds. More suitable for this purpose is a comprehensive 2D LC $\times$ LC technique, with two columns connected serially online via a modulator-switching valve interface collecting small-volume narrow fractions from the first-column effluent for fast reinjection onto the second-dimension column in multiple repeated alternating cycles in real analysis time (Schoenmakers et al., 2003). Figure 3.1 illustrates the function of a 10-port switching valve interface with two identical sample collecting loops (Figure 3.1A) or small trapping columns (Figure 3.1B). While one loop (the trapping column) is collecting a fraction from the first dimension, the previous fraction contained in the other loop is released and analyzed in the second dimension (1). When the second-dimension separation is finished, the modulator valve is switched into the other position and the function of the loops is interchanged (2). In this way, the entire sample is subject first to separation in the first dimension and then in subsequent fractions on the second-dimension column in alternating cycles.

A detector connected to the second column records continuously chromatograms of subsequent fractions (see the example of seven consecutive second-dimension analyses of first-dimension effluent fractions containing five compounds, with a fraction collection time of 25 s, in Figure 3.2). Every sample

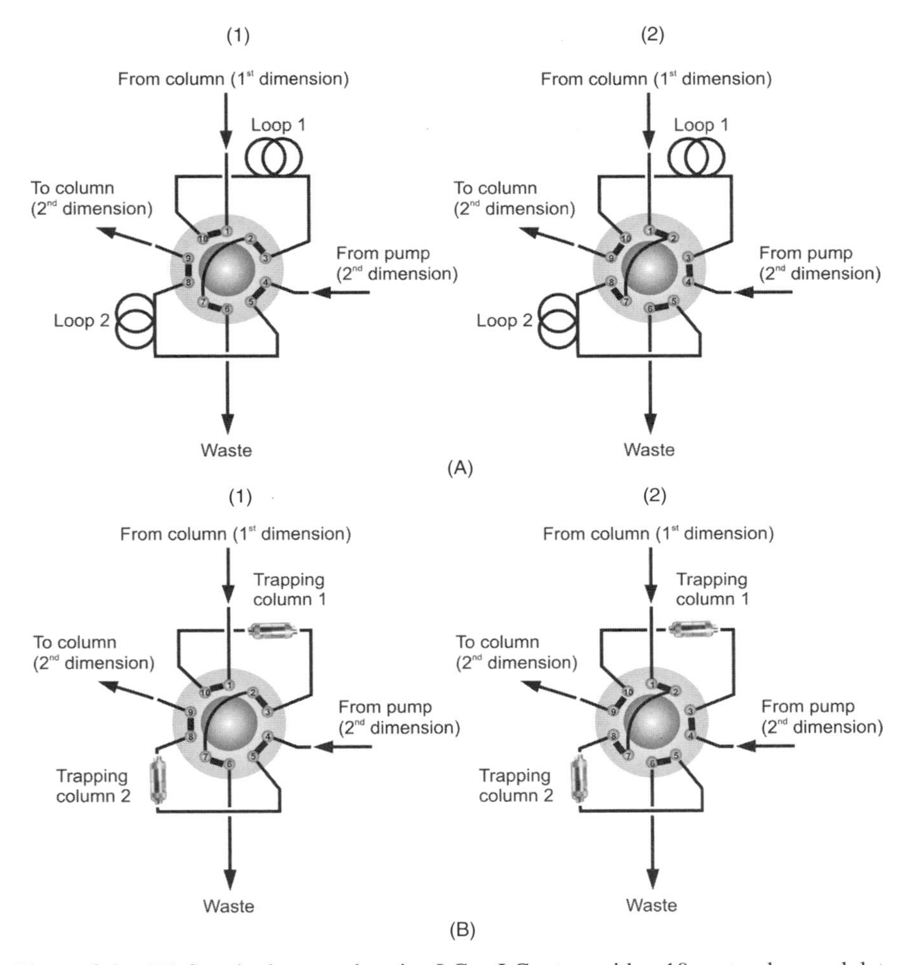

Figure 3.1 (A) Standard comprehensive LC × LC setup with a 10-port valve modulator interface and two collecting loops, operating in alternating cycles (left to right); (B) comprehensive LC × LC setup with a 10-port valve modulator interface and two trapping columns (SPE traps) substituting collecting loops, operating in alternating cycles (left to right).

compound is contained in several consecutive chromatograms of collected fractions (five in Figure 3.2) spanning over the bandwidth of a first-dimension peak. The end of each modulation period is marked automatically to allow the continuous signal from the detector at the end of the second-dimension column to be sorted out and attributed to consecutive second-dimension fractions. The individual compounds in a series of consecutive second-dimension chromatograms stacked side by side must be identified and distinguished from each other using appropriate software, such as a bilinear interpolation algorithm transferring the peaks and valleys read from the data matrix file onto a retention

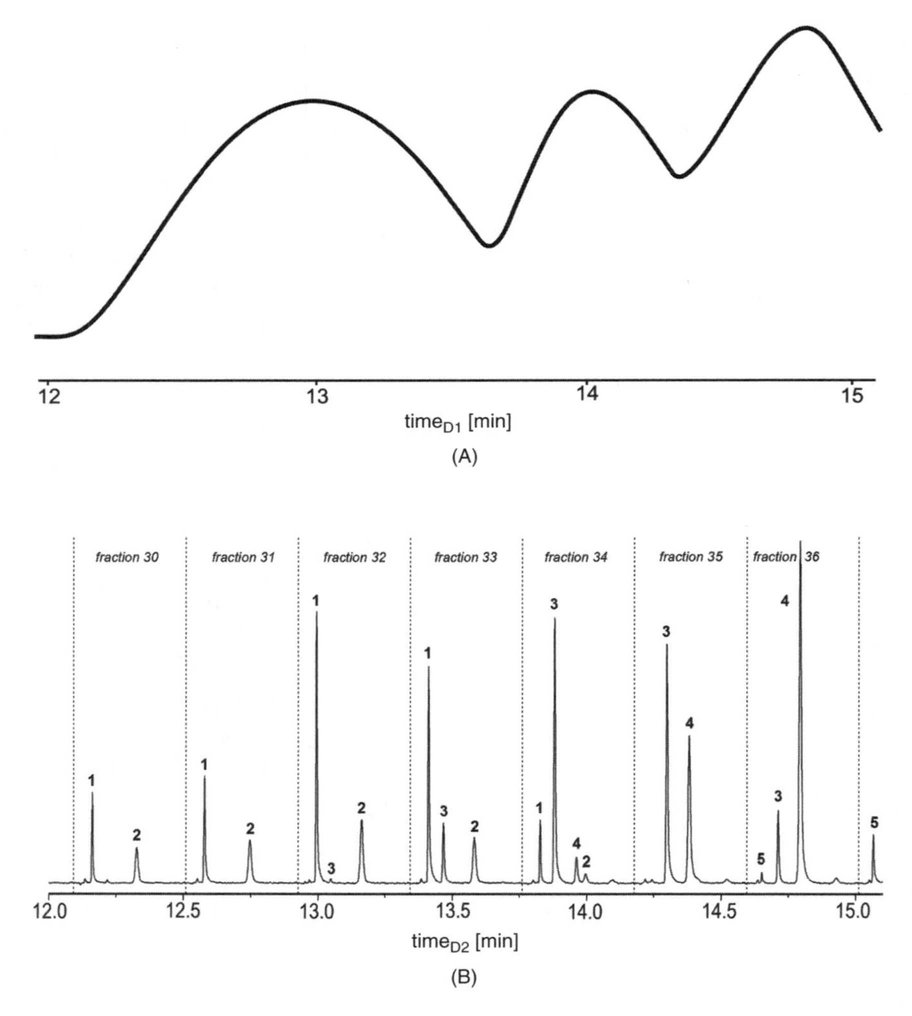

Figure 3.2 Concentration profiles in the first (A) and second (B) dimensions in a comprehensive LC × LC setup, before data processing and transformation (example). (A) Cutout of the first-dimension chromatogram (not recorded); (B) corresponding output signal of an ultraviolet detector connected to the second-dimension column: sequence of seven consecutive 25-s fractions containing five compounds.

plane with the separation times in the first and second dimensions as the coordinates (Pierce et al., 2008).

The second-dimension analysis in comprehensive LC × LC should be done in a short time, equal to the period of fraction collection, including the time necessary for the fraction transfer, usually 1 or 2 min or even less. The time constraint seriously limits the number of compounds that can be separated during one second-dimension analysis. In the stop-and-go technique, the first- and second-dimension columns are connected online via a six-port switching valve (Blahová et al., 2006). In one position, the first column operates in series with the second column, to which a fraction of the effluent is directed until the fraction volume desired has passed onto the second column. Then the six-port valve is switched to the second position, the flow of the mobile phase through the first column is stopped, and the fraction transferred is separated on the second-dimension column. When the separation is finished, the valve is switched back to the original position and mobile-phase delivery onto the first column is resumed. The entire procedure is repeated as many times as necessary. The second-dimension analysis may be limited to selected fractions presumably containing the analytes of interest, while the rest of the first-column effluent is diverted to waste, in a variation of the online heart-cutting technique. The stop-and-go technique allows increasing the number of compounds resolved with respect to continuous comprehensive LC × LC, by using a longer second-dimension column with a higher plate number separation period, at a cost of increased total 2D analysis time per sample.

As the sample components are distributed over several consecutive fractions in comprehensive LC × LC (Figure 3.2), quantitative evaluation of 2D chromatograms is not a simple task. Assuming highly reproducible control of a 2D LC × LC system, a group of peaks belonging to each target compound in the subsequent first-column effluent sampled fractions can be included in a box, within which all peak areas are summed for quantitative evaluation. This procedure requires reliable identification of peaks that should be included in the summation, preferably using additional mass spectrographic (MS) or other spectral information.

3.2 PEAK CAPACITY IN HPLC: ONE- AND MULTIDIMENSIONAL SEPARATIONS

The suitability of the LC separation system for resolving complex samples can be characterized by the theoretical peak capacity, n_c, which determines the maximum number of peaks that can be accommodated side by side in the chromatogram at a desired degree of resolution (e.g., $R_s = 1$). The separation time can be specified as the retention time interval $\Delta t_R = t_{R,z} - t_{R,1}$, between the first (1) and the last (z) sample peaks. More general characterization of the separation system is possible by selecting the time interval Δt_R between the column hold-up time and the peak of a compound with a predetermined retention factor (isocratic conditions), or between the peaks at the start and end of a gradient. Peak capacity, n_c, can be estimated from the average bandwidth, w_{av}, in an experimental chromatogram of a simple mixture, such as a mixture of homologous compounds, providing well-resolved individual peaks regularly distributed over the entire retention window, $\Delta t_R : n_c = \Delta t_R / w_{av}$. However, this approach is more suitable for gradient-elution separations, yielding similar bandwidths of sample compounds over a broad retention-time interval, than in isocratic experiments, where the bandwidths regularly increase proportionally to the retention times.

The peak capacity depends strongly on the elution mode. Under isocratic conditions, the baseline bandwidths of the sample solutes, w_i, increase proportionally to increasing retention times at a constant column efficiency for all sample compounds (number of theoretical plates, N). Assuming sufficient column efficiency ($N > 1000$), Giddings (1967) derived a simplified equation that can be used to estimate the theoretical peak capacity, n_c, under isocratic or isothermal conditions at constant resolution, $R_s = 1$:

$$n_c = \frac{t_{R,z} - t_{R,1}}{\overline{w_t}} + 1 = \frac{\sqrt{N}}{4} \ln \frac{t_{R,z}}{t_{R,1}} + 1 = \frac{\sqrt{N}}{4} \ln \frac{k_z + 1}{k_1 + 1} + 1 \qquad (3.1)$$

k_1 and k_z in Eq. (3.1) are the retention factors of the first (1) and last (z) eluting compounds, respectively [$k = (t_R - t_M)/t_M$]; $t_{R,1}$ is often set equal to t_M, the column hold-up time.

Gradient elution with programmed mobile-phase composition covers a broader range of retention factors, k; provides more narrow peaks and almost constant bandwidths, especially for strongly retained compounds (Giddings, 1967; Grushka, 1970); and consequently, higher peak capacity can be achieved in gradient-elution mode than in isocratic chromatography. Gradient elution is used primarily in reversed-phase LC, where the peak capacity at constant 4σ resolution, $R_s = 1$, can be described, to first approximation, by (Neue et al., 2001)

$$n_c \cong \frac{\sqrt{N}}{4} \left(\frac{t_{R,z}}{t_{R,1}} - 1 \right) + 1 \cong \frac{\sqrt{N}}{4} \frac{1}{t_M} \frac{\Delta t_R}{1 + k_e} + 1 \qquad (3.2)$$

k_e is the retention factor at the time of elution of peak maxima, which is considered approximately independent of the elution time in gradient elution. (It should be stressed that the plate number, N, cannot be determined *directly* as the ratio of the difference in the experimental gradient elution times and bandwidths, as the retention factors change during the elution.) Figure 3.3 illustrates significant increase in the peak capacity in gradient HPLC compared to isocratic conditions, especially at an increasing span in the retention times between the first, $t_{R,1}$, and the last, $t_{R,z}$, peaks. The time necessary for column regeneration of the column after the end of the analysis is not included. High-temperature operation also increases the peak capacity; due to the decreased viscosity of the mobile-phase, bandwidths are narrower and a higher flow rate can be used (Stoll and Carr, 2005).

As shown in both Eqs. (3.1) and (3.2), the peak capacity increases proportionally to the square root of the column plate number, N, in the isocratic as in the gradient elution mode. To a certain extent, the number of resolved peaks can be traded for the time of separation. Whereas N is directly proportional to the column length and, at a constant flow rate, to the time of analysis, the chromatographic resolution and the peak capacity increase in proportion to the square root of N, so that using a long column may increase the peak capacity only moderately, at the cost of a significantly longer separation time.

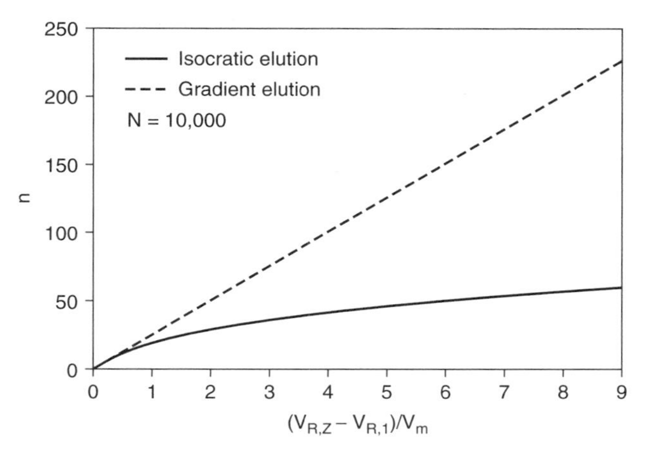

Figure 3.3 Theoretical peak capacity, n, in isocratic and programmed elution chromatography at various retention volume intervals between the elution volumes of the last, $V_{R,Z}$, and the first, $V_{R,1}$, compound, calculated using Eqs. (3.1) and (3.2). V_m is the column hold-up volume.

The theoretical peak capacity can never be achieved in practice and is even difficult to approach, since possibly except for some samples with repeat structural increments showing constant contributions to the retention, the eluting peaks do not regularly fill the entire separation time available. For one-dimensional separation of an m-component randomly distributed sample, the probability of resolving all sample components, P_m, increases with increasing peak capacity, n_c, while it decreases exponentially with the second power of the number of sample components, m:

$$P_m = e^{-(m^2/n_c)} \tag{3.3}$$

This means that the peak capacity required for separation increases with the second power of the real number of sample components (Davis and Giddings, 1983). In poorly resolved one-dimensional chromatograms of samples containing m components, the expected average number of singlets (S), overlapping doublets (D), and triplets (T) can be predicted using statistical overlap theory, assuming constant peak sizes and a Poisson distribution of the retention in complex samples (Davis, 1999):

$$S = \lambda n_c e^{-2\lambda} \tag{3.4}$$

$$D = \lambda n_c e^{-2\lambda}(1 - e^{-\lambda}) \tag{3.5}$$

$$T = \lambda n_c e^{-2\lambda}(1 - e^{-\lambda})^2 \tag{3.6}$$

where $\lambda = m/n_c$ is the saturation factor of the column peak capacity (i.e., the fraction of the maximum peak capacity really occupied by sample compounds).

The number of singlet peaks expected is maximum at $\lambda = 0.5$ and decreases rapidly at further increasing saturation factor. For example, a 100-component sample can be expected to provide only 37 clearly distinguished nonoverlapped peaks on a column with peak capacity $n_c = 200$ (Guiochon et al., 2008).

The peak capacity increases significantly in two-dimensional chromatography, where the sample is subject to separation in two different coupled separation systems. The total theoretical peak capacity of a 2D chromatographic system, $n_{c,2\mathrm{D}}$, should ideally be equal to the product of the peak capacities of the two columns, n_{c1} and n_{c2} (Giddings, 1984):

$$n_{c,2\mathrm{D}} = n_{c1} n_{c2} \tag{3.7}$$

The real 2D peak capacity in the online (comprehensive) setup is always significantly lower than that predicted from Eq. (3.7). In the practice of comprehensive LC × LC, the time for second-dimension separation is strongly limited, which means that the peak capacity in the second dimension is much lower than the peak capacity in the first dimension.

As in one-dimensional separation systems, the theoretical 2D peak capacity is much larger than the number of peaks that can really be separated in samples with randomly distributed retention. Based on statistical theory, the maximum fraction of the 2D peak capacity that can be expected to resolve the sample compounds is significantly lower than that of one-dimensional separation (Oros and Davis, 1992).

$$S = \lambda n_c e^{-4\lambda} \tag{3.8}$$

Some loss in the 2D peak capacity is caused by remixing the compounds in the fractions collected from the first dimension during transfer to the second dimension. Generally, both the first-dimension column dispersion, σ_1, and the dispersion in the interface between the first (D1)- and second-dimension (D2) columns, σ_M, contribute to the dispersion evaluated from the contour plot along the first-dimension time axis, σ_{D1}:

$$\sigma_{\mathrm{D1}}^2 = \sigma_1^2 + \sigma_M^2 \tag{3.9}$$

The first-dimension peak dispersion before the sampling, σ_1, is always smaller than σ_{D1}, by the band-broadening factor, $\beta' = \sigma_1/\sigma_{\mathrm{D1}}$, which diminishes the theoretical 2D peak capacity predicted using Eq. (3.7), especially at longer sampling times:

$$n_{c,2\mathrm{D}} = \frac{n_{c1} n_{c2}}{\beta'} \tag{3.10}$$

The average band-broadening factor, β' in the interface decreases at shorter sampling times, $t_S{}'$ (Horie et al, 2007):

$$\beta' = \sqrt{\kappa \left(\frac{t_S{}'}{\sigma_1}\right)^2 + 1} \quad (3.11)$$

Here, the parameter $\kappa = 0.21$ over the practical range of sampling times.

3.3 ORTHOGONALITY IN TWO-DIMENSIONAL LC–LC SYSTEMS

Maximum 2D peak capacity can be obtained with *orthogonal separation systems*, which employ completely different separation mechanisms in the first and second dimensions (Giddings, 1984; Guiochon et al., 2008i; Jandera, 2006). Orthogonal systems provide statistically independent retention times, corresponding to regular coverage of the available 2D retention space. In a recent study (Gilar et al., 2005), the degree of orthogonality (O) in a 2D system was defined as the normalized area covered by the eluting peaks in a separation plane divided into a discrete number of space elements (bins), which may contain one sample compound each. The degree of orthogonality is estimated as the percent coverage of the bins using an appropriately large set of a wide range of compounds. For this purpose, the retention times of representative sample components, $t_{R,i}$, are normalized within the available separation time in between the retention times of the most and least retained compounds, respectively, $t_{R,\max}$ and $t_{R,\min}$, as $t_R(\text{norm}) = (t_{R,i} - t_{R,\min})/(t_{R,\max} - t_{R,\min})$ to evaluate the coverage of the practically useful area in 2D space. An ideally orthogonal separation, in which the bins are randomly occupied by sample compounds, was considered to provide 63% coverage of a square retention space, assuming equal one-dimensional peak capacities in both dimensions. Hence, the maximum 2D peak capacity, $n_{c,\max}$, can be determined from

$$O = \frac{\sum \text{bins} - \sqrt{n_{c,\max}}}{0.63 n_{c,\max}} \quad (3.12)$$

This concept was later criticized, as in practice the first-dimension peak capacity is usually much higher than the second-dimension peak capacity; further, the value of 0.63 in the denominator of Eq. (3.12) is approached as a limiting value only at high one-dimensional peak capacity in each dimension (Watson et al., 2007).

Various LC modes can be combined in 2D LC × LC systems to suit specific separation problems. The selectivity of separation is based principally on the differences in specific and nonspecific interactions, depending on the properties of the separation system and on the structure of the sample molecules [i.e., on the size, polarity and shape, acidity–basicity, or the charge (ionic compounds)]. Very frequently, various samples are separated according to lipophilicity in one dimension, usually using a C_{18} column and gradient elution with increasing concentration of acetonitrile or methanol, providing high resolution and peak capacity, while a selective separation system (column, mobile phase) suited to the specific sample character is used in the second dimension. For example, cation-exchange chromatography coupled to reversed-phase HPLC is the most used combination in the separation of peptides according to their charge and hydrophobicity (Delmotte et al., 2009).

Both reversed-phase (RP) and normal-phase (NP) separations are based principally on the differences in polarities of the analytes. RP selectivity depends mainly on nonspecific (nonpolar) interactions with the hydrocarbon parts of sample molecules. Differences in the selective dipole–dipole or proton donor–acceptor interactions may cause major selectivity differences for various isomers in NP systems, whereas differences in the charge, acid–base properties, and nonionic interactions are the principal sources of selectivity in ion-exchange chromatography (IEC) or ion-pairing systems. Differences in molecular size are the basis of separation in size-exclusion chromatography (SEC), but may contribute more or less significantly to the separation selectivity in RP systems (and in some cases also in NP systems). In real separation systems, a "cocktail" of various interactions may affect the selectivity, which depends on both the stationary and the mobile phase and on the operation conditions, such as in hydrophilic interaction liquid chromatography (HILIC; aqueous–organic NP systems) (Jandera, 2008).

To select an orthogonal 2D system, the appropriate chemistry of the stationary phases, mobile phase, and wherever possible the operation temperature should be selected in each separation system. Usually, the differences in nonspecific interactions of samples with the stationary and mobile phases are used widely as the basis of separation in one dimension, whereas separation in the other dimension is based mainly on specific interactions, characteristic for the target classes of compounds, which, however, are often combined to some extent with nonspecific interactions, causing more or less significant correlations between systems.

Unfortunately, even careful experiment planning often cannot completely prohibit some correlations between the separation selectivities in the first and second LC–LC dimensions. When partially correlated chromatographic systems with some degree of similarity are used, the number of compounds that can be separated in a 2D setup is lower than the product of the individual peak capacities in Eq. (3.7) or (3.10). The 2D peak capacity for a nonorthogonal 2D separation decreases with increasing similarity of retention on the coupled columns.

The peak capacity in partially correlated 2D separation systems can be estimated as the weighted average of the two limiting cases: (1) a completely orthogonal 2D system and (2) two fully identical systems, 1 and 2, connected in series (Jandera et al., 2004):

$$n_{c,2D} = n_{c,1} n_{c,2}(1 - R) + \sqrt{n_{c,1}^2 + n_{c,2}^2}\,R \tag{3.13}$$

The weighting factor, R, in Eq. (3.13) is a measure of the degree of similarity (correlation) of the separation systems (i.e., the ratio of differences in free energy of the retention in the two dimensions). With noncorrelated (fully orthogonal) systems $R = 0$ and Eq. (3.13) becomes identical with Eq. (3.7). However, it is difficult to determine the appropriate correlation criterion R for practical use of Eq. (3.13).

In some 2D LC × LC systems, the retention increases in the first dimension, but at the same time, decreases in the second dimension. Such systems are inversely correlated. Figure 3.4 A illustrates this effect on the example of retention of ethylene glycol–propylene glycol (EO–PO) copolymer surfactants in partially correlated NP × HILIC systems. A bonded aminopropyl silica column is used in the two normal-phase systems, with propanol–hexane mobile phase in the normal-phase system (NP1) and ethanol–dichloromethane–water mobile phase in the HILIC system (NP2). The retention increases with an increasing number of both oxyethylene (EO) and oxypropylene (PO) repeat monomer units. In Figure 3.4B, a reversed-phase system with a C_{18} column and acetonitrile–water mobile phase (RP) and a HILIC (NP2) system are (almost) orthogonal with respect to EO units, as the RP retention is independent of the number of EO units, but the retention increases with increasing number of EO units in the NP2 system. The RP and NP2 systems show partial inverse correlation with respect to the selectivity for PO units, because the RP retention increases but the NP2 retention decreases as the number of PO units increases (Jandera et al., 2000).

3.4 SAMPLE DIMENSIONALITY AND STRUCTURAL CORRELATIONS

The statistical treatment of peak capacity is based on the assumption that the retention is a random process (Davis and Giddings, 1983). However, real samples usually contain some structurally similar analytes, which are not randomly distributed in the chromatogram. Hidden structurally based relationships providing ordered retention patterns in complex chromatograms can be decoded to some extent using the autocovariance function (Fraga et al., 2001; Marchetti et al., 2004). If the nature of structural similarities of the groups of compounds present in the sample is known, the separation conditions can often be adjusted to enhance the separation selectivity and to decrease the necessary practical peak capacity in comparison to randomly distributed noncorrelated samples.

The *chemical variance function* is based on the fact that the additivity of structural contributions to the retention can distinguish the contributions of different families of compounds in multidimensional samples to 2D retention data sets treated as matrices. It can be used to improve the resolution of overlapping peaks. This approach has been applied, for example, to the characterization of functional polymers (Vivó-Truyols and Schoenmakers, 2006).

Two-dimensional LC–LC methods are excellent tools for the characterization and separation of structurally correlated (ordered) samples with regular distribution of one or more types of characteristic structural units. Such samples show ordered peak appearance in chromatograms, which Giddings (1995) called *sample dimensionality*. For example, such behavior is observed with (poly)aromatic hydrocarbons containing different numbers of aromatic rings in petroleum products, in homologs with different alkyl lengths, and in oligomers, (homo)polymers,

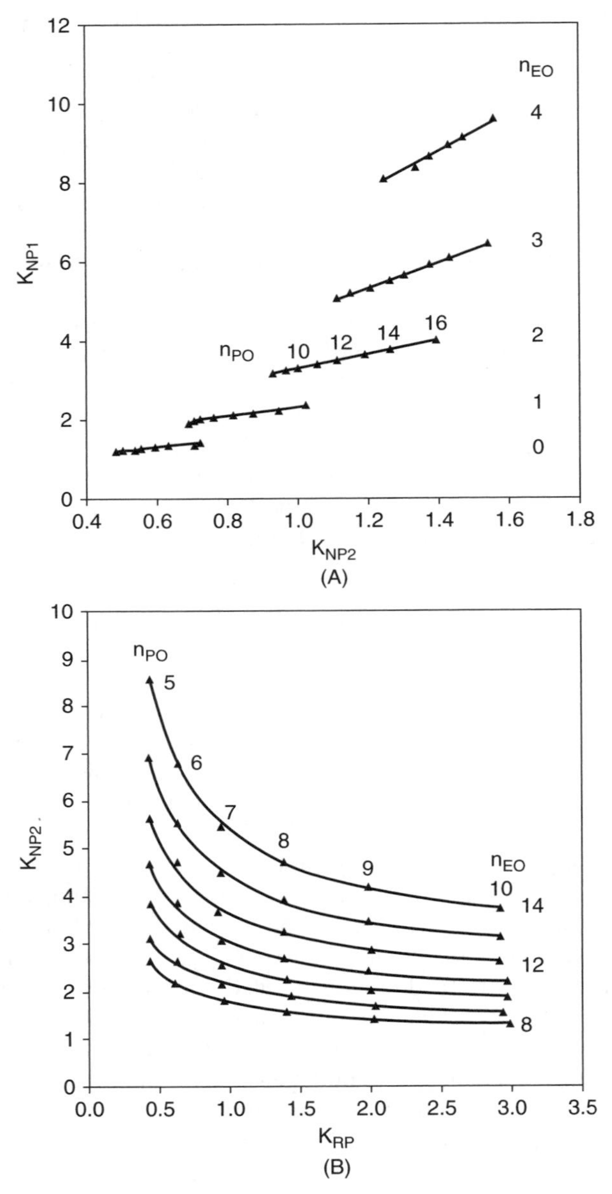

Figure 3.4 Experimental retention factors of individual oxyethylene–oxypropylene (EO–PO) (co)oligomers with varying numbers of the repeat EO and PO units in a commercial surfactant sample. (A) Partial correlation between two normal-phase systems based on an aminopropyl silica column, NP1 (propanol/hexane, 30 : 70) and NP2 (HILIC, acetonitrile/dichloromethane/water, 39.6 : 60 : 0.4); (B) absence of correlation between the NP2 HILIC system and the RP system with a bonded C_{18} column (acetonitrile/water, 50 : 50).

or (co)polymers with different types and numbers of repeat monomer units, representing regular molar mass differences (Cho et al., 2003; Jandera et al., 2000; Van der Horst and Schoenmakers, 2003), different numbers of double bonds in fatty acid esters (Dugo et al., 2006a) and carotenoids (Dugo et al., 2006b).

In many reversed-phase and normal-phase LC systems, repeat structural units show a constant additive contribution to the Gibbs free energy of retention, $-\Delta G$, which is directly proportional to the logarithm of the retention factor, k. In this case, the retention of homologs or oligomers containing n repeat structural groups can be described by (Jandera, 1984)

$$\log k = \log \beta + n \log \alpha \tag{3.14}$$

where $\alpha = k_j/k_i$ is the separation factor characterizing the repeat unit separation selectivity for two compounds, i and j, differing by one repeat structural unit, $\Delta n = 1$, with the retention factors k_i and k_j, respectively (i.e., the homologous or oligomer selectivity). The constant β in Eq. (3.14) characterizes the contribution of the nonrepeat moieties (end groups) to the retention.

Various samples can be separated into groups with the same number n of repeat structural units if the separation conditions are optimized to provide maximum repeat unit selectivity α and to suppress the effects of the differences in the nonrepeat structural elements (β) on the retention. In 2D systems, polymer species differing in molar masses are usually separated according to the molar mass distribution on an SEC column in one dimension and according to structural differences in the end groups in the other dimension in the RP or NP mode, depending on the sample lipophilicity (Van der Horst and Schoenmakers, 2003). Some oligomers containing polar or lipophilic repeat monomer units can often be separated according to the molar mass distribution (often with better resolution than in SEC) in NP or RP systems under conditions enhancing the repeat group selectivity, while the effects of the end groups on the retention are suppressed (Cho et al., 2003; Jandera et al., 2001).

The retention of ordered "two-dimensional" samples with dual structural distribution of the repeat groups A and B, contributing to the retention, can be described in a way similar to the retention of one-dimensional samples. Assuming constant regular contributions of each of the two repeat groups to the energy of retention, Eq. (3.14) can be extended to characterize the effect of dual structural distribution on the retention factor, k (Jandera et al., 2001):

$$\log k = \log \beta + n_A \log \alpha_A + n_B \log \alpha_B \qquad (k = \beta, {\alpha_A}^{n_A}, {\alpha_B}^{n_B}) \tag{3.15}$$

Here, α_A and α_B are the selectivity factors of solutes differing by one repeat structural unit A or B, respectively, n_A and n_B are the numbers of the two types of repeat structural units in the sample, and the constant β characterizes the contribution to the retention of the nonrepeat structural moieties (such as the endgroups in copolymers or co-oligomers). Equation (3.15) enables predicting the coelution of samples with various combinations of the two repeat units, n_A

and n_B, providing the same retention (k) in the separation systems with more or less correlated repeat unit separation selectivities, α_A and α_B (Jandera et al., 2005):

$$\log \alpha_A = C' \log \alpha_B \tag{3.16}$$

The correlation constant C' in Eq. (3.16) depends on the structural elements and on the stationary and mobile phases.

Two-dimensional HPLC is better suited than one-dimensional systems for group-specific separations of "two-dimensional" samples (Giddings, 1995). If Eq. (3.15) describes the retention in both the first and second dimensions, 1 and 2, the correlation between the retention factors k_1 and k_2 of the molecules with the same numbers of the A units in the two systems can be characterized by (Jandera et al., 2005)

$$\begin{aligned}\log k_2 = \log k_1 \frac{\log \alpha_{A,2}}{\log \alpha_{A,1}} + \log \beta_2 - \log \beta_1 \frac{\log \alpha_{A,2}}{\log \alpha_{A,1}} \\ + n_B \left(\log \alpha_{B,2} - \log \alpha_{B,1} \frac{\log \alpha_{A,2}}{\log \alpha_{A,1}} \right) = a + b \log k_1\end{aligned} \tag{3.17}$$

Equation (3.17) describes a set of plots with correlated selectivities, $\alpha_{i,j}$, for 2D samples with various numbers of repeat structural units A (n_A) and B (n_B), depending on the degree of correlation between the selectivities for repeat units A and B in the first ($\alpha_{A,1}, \alpha_{B,1}$) and second ($\alpha_{A,2}, \alpha_{B,2}$) dimensions.

For separations of 2D samples into groups with different numbers of A and B units in a 2D plane, LC systems yielding high α_A and low α_B in one dimension and vice versa in the second dimension should be found. In other words, the conditions should be selected for separation of compounds with different numbers of repeat unit A in the first dimension and of the species differing in the distribution of the other repeat unit, B, in the second dimension. If system 2 does not distinguish compounds with different numbers of the repeat group A, $\alpha_{A,2} = 1$ and $b = 0$ in Eq. (3.17), the retention in the second dimension, k_2, is independent of the retention in the first dimension, k_1, and of the number of repeat units A. Otherwise, the individual $k_1 - k_2$ plots for compounds with different n_B are shifted along the k_1 axis, depending on the degree of correlation between the selectivities for the repeat unit A in the two dimensions $\alpha_{A,1}$ and $\alpha_{A,2}$.

Two examples of sets of plots for EO–PO (co)oligomers used in commercial surfactant products described by Eq. (3.17) are shown in Figure 3.4. The RP–NP2 two-dimensional LC–LC system is orthogonal with respect to EO (oxyethylene) units and provides highly regular coverage of the 2D retention space (Figure 3.4B). In the NP1–NP2 system in Figure 3.4A, the separation selectivities for the EO and PO (oxypropylene) units are correlated and the compounds differing in the number of PO units do not completely fill the retention space available. Hence, the resolution of sample compounds is incomplete, and the useful 2D peak capacity for "two-dimensional" samples is significantly less than the maximum system theoretical peak capacity (Jandera et al., 2005). However, the individual classes of compounds with common structural groups can often be more or less distinguished, even in systems with partially correlated separation selectivities.

Another example of noncorrelated 2D HPLC–HPLC analysis of 2D samples reported in the literature is reversed-phase separation of the diastereomers of low-molecular-weight polystyrenes (Gray et al., 2004). In the first dimension, a C_{18} column was employed, which shows high hydrophobic selectivity for oligomers but almost no selectivity for diastereomers. The isomers were separated in the second dimension on a column packed with zirconia carbon-clad adsorbent.

3.5 SEPARATION SELECTIVITY AND SELECTION OF PHASE SYSTEMS IN TWO-DIMENSIONAL LC–LC

Generally, 2D HPLC systems should have different selectivities in the first and second dimensions. The selectivity for any two compounds, i and j, is defined by their separation factor, $\alpha = k_j/k_i$ (with $k_j > k_i$). As similar specific polar and nonspecific dispersion interactions contribute to the retention in various chromatographic systems combined in 2D HPLC, a more or less significant retention correlation is usually present not only with "ordered" but in almost all complex samples, even though structural similarities between the sample components may not be apparent at first glance. In practice, sample molecules are always subject to dispersion interactions, which may be more or less suppressed in some separation systems (Turowski et al., 2001). Even size-exclusion chromatography is partially correlated with reversed-phase HPLC, where the retention also depends on the molecular weight. Various specific effects influencing the polar contributions to retention may depend strongly on the mobile phase, adequate selection of which is as important as the selection of suitable chemistry for the stationary phase in a 2D LC–LC system with low selectivity correlations. Two-dimensional HPLC separation systems usually employ two different stationary phases, but successful separations were also reported with one stationary phase and two different mobile phases (Ikegami et al., 2006). Gradient separation of peptides on a C_{18} column at a low pH in one dimension and at a high pH in the other (François et al., 2009) is a nice example of this approach.

There is no universal "best" 2D LC–LC phase system generally suitable for all separations. The selection of a suitable two-dimensional phase system depends strongly on the type of sample, often relies on previous experience or on the information in the available literature and usually involves some trial-and-error experiments. A systematic approach to the development of 2D LC–LC separations requires the selection of a set of test compounds adequately representing a variety of components that can be expected in particular sample types. If adequate standards are not available, the sample itself can be used, especially if the separation is aimed at acquiring the information from the "fingerprint chromatogram of a complex sample." A sufficient amount of data should be collected in potentially suitable separation systems to allow a comparison of separation selectivity by plotting the retention data of all compounds in one dimension versus the data in the second dimension. Retention factors, separation factors, or dimensionless retention parameters $\xi = (t_i - t_0)/(t_z - t_1)$, measuring the distribution of the sample components in the separation space available between the retention times of the most, t_z, and the least, t_1, retained sample components, may be used for this purpose (Neue et al., 1999). Dissimilar chromatographic systems with large selectivity differences for sample compounds in the potential first- and second-dimension systems can be evaluated based on correlation coefficients, or more advanced chemometric methods, such as principal components analysis, factor analysis, chemical variance, similarity dendrograms, or the orthogonal projection approach (Dumarey et al., 2008). This strategy was used, for example, to select suitable combinations of reversed-phase columns for 2D separations of some pharmaceuticals (Van Gyseghem et al., 2003), or for orthogonal separations of phenolic acids and flavones [a bonded poly(ethylene glycol) column in the first dimension and a C_{18} column in the second dimension] (Cacciola et al., 2006; Jandera et al., 2008). A similar approach was used to select anion-exchange columns with large selectivity differences for the separation of inorganic anions in comprehensive 2D ion chromatography (Shellie et al., 2008).

Instead of direct comparison of the retention data, chromatographic columns can be classified based on quantitative structure–retention relationship (QSRR) concepts, which distinguish selective contributions of various polar interactions to the retention in liquid chromatography. QSRR concepts are statistically derived relationships between chromatographic retention parameters (usually, $\log k$ or $\log \alpha$) and molecular descriptors characterizing the structures of the test compounds (Kaliszan, 1997). Chromatographic retention is treated as a linear function of a number of solute–column–mobile phase interactions, described by a set of quantum chemical indices and molecular descriptors provided by calculation chemistry.

The linear solvation energy relationship (LSER) strategy is based on a QSRR equation using multiple correlations between the retention and solvatochromic molecular descriptors, characterizing the volume of solvated solute, V_X; polarity, π_2^*; hydrogen-bonding basicity, β_2; and hydrogen-bonding acidity, α_2 (Abraham et al., 1997):

$$\log\ k = (\log\ k)_0 + \frac{m_1 V_X}{100} + s_1\pi_2^* + a_1\alpha_2 + b_1\beta_2 \tag{3.18}$$

The retention data of a set of representative test compounds are subject to multiple regression versus the molecular descriptors, to obtain the parameters m_1, s_1, a_1, and b_1, which characterize the retention response of the chromatographic system to various structural sample characteristics via selective polar and nonpolar interactions. Suitability of various HPLC columns (mainly reversed-phase columns) for 2D LC–LC separations is evaluated by comparing the parameters m_1, s_1, a_1, and b_1 of the individual separation systems using principal components or cluster analysis (Jandera et al., 2008).

The LSER model was later modified to the hydrophobic subtraction model (HSM), disregarding the hydrophobic contribution to the retention, by subtracting the retention factor of ethylbenzene, k_{EB} (the reference nonpolar solute), which makes it possible to distinguish the relative contributions of polar interactions to the selectivity, α, in reversed-phase LC (Snyder et al., 2004):

$$\log\ \alpha \equiv \log\ \frac{k}{k_{EB}} = \eta' H - \sigma' S^* + \beta' A + \alpha' B + \kappa' C \quad (3.19)$$

Here, k is the retention factor of a test solute, measured under the same conditions as k_{EB}. In the HSM model, the selectivity-related symbols on the right-hand side of Eq. (3.19) represent the eluent- and temperature-dependent properties of the solute (η', hydrophobicity; σ', molecular size; β', hydrogen-bonding basicity; α', hydrogen-bonding acidity; κ', partial charge, positive or negative), and the eluent- and temperature-independent properties of the stationary phase. H is the hydrophobicity, S^* the steric resistance to penetration of the analyte molecule into the stationary phase, A the column hydrogen-bonding acidity, attributed to residual nonionized silanols, B the column hydrogen-bonding basicity due to the water adsorbed in the stationary phase, and C the column cation-exchange activity due to ionized silanols.

The stationary-phase characteristics (parameters) of Eq. (3.19) have been measured for more than 60 various test compounds on different columns from a 400-column database, including silica gel supports with bonded alkyl-, cyanopropyl-, phenylalkyl-, and fluoro-substituted stationary phases and columns with embedded or end-capping polar groups (Wilson et al., 2002). Based on the differences in the column characteristics of Eq. (3.19), $\Delta H = H_1 - H_2$, $\Delta S = S_1 - S_2$, $\Delta A = A_1 - A_2$, $\Delta B = B_1 - B_2$, and $\Delta C = C_1 - C_2$, the column-selectivity comparison function, F_S, was introduced to select the pairs of columns 1 and 2 with the lowest retention correlations for general applications in 2D reversed-phase systems. The relative importance of the individual selectivity contributions is attributed to empirical weighting factors:

$$F_S = \sqrt{(12.5\Delta H)^2 - (100\Delta S)^2 + (30\Delta A)^2 + (143\Delta B)^2 + (83\Delta C)^2} \quad (3.20)$$

3.6 PROGRAMMED ELUTION IN TWO-DIMENSIONAL HPLC

Programmed elution techniques provide significantly improved peak capacity with respect to isocratic elution (see Figure 3.3) and hence should be used in 2D liquid chromatography wherever possible, especially in reversed-phase systems, but also in ion-exchange systems and in nonaqueous normal-phase separation mode on polar adsorbents. Gradient elution covers a broad range of retention factors, k; hence it enables the separation of samples with widely differing retention in complex samples. There is an additional reason for using gradient elution in the first dimension of comprehensive LC $\times$ LC: Unlike isocratic HPLC, gradient conditions provide approximately constant bandwidths for samples with different retention times, which enables collecting the same number of fractions per bandwidth for second-dimension analysis at a constant sampling rate. Gradient elution is also useful for increasing the peak capacity in the second dimension; however, there are severe restrictions connected with the short separation time available, to which the gradient program should be adapted. Approaches to solving this problem are discussed at the end of the section.

In reversed-phase LC, the effect of the volume fraction of the organic solvent, ϕ, in an aqueous–organic mobile phase on the retention factors, $k = (V_R - V_m)/V_m$, in reversed-phase LC can often be characterized—at least over a more or less limited range of φ—using the well-known equation (Jandera and Churáček, 1974a; Snyder and Dolan, 1998)

$$\log k = a - m\varphi \tag{3.21}$$

where the constants a and m depend on the analyte and separation system.

Assuming the validity of Eq. (3.21), the elution volume, V_R, in reversed-phase chromatography with a linear gradient, $\phi = A + BV$, can be calculated as (Jandera and Churáček, 1974b; Jandera et al., 1979; Snyder et al., 1979)

$$V_R = \frac{1}{mB}\log[2.31mB(V_m \cdot 10^{a-mA} - V_1) + 1] + V_m + V_1 \tag{3.22}$$

Here, V is the volume of the column effluent since the start of the gradient; A is the initial volume fraction of the organic solvent, φ, at the start of the gradient and ϕ_f at the end of the gradient; and $B = (\phi_f - A)/V_G$ is the gradient steepness parameter (i.e., the change in the volume fraction of the organic solvent in the mobile phase, from A to ϕ_f, in the gradient volume, V_G). V_m is the column hold-up volume and V_1 is the volume of the first, isocratic segment with $\varphi = A$, equal to or including the instrumental gradient dwell volume, V_D (i.e., the volume in between the gradient mixer and the column top, filled with the starting mobile phase at the beginning of elution).

The bandwidths in the RP gradient chromatography, w_g, can be estimated as approximately equal to the bandwidths under isocratic conditions at the elution time of the peak maximum, with the instantaneous retention factor k_e (Jandera and Churáček, 1974b; Snyder and Dolan, 1998):

$$w_g = \frac{4V_m}{\sqrt{N}}(1 + k_e) = \frac{4V_m}{\sqrt{N}}\left(1 + \frac{1}{2.31mBV_m + 10^{mA-a}}\right) \tag{3.23}$$

The peak capacity in gradient HPLC is described by Eq. (3.2) (Grushka, 1970). It is advantageous to design one dimension of a two-dimensional LC × LC system to separate sample components on the basis of differences in lipophilicity, which, when using gradient elution, can be characterized by the *gradient lipophilic capacity*, P_l, defined as the number of peaks that can be stacked side by side in the time interval between the retention times of homologous alkylbenzenes differing by one methylene unit, assuming that $R_s = 1$. Based on Eqs. (3.2) and (3.14), P_l can be estimated as (Jandera et al., 2003)

$$P_l \cong 1 + \frac{\sqrt{N}}{4} \frac{\log\ \alpha_A}{m\Delta\varphi(V_m/V_G)(1+k_e)} \tag{3.24}$$

The gradient lipophilic capacity depends on the reversed-phase stationary phase employed. Equation (3.24) shows that P_l increases at a higher methylene selectivity, α_A [Eq. (3.14)], for more efficient columns (a higher number of theoretical plates based on the column hold-up volume) and for less steep gradients (lower B), starting at a lower concentration of the organic solvent. However, these parameters have the opposite effect on the number of homologous alkylbenzenes, Δn, eluting in a fixed gradient time period:

$$\Delta n \cong \frac{m\Delta\varphi}{\log\ \alpha_A} \tag{3.25}$$

The following discussion shows the impact of the type of gradient profile on 2D HPLC separation, illustrated by three examples of gradient separation of alkylbenzenes on a short porous-shell C_{18} column. From Figure 3.5A it can be seen that benzene and nine alkylbenzenes can be eluted within the gradient time of 1.5 min using a gradient covering the full range of acetonitrile concentration (2 to 100%) in water, even though only five alkylbenzenes eluted in the first 1.2 min are shown. A moderately steep gradient covering a narrower concentration range and starting at a higher initial concentration of acetonitrile [i.e., at a lower Δ_φ (15 to 50%)] (Figure 3.5B) shows a higher lipophilic peak capacity, P_l, in broader time intervals between the peaks, but a lower number of homologs (benzene to ethylbenzene) eluted during the gradient time. A chromatogram with a shallow gradient starting at a relatively high concentration of acetonitrile (48.5 to 50%) shows a pattern similar to those of isocratic separations, with markedly increasing bandwidths of later-eluting compounds.

Monolithic columns and columns packed with sub-2-μm or porous shell particles provide fast and efficient separations at high flow rates, so that they are excellently suited for the second dimension of comprehensive LC × LC systems. Such columns, with bonded C_{18} or C_8 stationary phases, can be used for fast RP gradient elution in the second dimension. Figure 3.6A to C shows three examples of 2D contour plot chromatograms with gradients of acetonitrile in ammonium acetate aqueous buffer run simultaneously on a porous shell Kinetex C_{18} column (50 mm × 3.0 mm i.d.) in the second dimension (solid lines) and on a bonded poly(ethylene glycol) microbore column (150 mm × 2.1 mm i.d.) in the first dimension (dashed lines) at a cycle time of 1.5 min.

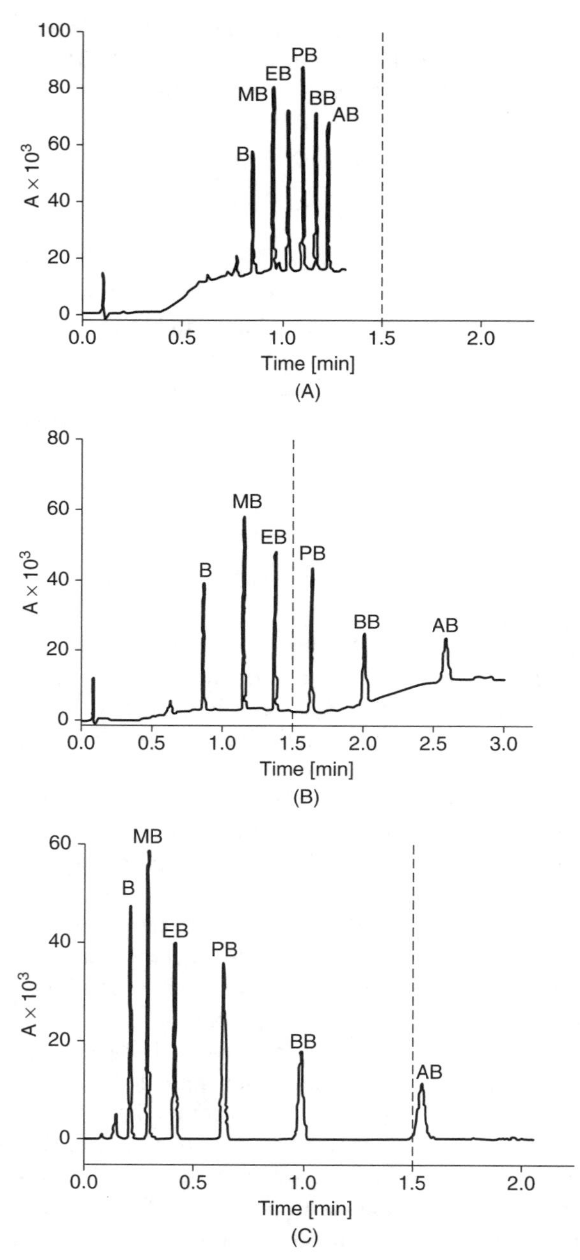

Figure 3.5 Separation of alkylbenzenes on a porous-shell Kinetex C_{18} 100A column (50 mm × 3.0 mm, 2.6 μm). Gradients of acetonitrile in water, gradient time 1 min + postgradient isocratic elution, $F_m = 2.5$ mL/min, $T = 50$°C. (A) 2 to 100% ACN; (B) 15 to 50% ACN; (C) 48.5 to 50% ACN. B, benzene; MB, methylbenzene; EB, ethylbenzene; PB, propylbenzene; BB, butylbenzene; AB, amylbenzene.

1. A very steep second-dimension gradient run in each fraction over a broad range of acetonitrile concentrations (2 to 50%) offers high bandwidth suppression; however, the time necessary for postgradient column re-equilibration to the starting conditions (0.5 min) diminishes the actual time available for separation and the actual peak capacity (Figure 3.6A). The compounds do not regularly fill the available 2D retention plane and tend to cluster around the diagonal connecting the lower left corner with the upper right corner, due to partial correlation of retention in the two RP systems. Finally, with very steep gradients it is more probable that some strongly retained compounds, which do not have enough time to elute within one fraction, may be carried over to the next fraction(s), eluting as broad peaks (such as compounds 15 and 22).
2. A less-steep second-dimension in-fraction gradient with two segments covering different ranges of acetonitrile concentration at lower and higher acetonitrile concentrations in the first dimension (the first with 2 to 30% and the second with 15 to 50%) (Figure 3.6B) provides even narrower bandwidths than the gradient in Figure 3.6A and improved resolution, despite the same postgradient re-equilibration time in each fraction. The sample components are more widely spread in the 2D retention plane and the carryover to the next fractions is less probable, because the concentration ranges of the organic solvent in the individual segments can be better adjusted to the partially correlated retention in the two dimensions (compounds 15 and 22 elute at high second-dimension times but are not carried over to the next fractions).
3. Continuous second-dimension gradients with two segments of different steepness and an isocratic hold segment in between (Figure 3.6C) covering the entire 2D separation time in parallel with the first-dimension gradient show slightly larger second-dimension bandwidths than those of the chromatograms in Figure 3.6A and B, because of shallow gradient conditions. However, the second-dimension space is covered more regularly with samples showing partially correlated retention in the two dimensions than in the other two gradient types. Further, this type of gradient offers a longer time period for the second-dimension separation, as column re-equilibration is not necessary within each individual fraction cycle time. Consequently, the total 2D separation time is shorter. The carryover of compounds retained from one fraction to the next in the second dimension can be avoided by careful adjustment of the parallel gradient profiles. For this purpose, window diagram optimization can be used for maximum resolution within the fixed separation time (Česla et al., 2009). However, as the gradient conditions gradually change slightly between subsequent fractions, minor shifts of the second-dimension elution times may be observed between the fractions, which can be corrected by software that recalculates the elution times automatically using Eq. (3.22) (Česla et al., 2009).

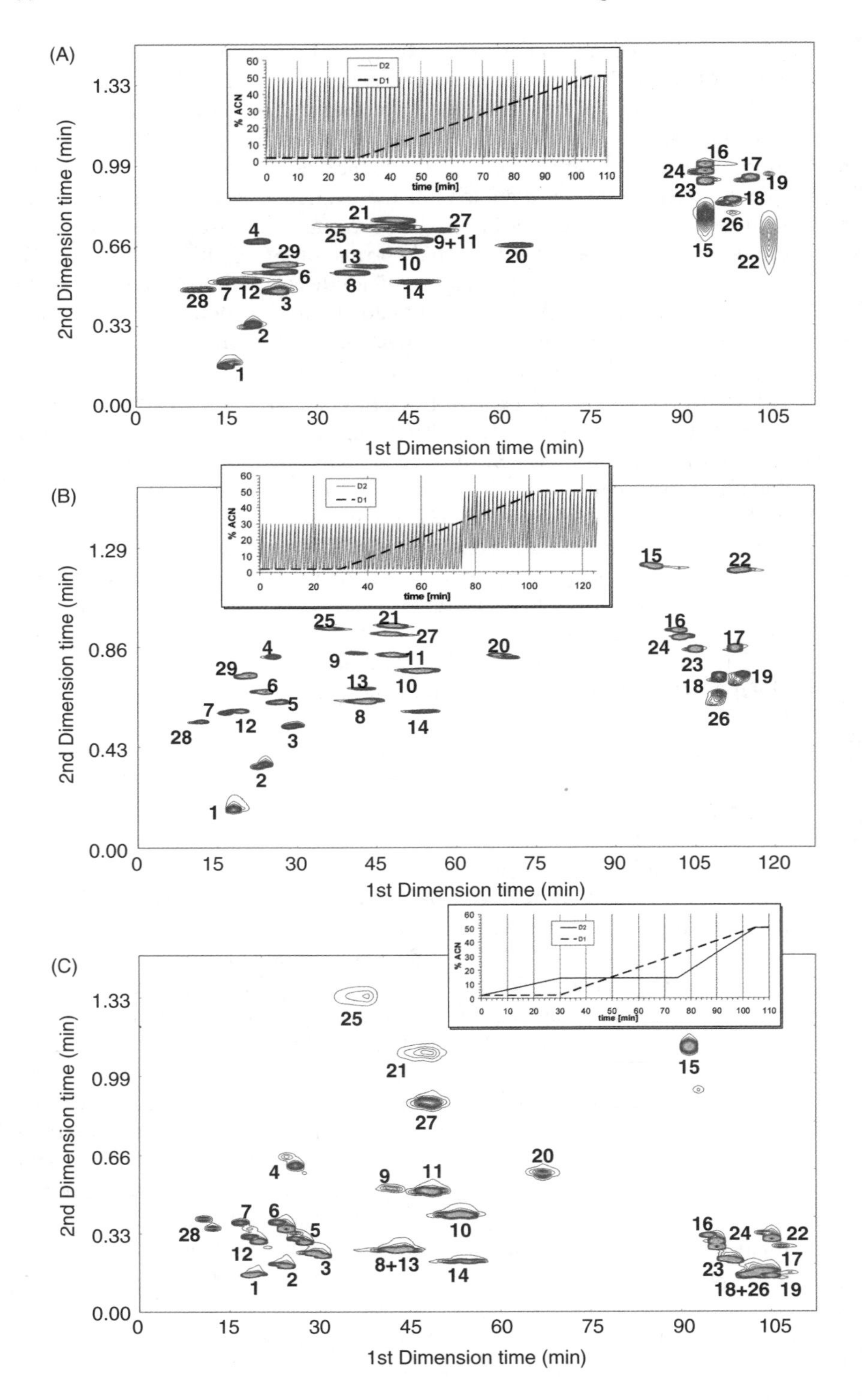
(A)
2nd Dimension time (min)
1st Dimension time (min)
% ACN
time [min]
D2
D1
(B)
2nd Dimension time (min)
1st Dimension time (min)
(C)
2nd Dimension time (min)
1st Dimension time (min)

3.7 FRACTION TRANSFER MODULATION IN COMPREHENSIVE LC x LC: ADDITIONAL BAND BROADENING

In comprehensive LC × LC, the frequency of the fraction transfer cycle from the first-dimension column into the second dimension depends on the width of peaks sampled from the first dimension and on the time available for separation in the second dimension. Second-dimension separation must operate at a much higher flow rate than that used in the first dimension, so that a series of several tens or hundreds of second-dimension chromatograms is produced over the entire period of a first-dimension run, generally leading to longer separation times compared to those of common one-dimensional separations. The maximum acceptable second-dimension separation time limits the modulation time, the period used to store the effluent from the first column before releasing it to the second column. The sampling rate of the first-dimension effluent fractions is characterized by the modulation ratio, M_R, defined as the ratio of the width of a peak at baseline, $w_b = 4\sigma$, to the modulation period selected, P_M (Khummueng and Marriott, 2009):

$$M_R = \frac{w_b}{P_M} = \frac{4\sigma}{P_M} \tag{3.26}$$

P_M determines the duration of a complete modulation cycle in comprehensive LC × LC mode: that is, the time between the two successive fraction transfers into the second column, which limits the time for second-dimension separation. The modulation period should be longer than the retention time of any component on the second-dimension column, to avoid carryover of a compound from the previous to the next fraction in the second dimension. The second-dimension separation time should match the modulation ratio $M_R \sim 3$ to 4, with three or four

Figure 3.6 Two-dimensional comprehensive LC × LC separation of phenolic acids and flavones with simultaneous gradients of acetonitrile in buffered aqueous organic mobile phase (0.01 mol/L CH_3COONH_4, pH 3.1). First dimension: a poly(ethyleneglycol) column (5 μm, 150 mm × 2.1 mm i.d.), 65 μL/min; second dimension: a Kinetex C_{18} column (2.6 μm, 50 mm × 3.0 mm i.d.), 2.5 mL/min. (A) Fast second-dimension gradient with a broad acetonitrile concentration range; (B) fast second-dimension gradients with two segments of acetonitrile concentration range; (C) segmented second-dimension gradient parallel to the first-dimension gradient with shallow acetonitrile concentration range. The acetonitrile concentration profiles are shown in the insets (dashed line for the first dimension, solid line for the second dimension). Sample compounds: 1, gallic acid; 2, protocatechuic acid; 3, *p*-hydroxybenzoic acid; 4, salicylic acid; 5, vanillic acid; 6, syringic acid; 7, 4-hydroxyphenylacetic acid; 8, caffeic acid; 9, sinapic acid; 10, *p*-coumaric acid; 11, ferulic acid; 12, chlorogenic acid; 13, (−)-epicatechine; 14, (+)-catechine; 15, flavone; 16, 7-hydroxyflavone; 17, apigenine; 18, lutheoline; 19, quercetine; 20, rutine; 21, naringine; 22, biochanin A; 23, naringenine; 24, hesperetine; 25, hesperidine; 26, morine; 27 : 4-hydroxycoumarine; 28, esculine; 29, vanilline.

samplings across the time equivalent of a first-dimension peak width (Murphy et al., 1998), so that the peaks resolved in the first dimension remain so throughout the entire 2D separation time (Murphy et al., 1998). If the separation on the second-dimension column is too slow relative to the first dimension (such as in undersampling modulation), the resolution obtained for the first column is degraded, whereas too fast a second-dimension separation does not provide enough time for adequate resolution (oversampling modulation).

The first-dimension column should have a high peak capacity and high efficiency at low flow rates and should provide low peak volumes, whereas the second-dimension column should have high permeability and efficiency at the high flow rates necessary for fast separations. Hence, most frequently used experimental arrangements for comprehensive LC $\times$ LC employ a relatively long microcolumn (15 to 20 cm) in the first dimension and a short packed or monolithic second-dimension column (2 to 5 cm). Band broadening during fraction transfer modulation should be suppressed to improve the chromatographic resolution and peak capacity [Eq. (3.10)] and to enhance the sensitivity and improve the detection limits. In addition to the system orthogonality, suppressing the fraction transfer contribution to the band broadening is another criterion that should be taken into account when selecting suitable combinations of stationary and mobile phases for comprehensive LC $\times$ LC systems. The second-dimension separation system should preferably provide stronger sample retention with respect to the first dimension, so that the mobile phase used for sample transfer from the first dimension has a lower elution strength in the second dimension. In this case, the analytes from the transferred fraction are adsorbed in a compressed narrow zone on the top of the second column, before elution with a stronger second-dimension mobile phase. Using this well-known on-column sample focusing principle results in the suppression of band broadening, improved resolution, and peak capacity in the second dimension, or can be used to increase the volume of the fraction transferred and the fraction transfer switching time. If a sample compound has higher retention on the second-dimension column in the mobile phase transferring the fraction (k_a) than in the mobile phase used subsequently for elution (k_b), $k_a > k_b$, the original fraction volume is reduced due to sample focusing on the second column by the compression factor z:

$$z = \frac{1 + k_b}{1 + k_a} \tag{3.27}$$

An even stronger sample focusing effect can be achieved by using two short trapping columns (SPE traps), packed with a suitable highly retentive sorbent instead of sampling loops in the switching valve modulation interface between the first and second dimensions (SPE trap, Figure 3.1B). The fractions are focused alternately in narrow zones on one of the two trapping columns, while the sample zone adsorbed is back-flushed from the other trapping column onto the second-dimension separation column in a very low volume of the second-dimension mobile phase (Cacciola et al., 2006). This setup may significantly reduce the volume of the fractions transferred to the second dimension and represents in principle a simple example of a three-dimensional LC × LC × LC setup.

Mobile-phase compatibility may present a serious problem when transferring fractions between reversed-phase systems with aqueous–organic mobile phases and normal-phase systems using purely organic mobile phases immiscible with water, and vice versa. Poor miscibility of mobile phases and desactivation of polar adsorbents such as bare silica gel may destroy the separation completely. Therefore, RP–NP systems are as a rule connected off-line (Dugo et al., 2004). A vacuum evaporative interface was introduced to evaporate volatile organic mobile phase from the first-dimension NP system collected before transfer into the second, RP dimension, often at the cost of incomplete recovery of sample compounds with low boiling points (Tian et al, 2006). Lipophilic compounds such as fats and oils can often be separated online using a 2D combination of normal-phase silver-ion chromatography in the first dimension with nonaqueous reversed-phase (NARP) chromatography in the second dimension (NP × NARP LC) (Dugo et al., 2006b).

Online RP × NP separation of polar compounds can be solved theoretically by combining reversed-phase and hydrophilic interaction liquid chromatography, which is essentially normal-phase chromatography with mobile phases containing low concentrations of water in an organic solvent such as acetonitrile (Jandera, 2008). RP chromatography often shows separation selectivity complementary to HILIC; unfortunately, the mobile phase used in one separation mode is a very strong eluent in the other mode, with detrimental effects on separation, unless the mobile-phase composition and the volume of fractions transferred are carefully optimized (Jandera et al., 2006). An online HILIC–solid-phase extraction (SPE)–RP system was developed for the separation of tryptic digest peptides to solve the problem of mobile-phase compatibility by diluting the HILIC fractions from the first dimension with water and trapping on one of 18 separate (RP) SPE trapping columns arranged in a column selector interface, before back-flushing the trapped HILIC fractions to a polymeric C_{18} column for second-dimension separation (Wilson et al., 2007).

3.8 FUTURE PERSPECTIVES

Two-dimensional HPLC has proven its usefulness as a powerful tool for the separation of complex samples. Two-dimensional separations of several hundred compounds have become quite common in the last decade; theoretically, separation of samples containing several thousand analytes could be possible in the future. We can expect a rapid increase in the number of new applications for various sample types when a new generation of instrumentation for 2D HPLC becomes commercially available. The theoretical concepts of multidimensional separations are reasonably well understood; however, the reliable determination of practical peak capacity still has not been solved satisfactorily. Further, the orthogonality concept in LC × LC separations surely needs improvement in closer relation to the separation selectivity in two dimensions. Great effort is focused on practical selection of separation systems with low-selectivity correlations in the two dimensions, and development in this direction will progress with future introduction of new stationary phases. Progress in column technology will result in the production of new types of columns with high efficiency, temperature stability, and permeability, enabling wider use of high-pressure and high-temperature fast second-dimension operation, which will have a major impact on the overall peak capacity and speed in comprehensive LC × LC.

Three-dimensional separation techniques offering analysis of samples containing 10,000 or more compounds in three dimensions still remain subject to theoretical considerations for future development (Guiochon et al., 2008), as the necessary separation speed in the third dimension and the complexity of data handling present problems far beyond the possibility of practical solution at the present state of the art in separation techniques. However, combinations of 2D HPLC–HPLC with mass spectrometry (MS, MS–MS), providing structural information for sample identification, can be classified as three-dimensional analytical techniques.

REFERENCES

Abraham MH, Rosés M, Poole CF, Poole SK. *J. Phys. Org. Chem*. 1997; 10:358–368.

Blahová E, Jandera P, Cacciola F, Mondello L. *J. Sep. Sci*. 2006; 29:555–566.

Bushey MM, Jorgenson JW. *Anal. Chem*. 1990; 62:161–167.

Cacciola F, Jandera P, Blahová E, Mondello L. *J. Sep. Sci*. 2006; 29:2500–2513.

Česla P, Hájek T, Jandera P. *J. Chromatogr. A* 2009; 1216:3443–3457.

Cho D, Hong J, Park S, Chang T. *J. Chromatogr. A* 2003; 986:199–206.

Davis JM. *J. Chromatogr. A* 1999; 831:37–49.

Davis JM, Giddings JC. *Anal. Chem*. 1983; 55:418–424.

Delmotte N, Lasaosa M, Tholey A, Heinzle E, van Dorsselaer A, Huber CG. *J. Sep. Sci*. 2009; 32:1156–1164.

Dugo P, Favoino O, Tranchida PQ, Dugo G, Mondello L. *J. Chromatogr. A* 2004; 1041:135–142.

Dugo P, Kumm T, Crupi ML, Cotroneo A, Mondello L. *J. Chromatogr. A* 2006a; 1112:269–275.

Dugo P, Škeříková V, Kumm T, Trozzi A, Jandera P, Mondello L. *Anal. Chem*. 2006b; 78:7743–7750.

Dumarey M, Put R, Van Gyseghem E, Heyden YV. *Anal. Chim. Acta* 2008; 609:223–234.

Erni F, Frei RW. *J. Chromatogr*. 1978; 149:561–569.

Fraga CG, Bruckner CA, Synovec RE. *Anal. Chem*. 2001; 73:675–683.

François I, Cabooter D, Sandra K, Lynen F, Desmet G, Sandra P. *J. Sep. Sci*. 2009; 32:1137–1144.

Giddings JC. *Anal. Chem*. 1967; 39:1027–1028.

Giddings JC. *Anal. Chem*. 1984; 56: 1258A–1270A.

Giddings JC. *J. Chromatogr. A* 1995; 703:3–15.

Gilar M, Olivová P, Daly AE, Gebler JC. *Anal. Chem*. 2005; 77:6426–6434.

Gray MJ, Dennis GR, Slonecker PJ, Shalliker RA. *J. Chromatogr. A* 2004; 1041:101–110.

Grushka E. *Anal. Chem*. 1970; 42:1142–1147.

Guiochon G, Marchetti N, Mriziq K, Shalliker RA. *J. Chromatogr. A* 2008; 189:109–168.

Horie K, Kimura H, Ikegami T, Iwatsuka A, Saad N, Fiehn O, Tanaka N. *Anal. Chem*. 2007; 79:3764–3770.

Ikegami T, Hara T, Kimura H, Kobayashi H, Hosoya K, Cabrera K, Tanaka N. *J. Chromatogr. A* 2006; 1106:112–117.

Jandera P. *J. Chromatogr*. 1984; 314:13–36.

Jandera P. *J. Sep. Sci*. 2006; 29:1763–1783.

Jandera P. *J. Sep. Sci*. 2008; 31:1421–1437.

Jandera P, Churáček J. *J. Chromatogr*. 1974a; 91:207–221.

Jandera P, Churáček J. *J. Chromatogr*. 1974b; 91:223–235.

Jandera P, Churáček J, Svoboda L. *J. Chromatogr*. 1979; 174:35–50.

Jandera P, Holčapek M, Kolářová L. *J. Chromatogr. A* 2000; 869:65–84.

Jandera P, Holčapek M, Kolářová L. *Int. J. Polym. Anal. Charact*. 2001; 6:261–294.

Jandera P, Halama M, Novotná K, Bunčeková S. *Chromatographia* 2003; 57: S153–S161.

Jandera P, Novotná K, Kolářová L, Fischer J. *Chromatographia* 2004; 60: S27–S35.

Jandera P, Halama M, Kolářová L, Fischer J, Novotná K. *J. Chromatogr. A* 2005; 1087: 112–123.

Jandera P, Fischer J, Lahovská H, Novotná K, Česla P, Kolářová L. *J. Chromatogr. A* 2006; 1119:3–10.

Jandera P, Vyňuchalová K, Hájek T, Česla P, Vohralík G. *J. Chemometr*. 2008; 22: 203–217.

Kaliszan R. *Structure and Retention in Chromatography: A Chemometric Approach*. Chichester, UK: Horwood Academic Publishers, 1997.

Khummueng W, Marriott PJ. *LC-GC Eur*. 2009; 22:38–45.

Little EL, Jeansonne MS, Foley JP. *Anal. Chem*. 1991; 63:33–44.

Marchetti N, Felinger A, Pasti, L, Pietrogrande MC, Dondi F. *Anal. Chem*. 2004; 76:3055–3068.

Murphy RE, Schure MR, Foley JP. *Anal. Chem*. 1998; 70:1585–1594.

Neue UD, Alden BA, Walter TH. *J. Chromatogr. A* 1999; 849:101–116.

Neue UD, Carmody JL, Cheng YF, Lu Z, Phoebe CH, Wheat TE. *Adv. Chromatogr*. 2001; 41:93–136.

Oros FJ, Davis JM. *J. Chromatogr*. 1992; 591:1–18.

Pierce KM, Hoggard JC, Mohler RE, Synovec RE. *J. Chromatogr. A* 2008; 1184: 341–352.

Schoenmakers PJ, Mariott P, Beens J. *LC-GC Eur*. 2003; 16:335–339.

Shellie RA, Tyrrell E, Pohl CA, Haddad PR. *J. Sep. Sci*. 2008; 31:3287–3296.

Snyder LR, Dolan JW. *Adv. Chromatogr*. 1998; 38:115–187.

Snyder LR, Dolan JW, Gant Jr. *J. Chromatogr*. 1979; 165:3–30.

Snyder LR, Dolan JW, Carr PW. *J. Chromatogr A* 2004; 1060:77–116.

Stoll DR, Carr PW. *J. Am. Chem. Soc*. 2005; 127:5034–5035.

Tian H, Xu J, Xu Y, Guan Y. *J. Chromatogr A* 2006; 1137:42–48.

Turowski M, Morimoto T, Kimata K, Monde H, Ikegami T, Hosoya K, Tanaka N. *J. Chromatogr. A* 2001; 911:177–190.

Van der Horst A, Schoenmakers PJ. *J. Chromatogr. A* 2003; 1000:693–709.

Van Gyseghem E, Van Hemelryck S, Daszykowski M, Questier F, Massart DL, Vander Heyden Y. *J. Chromatogr. A* 2003; 988:77–93.

Vivó-Truyols G, Schoenmakers PJ. *J. Chromatogr. A* 2006; 1120:273–281.

Watson HE, Davis JM, Synovec RE. *Anal. Chem.* 2007; 79:7924–7927.

Wilson NS, Nelson MD, Dolan JW, Snyder LR, Wolcott RG, Carr PW. *J. Chromatogr. A* 2002; 961:171–193.

Wilson SR, Jankowski M, Pepaj M, Mihailova A, Boix F, Truyols GV, Lundanes E, Greibrokk T. *Chromatographia* 2007; 66:469–474.

4

HISTORY, EVOLUTION, AND OPTIMIZATION ASPECTS OF COMPREHENSIVE TWO-DIMENSIONAL GAS CHROMATOGRAPHY

AHMED MOSTAFA AND TADEUSZ GÓRECKI
University of Waterloo, Waterloo, Ontario, Canada

PETER Q. TRANCHIDA AND LUIGI MONDELLO
University of Messina, Messina, Italy

Chromatography is the technique used most widely to separate and yield quantitative information about components of complex mixtures. Gas chromatography (GC) using modern capillary columns offers higher peak capacities than those available with other chromatographic techniques. However, it fails to separate all individual sample constituents when the matrix is very complex. One way to improve the separation power of a GC system is to subject the sample separated by a given GC column to an additional separation using a different separation mechanism. This method, referred to as *multidimensional separation*, was introduced more than 50 years ago (Simmons and Snyder, 1958). In 1984, Giddings discussed at length the idea of subjecting a sample to multiple types of separation to get improved resolution and separation power (Giddings, 1984). He concluded that the best results are obtained when the two separation mechanisms are independent. However, within a class of similar compounds there is often a

Comprehensive Chromatography in Combination with Mass Spectrometry, First Edition.
Edited by Luigi Mondello.

correlation between the separation mechanisms, giving rise to diagonal lines on the retention plane (Giddings, 1984).

Multidimensional gas chromatography can be subdivided into two categories: heart-cut multidimensional gas chromatography (GC–GC), where a single fraction (or a few specific fractions at most) of the first-dimension effluent is introduced into the second dimension for further separation (Cortes, 1992), and comprehensive multidimensional gas chromatography (GC × GC), where the entire sample (or at least a representative fraction of each sample component) eluting from the first dimension is introduced into the second dimension for further separation (Liu and Phillips, 1991).

Although combining two (or more) different separation mechanisms is the best way to improve resolution and separation power in chromatography, it is not always a trivial task. The idea can be implemented easily only in planar chromatography (e.g., thin-layer chromatography), where the chromatographic plate can be physically rotated by 90° after the first separation and then developed with a second, different solvent. For obvious reasons, such an approach is not practical in column chromatography (including GC × GC). The fundamental requirement of comprehensive two-dimensional (2D) separation is that components separated in the first dimension must remain separated after passing through the second dimension. This precludes the use of two columns connected directly in series for this purpose, as explained below. It was not until 1991 that Liu and Phillips realized the vision of Giddings and developed the first GC × GC system using a special modulator (Liu and Phillips, 1991).

4.1 FUNDAMENTALS OF GC x GC

As mentioned above, there are two types of multidimensional gas chromatography: heart-cut GC (GC–GC) and comprehensive multidimensional GC (GC × GC). In GC–GC (Figure 4.1A) two different columns are used, but only a small portion of the material eluting from the first column ("heart cut") is introduced for further separation into the second column. The number of heart cuts can be increased provided that the time allowed for separation of the cuts in the second dimension is reduced proportionally (Figure 4.1B). When the number of heart cuts gets high enough (and the time for their separation short enough), one accomplishes a comprehensive separation (Figure 4.1C), in which the entire sample (or a representative fraction of each sample component) is subjected to separation in both dimensions. Consequently, one can say that GC × GC is in essence an extension of conventional heart-cut GC.

In true multidimensional separations, two different separation mechanisms should be applied to the entire sample to achieve separation orthogonality (Marriott and Shellie, 2002; Venkatramani et al., 1996). Figure 4.2 illustrates the basic layout of a GC × GC system used to accomplish this goal. The sample is first introduced and separated on the first capillary column. However, rather than being sent to the detector, the effluent is introduced into a second capillary column coated with a different stationary phase for further separation. The two

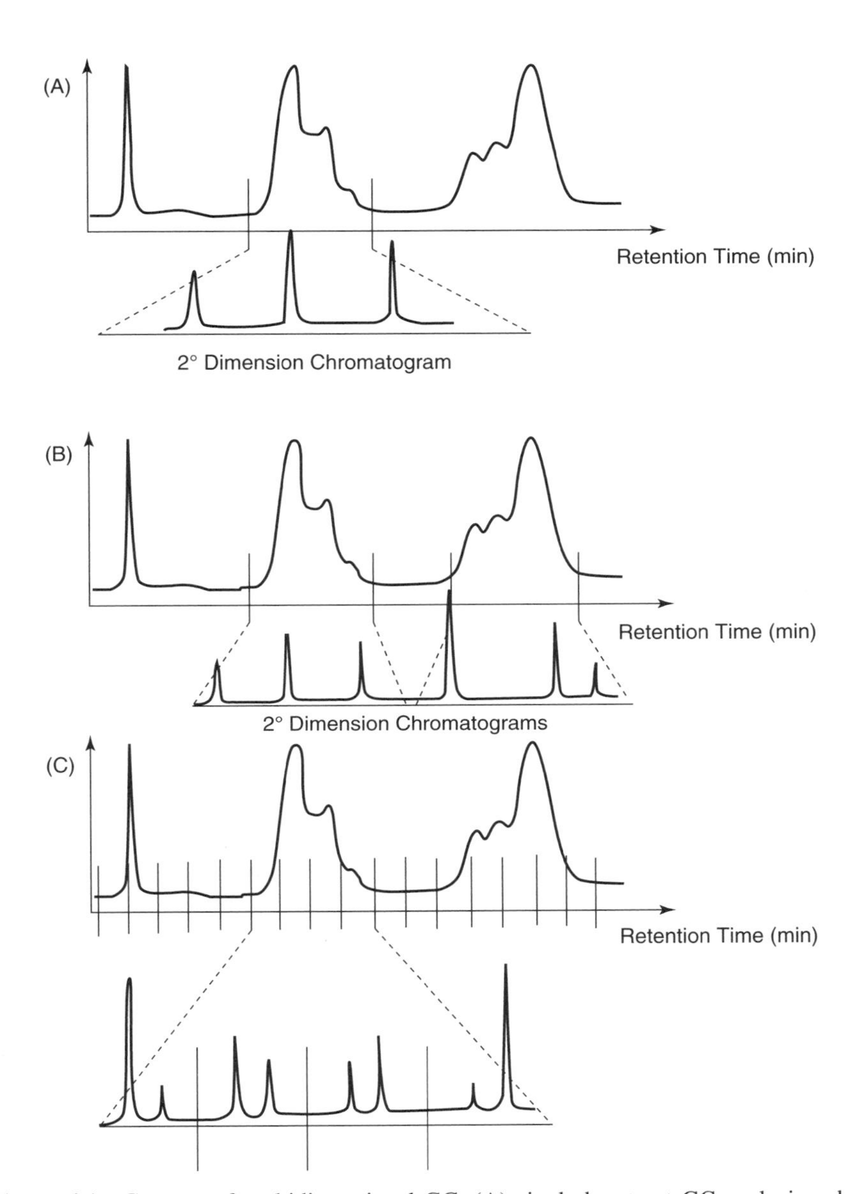

Figure 4.1 Concept of multidimensional GC. (A) single heart-cut GC analysis, where a large portion of the effluent from the primary column with coelutions is diverted to the second-dimension column and separated over an extended period of time. (B) Dual heart-cut GC analysis, where two regions with coelutions are diverted to the second-dimension column, but with less time to perform each separation. (C) Comprehensive two-dimensional GC analysis occurs when the size of the sequential heart cuts is very short, as are the second-dimension chromatograms. [From Górecki et al. (2004), with permission. Copyright © 2004 by Wiley-VCH Verlag GmbH & Co. KGaA.]

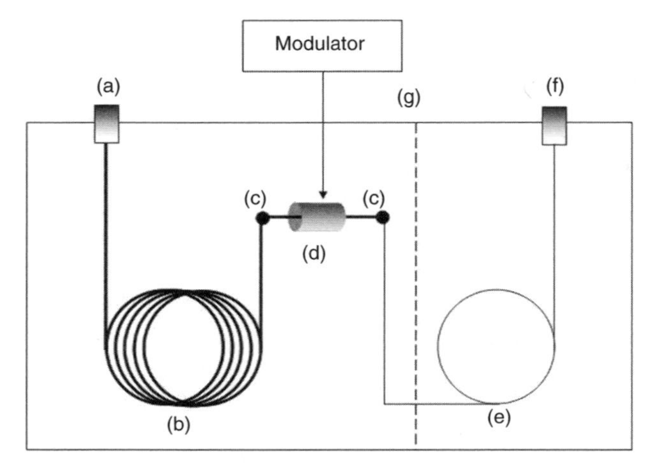

Figure 4.2 Block diagram of a GC × GC system: (a) injector; (b) primary column; (c) column connectors (at least one is necessary; more may be required depending on exact configuration); (d) GC × GC interface; (e) secondary column; (f) detector; (g) optional division for secondary oven. [From Górecki et al. (2004), with permission. Copyright © 2004 by Wiley-VCH Verlag GmbH & Co. KGaA.]

columns are connected through a special interface called a *modulator*. The primary column typically contains nonpolar stationary phase; therefore, separation in this column is based primarily on analyte volatility (applies strictly to nonpolar analytes only). The secondary column, which is much shorter and often narrower than the primary column, is usually coated with a more polar stationary phase to achieve orthogonality (a more detailed discussion of the column combinations used in GC × GC is presented later in this chapter). The separation in the secondary column has to be extremely fast (a few seconds) to make sure that fractions of the first-dimension effluent are sampled frequently enough to preserve the separation accomplished in the first dimension. The effluent from the second column is directed to the detector.

4.2 MODULATION

The modulator is arguably the most important component of any GC × GC system. Figure 4.3 explains the need for and the role of the modulator. Direct serial connection of two different columns without a modulator results in a one-dimensional (1D) separation because analytes separated on the first column are not prevented from coelution at the exit of the second column (Figure 4.3B). Their elution order might even be reversed (Figure 4.3C). The modulator allows the flow of analytes from the first column to the second column to be controlled, which changes the nature of the separation fundamentally, as illustrated in Figure 4.3D–G. Following the same separation in the first column (Figure 4.3D), the modulator traps and focuses the first band (black in Figure 4.3E), and then

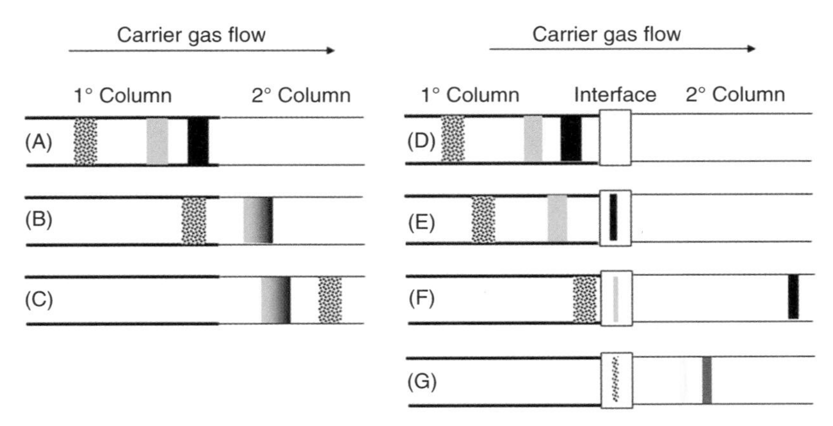

Figure 4.3 Need for a GC × GC interface: (A–C) bands separated on one column can recombine or change elution order on a second column if they flow uncontrolled from one column to the other; (D–G) the interface traps material from the primary column and then allows discrete bands to pass to the second-dimension column while trapping other fractions. [From Górecki et al. (2004), with permission. Copyright © 2004 by Wiley-VCH Verlag GmbH & Co. KGaA.]

injects it into the second column while collecting the following band (gray in Figure 4.3F). The gray band is injected into the second column only after the black band had eluted from it. The gray band is then separated into two bands on the second column, while the spotted band is collected by the modulator (Figure 4.3G). This sequence of events assures that separation achieved in the first-dimension column is preserved and that additional separation in the second column is possible.

To sum up, the role of the modulator is to trap, refocus, and inject the primary column heart cuts sequentially into the second dimension (although it should be pointed out that instead of being trapped, the effluent from the first dimension might also be sampled, as explained later in the chapter). The time taken to complete a single cycle of events is called the *modulation period*. The preservation of the first-dimension separation can be accomplished only if every peak eluting from the primary column is sampled at least three times (Murphy et al., 1998), although 2.5 times has also been proposed as the optimal value (Blumberg, 2003). Thus, if, for example, a 12-s-wide peak elutes from the primary dimension, the modulation period should be no longer than 4 s. Górecki et al. (2004) illustrated the effect of the length of the modulation period on the preservation of the primary column separation in a review in 2004.

4.3 GC × GC DATA INTERPRETATION

A conventional 1D GC chromatogram is a two-dimensional plot of detector signal intensity versus retention time. A GC × GC chromatogram, on the other hand, is a three-dimensional plot with two retention times and signal

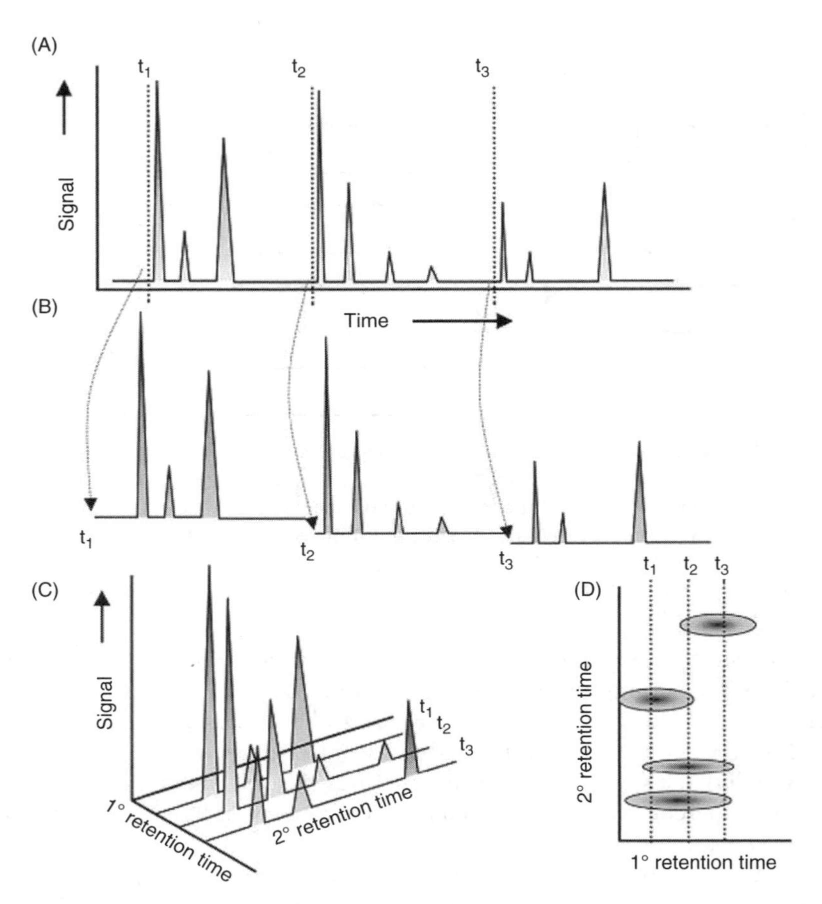

Figure 4.4 Interpretation of GC × GC data and generation of contour plots. (A) The raw GC × GC chromatogram consisting of a series of short second-dimension chromatograms; t_1, t_2, and t_3 indicate the times when injections to the second-dimension column occurred. The computer uses these injection times to slice the original signal into a multitude of individual chromatograms (B). These are then aligned on a two-dimensional plane with primary retention and secondary retention as the x- and y-axes, and signal intensity as the z-axis (C). When viewed from above, the peaks appear as rings of contour lines or color-coded spots (D). [From Górecki et al. (2004), with permission. Copyright © 2004 by Wiley-VCH Verlag GmbH & Co. KGaA.]

intensity as the axes. However, the detector positioned at the outlet of the second column records a continuous linear signal, being in fact a series of second-dimension chromatograms produced by each modulation cycle (Figure 4.4A). Interpretation of such a chromatogram is very difficult, especially considering that each component eluting from the first column might be present in several secondary chromatograms. Consequently, the data have to be converted into a three-dimensional representation before they can be analyzed effectively. This task is handled by appropriate computer software. The construction of such a plot is illustrated in Figure 4.4. The software utilizes the times of injection of the fractions of the first-dimension effluent into the second column (t_1, t_2, and t_3 in Figure 4.4) to "slice" the continuous chromatogram into the individual second-dimension chromatograms (Figure 4.4B). The times when the injection into the second column took place provide the first-dimension retention times for all peaks eluting in a given modulation period. The second-dimension retention time of each peak is its absolute retention time minus the injection time for the modulation cycle (Harynuk et al., 2002). As illustrated in Figure 4.4C, after arranging the individual second-dimension chromatograms side by side, the x-axis becomes the primary retention time, the y-axis the secondary retention time, and the z-axis represents signal intensity (Harynuk et al., 2002). In such representation, second-dimension peaks of the same component appear in several consecutive slices with the same secondary retention times (recall that each primary dimension peak should be sampled at least three times; therefore, component(s) of this peak might show up in several consecutive second-dimension chromatograms). The peaks of a given analyte observed in the individual secondary chromatograms are then merged into a single component peak. The approach described requires that the modulation periods be both known precisely and reproducible, which can be accomplished using modern computer-controlled hardware.

Quantification of GC × GC data is similar to that encountered in 1D GC, except that instead of integrating a single peak for a single analyte, multiple peaks for each analyte have to be taken into consideration. In the simplest approach, each peak in the second-dimension chromatogram is integrated separately, and the areas of the peaks belonging to the same analyte are summed. Although different approaches to analyte quantitation in GC × GC have been proposed, the simple summation of the peak areas seems to work best (Bertsch, 2000). Quantitative methods employed in GC × GC were reviewed by Dallüge et al. (2003), Górecki et al. (2006), Adahchour et al. (2006a), and Amador-Muñoz and Marriott (2008).

In early GC × GC history, most research groups had to develop their own in-house-written GC × GC analysis software (Beens et al.,1998a; Frysinger et al., 1999; Harynuk and Górecki, 2003; Penet et al., 2006). These GC × GC data-handling packages usually allowed visualization of the multidimensional data, data preprocessing, and (less often) peak detection and quantification. Harynuk et al. (2002) demonstrated the steps of data manipulation that permit conversion of raw data to the contour plots generated in GC × GC. They explained the challenges associated with accurate conversion of GC × GC data, based on their observations when developing GC × GC data analysis software. Reichenbach et al. (2003, 2004) at the University of Nebraska–Lincoln developed software called GC Image that produced background-free peaks in GC × GC chromatograms. The software was available for quantification purposes. The same group later developed software with added support for mass spectrometry (GC × GC–MS), three-dimensional visualization, and many other features (Reichenbach et al.,

2005). Techniques for peak alignment were also developed, which significantly improved comparative analysis of GC × GC data (Hollingsworth et al., 2006). GC Image is available commercially from Zoex Corporation (Reichenbach et al., 2004). Another GC × GC software package, HyperChrom, developed by Thermo Fisher Scientific, was paired with the TRACE GC × GC instrument sold by the same company. It has been used extensively by some researchers (e.g., Korytár et al., 2005a, 2005b). The leading GC × GC–mass spectrometry (MS) software package, ChromaTOF, is available from LECO together with their Pegasus 4D GC × GC–time of flight (TOF) MS system. The software, which also works with flame ionization detector (FID) data, incorporates perhaps the most sophisticated data-processing capabilities for GC × GC.

4.4 GC x GC INSTRUMENTATION

GC × GC utilizes much of the basic instrumentation used for 1D GC. For example, injectors used in GC × GC play the same role as in 1D GC. Consequently, injectors and injection techniques used for 1D GC can in principle also be used for GC × GC analyses. The primary column is usually long, with typical dimensions of 15 to 30 m × 0.25 mm. The stationary-phase film thickness in the primary column is usually in the range 0.25 to 1.0 μm. These columns allow for the generation of peaks with widths of 10 to 20 s, which are required for typical modulation periods (3 to 6 s). The second-dimension column has to be very short and efficient, as each individual separation in this column should be finished in a time shorter than the modulation period. Typical secondary column dimensions range from 0.5 to 1.5 m in length and 0.1 to 0.25 mm in diameter. The primary columns are typically coated with nonpolar stationary phases (although other options are explored increasingly often); some common choices include 100% poly(dimethylsiloxane) or 95 : 5 methyl/phenyl polysiloxane. The stationary phase chosen for the second-dimension column should offer a different separation mechanism than that in the primary column; hence, it should be distinctly different from the latter. With nonpolar coating in the first column, the secondary columns are usually coated with polar stationary phases. Typically, the stationary phases are 50 : 50 phenyl/methyl polysiloxane or “wax” [poly(ethylene glycol)]-based; sometimes, though, good results can be achieved with liquid crystal, fluorinated, or chiral phases.

The two columns used in GC × GC systems can be housed in one oven, or a second optional oven can be used for the second column. In fact, most commercial systems use two separate ovens. This option provides more flexibility in method development and allows better control over secondary column separation. In contrast, in most homemade GC × GC systems, both columns are housed in one oven to avoid making the system too complex. Satisfactory separations can usually be achieved with both setups.

The detectors used for 1D GC can be used for GC × GC as well, but with an additional requirement that their data acquisition rates must be high. Peaks eluting from the secondary column are typically very narrow, with widths of 100 to 500 ms at the base (Beens et al., 2001; Kinghorn and Marriott, 1999; Oldridge et al., 2008; Ong and Marriott, 2002). To get reliable and reproducible determination of a peak area, at least 10 data points should be collected along the peak profile (Beens et al., 1998a). Consequently, a detector that is capable of collecting data at a rate of at least 50 Hz is required. Thus far, flame ionization detectors (FIDs) have been the most popular choice in GC × GC, followed by mass spectrometers. Micro-electron capture detectors (Bordajandi et al., 2005; Marriott et al., 2003), atomic emission detectors (van Stee et al., 2003), nitrogen chemiluminescence detectors (Wang et al., 2004), miniaturized pulsed discharge detectors (Winniford et al., 2006), and sulfur chemiluminescence detectors (Hua et al., 2004) have also been used for GC × GC work.

As we noted in Section 4.2, the most critical and important component of any GC × GC system is the modulator. Two main classes of modulators are used: thermal modulators and valve-based modulators. Thermal modulators can be subdivided further into heater-based (modulation via an increase in temperature), and cryogenic (modulation via a decrease in temperature). The next section is a brief summary of the history of interface design. For a more thorough analysis of interface technology, readers are advised to refer to reviews by Górecki et al. (2004), Bertsch (2000), and Adahchour et al. (2006a).

4.5 THERMAL MODULATORS

4.5.1 Heater-Based Modulators

Liu and Phillips introduced the first GC × GC modulator in 1991 (Figure 4.5). The interface, consisting of a segment of thick-film capillary column painted with conductive paint, was positioned between the primary and secondary columns (Liu and Phillips, 1991). Analytes from the primary column were focused in the thick film of the stationary phase (Figure 4.5A). Application of an electrical pulse to the conductive capillary tube resulted in heating of the stationary phase and drove the analytes back into the gas phase. They were then introduced by the flow of carrier gas into the second column (Figure 4.5B). Trapping was restored again by turning off the electrical pulse and allowing the capillary to cool down. It was quickly realized that when the entire trapping capillary was heated at once, some of the analytes exiting the primary column passed through the interface without being focused. This resulted in broad injection bands onto the second column and peak shape irregularities due to analyte breakthrough (Figure 4.5C). To solve this problem, a dual-stage modulator was proposed consisting of two independently heated trapping segments. At the beginning of a cycle, the first segment trapped the primary column effluent (Figure 4.5D). When this segment was heated, the trapped material was injected into the second segment and re-trapped there, together with any material from the primary column that passed through the hot first stage (Figure 4.5E, F). Once the first stage cooled back to the trapping temperature, the second segment was heated, resulting in the injection

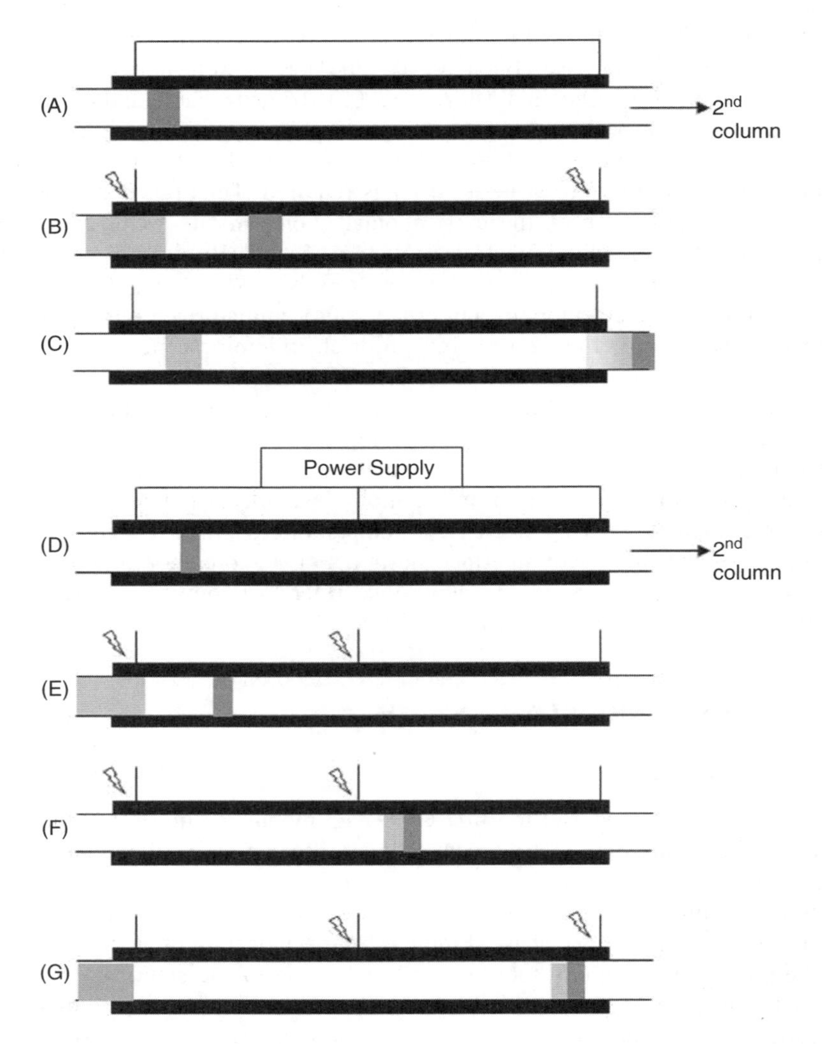

Figure 4.5 Operation of the original Phillips modulator. The heavy lines denote the thick-filmed trapping capillary coated in metallic paint. An analyte band enters the modulator and is focused by the thick-filmed column (A). Upon passing electric current through the metallic coating, the modulator is heated and the focused band is released (B). Any analytes entering the modulator at this time pass through the capillary until it starts to cool. This breakthrough enters the second-dimension column as a broad, unfocused zone (C). Using a dual-stage trap (D) avoids this problem. As the first zone is heated (E), analytes are passed to the second stage, which is cool and can trap them. Any material that breaks through the first stage before it cools down is trapped by the second stage (F). With the first stage cool, the second stage can launch analytes to the second column without breakthrough from the first stage (G). [From Górecki et al. (2004), with permission. Copyright © 2004 by Wiley-VCH Verlag GmbH & Co. KGaA.]

of a narrow band of the trapped material into the second column (Figure 4.5G). With the first stage collecting material while the second stage was injecting its contents into the second column, analyte breakthrough in the modulator was prevented. Although good results were obtained using those thermal modulators (Liu and Phillips, 1994; Phillips and Xu, 1995; Venkatramani and Phillips, 1993), they appeared to have significant disadvantages, mostly due to the fact that the modulator capillaries and the paint coating were not very robust or reproducible. However, the early thermal modulators did demonstrate that GC × GC was a workable technique and formed the foundation for many modulators that followed. The idea of dual-stage modulation in GC × GC is used in almost all modern thermal modulators.

The first commercial modulator, the rotating thermal modulator (Figure 4.6), was also developed by Phillips et al. (Phillips and Ledford, 1996; Phillips et al., 1999). Briefly, a thick-film capillary housed in the GC oven was used to trap and focus the primary column effluent. Desorption of the trapped material and its reinjection into the secondary column were accomplished by moving a slotted heater along the trapping capillary, heating it locally to a temperature higher than the oven temperature by about 100°C. As the heater passed over the trap, moving in the same direction as the carrier gas, any material sorbed by the stationary phase in the heated region partitioned back into the gas phase and was "swept" toward the end of the trap and focused into a narrow band before entering the second column, as shown in Figure 4.6. This modulator worked satisfactorily in many different applications (Beens et al., 1998a, 1998b; Frysinger and Gaines, 1999; Frysinger et al.,1999; Phillips et al., 1999); however, the moving parts caused problems when the alignment of all the modulator parts was not perfect. In addition, the modulator was incapable of collecting volatile compounds at conventional oven temperatures, and the maximum oven temperature had to be kept about 100°C lower than the maximum operating temperature of the stationary phase in the modulator, which significantly limited the range of analyte volatilities. In light of these disadvantages, the rotating thermal modulator concept was abandoned and the device is no longer available commercially.

Harynuk and Górecki (2002) developed a different dual-stage thermal modulator based on the early concepts of Phillips et al. It consisted of a pair of in-line microsorbent traps housed in deactivated Silcosteel capillaries that could be resistively heated in order to desorb and reinject the trapped analytes. Very rapid heating was accomplished by means of a capacitive discharge power supply. This modulator design had no moving parts and a much higher capacity for analyte trapping than that of the thick-film modulator capillaries used in other modulators. The main disadvantage was the limited thermal stability of the sorbent, which made the modulator poorly suitable for higher-boiling compounds (similarly to other heated modulator designs).

Another heated modulator was developed by Burger et al. (2003). This modulator used rapid resistive heating of consecutive segments of a stainless-steel capillary housing a thick-film column inside, where the primary column effluent was trapped. The steel capillary had multiple electrical contacts that allowed

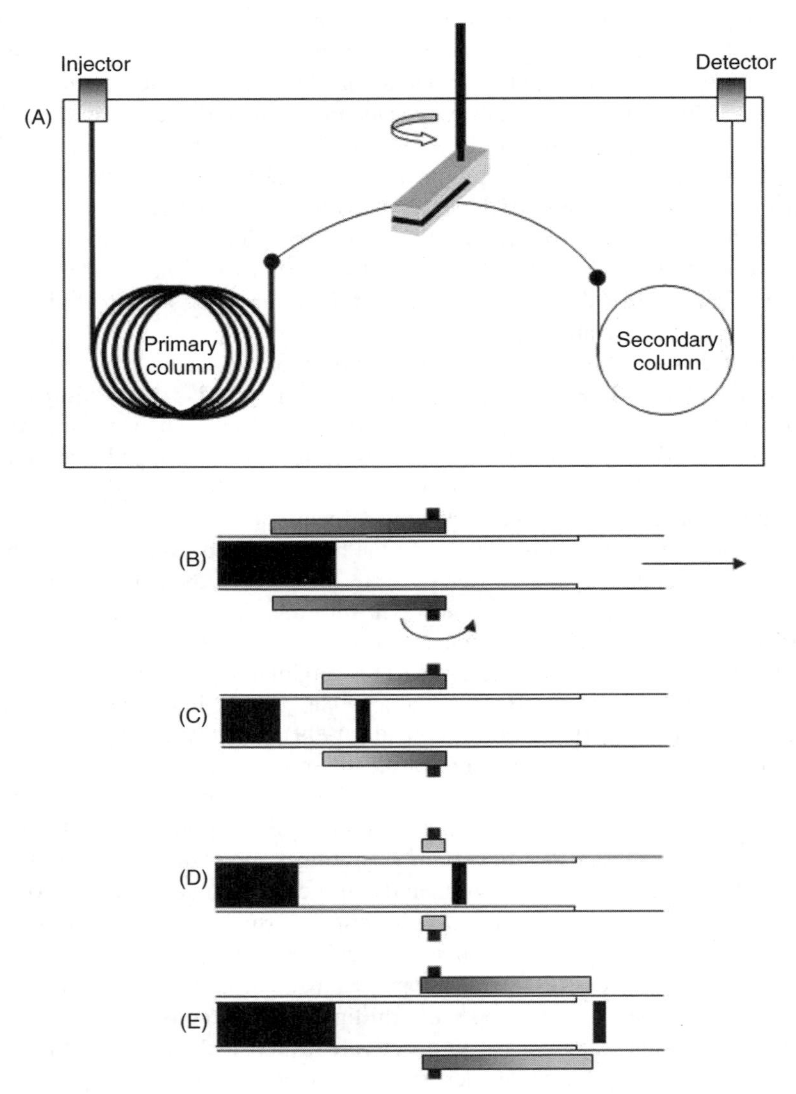

Figure 4.6 Operation of the rotating thermal modulator (A). As an analyte band travels along the trapping capillary (B), the slotted heater heats a portion of the column, speeding up the material in this region while leaving the rest of the analyte band traveling slowly (C). The heater then moves along the column in the direction of the carrier gas flow (D), focusing the analyte band and launching it into the second-dimension column (E). [From Górecki et al. (2004), with permission. Copyright © 2004 by Wiley-VCH Verlag GmbH & Co. KGaA.]

small regions to be heated in sequence for brief periods of time to shuttle bands of analytes through the modulator, keeping them focused. Lack of moving parts was the main advantage of this design, but it suffered from the same disadvantages as those of other heated modulators when it came to the modulation of very volatile analytes (up to approximately n-C_8) and thermal stability of the coating. Today, the sweeper and/or related modulators are rarely used. Their low efficiency in trapping volatile compounds at conventional oven temperatures and the limited range of analyte volatilities were the main reasons why they were largely replaced by cryogenic modulators. One exception to this rule is the thermal modulator developed recently by Górecki and co-workers for the determination of the composition of the semivolatile fraction of air particulate matter ($PM_{2.5}$) in the field using a thermal desorption aerosol GC (2D-TAG) system (Goldstein et al., 2008). The design of this modulator was based on the original idea of Liu and Phillips (1991), except that deactivated stainless-steel Silcosteel® capillary tubing was used for the modulator rather than painted fused-silica capillary. The tubing used for the modulator was either coated with a 1-μm layer of poly(dimethylsiloxane) stationary phase, or just deactivated. The modulator was mounted outside the GC oven, allowing for continuous forced-air cooling. The effluent was trapped in the interface at ambient temperature. Desorption was performed through pulsed resistive heating of one segment of the trapping capillary using a capacitive discharge power supply, while the other segment was trapping and focusing another portion of the effluent. This modulator proved to be robust, had no moving parts, and did not require any consumables, making it suitable for an automated in-situ field instrument, as well as for laboratory work. Very recently, the same group introduced a modified version of the modulator using single-stage rather than dual-stage thermal modulation. Other modifications included the use of a vortex cooler to allow cooling with subambient temperature air at the beginning of the run and temperature programming of the cooling air (Górecki et al., 2009).

4.5.2 Cryogenic Modulators

The first cryogenic modulator, the longitudinally modulated cryogenic system (LMCS), was developed in Australia by Kinghorn and Marriott (1998). This modulator worked on a principle similar to that of heated modulators. However, rather than trapping the effluent at the oven temperature and desorbing it by increasing the temperature above the oven temperature, trapping was performed at temperatures significantly below that of the GC oven, and reinjection was accomplished at the oven temperature. A schematic diagram of the LMCS is presented in Figure 4.7. The interface consisted of a movable, liquid CO_2-cooled trap installed around the end segment of the primary column. As the effluent exited the primary column, analytes were cold-trapped in a small region of the column marked “R”. The cold trap was then moved to a region of the column downstream of the trapping position (“T”). This resulted in the cold spot being rapidly heated back to the oven temperature, which caused the analytes trapped in the stationary phase to partition back into the gas phase. As the desorbed band moved along the column carried by the carrier gas, it was re-trapped in

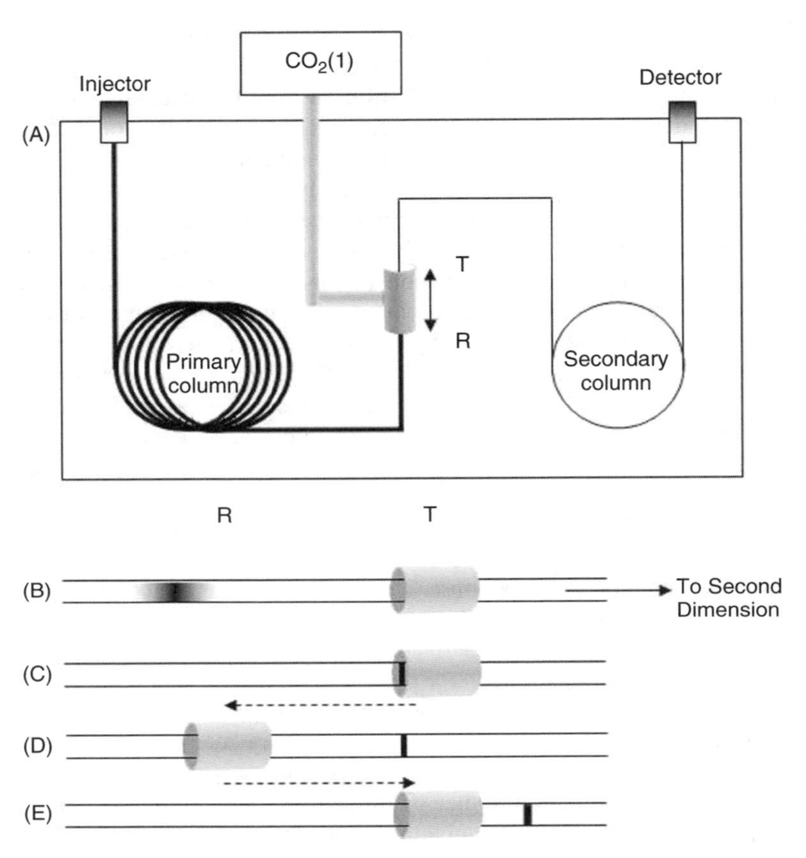

Figure 4.7 Operation of the LMCS (A). “T” and “R” denote the trap and release positions of the cryotrap as it moves along the column. As an analyte band moves down the column (B), it encounters the trap at the T position and is focused and held in place (C). When the trap moves to the R position, the band that was trapped at the T position is released and launched to the second-dimension column, while the cryotrap prevents breakthrough of material from the first column (D). After the focused band has left the trapping zone, the cryotrap returns to the T position (E) and the cycle repeats itself. [From Górecki et al. (2004), with permission. Copyright © 2004 by Wiley-VCH Verlag GmbH & Co. KGaA.]

the downstream segment where the cryotrap was positioned (“T”). In this way, analyte breakthrough was prevented. The cryotrap was then moved back to its initial position (“R”), releasing the analyte band into the second column, while a new portion of the primary column effluent was trapped in the upstream position. Modulation was accomplished by repeating this cycle throughout the entire run. The dual-stage modulation in LMCS brought about the same advantages as in heater-based modulation: elimination of analyte breakthrough and excessive band broadening in the second dimension.

The cryogenic modulation introduced with LMCS offered significant advantages over heated modulation. First, the column segment used for analyte trapping only needed to be raised to the GC oven temperature for desorption, not to a temperature above oven temperature. Consequently, higher final oven temperatures could be used during separation than those for heated modulators. Second, the modulator was capable of trapping volatile analytes much more efficiently than were heated modulators, owing to the low trapping temperature. On the other hand, there were some limitations to this approach. The first was the use of a moving trap, which could damage the column (in the worst-case scenario) or cause other problems. Another limitation was the use of liquid CO_2 as the cryogenic agent, which allowed the trap to be cooled to about $-50^{\circ}C$. This temperature is not sufficiently low to trap highly volatile analytes (up to approximately n-C_6) Generally speaking, though, cryogenic modulators had fewer limitations than heated modulators. Today, cryogenic modulators have replaced the latter almost completely.

The next goal in the development of cryogenic modulators was the elimination of moving parts inside the GC oven. The first modulator of this type was reported by Ledford (2000). In this design, shown in Figure 4.8A, dual-stage modulation was accomplished with the use of two cold CO_2 jets (C_1 and C_2 in Figure 4.8A) for analyte trapping and focusing and two hot air jets (H_1 and H_2 in Figure 4.8A) for desorption. When the upstream cryojet (C_1) was turned on, the effluent from the first column was trapped in the upstream position. Cryojet C_1 was then turned off and the upstream warm jet (H_1) was activated together with the downstream cryojet (C_2). The effluent collected at position C_1 along with any effluent that arrived at the trap while the warm jet (H_1) was on were then carried to the downstream cold spot (C_2), where they were refocused. In the next part of the cycle, the upstream warm jet (H_1) was turned off and the upstream cryojet (C_1) was activated, while the opposite happened at the downstream position. This prevented analyte breakthrough when the downstream cryojet (C_2) was turned off and the downstream warm jet (H_2) was turned on to inject the narrow analyte band onto the second-dimension column. This modulator generally worked very well. Its main limitation was the somewhat complicated design, including four jets that needed to be controlled in sequence. It did provide the basis for the design of most of the new cryogenic modulators that have been developed recently, however. A commercial version of this modulator utilizing liquid nitrogen as the cooling agent is now available on GC $\times$ GC instruments from the LECO Corporation.

Shortly after the quad-jet modulator was introduced, Harynuk and Górecki (2001, 2002) reported the development of another cryogenic modulator with no moving parts. The interface consisted of two empty deactivated Silcosteel capillaries connected in series and mounted inside a cryochamber cooled with liquid nitrogen. In this modulator, trapping was performed through freezing rather than partitioning into the stationary phase as in other cryogenic modulator designs. Launching of the trapped analytes into the second column was accomplished through resistive heating of the trap using an electrical pulse. This modulator

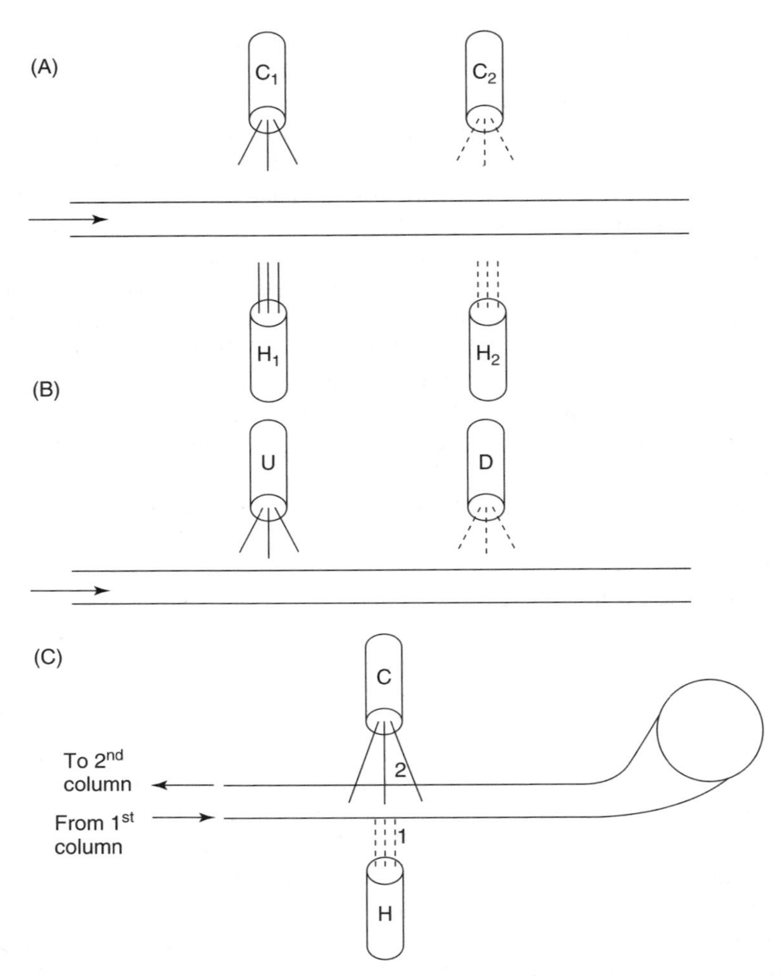

Figure 4.8 Selected dual-stage cryogenic modulators: (A) quad-jet cryogenic modulator developed by Ledford and Billesbach (2002), (B) dual-jet cryogenic modulator developed by Beens et al. (2001); (C) single-cryojet dual-stage modulator with a delay loop. [From Górecki et al. (2006), with permission. Copyright © by Taylor and Francis Group LLC.]

was able to trap highly volatile analytes (including propane), which was one of its major advantages in addition to its ability to control the injection timing very precisely. The main drawback of this modulator were the occasional leaks on the seals between the Silcosteel capillaries and the cryochamber, which led to the development of cold spots, resulting in band broadening (Harynuk and Górecki, 2002).

Utilizing liquid nitrogen instead of liquid CO_2 for analyte trapping via partitioning into the stationary phase allows much lower trapping temperatures, which makes modulation of highly volatile components possible. On the other hand, it has some limitations. The heating required for analyte desorption limits the maximum GC oven temperature. In addition, liquid nitrogen is not easily available in all laboratories and requires bulky insulation to be transported through tubing (Beens et al., 2001). Beens et al. simplified Ledford's quad-jet design by using two liquid CO_2 cryojets for trapping the analytes, which were then remobilized by the heat from the oven air. In this way, the two hot jets of the quad-jet design could be eliminated (Figure 4.8B) (Beens et al., 2001). The commercial version of this modulator is offered on GC × GC instruments from Thermo Fisher Scientific. A further simplification of this modulator was introduced by Adahchour et al. (2003). In this version only a single jet was used to perform single-stage modulation (Figure 4.9). The primary advantage of this technique was the simplicity of the instrumentation and a decrease in CO_2 consumption by more than

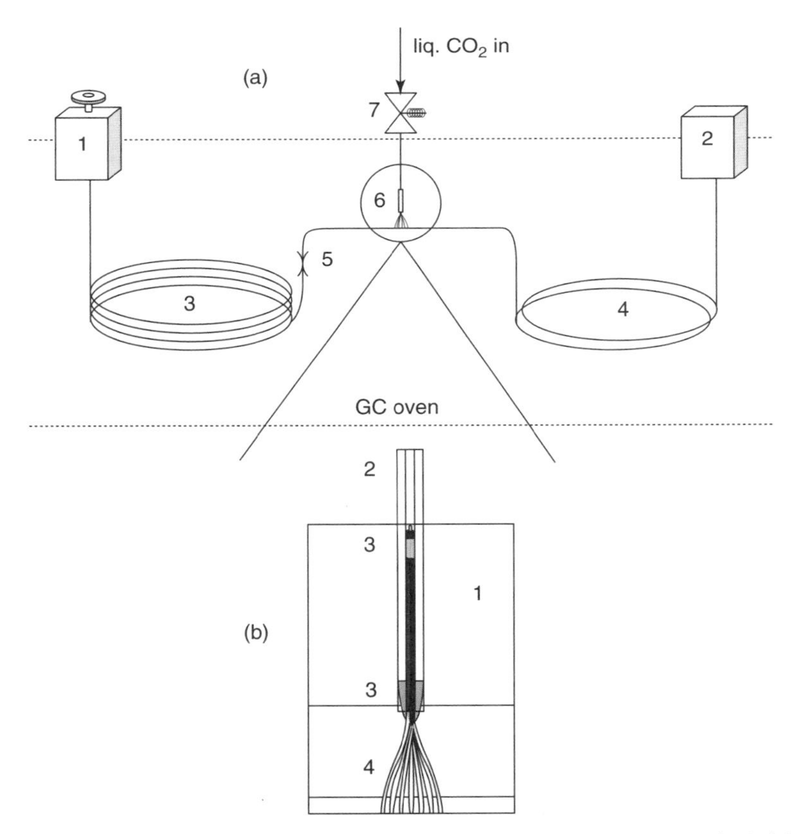

Figure 4.9 (a) Single-jet cryogenic modulator developed by Adahchour et al.: 1, injector; 2, detector; 3, first column; 4, second column; 5, column connection (press-fit); 6, CO_2 nozzle; 7, CO_2 valve. (b) CO_2 nozzle: 1, brass block; 2, stainless-steel (SS) connecting capillary; 3, soldering; 4, seven SS spraying capillaries, 0.11 mm i.d. [From Adahchour et al. (2003), with permission of The Royal Society of Chemistry.]

37% with no significant effect on modulation performance (Adahchour et al., 2003). The disadvantage of this approach was that when using only one trapping zone, the timing of the jet and the tuning of the instrumental parameters had to be done very carefully so as to minimize breakthrough from the first dimension while the trap was hot. In another development, Ledford and co-workers (2002) simplified the concept of the original quad-jet dual-stage modulator by introducing a single cryojet interface capable of dual-stage modulation with the use of a delay loop (Figure 4.8C). In this design, gaseous nitrogen cooled with liquid nitrogen was used as the cryogenic agent. The single jet cooled two segments of a coiled trapping capillary simultaneously. The effluent from the primary column was first trapped at the upstream cold spot when the cryojet was turned on. Turning the hot jet on caused the cold spot to warm up quickly and launched the material collected there to the delay loop, along with any breakthrough from the first column. The hot jet was then turned off, which led to cooling of both cold spots, so that the effluent from the primary column began to be collected again in the first (upstream) cold spot, while the material in the loop was re-trapped in the second (downstream) cold spot. When the hot jet was turned on again, material from the first cold spot was injected into the loop, while the band collected in the second cold spot was injected to the second-dimension column. This modulator, available commercially from Zoex Corporation, represents one of the simplest dual-stage cryogenic designs. The main drawback of this modulator is that the length of the loop and the velocity of the carrier gas have to be carefully adjusted whenever the chromatographic conditions change. If the flow of the carrier gas is not adjusted properly, the band traveling through the delay loop might not reach the trapping spot at a time when it is cold, and therefore it might not be refocused.

Harynuk and Górecki (2003) developed another modulator based on the idea of dual-stage single-jet modulation with a delay loop (Figure 4.10). The main difference between this design and the design developed by Ledford et al. was that liquid nitrogen was used as a cryogen rather than cold nitrogen gas; thus, highly volatile analytes (propane and up) could be trapped efficiently. In addition, trapping was performed in uncoated fused-silica capillaries, thus eliminating potential problems occurring when coated capillaries were used for trapping (e.g., pre-separation of the analytes in the trap due to different times required to desorb the analytes from the stationary phase). In addition, the consumption of liquid nitrogen was decreased to about 30 L/day, owing to careful cooling agent management (Harynuk and Górecki, 2003), whereas commercially available modulators usually consume 50 to 100 L/day (Ledford et al., 2002).

A rather different interface design introduced by Górecki's group was based on stop-flow GC × GC (Harynuk and Górecki, 2004, 2006). In this new design, the flow in the primary column was stopped periodically for short periods of time using a six-port valve while supplying carrier gas to the second column from an auxiliary source. Thus, the two columns were decoupled, which allowed more time for separation in the second dimension. As a result, both columns could be operated simultaneously under optimal flow conditions, and the second column

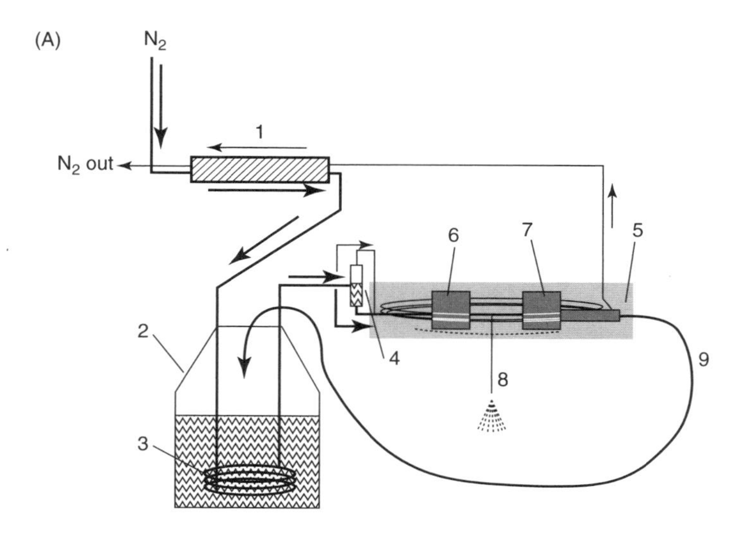

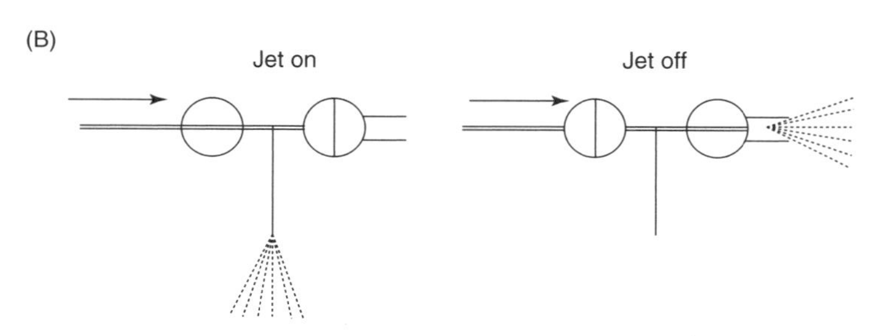

Figure 4.10 (A) Cryogen supply system. Bold solid arrows denote the primary flow path of nitrogen from the high-pressure supply to the cryojet nozzle and back to the LN_2 dewar. Dashed arrows denote the secondary flow path of cold gaseous nitrogen and excess liquid nitrogen from the phase separator through the cooling coils around the valves to the heat exchanger and finally to the atmosphere. 1, Heat exchanger; 2, Dewar with liquid nitrogen; 3, cooling coils; 4, phase separator; 5, modulator; 6, upstream solenoid valve; 7, downstream solenoid valve; 8, cryojet nozzle; 9, cryojet vent with liquid nitrogen return to the LN_2 dewar. (B) Use of two on–off solenoid valves to control the flow of liquid nitrogen through the cryojet. The arrows denote the direction of liquid nitrogen flow from the phase separator. [From Harynuk and Górecki (2003), with permission. Copyright © 2003 by Elsevier.]

did not have to be short and/or narrow. The stop-flow interface was followed by a cryogenic modulator. The utilization of valves in this design had a number of drawbacks. High-temperature rotary valves did not last long when switched continuously, and pneumatic valves had low upper temperature limits. In addition, artifact peaks appeared in the second dimension, as polymeric materials used in the valves tended to off-gas at high temperatures (Oldridge et al., 2008).

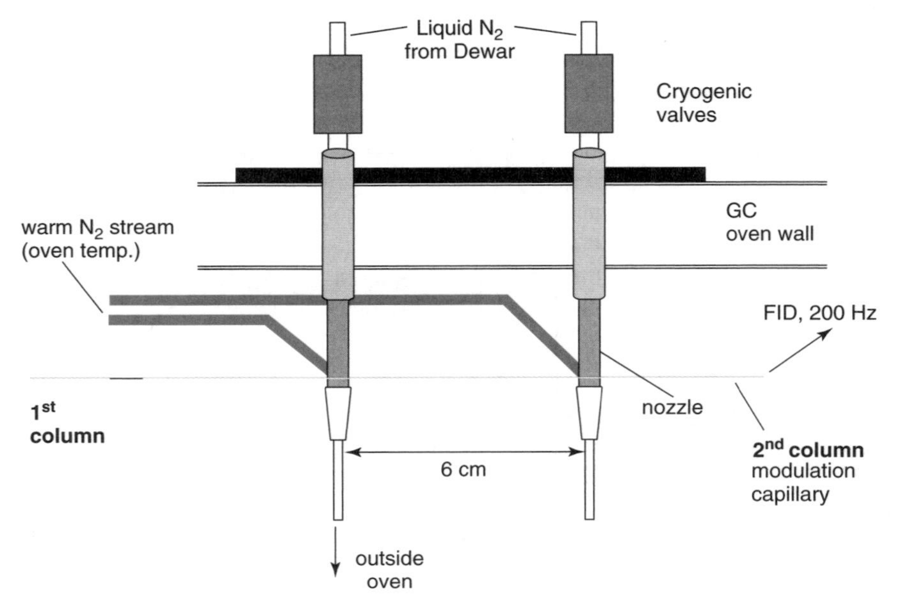

Figure 4.11 Design of two-jet modulator capable of pulsing liquid N_2. [From Pursch et al. (2003), with permission. Copyright © 2003 by Elsevier.]

Another cryogenic modulator, developed by Pursch et al. (2003) (Figure 4.11), used two jets supplying liquid nitrogen directly from a Dewar flask for modulation. The amount of liquid nitrogen consumed per day was about 30 to 40 L (Pursch et al., 2003).

Libardoni et al. (2005) developed an air-cooled resistively heated single-stage thermal modulator with no moving parts. The modulator utilized refrigerated air for trapping and resistive heating for desorption. The main disadvantage was the single-stage design, which resulted in analyte breakthrough during the desorption cycle of the trap. The same group developed another modulator based on the same concept but using liquid ethylene glycol rather than air for cooling (Libardoni et al., 2006). The most significant limitation in both designs was the higher trapping temperature (−30°C) relative to the temperatures attainable with liquefied gases used as cryogenic agents. As a result, both designs were incapable of trapping volatile analytes. On the other hand, they were characterized by very low operating costs. A cryogen-free modulator based on this research has recently been commercialized by the LECO Corporation.

A different cryogenic modulator developed by Hyötyläinen et al. (2002) was based on two-step cryogenic trapping with continuously delivered CO_2 and thermal desorption with electric heating. In this design, two nozzles were mounted on a tube at an angle of 45° to each other. When the tube was rotated, liquid CO_2 from the nozzles would alternately hit a trapping capillary and create two trapping zones. The major problems with this design were the high consumption of CO_2 and the fact that occasional overheating tended to burn the stationary phase. The same group then simplified the design of this modulator to one that used a single CO_2 nozzle mounted on a disk that rotated by 180° back and forth to provide two spots on the column that were alternately cooled for trapping and focusing (Kallio et al., 2003a). Recently, the same group developed a third version of the semirotating modulator to allow more reliable and rugged performance and operation (Kallio et al., 2008). In this version, a separate modulator control program was replaced with a preprogrammed microcontroller equipped with a 4-MHz quartz crystal to control the movement of the modulator, which led to improved repeatability of retention times (Kallio et al., 2008). In addition, the new version was simpler and lighter, and the modulator was easy to install in any commercial GC system.

Although the semirotating modulator was based on the same principle as that of any two-stage cryogenic modulator, the main advantage came from the fact that with the single jet rotating from one position to the other, there was no need for valves to control the flow of the cryogen. Moreover, this design did not cause any risk of column breakage during modulation as was found with the rotating thermal modulator and the early implementations of LMCS (Górecki et al., 2004).

4.5.3 Valve-Based Modulators

Diaphragm Valves The sometimes limited availability and large consumption of liquefied gas cryogenic agents (LCO_2 and LN_2) were the main limitations of cryogenic modulation. In addition, instrument portability was very limited. This led to the parallel development of valve-based modulators requiring no consumables and assembled using inexpensive off-the-shelf components. These modulators are reviewed only briefly here, as they are discussed more extensively in Chapter 5.

The first valve-based modulator based on a fast-switching diaphragm valve was developed by Bruckner et al. (1998). Early valve-based modulators operated through the diversion of portions of the effluent exiting the primary column into the secondary column (Seeley et al., 2000). In fact, a controversy developed initially among the GC × GC community as to whether such an approach really produced a comprehensive multidimensional separation, considering the fact that not all of the sample was subjected to additional separation in the second dimension. Following extensive discussions at the First International Symposium on Comprehensive Multidimensional Gas Chromatography held in 2003 in Volendam (Holland), the community agreed that this was indeed the case, as the resulting chromatogram was representative of the entire sample. A significant limitation of the diaphragm valve–based modulator design was the low maximum operating temperature of the valve, which restricted the range of applications to compounds eluting from the GC oven at temperatures not exceeding the maximum operating temperature of the valve (ca. 180°C).

Hamilton et al. (2003) developed a different valve-based modulator using a design similar to that of Bruckner et al. (1998). This valve modulation system utilized a rotary valve that transferred only small fractions of the primary column effluent to the second-dimension column. Mohler et al. (2006) developed a diaphragm valve–based modulator that accomplished total transfer of the compounds from the primary column to the secondary column. The design was based on a sample loop that was closed at one end during sampling, with the head pressure on the primary column higher than the head pressure on the secondary column, ensuring compression instead of diffusion of the primary column effluent into the loop.

Flow Modulation Systems Bueno and Seeley (2004) developed a modulator based on the idea of differential flow modulation without the need for diaphragm valves incorporated into the flow path of the sample. The interface used two sample loops and a pneumatic switching system. The direction of the flow of the first column effluent was controlled by a three-port solenoid valve located outside the GC oven and not in the sample flow path (see Chapter 5 for details). This was the first valve-based modulator that launched 100% of the primary column effluent into the second-dimension column for separation (Bueno and Seeley, 2004). In addition, the approach eliminated the temperature limitations of diaphragm valve–based modulators.

Seeley et al. (2006) later simplified their design and introduced a simple in-line fluidic modulator. Wang (2007, 2008) proposed a different differential flow modulation unit design based on two four-port two-position switching valves for complete sample transfer from the first to the second column.

Górecki's group introduced a different interface design, stop-flow GC × GC with pneumatic switching, which was a new version of their previous stop-flow interface (Harynuk and Górecki, 2004, 2006). The flow in the primary column was stopped pneumatically, with no valves in the pathway of the carrier gas (by applying pressure pulses at the junction between the two columns). Figure 4.12 shows a detailed setup of the pneumatic switching system. It used a solenoid valve placed outside the oven. When the valve was opened, pressure was applied at the midpoint junction at the outlet of the primary column, which stopped

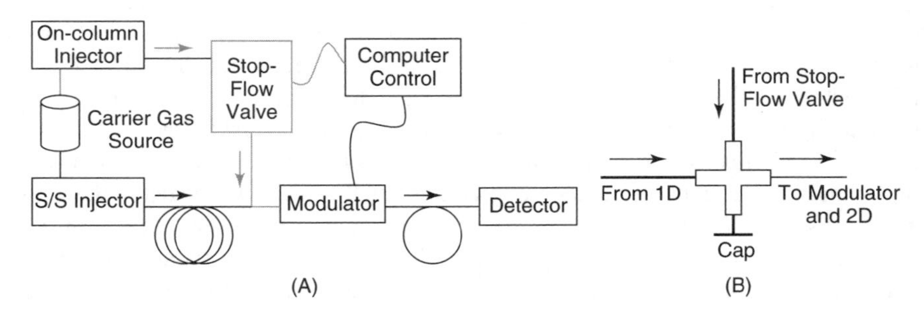

Figure 4.12 (A) Pneumatic stop-flow interface. Items in gray are additions to the conventional GC × GC column train. Arrows indicate direction of gas flow. (B) Midpoint junction connector. A four-port junction (cross) was used, with one end capped. [From Oldridge et al. (2008), with permission. Copyright © 2008 by Wiley-VCH Verlag GmbH & Co. KGaA.]

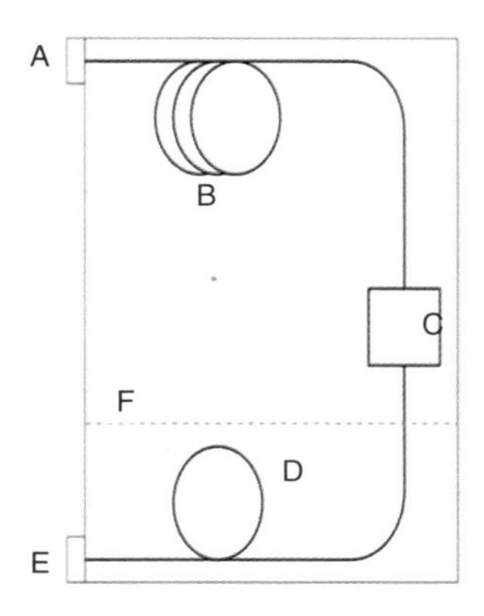

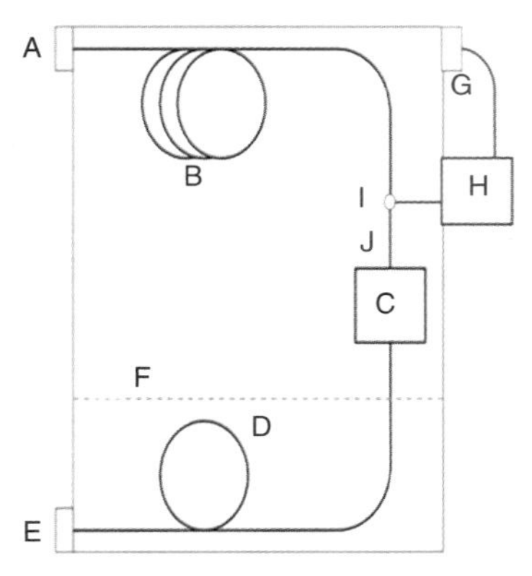

Figure 4.13 Conventional GC × GC system (left) and stop-flow GC × GC system with pneumatic switching (right). Common instrumental components include (A) injector, (B) 1D column, (C) modulator, (D) 2D column, (E) detector, and (F) optional second oven. The stop-flow system utilizes (G) a cool on-column injector as an auxiliary gas pressure source, (H) a solenoid valve, (I) a three-port junction, and (J) a restricting transfer capillary from the junction to the modulator. [From Oldridge et al. (2008), with permission. Copyright © 2008 by Wiley-VCH Verlag GmbH & Co. KGaA.]

the primary column flow. It is important to note that in this approach a thermal modulator was used to achieve effective modulation (Figure 4.13). Elimination of mechanical valves from the column train improved the robustness of the system and eliminated the temperature limitations and artifacts related to off-gassing of the valve components (Oldridge et al., 2008). The main disadvantage of the design was the complexity of the instrumentation and the longer total analysis time (Adahchour et al., 2006a).

Recently, Agilent Technologies introduced a new modulator utilizing the company's capillary flow technology hardware and Seeley's differential flow modulation concepts. The modulator utilizes a low-thermal-mass deactivated stainless-steel capillary flow device with no moving parts for switching purposes. The pneumatics are controlled via a three-way solenoid valve installed on the top of the GC oven.

4.6 COMPREHENSIVE TWO-DIMENSIONAL GC METHOD OPTIMIZATION

The introduction of comprehensive two-dimensional (2D) GC has opened a new era of interest and excitement in the GC field. GC × GC has certainly enabled a much deeper insight into the composition of many GC-amenable samples, defined as well known up until a decade ago. However, though the usefulness and analytical power of GC × GC have been fully demonstrated, the technique is still far from being well established. Apart from a natural scepticism toward new methodologies (old habits are hard to shake!), one of the main reasons behind this reluctance toward innovation is related to optimization issues.

In general, the main advantages of GC × GC over conventional GC methods are essentially five:

1. *Speed*: the number of peaks resolved per unit of time
2. *Selectivity*: the use of two stationary phases of different selectivity
3. *Separation*: increased resolving power
4. *Sensitivity*: thermal modulation normally achieved using cryogenic gases, generating a band compression effect (additionally, the isolation of chemical noise has a considerable influence on sensitivity)
5. *Spatial order*: the contour plot formation of chemically similar compound patterns

Comprehensive 2D GC method optimization is generally directed toward the maximization of separation power and sensitivity; spatial order is highly important only when homologous series of compounds (e.g., fatty acid methyl esters, alkanes) are present in the sample. As mentioned earlier the construction of a comprehensive 2D GC system is quite simple and can be achieved on practically any commercially available GC instrument, because GC × GC is essentially a tandem-column methodology with a modulation system usually located at the head of the secondary column. Thus, all that is required to produce a double-axis chromatogram is a modulator mounted inside a traditional GC oven. Injection features are no different from those of single-column GCs, inlet pressure requirements are within the capabilities of a standard gas chromatograph, and the temperature-program ramp is generally "slowish" (2 to 3°C/min). Considering detection, present-day flame ionization detectors are characterized by low internal volumes and are more than fast enough for the rapid peaks generated by modulation. Moreover, a number of commercially available selective detectors can easily be used in a GC × GC experiment. It must be added, however, that dedicated software is necessary to process GC × GC data.

The first GC × GC analysis (Liu and Phillips, 1991) was carried out using a Varian Model 3700 gas chromatograph and a low-cost thermal-desorption modulator; a polar first dimension (21 m × 0.25 mm i.d. × 0.25 μm d_f) and a nonpolar second dimension (1 m × 0.10 mm i.d. × 0.5 μm d_f) were used as a column set. The first GC × GC stationary-phase combination is worthy of note because it is defined as "reversed" and is not considered as orthogonal in nature. The apolar–polar capillary combination is universally recognized as the most orthogonal; overlapping compounds at the primary column outlet (boiling-point separation) are characterized by the same or similar vapor pressures and can ideally be separated on a secondary column on a polarity basis. It will be seen that reversed column sets can provide very good GC × GC results for specific sample types. Another issue worthy of emphasis from that first paper is the use of a splitter (10 cm × 0.25 mm i.d. fused-silica tubing), connected at the union point between the two capillaries, to adjust the secondary-column gas flow.

Throughout the following years, employment of the afore-described approach disappeared from the scene. Nowadays, modulation is generally carried out with cryogenic gases, with excessively high costs per analysis. Moreover, the use of a splitter gradually declined. The long conventional + the short fast column combination (from now onward defined as conventional GC × GC) has always been by far the most preferred by researchers in this field. However, it will be shown that most conventional GC × GC separations are carried out at gas velocities that are ideal (or nearly ideal) in the first dimension and excessively high in the second dimension. It will also be demonstrated that the direction of an appropriate volume of primary column effluent to waste enables the generation of optimum gas flows in both dimensions.

If the transformation of a 1D GC instrument into a 2D GC instrument is straightforward, things become slightly more complicated if method development is considered. Comprehensive 2D GC method optimization can be a painstaking issue and is, in the opinion of many, one of the main reasons behind the still rather limited employment of the technique. The scenario is much more complex than conventional GC, because the two dimensions are intimately related. A solid knowledge of chromatography basic theory and acquired experience in the field of conventional, classical multidimensional, very fast microbore column, and high-speed megabore column low-pressure GC is of great help. Apart from modulation parameters (period and temperature), the main operational conditions that must be considered are the stationary-phase chemistries, capillary column dimensions, gas flow, temperature programs, outlet pressure conditions, and detector settings.

4.6.1 Stationary-Phase Combinations

Analyte vapor pressures play a major role in all GC separations, irrespective of the liquid stationary phase employed. In every GC × GC experiment, there is always a certain degree of correlation between the two dimensions, inhibiting full exploitation of the bidimensional space. Partial correlation is one of the main reasons that the much acclaimed $n_{c1} \times n_{c2}$ peak capacity value is an excessive estimation of the real value. A high number of published GC × GC chromatograms are characterized by fan-shaped analyte bands, extending across different lengths of the space plane, with a left-to-right upward inclination and with plenty of unexploited space. Due to the lack of entirely dissimilar separation mechanisms, it is very rare to find an early-eluting first-dimension compound with a high second-dimension retention time, and vice versa. A typical example of low exploitation of the available 2D chromatography space can be observed in the cod oil fatty acid methyl ester (FAME) chromatogram, illustrated in Figure 4.14. Although the chromatogram is well structured (C_{14} to C_{22} group-type patterns are evident), the amount of unoccupied 2D space is high; the FAME band was located inside a quadrilateral, the area of which was approximately 22.3% of the entire 2D space (Tranchida et al., 2007). This negative GC × GC feature was caused by two main factors: partial correlation between the two dimensions [a 5% diphenyl + 95% dimethyl polysiloxane–poly(ethylene glycol) combination was used] and an excessively high second-dimension gas linear velocity. The latter aspect is discussed in depth later.

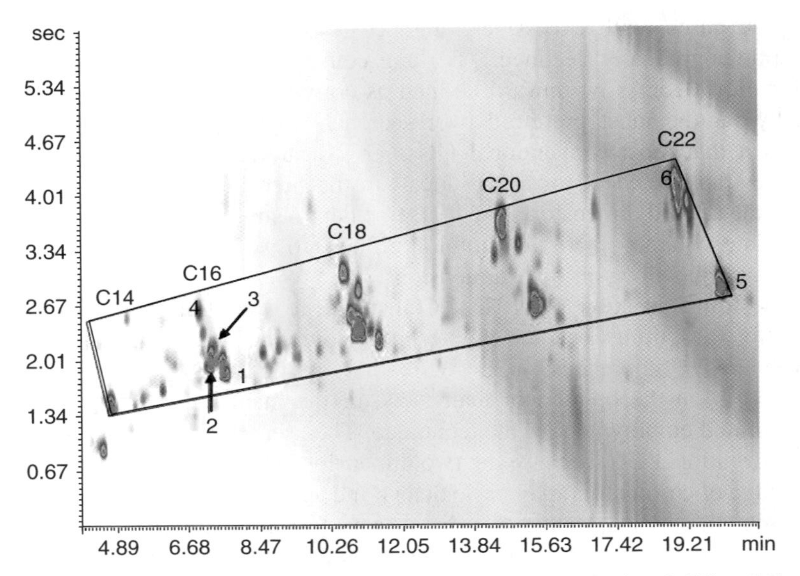

Figure 4.14 Bidimensional chromatogram relative to a conventional GC × GC application, achieved on an orthogonal column combination and carried out on cod oil FAME. Peak identification: 1. $C_{16:0}$; 2. $C_{16:1\omega7}$; 3. $C_{16:2\omega4}$; 4. $C_{16:4\omega3}$; 5. $C_{22:1\omega9}$; 6. $C_{22:6\omega3}$. [Reproduced from Tranchida et al. (2007), with permission. Copyright © 2007 by The American Chemical Society.]

Real-world samples are generally complex mixtures containing a wide range of volatiles with different chemical structures and polarities. Even if an analyst has an a priori knowledge of the composition of a sample, it is not easy to foresee which phases will provide the best results. Consequently, GC × GC stationary-phase optimization is usually achieved through trial-and-error testing. An apolar first dimension (100% dimethyl polysiloxane, 5% diphenyl + 95% dimethyl polysiloxane, etc.) and polar second dimension [100% poly(ethylene glycol), 50% diphenyl + 50% dimethyl polysiloxane, etc.] are generally tested before other combinations. However, the initial orthogonal choice does not always satisfy the analytical objective(s), namely: (1) separation of the highest number of sample volatiles, (2) isolation of target compounds, and (3) formation of ordered groups of structurally related compounds. A common occurrence when using an orthogonal column set is that polar compounds can undergo extensive wraparound. However, wraparound is commonly observed in many GC × GC applications and is acceptable as long as the objectives of the analysis are not lost.

Over the last two decades, hundreds of GC × GC applications have been published, with a wide variety of stationary-phase combinations. It is obvious that all experiments of interest, from a selectivity viewpoint, cannot be reported here. A description of some significant GC × GC experiments, highlighting stationary-phase selectivities, follows. Petrochemical fuels appear to be the perfect sample types for GC × GC analysis because (1) they are formed

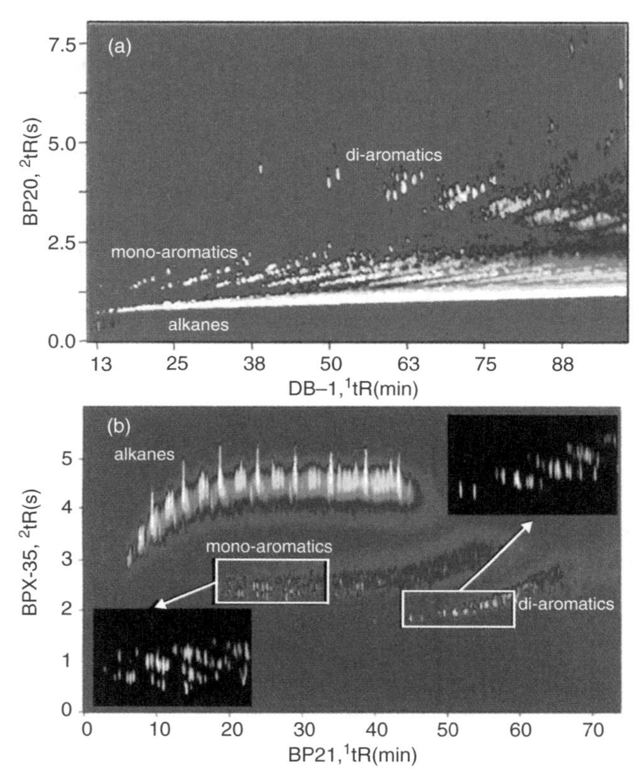

Figure 4.15 GC × GC–FID chromatograms of a diesel oil obtained on two different column sets: apolar–polar (top) and polar–medium polarity. [From Adahchour et al. (2004), with permission. Copyright © 2004 by Elsevier.]

of thousands of compounds, (2) they contain specific biomarkers, and (3) they are characterized by chemically related constituents. Figure 4.15 illustrates both an orthogonal [100% methyl polysiloxane–poly(ethylene glycol) and reversed [poly(ethylene glycol)–35% phenyl polysilphenylene-siloxane] GC × GC–FID analysis on diesel oil (Adahchour et al., 2004). The orthogonal combination enabled a primary column boiling-point separation with little interclass differentiation (e.g., there is no separation between monoaromatic and alkane groups), and a secondary column polarity-based separation, with satisfactory intergroup separation between the diaromatics and monoaromatics, and partial overlapping at the border between the monoaromatics and the alkanes. The quality of intraclass resolution is good for diaromatics and monoaromatics, while, as expected, the polar capillary showed insufficient selectivity for the alkanes. It must be added that a +20°C offset was employed in the second-dimension oven to maintain the diaromatics within the 8-s modulation time window. The reversed combination enabled a first-dimension polarity-based separation, showing some degree of intergroup separation (e.g., there is partial group isolation between diaromatics, monoaromatics, and alkanes), and a secondary-column medium polarity–based 2D analysis with good intergroup separation between all chemical classes. However, the quality of intragroup resolution cannot be defined as sufficient for any of the three chemical classes. A +40°C offset was applied in the second-dimension oven. Considering both GC × GC experiments, it can be concluded that the orthogonal approach enabled the resolution of the highest number of volatiles and would certainly be more suited for the analysis of target analytes. The reversed set was preferred for intergroup resolution and thus for the quantitation of specific chemical classes.

Seeley et al. (2001) proposed a dual-secondary column GC × 2GC system with valve-based modulation. A 15 m × 0.25 mm i.d. × 1.4 μm d_f DB-624 (6% cyanopropylphenyl + 94% dimethyl polysiloxane) column was connected to a 5 m × 0.25 mm i.d. × 0.5 μm d_f DB-210 (trifluoropropylmethyl polysiloxane) and a 5 m × 0.25 mm i.d. × 0.25 μm d_f DB-Wax [poly(ethylene glycol)]. The first column interacts mainly through dispersive forces, while the trifluoropropylmethyl polysiloxane phase shows strong dipole–dipole interactions and the poly(ethylene glycol) phase exhibits intense hydrogen-bonding interactions. A 55-compound mixture containing C_5–C_{13} *n*-alkanes, C_1–C_8 1-alcohols, C_3–C_8 2-alcohols, C_4–C_7 2-methyl-2-alcohols, C_3–C_8 and C_{10} acetates, C_3–C_{11} aldehydes, C_3–C_8 2-ketones, and C_6–C_{10} alkyl aromatics was subjected to analysis (Figure 4.16). It was observed that alkanes, as expected, had minimum retention on both columns; aromatics (i.e., highly polarizable molecules but with low dipole moments and no hydrogen-bonding functional groups) showed moderate retention on both capillaries; 1- and 2-alcohols (i.e., compounds with high degrees of hydrogen-bond acidity but moderate dipole moments) were highly retained on the DB-Wax capillary and only moderately on the DB-210 capillary; ketones and aldehydes (i.e., molecules with low hydrogen-bond acidity and large dipole moments, were characterized by high elution times on the DB-210 column and moderate retention on the DB-Wax second dimension.

In a much more recent study, Seeley et al. (2009) used the solvation parameter model to generate simulated GC × GC chromatograms for 54 compounds, on four different stationary-phase combinations: DB-1: poly(dimethylsiloxane) × DB-Wax; DB-1 × DB-210; DB-1 × DB-1701 poly (cyanopropylphenyldimethylsiloxane); and HP-50: poly (dimethyldiphenyl- siloxane) ×DB-1. Considering the DB-1 × DB-Wax combination, the model predicted that retention along the first and second dimensions would have been determined predominantly by solute size and hydrogen-bond acidity (e.g., alcohols interact intensively), respectively. With regard to the DB-1 × DB-210 combination, it was predicted that second-dimension elution times would have been much less dependent on hydrogen-bond acidity and highly influenced by dipolarity (e.g., nitroalkanes, aldehydes, ketones, and esters are more retained than alcohols). The model predicted comparable influence of dipolarity and hydrogen-bond acidity when using the DB-1701 column in the second dimension. Considering the HP-50 × DB-1 combination, it was predicted that analyte dipolarity, as well as solute size, would greatly influence the degree of primary column retention, while an increase in dipolarity

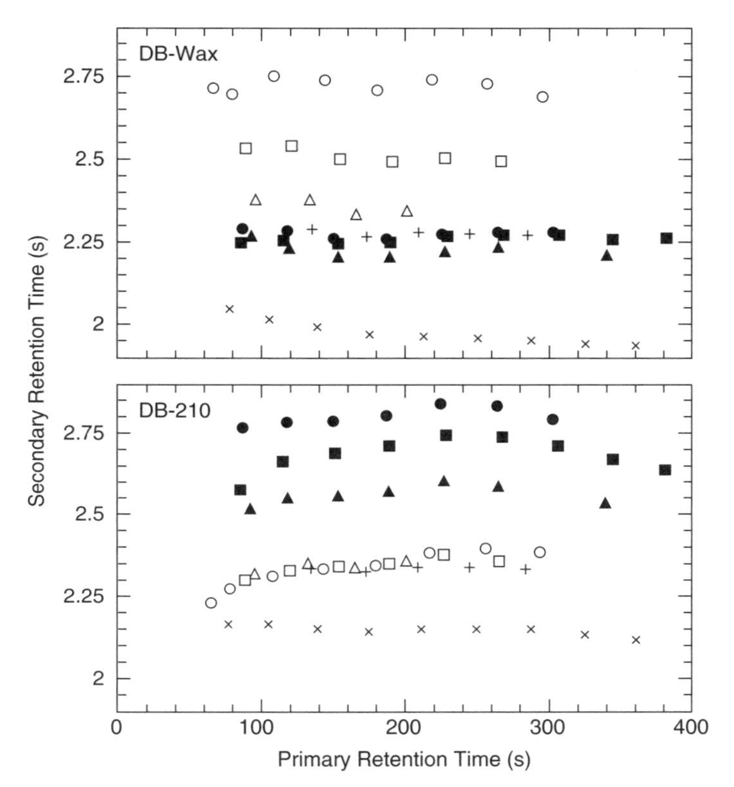

Figure 4.16 Excel-reconstructed plots relative to GC × 2GC analysis carried out with a low-polarity primary column, and DB-Wax/DB-210 secondary columns. Chemical class identification: ×, alkanes +, aromatics; ▲, acetates; ■, aldehydes; ●, ketones; △, 2-methyl-2-alcohols; □, 2-alcohols; ○, 1-alcohols. [From Seeley et al. (2001), with permission. Copyright © 2001 by Wiley-VCH Verlag GmbH & Co. KGaA.]

and hydrogen-bond acidity would decrease secondary elution times. In general, very good agreement was observed between the theoretical and experimental chromatograms.

If GC selectivity is considered, there is nothing more selective than a chiral stationary phase. In particular, cyclodextrin derivatives have been employed successfully in many applications, particularly in the analysis of essential oil chiral constituents (Bicchi et al., 1999). Enantioselective GC (enantio-GC) using cyclodextrin chiral selectors can be very useful for the determination of genuineness, origin, and quality. However, the complexity of essential oils often exceeds the peak capacity of a single chiral column, and the use of classical heart-cutting multidimensional GC (MDGC) is advisable. Enantio-MDGC experiments are typically achieved by using an apolar–chiral stationary phase combination.

The concept of enantio-MDGC was extended to GC × GC, in the analysis of tea tree oil, by Shellie et al. (2001). The authors reported the use of a β-cyclodextrin 25 m × 0.25 mm i.d. × 0.25 μm d_f primary column and a polar

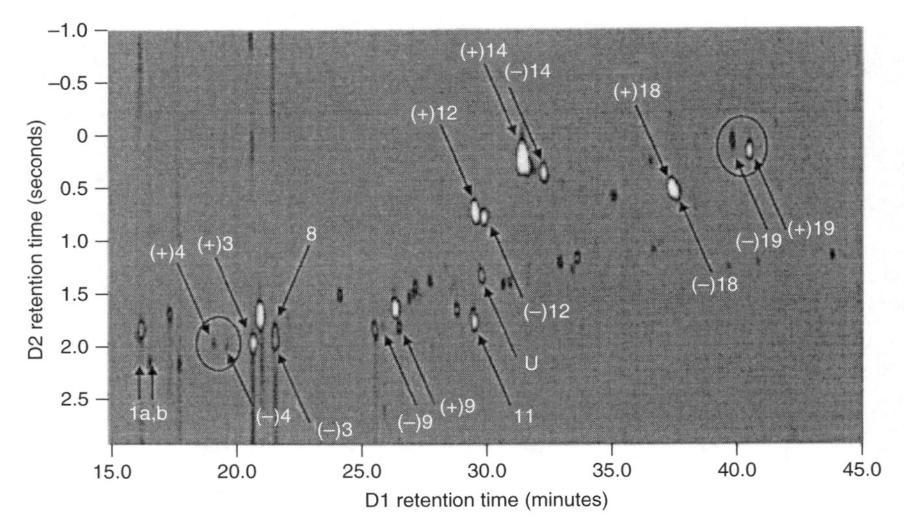

Figure 4.17 Enantio-GC × GC–FID chromatogram of tea tree oil. Peak identification: 1, α-thujene; 3, sabinene; 4, β-pinene; 8, *p*-cymene; 9, limonene; 11, γ-terpinene; 12, *trans*-sabinene hydrate; 14, *cis*-sabinene hydrate; 18, terpinen-4-ol; 19, α-terpineol. Optical isomers are differentiated by a + or a − symbol; the letters a and b are used where the assignment of the optical isomers was not confirmed. U, unidentified compound. [From Shellie et al. (2001), with permission. Copyright © 2001 by Wiley-VCH Verlag GmbH & Co. KGaA.]

0.8 m × 0.1 mm i.d. × 0.1 μm d_f second dimension. Consequently, the aim of the enantio-GC × GC experiment was to separate target optic isomers in the first dimension and free them from interference in the second dimension. Figure 4.17 shows a tea tree oil enantio-GC × GC-FID chromatogram, characterized by a nice series of entirely isolated chiral volatiles, except for +/− terpinen-4-ol (peak 18), which partially overlapped in both dimensions. It must be emphasized that in such applications (i.e., where the resolution achieved in the first dimension is of the utmost importance), each peak must be modulated a sufficient number of times (minimum three, better four).

4.6.2 Gas Linear Velocities

If one browses through the GC × GC literature, it can be observed that the many impressive separations reported, demonstrating the unrivaled peak capacity of this approach, hide the fact that the full potential of GC × GC is rarely expressed. Apart from the correlation between stationary phases discussed previously, most GC × GC separations are carried out at gas velocities that are ideal (or nearly ideal) in the first dimension and far from optimum in the second dimension (Beens et al., 2005; Shellie et al., 2004; Tranchida et al., 2007). The reason for such a gas velocity condition is related to the fact that a single pressure source supplies gas to a long conventional and a short microbore twin-column system. Under optimum gas linear velocity conditions, we intend that a minimum theoretical plate height value be generated.

In an interesting paper, Shellie et al. (2004) reproduced equal gas linear velocity conditions in GC × GC–FID and GC × GC–TOF MS experiments. It was demonstrated that matching GC × GC–TOF MS chromatograms can be generated if an auxiliary gas is supplied between the column outlet and the MS interface. The authors reported the use of a 30 m × 0.25 mm i.d. primary column and a 0.5 m × 0.1 mm i.d. secondary capillary, with linear velocities of about 34 and 295 cm/s in the first and second dimensions, respectively. Although the main objective of the experiment was not gas velocity optimization, the authors concluded that the linear velocity in the second dimension was about three times higher than the optimum value.

In terms of gas velocity optimization, different options exist:

1. A head pressure reduction will obviously generate lower linear velocities in both dimensions. Besides extending analysis times (not a foremost worry in comprehensive 2D GC), such a choice can have a negative effect on first-dimension resolution. Moreover, analytes will elute from the first dimension at higher temperatures, reducing the benefits of lower second-dimension gas velocities.
2. The use of a longer secondary column will also enable the generation of lower gas velocities in the second dimension. However, if the second-dimension length is doubled, so are elution times, while a factor of 1.4 gain in resolution should be expected. Furthermore, the extended retention times would probably require a greater modulation period, which could lead to a higher degradation of primary column resolution.
3. The employment of a 0.15 to 0.18-mm i.d. secondary column, in combination with a primary 0.25-mm i.d. column, would enable the operation of both dimensions under nearly optimum conditions. However, the price to pay for such an option is reduced efficiency compared to a 0.1 mm i.d. capillary.

An additional option exists, and it can be derived from the first GC × GC experiments (Liu and Phillips, 1991; Venkatramani and Phillips, 1993), where part of the flow was diverted to waste prior to modulation: a T-union was used to connect the two analytical columns to a short capillary segment, enabling the diversion of about 30% of the primary column flow. It was also stated that such a column configuration would produce ideal flows in both dimensions and could reduce overloading in the second dimension. Unfortunately, in the following years the employment of a splitter between the two dimensions disappeared from the GC × GC scene.

In 2007, Tranchida et al. developed a GC × GC method defined as split-flow comprehensive 2D GC. The novel system, illustrated in Figure 4.18, was characterized by a primary apolar 30 m × 0.25 mm i.d. column, linked to a 1 m × 0.10 mm i.d. polar column and to a 30 cm × 0.10 mm i.d.

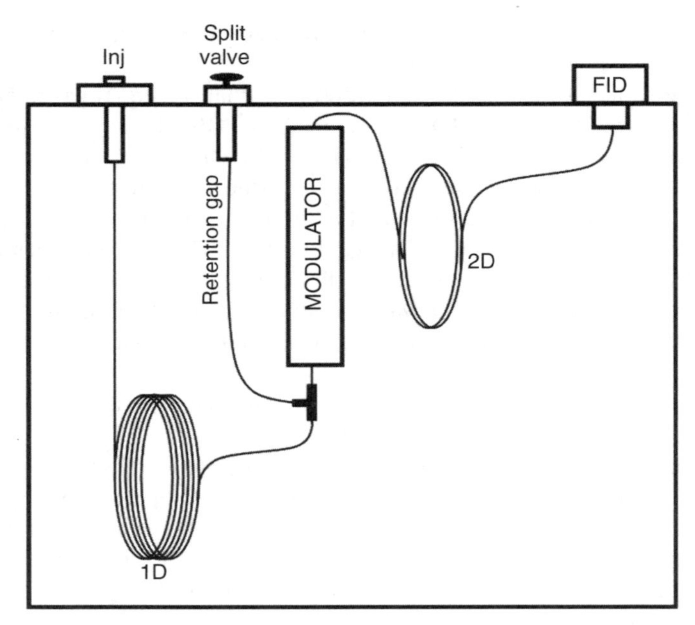

Figure 4.18 Split-flow single-oven GC × GC–FID system. (From Tranchida et al. (2007), with permission. Copyright © 2007 by The American Chemical Society.]

uncoated capillary, using a Y press fit. The uncoated column was connected to a manually operated split valve, while the polar capillary was passed through a cryogenic modulator and then connected to an FID. The hydrogen linear velocities in both dimensions were regulated by manual operation of the split valve.

The cod oil methyl ester chromatogram, illustrated in Figure 4.14, was attained by closing the split valve and thus can be considered as a conventional GC × GC–FID experiment. The applied head pressure (194.9 kPa) generated linear velocities of about 35 and 333 cm/s in the first and second dimensions, respectively. These approximate values were attained by calculating the equivalent dimensions of the twin column set, through the evaluation of flow relationships between columns of different lengths and i.d.s (Shellie et al., 2004). With regard to the 35-cm/s H_2 velocity in the primary column, this can be considered as a nearly optimum operational parameter, while the second-dimension 333-cm/s gas velocity is a far-from-ideal experimental condition.

Although the gas velocity conditions generated a well-structured chromatogram plot (C_{14} to C_{22} group-type patterns are evident), there is a large amount of unoccupied 2D space. The width of the C_{16} group along the y-axis was 0.688 s and can be considered as the difference in the elution times between peaks 1 and 4, while the width of the C_{22} group was 1.176 s, considering peaks 5 and 6. A single high-speed 2D chromatogram, which shows the separation of peaks 2 and 3, is reported in Figure 4.19a; the R_s value for these compounds is 0.8.

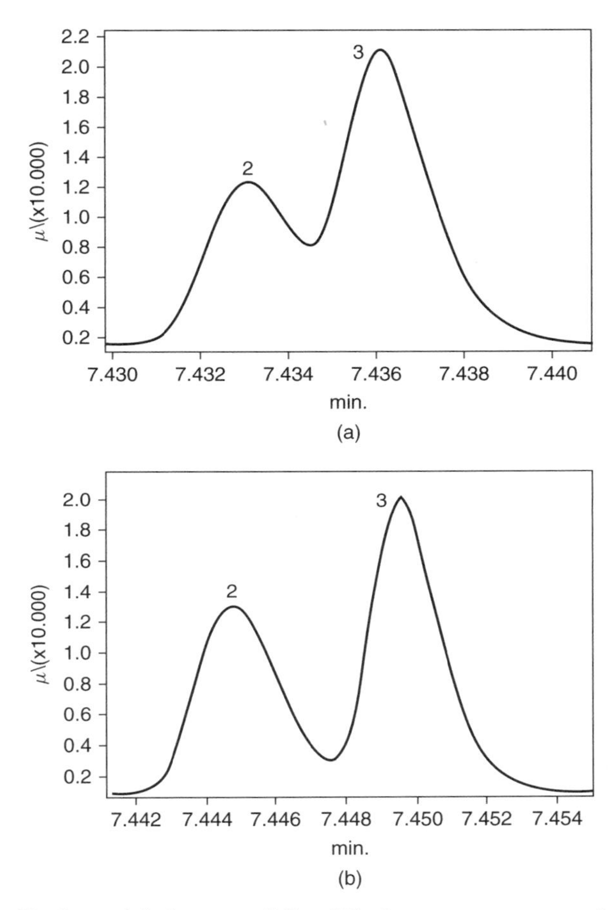

Figure 4.19 Single-modulation raw GC × GC chromatogram expansions relative to conventional (a), and 35 : 65 split-flow (b) applications for compounds 2 and 3. Peak identification as in Figure 4.14. [From Tranchida et al. (2007), with permission. Copyright © 2007 by The American Chemical Society.]

The split-flow setup was used for analysis of the cod oil FAME, with the objective of increasing efficiency in the second dimension while preserving the group structures. The best result was attained using a 146.3-kPa head pressure and a 35 : 65 (FID) split-flow ratio, which generated gas velocities of about 35 and 213 cm/s in the first and second dimensions respectively. As can be seen in the split-flow 2D chromatogram shown in Figure 4.20, the structure of the chromatogram is maintained. Moreover, the secondary microbore column provided an improved separation performance. The widths of the C_{16} and C_{22} groups were 1.104s (+ 60%) and 1.728s (+ 47%), respectively, while resolution between peaks 2 and 3 (Figure 4.19b) increased to 1.2 (+ 50%). Considering the quadrilateral surrounding the FAMEs band, this occupied approximately 32.0% of

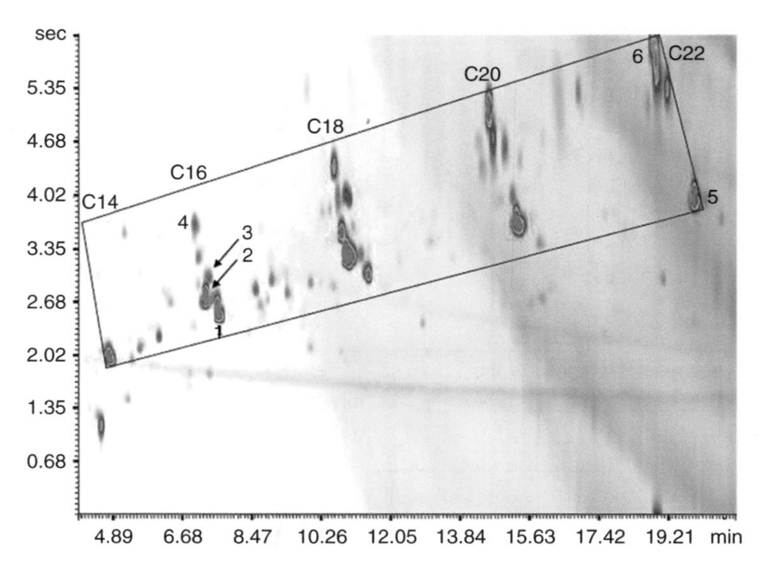

Figure 4.20 Two-dimensional chromatogram relative to a 35 : 65 split-flow GC × GC application, carried out on cod liver oil FAME. Peak identification as in Figure 4.14. [From Tranchida et al. (2007), with permission. Copyright © 2007 by The American Chemical Society.]

the entire contour plot. A further advantage of the split-flow method, emphasized by the authors, was that overloading of the second dimension can be reduced.

Recently, a 50- μm i.d. secondary column was employed under optimized gas velocity conditions using a twin-oven split-flow GC × GC–FID system (Tranchida et al., 2009). An apolar 30 m × 0.25 mm i.d. capillary was connected, through a Y-union, to a high-resolution 1 m × 0.05 mm i.d. polar capillary and to a 20 cm × 0.05 mm i.d. uncoated column segment. The secondary analytical column was connected to an FID, while the uncoated column was linked to a manually operated split valve located on top of the second GC. Various split-flow GC × GC–FID experiments were carried out by manual regulation of the split valve. The best operational conditions were attained with the split valve completely opened (83.4% waste); linear velocities of about 20 and 110 cm/s were calculated in the first and second columns, respectively. The optimized GC × GC–FID method was exploited for the analysis of a diesel sample. The results attained were compared directly with those derived by using what the authors defined as a classical GC × GC split-flow column set. The same first dimension was connected to an FID-linked 1 m × 0.10 mm i.d. polar column and to a 30 cm × 0.10 mm i.d. uncoated column. Both column sets were used under practically ideal chromatography conditions. The experimental results demonstrated the full benefits of using a 50-μm i.d. secondary column in a split-flow configuration.

On the basis of the experimental results, the authors concluded that the optimized employment of a 50-μm i.d. secondary column in a GC × GC–FID instrument could be achieved by using (1) two GC ovens and a suitable uncoated capillary segment for flow splitting or (2) a single GC oven and an external split valve. Considering option 1, it is important to use an adequate uncoated capillary segment length in relation to the second-dimension analytical column. The generation of a hydrogen linear velocity of about 20 cm/s in the first dimension and 100 to 130 cm/s in the second dimension is highly advisable. Once the linear velocities are optimized, the GC × GC analyst can regulate the temperature programs to attain a satisfactory result. For example, if wraparound occurs and needs to be avoided, a positive temperature offset can be applied in GC2. Considering option 2, gas flow regulation using the split valve would often be a necessity. Again, it would be advisable to use a suitable length of uncoated column to generate ideal velocities in both dimensions. Generally, there would be no need to alter the first-dimension linear velocity (ca. 20 cm/s is a good choice), unless particular conditions are required (e.g., a faster analysis, the generation of more broadened bands, etc.). If the second-dimension gas velocity is considered, the ideal value may not produce the best result. For example, if wraparound occurs, causing a loss of order in chromatogram group-type patterns, an increase in secondary column gas velocity (through spilt-valve regulation) would certainly overcome this problem. It must be added that the first of the two options appears to be the best choice, due to the easier use and greater flexibility. At the end of the paper the authors surmised that the use of a 0.15 mm i.d. + 0.05 mm i.d. column combination, in a split-flow GC × GC configuration, would appear to be of great interest.

As mentioned previously, an innovative comprehensive 2D GC technique, defined as stop-flow GC × GC, was recently described (Oldridge et al., 2008). A schematic of the stop-flow instrument, which is based on the stop-flow tandem-column system developed by Veriotti and Sacks (2001), is shown on the right-hand side of Figure 4.13, while a conventional GC × GC is shown on the left. Stop-flow conditions are generated by an auxiliary gas flow, derived from an on-column injector (G). The latter is connected to a solenoid valve (H), located outside the oven, and which, in turn, is linked to a three-port connector (I) via a 250-μm i.d. fused-silica capillary. The remaining two ports are connected to the primary column outlet and to a 100-μm i.d. fused-silica capillary, which is in turn connected to the modulator (Figure 4.12). The flow is stopped in the first dimension when the solenoid valve is opened and a specific pressure is directed to the junction point. A conventional GC × GC application on a diesel sample, achieved by maintaining the solenoid valve in the closed position, is illustrated in Figure 4.21. A constant gas flow of 1.2 mL/min and a 6-s modulation period were utilized. The first dimension was a nonpolar 30 m × 0.25 mm i.d. × 1 μm column, while the second column was a 2.1 m × 0.25 mm i.d. × 0.25 μm SolGel-WAX, both temperature-programmed from 100 to 260°C (7 min) at 3°C/min. The stop-flow GC × GC application on the diesel sample is illustrated in Figure 4.22. A constant gas flow of 1.2 mL/min was generated in the first dimension, while a 6-s modulation period was applied with a 3-s first-dimension flow followed by a 3-s stop flow. The number of cuts per peak were, in this way, doubled, leading to

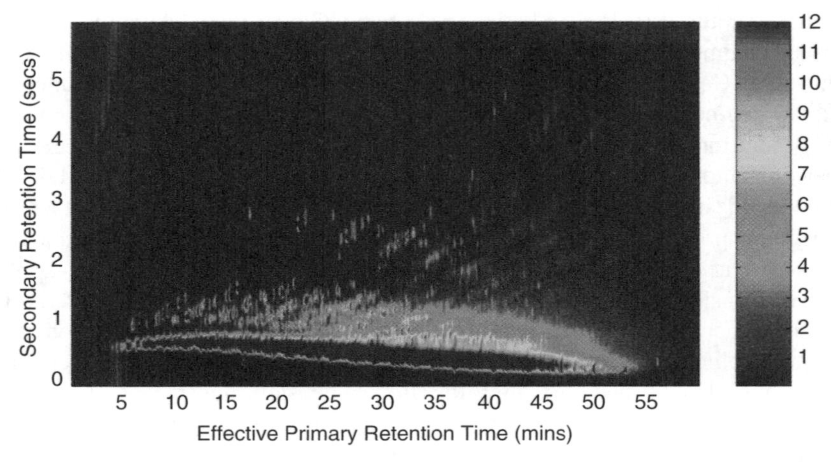

Figure 4.21 Conventional GC × GC analysis of diesel. [From Oldridge et al. (2008), with permission. Copyright © 2008 by Wiley-VCH Verlag GmbH & Co. KGaA.]

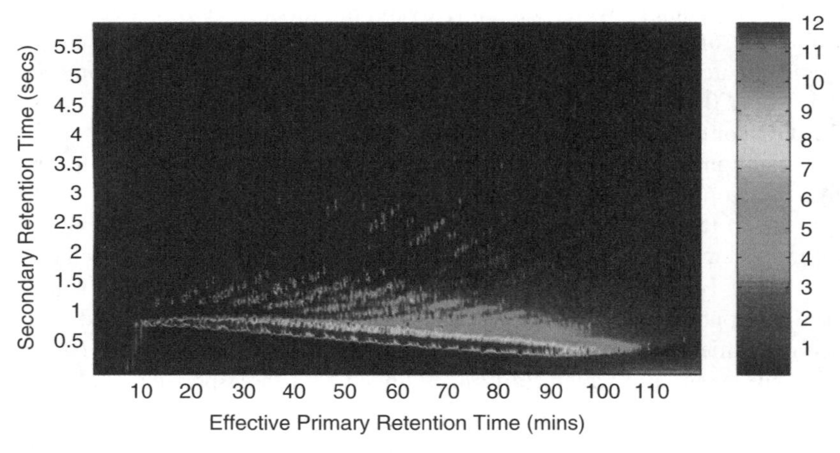

Figure 4.22 Stop-flow GC × GC analysis of diesel. [From Oldridge et al. (2008), with permission. Copyright © 2008 by Wiley-VCH Verlag GmbH & Co. KGaA.]

better preservation of the primary column resolution. The first dimension was the same as in conventional analysis, while the second column was a 5.7-m piece of SolGel-WAX. The temperature program, which was modified to keep the same elution temperatures, was from 100 to 260°C (13 min) at 1.5°C/min. The on-column injector (G) was operated under pressure gradient conditions, enabling constant stop-flow periods throughout the analysis. The authors demonstrated that the overall first-dimension separation benefited greatly from the higher number of modulations per peak. Moreover, the resolution in the second dimension was also improved, as a consequence of the use of a longer column. The results attained by using a stop-flow GC × GC system can be considered of great interest in terms of method optimization. The extended analysis times can be considered a minor drawback also because the authors used a novel modulation system, with no cryogenic gas requirements.

4.6.3 Temperature Program(s)

The temperature gradient is another aspect of outstanding importance in GC × GC, requiring careful optimization. Comprehensive 2D GC experiments are carried out using either a single- or a twin-oven configuration: the latter option certainly being the better solution, as it guarantees a high degree of flexibility. Two main disadvantages derive from using a single GC oven: the maximum column-set operational temperature is that of the less thermally stable capillary (e.g., 260 to 280°C if a poly(ethylene glycol) phase is employed), and the second-dimension analysis temperatures are entirely dependent on the first-dimension elution temperatures. Primary capillary temperature ramps are rather slow, usually in the range 1 to 3°C/min, generating broadened peaks prior to modulation. Using such conditions in a single-oven instrument, modulated fractions reach the secondary column at relatively low temperatures, overlapping components interact more intensely with the 2D stationary phase, resolution improves (up until a specific k value), and more of the bidimensional space is occupied. However, an excessively slow temperature rate can generate wraparound and a sensitivity reduction; if required, both of these negative effects can be avoided by using a second oven. Second-dimension separations are run under nearly isothermal conditions; for example, if a 2°C/min temperature gradient is used with an 8-s modulation time, a 0.27°C increase will be observed during each modulation. As a consequence, secondary-column retention times, relative to the same solute, will undergo a slight reduction from one modulation to the next. The main effect relative to the latter phenomenom is that 2D spots can present a downwardly inclined direction.

The effects of temperature conditions on peak distribution were studied in the years immediately following the invention of GC × GC, using a twin-oven configuration. Venkatramani et al. (1996) analyzed a 19-compound mixture (four alkanes, four alkenes, and five alcohols, plus six chemically dissimilar volatiles) on a nonpolar 3 m × 0.53 mm i.d. × 5 μm first column and a polar 5 m × 0.25 mm i.d. × 0.25 μm second column. In the first experiment, the columns were operated isothermally at 95 and 160°C in the first and second dimensions, respectively. As shown in Figure 4.23, peaks are located along a diagonal across the 2D plane, with retention times of homologous compounds increasing in relation to the C number (e.g., OH1 to OH4). The GC × GC peak distribution, illustrated in Figure 4.23, is very far from desirable. In the second application, the temperature of the second column was increased in a linear mode (starting from 130°C, and increasing at 3°C/min), while the primary column was maintained at 95°C. As can be observed in Figure 4.24, the higher second-dimension temperatures have a greater effect on the less-volatile components (e.g., O4, P4, OH4). In the third experiment, the first dimension was subjected to a temperature program (starting at 40°C and increasing at 1.5°C/min), while the second

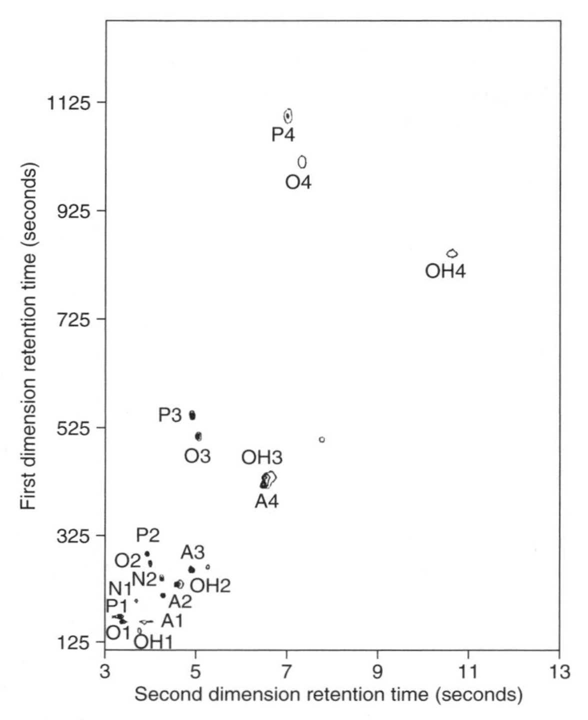

Figure 4.23 Isothermal GC × GC chromatogram of a test mixture; first dimension: 95°C, second dimension: 160°C. Peak identification: P1, P2, P3, P4: *n*-decane, *n*-dodecane, *n*-tridecane, *n*-tetradecane; OH1, OH2, OH3, OH4, OH5: 1-heptanol, 1-octanol, 1-nonanol, 1-decanol, 1-undecanol; O1, O2, O3, O4: 1-decene, 1-dodecene, 1-tridecene, 1-tetradecene; A1, A2, A3, A4: benzaldehyde, naphthalene, ethyl phenyl acetate, biphenyl; N1, N2: *tert*-butylbenzene, durene. [From Venkatramani et al. (1996), with permission. Copyright © 1996 by The American Chemical Society.]

was operated at a constant temperature of 160°C. As shown in Figure 4.25, such operational conditions generate a shortened-length diagonal (see Figure 4.23). When both columns are subjected to simultaneous temperature programming (primary capillary: starting at 40°C and increasing at 1.5°C/min; secondary capillary: starting at 120°C and increasing at 1.5°C/min), the sample components are separated on the primary apolar column according to their volatility. With regard to the secondary column, the temperature increase compensates for the progressive decrease in analyte volatility, leading to a nearly constant second-dimension t_R for members of each homologous series. The alkane and alkene peaks are aligned across the 2D plane and are characterized by low elution times on the moderately polar second dimension. The homologous series of alcohols, aligned along a distinct band, are characterized by high second-dimension elution times. In general, second-dimension retention is no longer influenced by the volatility of a molecule (Figure 4.26). The orthogonal nature of compound distribution is more visible in the chromatogram expansion illustrated in Figure 4.27.

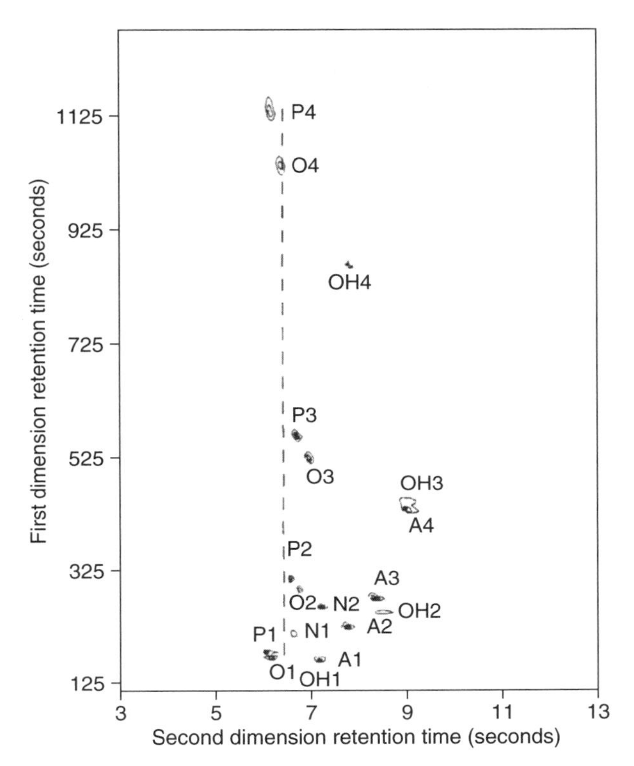

Figure 4.24 GC × GC chromatogram of a test mixture. First dimension, 95°C; second dimension, temperature program starting from 130°C and increasing at 3°C/min. Peak identification as in Figure 4.23. [From Venkatramani et al. (1996), with permission. Copyright © 1996 by The American Chemical Society.]

4.6.4 Thermal Modulation Parameters

The first GC × GC experiment was carried out by using a twin-stage thermal modulator, prepared on the head of the secondary nonpolar 1 m × 0.1 mm i.d. × 0.5 μm d_f column (Liu and Phillips, 1991). The modulator segment of column was looped outside the GC oven, at room temperature, and was coated with an electrically conductive film (gold paint). The modulator length was 15 cm, divided equally between the two stages. Current pulses of 20-ms duration were applied to each of the two stages with a 100-ms delay between stages, while the modulation period was of 2 s. Between frequent burnouts, the modulator produced some outstanding GC × GC results. Looking back at that revolutionary invention, the main modulator optimization issues would have been: modulation period, entrapment temperature, remobilization temperature, duration of the heating pulse, modulator gas velocity, and stationary-phase thickness. Although a variety of thermal modulators have been developed since 1991, the basic parameters that require fine tuning are generally quite similar.

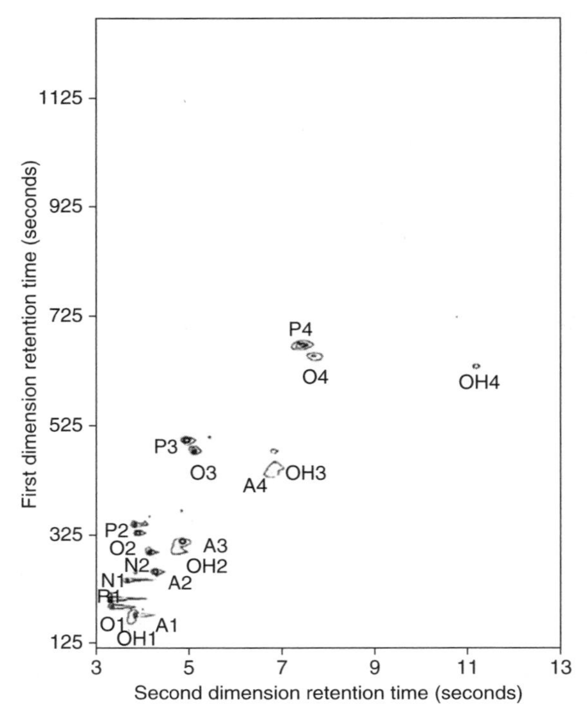

Figure 4.25 GC × GC chromatogram of a test mixture. First dimension, temperature program starting from 40°C and increasing at 1.5°C/min; second dimension, 160°C. Peak identification as in Figure 4.23. [From Venkatramani et al. (1996), with permission. Copyright © 1996 by The American Chemical Society.]

The modulation period is a general fundamental parameter and must be sufficiently low to maintain the resolution achieved on the primary column. Many GC × GC analysts would also add that the modulation period should be high enough to avoid wraparound. As aforementioned, wraparound represents a problem only in specific analytical situations (e.g., in the case of structured chromatograms). With regard to the preservation of primary column separation, in 1998 Murphy et al. described the effects of the modulation period on resolution in comprehensive 2D LC. The research results, which are valid for any type of comprehensive chromatography application, demonstrated that each first-dimension peak must be modulated at least three and four times in the case of in-phase and out-of-phase sampling, respectively. Considering GC × GC applications, it is very common to seek operational conditions that enable the generation of primary column peak widths between 15 and 20 s. As a consequence, modulation periods are usually in the range 4 to 6 s.

The entrapment temperature is a parameter that varies in relation to the type of thermal modulator used. For example, the first modulator entrapped analyte bands at ambient temperature on a rather thick layer of stationary phase.

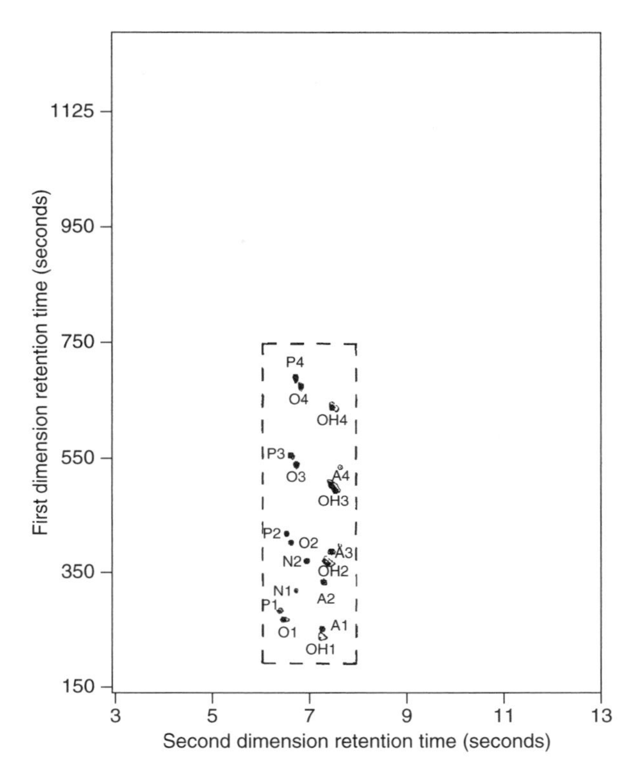

Figure 4.26 GC × GC chromatogram of a test mixture. First dimension, temperature program starting from 40°C and increasing at 1.5°C/min; second dimension, temperature program starting from 120°C and increasing at 1.5°C/min. Peak identification as in Figure 4.23. [From Venkatramani et al. (1996), with permission. Copyright © 1996 by The American Chemical Society.]

Hence, band isolation and compression were achieved by exploiting both low temperatures (in relation to the GC oven) and phase-ratio focusing; molecules as light as hexane were subjected to modulation. In 1999, a GC × GC thermal modulator, defined *thermal sweeper*, was introduced (Phillips et al., 1999). The system consisted of a slotted heater that was rotated mechanically over a modulator tube, generating chemical pulses in a four-step sequence: entrapment, isolation, compression, and reinjection. The modulator tube was characterized by a 10 cm × 0.1 mm i.d. × 3.5 μm d_f column, linked to a 0.1-mm i.d. segment of uncoated column. Entrapment occurred within the modulator tube at ambient GC oven temperature (phase-ratio focusing), while reinjection was achieved through thermal desorption by maintaining the sweeper temperature at least 100°C above the GC temperature.

The era of GC × GC cryogenic modulators began in 1998 (Kinghorn and Marriott), with the introduction of a longitudinally modulated cryogenic system. The LMCS device is a mobile trap, usually located at the head of the second

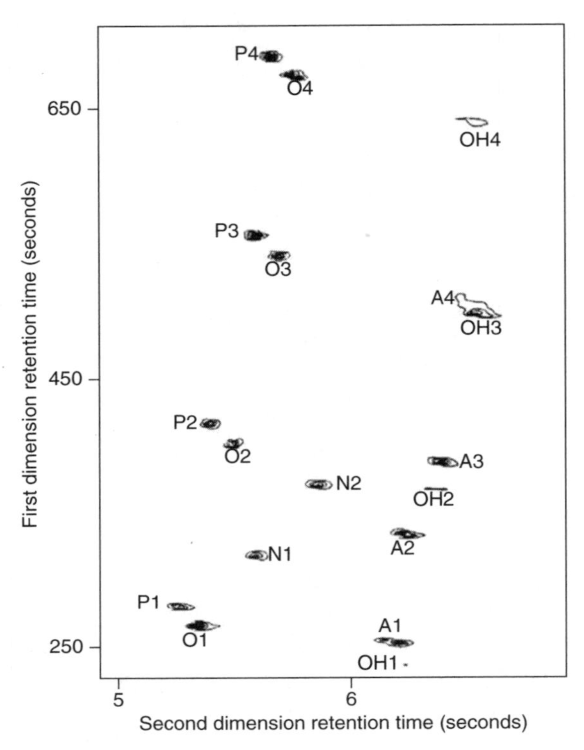

Figure 4.27 GC × GC expansion derived from the chromatogram illustrated in Figure 4.26. [From Venkatramani et al. (1996), with permission. Copyright © 1996 by The American Chemical Society.]

column. Analyte bands are entrapped and compressed through the freezing effect of an internal CO_2 stream. After entrapment, the transfer device is moved longitudinally along the column, exposing the previously cooled zone to the heat of the GC oven, thus ejecting a rapid pulse of entrapped components onto the secondary column. It was reported that peak widths on the order of 50 ms should be generated easily by the cryotrap (Kinghorn and Marriott, 1999). It is generally accepted that a LMCS temperature about -100°C lower than the GC elution temperature, for a specific analyte, is sufficient for effective entrapment. However, very volatile compounds (i.e., C_4 to C_6 alkanes) require even lower temperatures (-120 to -140°C). In general, attention must be paid to ensure that the most appropriate temperatures are used, because excessively low temperatures can retard analyte release.

With regard to the reinjection process, the LMCS system works fine with most volatiles; although problems can arise with the remobilization of high-molecular-weight compounds, these can be circumvented by using an uncoated column for trapping, by extending the release time, or even by using a thinner-walled column. The thermal sweeper and LMCS device have produced a great amount of excellent GC × GC research. However, both systems are characterized by a substantial disadvantage: the fact that both undergo continuous movement.

A loop-type modulator is a nonmoving transfer device that employs a hot and cold jet of gas (usually N_2) to effect dual-stage thermal modulation. The two stages are formed by looping a segment of capillary column (ca. 1 m) through the path of a single cold jet. The column between the two cold spots thus formed works as a delay loop (Ledford, technical note). Gaines and Frysinger (2004) carried out an interesting study on the thermal requirements of the loop modulator. It was found that the cold spot had to be 120 to 140°C lower than the analyte elution temperature to achieve efficient trapping. Additionally, it was observed that the N_2 flow required for efficient modulation decreased from 15.5 standard liters/min (SLPM) to 1.5 SLPM as the C number increased from C_4 to C_{40}. Excessively low temperatures can also be the cause of poor chromatography. Figure 4.28 shows the C_{24} to C_{36} part of two crude oil GC × GC chromatograms carried out under identical conditions apart from the cold-flow SLPM value, which was 17 in part (a) and programmed in (b). Under the high-SLPM conditions in (a), cooling temperatures are too low and analyte remobilization is not complete, thus generating long peak tails. Under programmed flow conditions, entrapment temperatures were adequate and band reinjection was satisfactory. However, an advantage of excessive cooling was the focalization of primary column bleed (indicated by the number 2 in Figure 4.28a), a phenomenon that did not occur under programmed conditions. In fact, the compression of column bleed isolates the chemical constituents of baseline noise.

The hot-jet pulse must be long enough to reduce analyte k values below a specific value, and short enough to avoid breakthrough. Furthermore, the hot spot should be at least 40°C higher than the elution temperatures. The authors reported

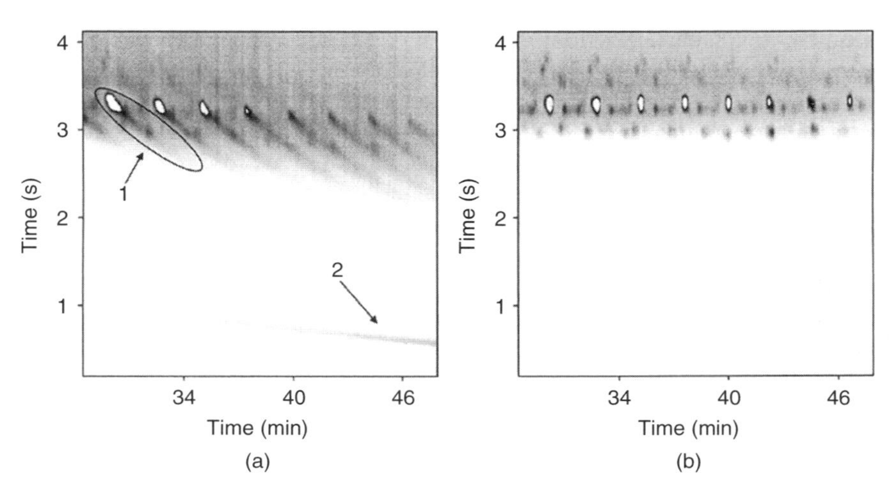

Figure 4.28 Effect of excessive cold jet flow on modulation. In (a) the flow was a constant 17.0 SLPM, while in (b), the flow was programmed. [From Gaines and Frysinger (2004), with permission. Copyright © 2004 by Wiley-VCH Verlag GmbH & Co. KGaA.]

optimum modulation using a hot-jet pulse of 300-ms duration and a heater block temperature 100°C above the column temperature.

A further issue, related to the loop modulater (not discussed by Gaines and Frysinger), is the length of the delay loop, which is usually about 1 m. If the delay loop is too short, breakthrough can occur during the hot-jet period; on the other hand, if the loop is too long, analyte bands do not reach the second cold spot before the end of cooling, causing inefficient trapping. An excessively long loop can even cause a lack of second-dimension analysis for every other injection onto the secondary column. Knowledge of the gas linear velocity inside the delay loop is important in order to define a correct length.

An interesting single-stage thermal modulator with air cooling and electric heating but with no cryogenic gas requirements has been described (Libardoni et al., 2005). The single-stage device consisted of an 8-cm segment of stainless-steel tubing located in an aluminum block containing a ceramic tube; the central part of the steel tube was air-cooled for sample entrapment, and its entire length was lined with stationary phase. The cold air, derived from a refrigeration unit, reached the interior of the modulator tube through holes in the ceramic tube, generating temperatures as low as −30°C. The modulator tube was heated by current pulses. The overall performance provided by the modulator was indeed promising: *n*-heptane was entrapped with no sign of breakthrough. It is the authors' opinion that such directions, related to the origins of GC × GC, appear to be worth taking.

4.6.5 Detection Parameters

Modulation and high second-dimension gas velocities generate very short chromatography bands, in both space and time. Gaines and Frysinger (2004) reported peak widths at half-height in the 40-ms range using loop-type modulation. Generally, GC × GC peak durations can vary in the range 100 to 500 ms, and are in relation to the modulator type, the second-dimension phase chemistry and film thickness, the temperature program, and the gas linear velocity. A GC × GC peak with a 150-ms width and at a 1-mL/min flow will occupy a volume of about 3 μL. Such rapid chromatography bands are similar to those produced in microbore column very fast GC experiments, and thus require detection systems characterized by micro internal volumes, low time constants, and high sampling frequencies (minimum 50 Hz). Such detector characteristics are necessary to avoid extra-column sources of band broadening and incorrect peak reconstruction (at least 10 data points per peak are needed). Detailed descriptions of the detectors used in comprehensive GC experiments have been reported in the literature (Adahchour et al., 2006a; von Mühlen et al., 2006).

In the first GC × GC experiment (Liu and Phillips, 1991), an FID was employed as a detection system and was operated at a 100-Hz sampling frequency. Since that initial GC × GC application, the FID has been the most commonly used detector in the field. The main disadvantages of using an FID, which is an otherwise perfect GC × GC detector, are the absolute lack of selectivity and structural information. Furthermore, the enhanced sensitivity of selective detectors compared to the FID is useful in trace analysis for specific molecules. While the combination of GC × GC and MS is described in depth in Chapter 6, a series of optimization aspects related to some selective detectors are discussed here.

A nitrogen–phosphorus detector (NPD) is characterized by a structure similar to that of the FID but with a heated ceramic bead containing an alkali metal salt, located above the jet. Negative ions are formed at the bead surface in a flameless, gaseous environment (H_2, air, makeup) and are directed onto a collector electrode that surrounds the bead. Peak tailing often occurs, due to temporary adsorption of decomposition products on the thermionic source; modification of the source temperature and/or gas flows can reduce such a disadvantage (Poole, 2003). The NPD is very convenient to use in the analysis of nitrogen- or phosphorus-containing molecules: in particular, pesticides and fungicides.

In 2007, von Mühlen et al. evaluated the geometry of the detector in a GC × GC–NPD study. The authors found that the combination of an extended jet with a standard collector electrode provided an improved response. The same detector geometry was employed in a GC × GC–NPD application directed to the analysis of fungicides (Khummueng et al., 2006). The best peak shapes and highest response were attained when NPD gas flows were 1.5, 7, and 100 mL/min for H_2, N_2, and air, respectively. Surprisingly, the conditions recommended by the manufacturer—3, 5, and 60 mL/min—produced a sensitivity reduced by more than 50%. GC × GC–NPD asymmetry factors measured for 10 fungicides showed a low degree of peak tailing (values were in the range 1.12 to 1.22). Finally, the NPD was operated at a 50-Hz sampling frequency, a sufficient value for peak widths at half-height, which ranged from 90 to 120 ms.

The great popularity of the electron capture detector (ECD) is due to its selectivity and sensitivity for high-electron-affinity components (Chen and Chen, 2004). Consequently, GC–ECD has been used widely in the analysis of volatile halogenated contaminants of various natures and sources. Conventional ECDs are characterized by rather high internal volumes, and, as such, are not really suitable for the detection of rapidly moving narrow-analyte bands. The ECD dead volume corresponds to the detector cell volume; for the FID, if the column exit protrudes into the base of the flame, the dead volume corresponds to the flame itself (Tijssen et al., 1987). Rather recently, however, a micro-ECD (μ-ECD) with an internal volume of 150 μL has become available commercially (Agilent Technologies).

In 2003, Kristenson et al. compared six different modulators (i.e., sweeper, LMCS, quad-jet, dual-jet CO_2, semirotating cryo, loop) and three commercially available ECDs with different internal volumes in the analysis of halogenated standard compounds. The authors found that ECD dead volumes played a fundamental, but not exclusive, role during GC × GC optimization. The ECDs were from Agilent (internal volume: 150 μL), Shimadzu (internal volume: 1.5 mL), and Thermo Finnigan (internal volume: 450 μL). GC × GC–ECD applications were achieved using the Agilent detector (μ-ECD) and a quad-jet modulator; the Shimadzu detector was used in combination with a dual-jet modulator.

In both GC × GC–ECD applications the first column consisted of a nonpolar 26.7 m × 0.32 mm i.d. × 0.25 μm d_f column, while the second was a polar 46 cm × 0.10 mm i.d. × 0.10 μm d_f column. The Thermo Finnigan ECD was employed under fast GC operational conditions. All the experimental ECD data were compared with FID-derived data. The authors tested the μ-ECD at different data acquisition frequencies (50, 100, and 200 Hz) and found no contribution toward band broadening using the lowest frequency. The makeup flow, on the contrary, had a considerable effect on band enlargement: when using the maximum makeup flow of 150 mL/min, widths at half-height were 185 to 200 ms. When an instrumental modification was made to increase the flow rate to 450 mL/min, peaks widths were reduced to 140 to 190 ms. Widths at half-height using the FID were approximately halved: namely, in the range 70 to 125 ms.

The Shimadzu ECD provided the best performance at a makeup flow of 200 mL/min: peak widths in the range 180 to 250 ms were attained, although substantial tailing occurred. The authors affirmed that the Shimadzu ECD was not suitable for GC × GC analysis, due to its high dead volume. The Thermo ECD performed poorly in fast GC applications on chloroalkanes; peak widths in the range 180 to 260 ms attained using a makeup flow of more than 2 L/min were much higher than those observed in the FID experiments (110 to 150 ms).

A sulfur chemiluminescence detector (SCD) is highly selective and sensitive toward sulfur-containing compounds, providing a linear and equimolar response. GC-SCD is probably the prime choice for the separation and detection of trace-amount sulfur constituents of complex samples (Yan, 2002). In 2004, Blomberg et al. used both an SCD and an FID, under fast GC conditions, in the analysis of dibenzothiophene. The authors found that the SCD-induced band broadening was caused by system electronics rather than by detector dead volumes. In fact, band broadening was reduced to that of the FID by using a modified electrometer intended for use with a flame-photometric detector. Such findings were in contrast with previous research (Hua et al., 2003). The fast SCD was employed in a series of GC × GC experiments directed to the analysis of sulfur-containing compounds in fuel products. Two single second-dimension traces, relative to a GC × GC–FID and GC × GC–SCD kerosene analysis, are illustrated in Figure 4.29. As can be seen, the modified SCD is highly selective and did not increase band broadening. It must also be noted that band widths are rather wide (ca. 1 s) for a GC × GC experiment, and the performance of the SCD on narrower peaks was not reported. A nicely structured GC × GC–SCD chromatogram of a crude oil is illustrated in Figure 4.30.

An atomic emission detector (AED) is a multielement system that combines well with gas chromatography. GC–AED is capable of simultaneous multichannel detection (up to four elements), wide-range linearity, excellent limits of detection and an element/C response ratio of about four to five orders of magnitude. Considering the simultaneous multichannel capabilities, AED optics are designed to achieve high-resolution emission differentiation; hence, during a single run, detection is possible in a 20- to 25-nm window, within a total range

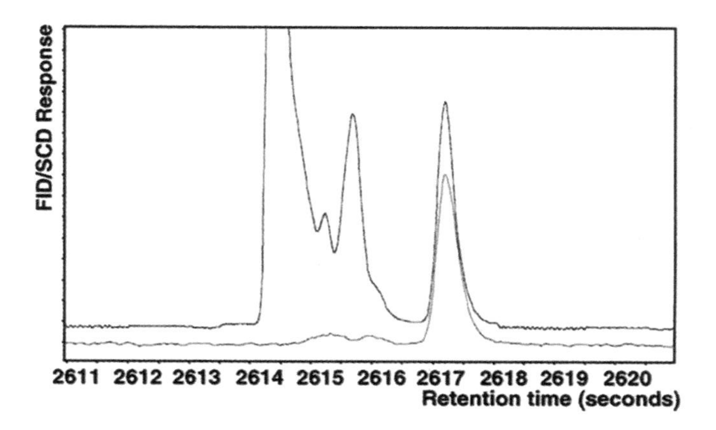

Figure 4.29 Second-dimension chromatograms relative to GC × GC–FID (upper trace) and GC × GC-SCD kerosene applications. [From Blomberg et al. (2004), with permission. Copyright © 2004 by Elsevier.]

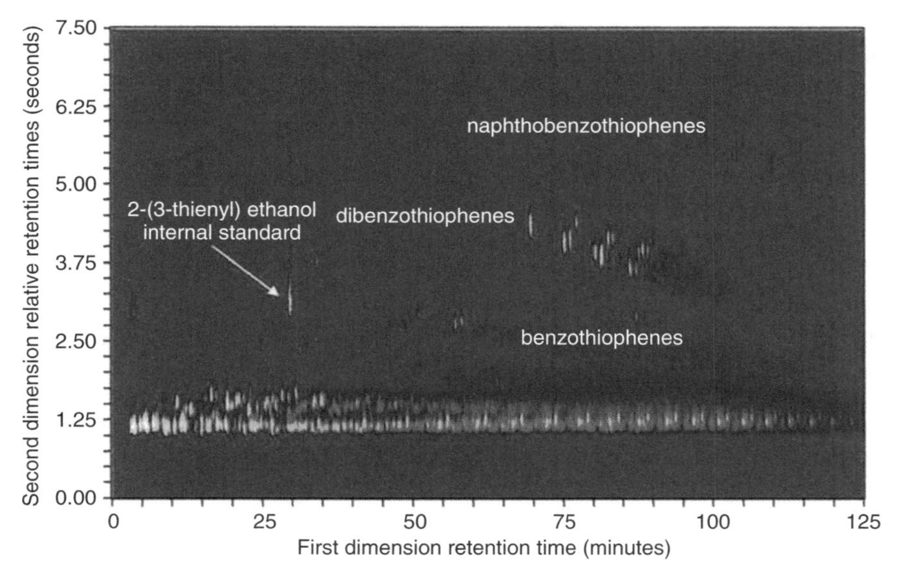

Figure 4.30 GC × GC–SCD chromatogram of a crude oil. [From Blomberg et al. (2004), with permission. Copyright © 2004 by Elsevier.]

of 160 to 800 nm. It is obvious that a contemporary analysis of more than one element is possible only if the emission lines fall within the same window (van Stee et al., 2002).

Although the AED is characterized by rather limited sampling frequencies, it has been used as a detection system in a GC × GC experiment (van Stee et al.,

2003). No structural modification was made to the AED. However, to meet an acquisition frequency of 10 Hz, the authors connected a 0.7 m × 0.25 mm i.d. uncoated column to the end of a 0.6 m × 0.10 mm i.d. × 0.10 μm polar secondary capillary to increase band enlargement. Such a combination of columns, as well as the detector, produced rather wide tailing chromatography bands, reconstructed with a sufficient number of data points for identification purposes. In this respect, three pesticide GC × GC–AED applications were performed to generate traces for (1) Br (478 nm), Cl (479 nm), H (486 nm), C (496 nm); (2) N (174 nm), S (181 nm), C (193 nm); and (3) P (178 nm). Notwithstanding the negative chromatography issues, the multielement selectivity of the GC × GC–AED experiments was a highly interesting aspect.

4.7 FINAL REMARKS

In the past decade, GC × GC became generally accepted as a powerful and widely applicable technique for the analysis of different types of complex samples. The number of papers devoted to GC × GC has been growing steadily over the years. In the first year after the birth of the technique, only four papers on GC × GC were published. By the end of 2002, the total number of publications reached nearly 100 (Dallüge et al., 2003). Three years later, an additional 150 papers appeared (Adahchour et al., 2006d). In the period from 2006 until the end of 2009, close to 250 new papers were published, indicating nearly exponential growth of the technique.

The focus of the papers shifted gradually over time from fundamental developments to applications. Of the close to 90 papers published in 2008 alone, nearly 70% were devoted to applications of GC × GC. Numerous examples of GC × GC applications are presented throughout this book, hence only a very brief sampling of application areas is noted below. Readers seeking a more thorough overview of GC × GC applications are encouraged to read earlier literature (e.g., Adahchour et al., 2006b, 2006c, 2008; Dallüge et al., 2003).

The first and still one of the most important application areas of GC × GC is the analysis of petroleum products. Typical petrochemical products, including gasoline, diesel, kerosene, and oils, contain thousands of components, making them some of the most complex samples known to analytical chemists. In 1993, kerosene was analyzed using GC × GC to demonstrate the separation power of the technique (Venkatramani and Phillips, 1993). Samples of petroleum products are still used for this purpose. As the technique evolved, it found a wide range of applications in areas including environmental analysis (e.g., Kallio et al., 2003b; Korytár et al., 2002), food (e.g., Rocha et al., 2007; Schurek et al., 2008), flavor (e.g., Zhu et al., 2007), fragrance (e.g., Rochat et al., 2007; von Mühlen et al., 2008), pharmaceuticals (e.g., Crimi and Snow 2008; Mitrevski et al., 2007), and metabolomics (e.g., Almstetter et al., 2009; Li et al., 2009). As the second decade of the development of the technique draws gradually to a close, one can expect that new and exciting applications of GC × GC will fuel the development of the technique for many years to come. The future of GC × GC looks bright.

REFERENCES

Adahchour M, Beens J, Brinkman UAT. *Analyst* 2003; 128(3):213–216.

Adahchour M, Beens J, Vreuls RJJ, Batenburg AM, Brinkman UAT. *J. Chromatogr. A* 2004; 1054:47–55.

Adahchour M, Beens J, Vreuls RJJ, Brinkman UAT. *Trends Anal. Chem*. 2006a; 25(6):540–553.

Adahchour M, Beens J, Vreuls RJJ, Brinkman UAT. *Trends Anal. Chem*. 2006b; 25(7):726–741.

Adahchour M, Beens J, Vreuls RJJ, Brinkman UAT. *Trends Anal. Chem*. 2006c; 25(8):821–840.

Adahchour M, Beens J, Vreuls RJJ, Brinkman UAT. *Trends Anal. Chem*. 2006d; 25(5):438–454.

Adahchour M, Beens J, Brinkman UAT. *J. Chromatogr. A* 2008; 1186(1–2):67–108.

Almstetter MF, Appel IJ, Gruber MA, Lottaz C, Timischl B, Spang R, Dettmer K, Oefner PJ. *Anal. Chem*. 2009; 81(14):5731–5739.

Amador-Muñoz O, Marriott PJ. *J. Chromatogr. A* 2008; 1184(1–2):323–340.

Beens J, Boelens H, Tijssen R, Blomberg J. *J. High Resolut. Chromatogr*. 1998a; 21(1):47–54.

Beens J, Tijssen R, Blomberg J. *J. High Resolut. Chromatogr*. 1998b; 21(1):63–64.

Beens J, Adahchour M, Vreuls RJJ, van Altena K, Brinkman UAT. *J. Chromatogr. A* 2001; 919(1):127–132.

Beens J, Janssen H-G, Adahchour M, Brinkman UAT. *J. Chromatogr. A* 2005; 1086:141–150.

Bertsch W. *J. High Resolut. Chromatogr*. 2000; 23(3):167–181.

Bicchi C, D'Amato A, Rubiolo P. *J. Chromatogr. A* 1999; 843:99–121.

Blomberg J, Riemersma T, van Zuijlen M, Chaabani H. *J. Chromatogr. A* 2004; 1050:77–84.

Blumberg LM. *J. Chromatogr. A* 2003; 985(1–2):29–38.

Bordajandi LR, Ramos L, Gonzalez MJ. *J. Chromatogr. A* 2005; 1078(1–2):128–135.

Bruckner CA, Prazen BJ, Synovec RE. *Anal. Chem*. 1998; 70(14):2796–2804.

Bueno PA Jr, and Seeley JV. *J. Chromatogr. A* 2004; 1027(1–2):3–10.

Burger BV, Snyman T, Burger WJG, van Rooyen WF. *J. Sep. Sci*. 2003; 26(1–2): 123–128.

Cortes HJ. *J. Chromatogr*. 1992; 626(1):3–23.

Chen ECM, Chen ES. *J. Chromatogr. A* 2004; 1037:83–106.

Crimi CM, Snow NH. *LC-GC N. Am*. 2008; 26(1):62, 64, 66, 68, 70.

Dallüge J, Beens J, Brinkman UAT. *J. Chromatogr. A* 2003; 1000(1–2):69–108.

Frysinger GS, Gaines RB. *J. High Resolut. Chromatogr*. 1999; 22(5):251–255.

Frysinger GS, Gaines RB, Ledford EB. *J. High Resolut. Chromatogr*. 1999; 22(4):195–200.

Gaines RB, Frysinger GS. *J. Sep. Sci*. 2004; 27:380–388.

Giddings JC. *Anal. Chem*. 1984; 56(12): 1258A–1270A.

Goldstein AH, Worton DR, Williams BJ, Hering SV, Kreisberg NM, Panić O, Górecki T. *J. Chromatogr. A* 2008; 1186(1–2):340–347.

Górecki T, Harynuk J, Panić O. *J. Sep. Sci*. 2004; 27(5–6):359–379.

Górecki T, Panić O, Oldridge NJ. *Liq. Chromatogr. Relat. Technol*. 2006; 29(7–8):1077–1104.

Górecki T, Goldstein AH, Kreisberg NM, Hering SV, Teng AP Worton DR, McNeish C, Eds. International Network of Environmental Forensics Conference, Calgary, Alberta, Canada, Aug. 31–Sept. 2, 2009, Abstracts, p. 30.

Hamilton JF, Lewis AC, Bartle KD. *J. Sep. Sci*. 2003; 26:578–584.

Harynuk J, Górecki T. Pittcon, New Orleans, LA, Mar. 4–9, 2001. Abstract 108.

Harynuk J and Górecki T. *J. Sep. Sci*. 2002; 25(5–6):304–310.

Harynuk J and Górecki T. *J. Chromatogr. A* 2003; 1019(1–2):53–63.

Harynuk J and Górecki T. *J. Sep. Sci*. 2004; 27(5–6):431–441.

Harynuk J and Górecki T. *J. Chromatogr. A* 2006; 1105(1–2):159–167.

Harynuk J, Górecki T, Campbell C. *LC-GC N. Am*. 2002; 20(9): 876, 878, 880, 882, 884, 886–890, 892.

Hollingsworth BV, Reichenbach SE, Tao Q, Visvanathan A. *J. Chromatogr. A* 2006; 1105(1–2):51–58.

Hua R, Li Y, Liu W, Zheng J, Wei H, Wang J, Lu X, Kong H, Xu G. *J. Chromatogr. A* 2003; 1019:101–109.

Hua R, Wang J, Kong H, Liu J, Lu X, Xu G. *J. Sep. Sci*. 2004; 27(9):691–698.

Hyötyläinen T, Kallio M, Hartonen K, Jussila M, Palonen S, Riekkola M. *Anal. Chem*. 2002; 74(17):4441–4446.

Kallio M, Hyötyläinen T, Jussila M, Hartonen K, Palonen S, Shimmo M, Riekkola M. *Anal. Bioanal. Chem*. 2003a; 375(6):725–731.

Kallio M, Hyötyläinen T, Lehtonen M, Jussila M, Hartonen K, Shimmo M, Riekkola M. *J. Chromatogr. A* 2003b; 1019(1–2):251–260.

Kallio M, Jussila M, Raimi P, Hyötyläinen T. *Anal. Bioanal. Chem*. 2008; 391(6):2357–2363.

Khummueng W, Trenerry C, Rose G, Marriott PJ. *J. Chromatogr. A* 2006; 1131:203–214.

Kinghorn RM, Marriott PJ. *J. High Resolut. Chromatogr*. 1998; 21(11):620–622.

Kinghorn RM, Marriott PJ. *J. High Resolut. Chromatogr*. 1999; 22(4):235–238.

Korytár P, Leonards PEG, de Boer J, Brinkman UAT. *J. Chromatogr. A* 2002; 958(1–2):203–218.

Korytár P, Parera J, Leonards PEG, de Boer J, Brinkman UAT. *J. Chromatogr. A* 2005a; 1067(1–2):255–264.

Korytár P, Parera J, Leonards PEG, Santos FJ, de Boer J, Brinkman UAT. *J. Chromatogr. A* 2005b; 1086(1–2):71–82.

Kristenson EM, Korytár P, Danielsson C, Kallio M, Brandt M, Mäkelä J, Vreuls RJJ, Beens J, Brinkman UATh. *J. Chromatogr. A* 2003; 1019:65–77.

Ledford EB. Presented at the 23rd Symposium on Capillary Gas Chromatography, Riva del Garda, Italy, June 2000.

Ledford EB, Billesbach C, Termaat J. Pittcon, Mar. 17–22, 2002, New Orleans, LA. Contribution 2262P.

Ledford EB Jr, Termaat JR, Billesbach CA. Technical Note KT030606-1. http://www.zoex.com.

Li X, Xu Z, Lu X, Yang X, Yin P, Kong H, Yu Y, Xu G. *Anal. Chim. Acta* 2009; 633(2):257–262.

Libardoni M, Waite JH, Sacks R. *Anal. Chem*. 2005; 77(9):2786–2794.

Libardoni M, Hasselbrink E, Waite JH, Sacks R. *J. Sep. Sci*. 2006; 29(7):1001–1008.

Liu Z, Phillips JB. *J. Chromatogr. Sci*. 1991; 29(6):227–231.

Liu Z, Phillips JB. *J. Microcol. Sep*. 1994; 6(3):229–235.

Marriott P, Shellie R. *Trends Anal. Chem*. 2002; 21(9–10):573–583.

Marriott PJ, Haglund P, Ong RCY. *Clin. Chim. Acta* 2003; 328(1–2):1–19.

Mitrevski B, Brenna JT, Zhang Y, Marriott PJ. *Chem. Aust*. 2007; 74(10):3–5.

Mohler RE, Prazen BJ, Synovec RE. *Anal. Chim. Acta* 2006; 555(1):68–74.

Murphy RE, Schure MR, Foley JP. *Anal. Chem*. 1998; 70(8):1585–1594.

Oldridge N, Panić O, Górecki T. *J. Sep. Sci*. 2008; 31(19):3375–3384.

Ong RCY, Marriott PJ. *J. Chromatogr. Sci*. 2002; 40(5):276–291.

Penet S, Vendeuvre C, Bertoncini F, Marchal R, Monot F. *Biodegradation* 2006; 17(6):577–585.

Phillips JB, Ledford EB. *Field Anal. Chem. Technol*. 1996; 1(1):23–29.

Phillips JB, Venkatramani CJ. *J. Microcol. Sep*. 1993; 5:511–516.

Phillips JB, Xu J. *J. Chromatogr. A* 1995; 703(1–2):327–334.

Phillips JB, Gaines RB, Blomberg J, Van Der Wielen FWM, Dimandja J, Green V, Granger J, Patterson D, Racovalis L, De Geus H, et al. *J. High Resolut. Chromatogr*. 1999; 22(1):3–10.

Poole C. *The Essence of Chromatography*. Amsterdam: Elsevier, 2003, pp. 225–257.

Pursch M, Eckerle P, Biel J, Streck R, Cortes H, Sun K, Winniford B. *J. Chromatogr. A* 2003; 1019(1–2):43–51.

Reichenbach SE, Ni M, Zhang D, Ledford EB. *J. Chromatogr. A* 2003; 985(1–2):47–56.

Reichenbach SE, Ni M, Kottapalli V, Visvanathan A. *Chemometr. Intell. Lab. Syst*. 2004; 71(2):107–120.

Reichenbach SE, Kottapalli V, Ni M, Visvanathan A. *J. Chromatogr. A* 2005; 1071(1–2):263–269.

Rocha SM, Coelho E, Zrostlíková J, Delgadillo I, Coimbra MA. *J. Chromatogr. A* 2007; 1161(1–2):292–299.

Rochat S, De Saint Laumer J, Chaintreau A. *J. Chromatogr. A* 2007; 1147(1):85–94.

Schurek J, Portoles T, Hajslová J, Riddellova K, Hernandez F. *Anal. Chim. Acta* 2008; 611(2):163–172.

Seeley JV, Kramp F, Hicks CJ. *Anal. Chem*. 2000; 72(18):4346–4352.

Seeley JV, Kramp FJ, Sharpe KS. *J. Sep. Sci*. 2001; 24(6):444–450.

Seeley JV, Micyus NJ, McCurry JD, Seeley SK. *Am. Lab*. 2006; 38(9):24–26.

Seeley JV, Libby EM, Hill Edwards KA, Seeley SK. *J. Chromatogr. A* 2009; 1216:1650–1657.

Shellie R, Marriott P, Cornwell C. *J. Sep. Sci*. 2001; 24:823–830.

Shellie R, Marriott P, Morrison P, Mondello L. *J. Sep. Sci*. 2004; 27:503–512.

Simmons MC, Snyder LR. *Anal. Chem*. 1958; 30:32–35.

Tijssen R, van den Hoed N, van Kreveld ME. *Anal. Chem*. 1987; 59:1007–1015.

Tranchida PQ, Casilli A, Dugo P, Dugo G, Mondello L. *Anal. Chem*. 2007; 79(6):2266–2275.

Tranchida PQ, Purcaro G, Conte L, Dugo P, Dugo G, Mondello L. *Anal. Chem*. 2009; 81:8529–8537.

van Stee LLP, Brinkman UAT, Bagheri H. *Trends Anal. Chem*. 2002; 21:618–226.

van Stee LLP, Beens J, Vreuls RJJ, Brinkman UAT. *J. Chromatogr. A* 2003; 1019(1–2):89–99.

Venkatramani CJ, Phillips JB. *J. Microcol. Sep*. 1993; 5(6):511–516.

Venkatramani CJ, Xu J, Phillips JB. *Anal. Chem*. 1996; 68(9):1486–1492.

Veriotti T, Sacks R. *Anal. Chem*. 2001; 73:3045–3050.

von Mühlen C, Zhummueng W, Alcarez Zini C, Bastos Caramão E, Marriott PJ. *J. Sep. Sci*. 2006; 29:1909–1921.

von Mühlen C, E. de Oliveira C, Morrison PD, A. Zini C, B. Caramão E, Marriott PJ. *J. Sep. Sci*. 2007; 30:3223–3232.

von Mühlen C, Zini CA, Caramão EB, Marriott PJ. *J. Chromatogr. A* 2008; 1200(1):34–42.

Wang FC-Y. Patents 2007106505, 20070920; PCT Int. Appl., 2007; U.S. Patent 2007/0214866 A1.

Wang FC-Y. *J. Chromatogr. A* 2008; 1188(2):274–280.

Wang FC-Y, Robbins WK, Greaney MA. *J. Sep. Sci*. 2004; 27(5–6):468–472.

Winniford BL, Sun K, Griffith JF, Luong JC. *J. Sep. Sci*. 2006; 29(17):2664–2670.

Yan X. *J. Chromatogr. A* 2002; 979:3–10.

Zhu S, Lu X, Ji K, Guo K, Li Y, Wu C, Xu G. *Anal. Chim. Acta* 2007; 597(2):340–348.

5

FLOW-MODULATED COMPREHENSIVE TWO-DIMENSIONAL GAS CHROMATOGRAPHY

JOHN V. SEELEY
Oakland University, Rochester, Michigan

This chapter covers the use of flow modulation with comprehensive two-dimensional gas chromatography (GC × GC). Flow modulation is a simple approach that allows GC × GC separations to be produced with a valve, a few fittings, and an auxiliary flow carrier gas. This places flow modulation in contrast to thermal modulation, which often requires large amounts of liquid cryogen. It is important to note that the simplicity and ease of implementation of flow modulation comes at the cost of lower resolving power along the secondary dimension and/or a smaller fraction of component transfer from the primary column to the secondary column. The chapter begins with an examination of the key requirements of modulation techniques. Then the flow modulation approaches that have been developed over the past 12 years are examined. Finally, the characteristics of flow modulation are compared to thermal modulation.

5.1 TIMING REQUIREMENTS OF GC × GC MODULATORS

A GC × GC separation can be viewed as an extension of a conventional single-column gas chromatographic (GC) analysis. The key difference is that instead

Comprehensive Chromatography in Combination with Mass Spectrometry, First Edition.
Edited by Luigi Mondello.

of sending the components to a detector as they emerge from the first column (i.e., the primary column), the components are put through a modulator and then through a second capillary column (i.e., the secondary column). The stationary phase in the secondary column is different from the primary stationary phase, and thus compounds that coelute on the primary column can potentially be separated by the secondary column. A schematic of a GC × GC instrument is shown in Figure 5.1. The modulator is the only additional piece of hardware required to generate a two-dimensional chromatogram. The modulator samples the effluent exiting the primary column and transfers it to the head of the secondary column as a pulse. This transfer process is repeated throughout the analysis at a constant interval known as the *modulation period*. The fraction of primary effluent sampled by the modulator is the duty cycle.

A hypothetical example of a primary column peak being modulated with a period of 1.0 s and a duty cycle of 0.2 is depicted in Figure 5.2. As the peak exits the primary column (Figure 5.2a), it is sampled for the first 20% of each modulation period (Figure 5.2b) as a result of the 0.2 duty cycle. Figure 5.2c shows the effluent as it exits the modulator. In this hypothetical case, the modulator increases the flux of the outgoing material by a factor of 2; thus, the pulses become twice as tall and half as wide (the mechanism for increasing component flux is discussed in Section 5.5). Ultimately, the modulator produces a series of 100-ms-wide pulses (i.e., 20% of 1.0 s divided by 2) separated by 900 ms of baseline "space." These pulses are passed through the secondary column, where they are separated on the secondary stationary phase. The pulses emerge from the secondary column shortly after injection (see Figure 5.2d). Only the concentration profile at the exit of the secondary column is actually monitored in a GC × GC separation. A two-dimensional chromatogram is generated by dividing the signal array into a series of segments that are as wide as the modulation

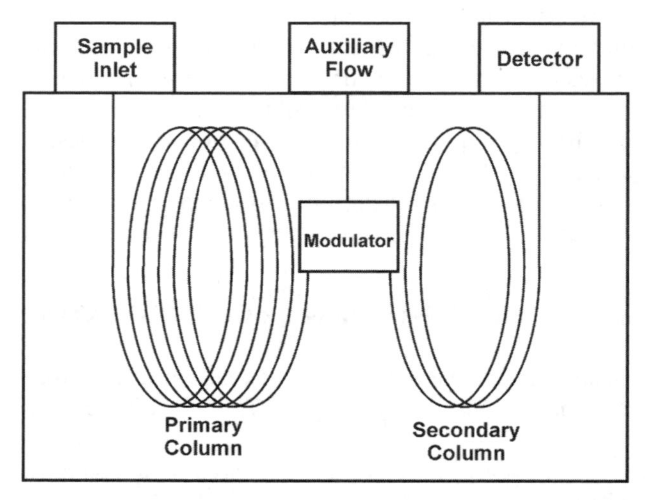

Figure 5.1 Flow modulation GC × GC system.

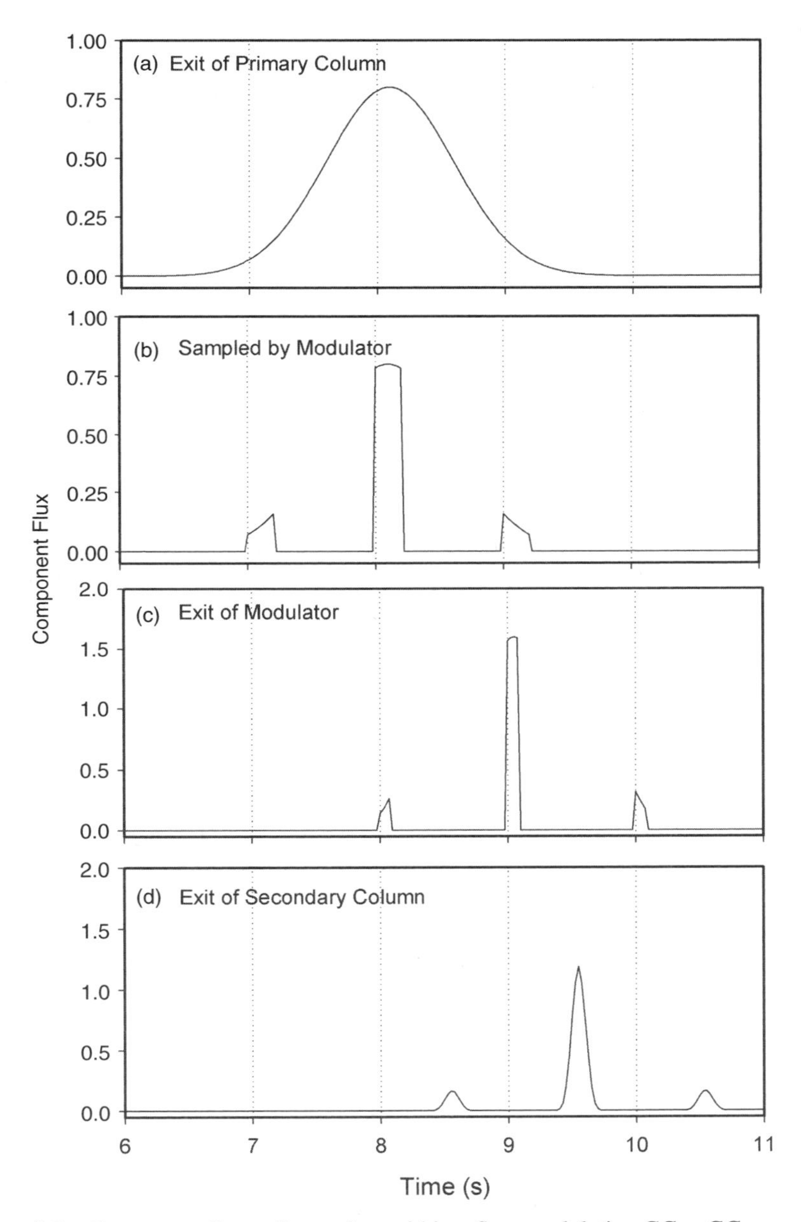

Figure 5.2 Component flux at four points within a flow modulation GC × GC apparatus. This theoretical example is of a compound that elutes with a primary retention time of 8.1 s (a), and is sampled with a 1.0-s modulation period and a duty cycle of 0.2 (b). The velocity of the sample material is increased by a factor of 2 within the modulator (c). The pulses broaden slightly within the secondary column and are detected at the exit of the secondary column (d).

period. These segments are stacked side by side to generate a two-dimensional chromatogram.

Effective modulation often requires that pulses be generated with modulation periods of a few seconds. One of the challenges of developing a new modulator is to find hardware that operates on a short time scale and can do so for a sustained period of time (e.g., over the course of several months) without a loss in performance. The source of this high-speed requirement is examined theoretically below. A more detailed account of these calculations can be found elsewhere (Seeley, 2002).

A central goal of most GC × GC analyses is to generate the secondary separations without significantly diminishing the primary separation. Experimental and theoretical studies have shown that the act of modulation can lead to a large loss of primary resolution if the modulation period is substantially longer than the widths of the peaks exiting the primary column. This effect was first studied by Murphy et al. (1998) for modulators with a duty cycle of 1.0 and then extended to all duty cycles by Seeley (2002). The amount of broadening depends on the modulation phase of the primary peak (i.e., the position of the peak relative to the timing of the modulation events). However, for most practical purposes, only the phase-averaged broadening is pertinent. The relationship between the average broadening along the primary dimension and the modulation period is shown in Figure 5.3a. In this figure the modulation period is represented in the form of the modulation ratio (Khummueng et al., 2006). The modulation ratio, M_R, is given by

$$M_R = \frac{w_b}{P_M}$$

where w_b is the baseline width of the peak (assumed to be four times the standard deviation of the peak eluting from the primary column, $w_b = 4\sigma$) and P_M is the modulation period. The modulation ratio is essentially the average number of pulses generated per primary peak.

Figure 5.3a clearly shows that peaks are broadened along the primary dimension as the modulation ratio is decreased and that the amount of broadening is dependent on the modulation duty cycle, with the degree of broadening decreasing as the duty cycle is decreased. However, if the modulation ratio is kept greater than 3.0, the peak widths increase along the primary dimension by less than 14%, regardless of the duty cycle (i.e., the broadening factor is less than 1.14).

It is clearly desirable for GC × GC analyses to retain the quantitative precision of conventional single-column GC. Inconsistency in the fraction of primary column effluent transferred to the secondary column is not a concern for modulators that have a duty cycle of 1.0 (e.g., thermal modulators and some flow modulators), as all of the primary column effluent is sampled and transferred to the secondary column; however, when duty cycles of less than 1.0 are employed, the proportion of the primary peak that is transferred to the secondary column is dependent on the relative position of the primary peak within the sequence of modulation events (i.e., the phase of modulation). This affects quantitative

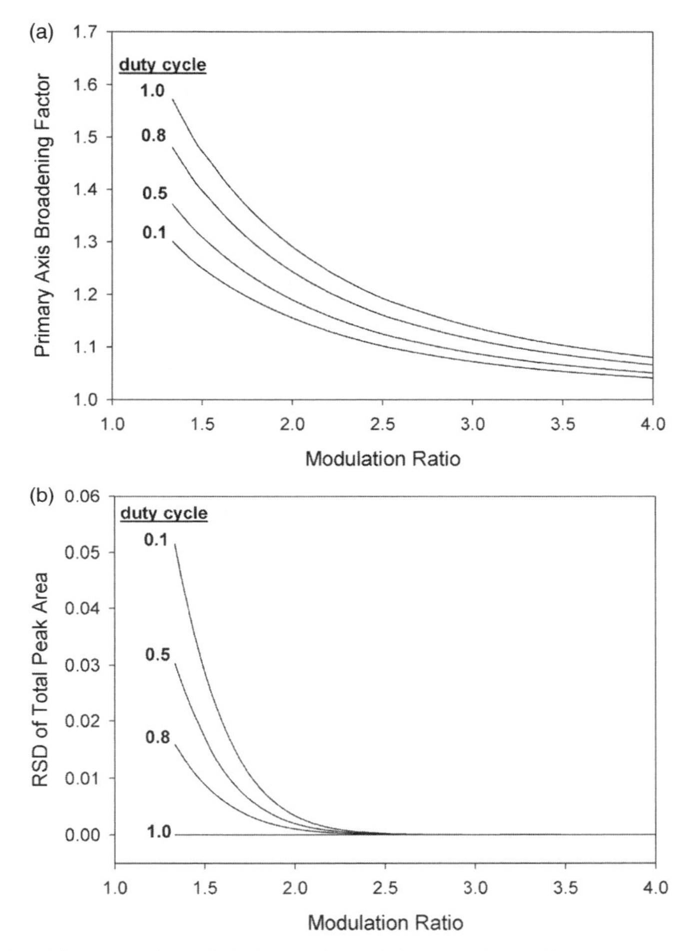

Figure 5.3 Influence of modulation ratio and duty cycle on instrument performance. Graph (a) shows how increasing the modulation ratio (i.e., increasing the number of modulated pulses per primary peak) decreases the broadening along the primary axis. Graph (b) shows how increasing the modulation ratio decreases the relative standard deviation of the total material transferred from the primary column to the secondary column.

precision, because small shifts in the primary retention time lead to changes in the modulation phase and hence generate a difference in the fraction of material transferred to the secondary column. The size of the phase-dependent variability is a function of duty cycle and modulation ratio. Figure 5.3b shows the relative standard deviation (RSD) of the amount of material transferred to the secondary column as a function of modulation number. The RSDs were calculated by considering all possible modulation phases. The RSDs calculated decrease as the duty cycle is increased. The RSDs also decrease as the modulation ratio is increased.

If the modulation ratio is greater than 3, the RSDs are less than 0.0001% for all duty cycles.

Throughout the history of GC × GC, practitioners have been warned to keep the modulation period small enough or to make the primary peaks wide enough to get "3 to 4 slices per component." This rule of thumb is the equivalent of stating that the modulation ratio should be keep above 3. The results in Figure 5.3 demonstrate that if this rule is obeyed, the act of modulation should lead to only a minor amount of broadening (<14%) along the primary dimension. Furthermore, under this rule, the quantitative precision of low-duty-cycle modulation should be similar to that produced by modulation with a duty cycle of 1.0. This conclusion assumes that the component of interest is present at levels well above the detection limit. As component concentrations are reduced to trace levels, modulators that employ a duty cycle near 1 have a distinct advantage, due to a larger fraction of the analyte being sent to the detector. The results shown in Figure 5.3 are especially important to flow modulation GC × GC, because unlike thermal modulators, many flow modulators employ duty cycles that are less than 1. It is the opinion of the author that the initial development of flow modulation was hindered by the frequently expressed belief that modulators with duty cycles below 1 were not "comprehensive" and not "quantitative." The fear was that the "missed" effluent would result in misleading two-dimensional chromatograms. The analysis depicted in Figure 5.3 demonstrates that if the standard rule of "3 to 4 slices per component" is followed, low-duty-cycle modulation should be as precise as thermal modulation.

5.2 CRITERIA FOR EVALUATING MODULATORS

Eleven flow modulator designs are discussed in this chapter. Thus, it is helpful to establish some of the important factors that distinguish an effective modulator.

Pulse Width Secondary separations are conducted under essentially isothermal conditions; thus, the widths of the peaks along the secondary dimension cannot be narrower than the widths of the pulses leaving the modulator. Ideally, the modulation pulses should be as narrow as possible and exhibit no tailing, to maximize peak capacity.

Duty Cycle High duty cycles are advantageous because they maximize the amount of material injected into the secondary column, and thus increase the sensitivity of the method. As shown in Section 5.1, high duty cycles also provide greater assurance that the modulation process will not decrease quantitative precision when operating at low modulation ratios (i.e., undersampled conditions). However, increasing duty cycles results in wider pulses for many flow modulators.

Flow Compatibility It is advantageous for a modulator to operate under flow rates that maximize the theoretical plates produced in both the primary and secondary columns. Ideally, the flow should be on the order of 1 mL/min so that a wide range of column diameters and lengths can be employed. Furthermore, operation of the modulator should not cause flow disturbances that substantially degrade the performance of the primary or secondary separations.

Modulation Speed The results described in Section 5.1 clearly show the benefits of working at modulation ratios greater than 3. There are two ways to achieve large modulation ratios: (1) generate wide primary peaks or (2) work with small modulation periods. Clearly, broadening the primary peaks is not as desirable as employing a modulator that can operate at high speeds. It is useful to perform a simple analysis to predict ideal modulation periods. Temperature-ramped single-column GC analysis with a standard capillary column (e.g., 30 m $\times$ 0.25 mm) and standard temperature ramping rates (e.g., 10°C/min) routinely produce peaks with baseline widths on the order of 6 s. If it is assumed that the primary column peaks will have similar widths, modulation periods near 2 s would be necessary to produce a modulation ratio of 3. Thus, it is very advantageous to have a modulator that can work easily on a time scale of 1 to 2 s.

Temperature Range It is best if the modulator does not impose additional temperature constraints in a chromatographic analysis. The biggest temperature constraint of most single-column analysis is the maximum temperature of the stationary phase. Polar stationary phases, such as poly(ethylene glycol) or substituted polysiloxanes, often have upper temperature limits ranging from 260 to 320°C. Ideally, a modulator can be used at temperatures above 320°C.

Robustness Numerous modulation cycles are generated during the course of a GC $\times$ GC analysis. For example, a 30-min chromatographic analysis with a 2.0-s modulation period involves nearly 1000 modulation events. This means that frequent use of a GC $\times$ GC instrument could easily result in 1 million modulation events in a year. It is advantageous for a modulator to be able to withstand numerous modulation events before requiring service.

Ease and Cost of Implementation Gas chromatography is often considered to be a straightforward technique to implement that does not require large amounts of costly consumables. It is advantageous that the addition of a modulator not drastically increase the ease or cost of analysis.

5.3 FORMS OF MODULATION

As described in Section 5.1, the main function of a modulator is to transform the peaks emerging from the primary column into a series of narrow pulses. There are three principal ways to perform this transformation: thermal modulation, single-stage flow modulation, and two-stage flow modulation. Thermal modulation is reviewed only briefly, as it is discussed extensively in Chapter 4.

Thermal modulation GC $\times$ GC was introduced by John Phillips in 1991 (Liu and Phillips, 1991) and it is the modulation technique used most extensively.

Thermal modulation operates by concentrating sample components as they emerge from the primary column in a retentive region. This is normally done by creating a cold spot in a capillary with a jet of cryogenically cooled gas. At the end of the modulation period, the retentive region is heated rapidly to desorb the collected components as a tight pulse. The concentration pulse can be directed to the head of the secondary column or, as is most often the case, subjected to another stage of thermal focusing and desorption. Thermal modulation provides outstanding performance in all of the evaluation criteria except the ease and cost of implementation. The liquid cryogen and other consumables associated with most thermal modulators lead to a significant increase in cost compared to single-column GC.

5.4 SINGLE-STAGE FLOW MODULATION

Single-stage flow modulation involves the precise addition of an auxiliary flow of carrier gas to generate pulses of primary effluent. Single-stage modulation does not involve the temporary storage of primary effluent. In 1998, Robert Synovec's research group published the first account of single-stage flow modulation GC × GC (Bruckner et al., 1998). Their method works by directing a small portion of the primary column flow to the head of the secondary column for a short period of time at the beginning of each modulation cycle. Synovec's original approach employed a multiport diaphragm valve, as shown in Figure 5.4. The valve is normally in the "bypass" state, where the primary column effluent is directed to an exhaust line. However, at the beginning of each modulation period, the valve is switched to the "inject" state, where the primary column flow is diverted to the head of the secondary column. The valve is then quickly returned to the "bypass" state. This rapid switching delivers a narrow pulse of primary effluent to the head of the secondary column.

Synovec's original work employed a 0.5-s modulation period and a 0.1 duty cycle. Thus, the valve was put in the inject state for 50 ms. The auxiliary flow was quite high (around 15 mL/min), but approximately two-thirds of the flow was split just upstream of the 0.85 m × 0.18 mm secondary column. Narrow pulses with widths near 50 ms were generated, and they were able to produce a two-dimensional group-type separation of the hydrocarbons in white gas in less than 75 s. A typical two-dimensional chromatogram is shown in Figure 5.5. Fully resolved compounds were found to have highly reproducible peak areas with RSDs near 1%. Clearly, the low duty cycle of modulation did not significantly degrade the analytical precision. The Synovec group followed this seminal publication with several other studies that focused on the high-speed separation of complex hydrocarbon mixtures, including a 4-min separation of jet fuel (Fraga et al., 2000) and a 5-min separation of naphtha (Prazen et al., 2001).

The diaphragm valve used by Synovec et al. placed a temperature constraint on the GC × GC separations. These valves have several internal parts that are not designed to be operated at temperatures above 175°C. This originally limited diaphragm valves for the separation of fairly volatile compounds; however,

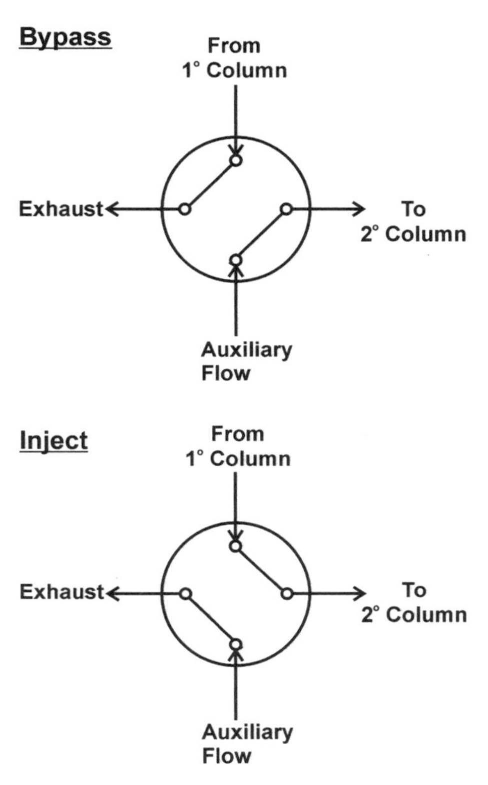

Figure 5.4 Two states of a single-stage modulator constructed from a four-port valve. The valve is normally in the "bypass" state shown at the top of the figure, but is briefly put into the "inject" state at the beginning of each modulation cycle.

techniques have been developed (Seeley et al., 2000; Sinha et al., 2003a) for protecting the temperature-sensitive portion of the valves while allowing the wetted portions of the valve to reach much higher temperatures.

In 2004, Cai and Stearns introduced a simple single-stage flow modulation technique for generating two-dimensional chromatograms. A Y-connector is placed between the primary and secondary columns. The auxiliary flow is mixed with the primary flow in the Y-connector and then the combined stream passes into the secondary column. A series of concentration spikes is generated by reducing the auxiliary flow for approximately 20 ms at the beginning of each modulation period. The key difference between this approach and all other published modulation techniques is that primary peaks are not fully converted into pulses; instead, sharply modulated spikes are observed on top of much broader unmodulated peaks. Cai and Stearns called this approach *partial modulation*. The detector signal array is digitally analyzed to separate the modulated spikes from the unmodulated peaks.

Despite the unorthodox nature of this approach, Cai and Stearns were able to produce highly resolved two-dimensional chromatograms of a wide variety of

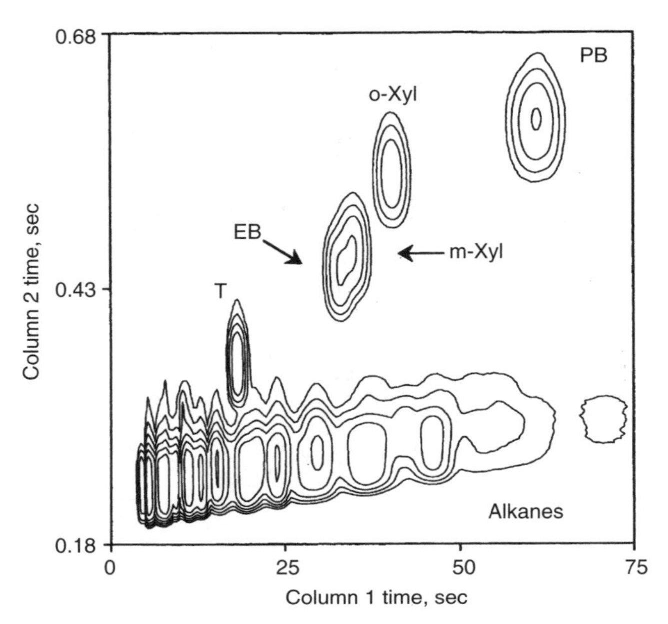

Figure 5.5 Typical high-speed two-dimensional chromatogram of white gasoline generated with a single-stage modulator constructed from a four-port valve. Individual peaks are denoted as follows: T, toluene; EB, ethylbenzene; m-Xyl, *m*-xylene; o-Xyl, *o*-xylene; PB, propylbenzene. [From Bruckner et al. (1998), with permission. Copyright © 1998 by The American Chemical Society.]

samples, including fuel oil. Peaks were very sharp along the secondary dimension, with widths ranging from 30 to 50 ms. They also reported that quantitative data with RSDs less than 1% could be obtained when the modulation ratio was kept above 4. The study represents the first published account of fluidic modulation. By keeping the valve out of the sample path and outside the oven, fluidic modulators can operate at higher temperatures. For example, Cai and Stearns were able to work at temperatures as high as 330°C. The two disadvantages of this approach are that the modulation duty cycle is a function of temperature (thus it changes during a temperature-ramped analysis) and the detector noise is increased when compounds coelute on the primary column.

Seeley et al. introduced the use of a Deans switch as a single-stage flow modulator (Seeley et al., 2007a). A Deans switch is a fluidic device that has been used for over 40 years in heart-cutting two-dimensional GC × GC (Deans, 1968). A Deans switch has two main components: a two-way three-port solenoid valve that is kept outside the GC oven, and an assembly of T-junctions that is kept inside the oven. A schematic of a Deans switch modulator is shown in Figure 5.6. The exit of the primary column is connected to the center T-junction, while the secondary column and an exhaust flow restrictor are connected to the peripheral T-junctions. The exhaust flow restrictor is often a fused-silica capillary with the same flow resistance as that of the secondary column. An auxiliary flow

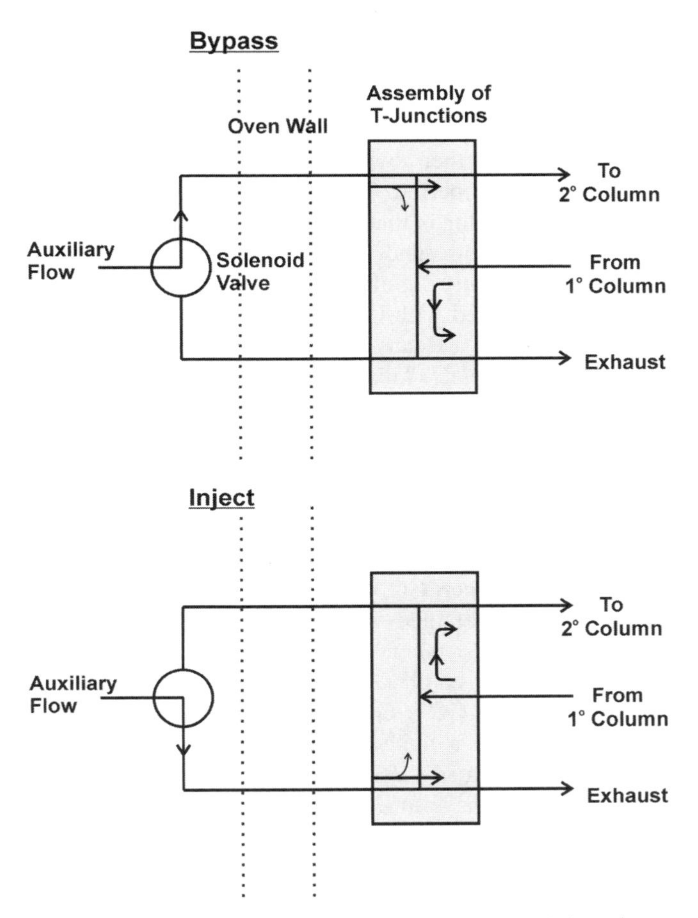

Figure 5.6 Two states of the Deans switch modulator. The modulator is normally in the "bypass" state shown at the top of the figure, but is put briefly into the "inject" state at the beginning of each modulation cycle.

that is larger than the incoming primary flow passes through the solenoid valve and then on to one of the two peripheral T-junctions. The state of the device is determined by the state of the solenoid valve. The "bypass" state is generated by having the solenoid valve direct the auxiliary flow to the T-junction connected to the secondary column (see the upper portion of Figure 5.6). In this state, the vast majority of the auxiliary flow goes to the secondary column, while a smaller portion moves toward the center T-junction, where it combines with the incoming primary column flow. This combined flow is directed to the T-junction connected to the exhaust flow restrictor. Thus, the primary column flow is prevented from entering the secondary column and, instead, passes through the exhaust restrictor. The "inject" state is generated by having the solenoid valve direct the auxiliary flow to the T-junction connected to the exhaust restrictor (see the lower portion of Figure 5.6). In this state, a small portion of the auxiliary flow goes to the center T-junction, combines with the primary column flow, heads to the opposite T-junction, and passes through the secondary column. Modulation is effected by placing the device briefly in the "inject" state at the beginning of each modulation period and then switching back to the "bypass" state for the remainder of the modulation period. The main advantage of this approach over the diaphragm valve modulator is that the T-assembly can easily be constructed from components that can withstand temperatures well above 300°C. Thus, the modulator does not impose any significant restrictions on the volatility range of analytes that can be monitored with GC × GC.

Seeley et al. (2007a) used a Deans switch constructed from a single manifold etched into a stainless-steel plate. This monolithic device, manufactured by Agilent Technologies, eliminated the need for three individual T-unions and resulted in a simple device that provided leak-free operation over a wide range of temperatures. They used the device with a 1.0-s modulation period and a 0.07 duty cycle. Peaks emerging from the secondary column had widths at half-maximum ranging from 40 to 60 ms. Seeley et al. first used Deans switch GC × GC to characterize the aromatic composition of finished gasoline. A typical chromatogram is shown in Figure 5.7. The Deans switch GC × GC system generated quantitative results that were similar to those obtained with GC–MS, and the RSDs for the peak areas were approximately 1% when a modulation ratio of 5.1 was employed. A subsequent study (Seeley et al., 2007c) demonstrated the high-temperature capabilities of Deans switch GC × GC. Petroleum-contaminated water and soil samples were analyzed with a method that had oven temperatures as high as 340°C. This study demonstrated that Deans switch GC × GC could fully separate the aliphatic hydrocarbons with sizes up to C_{36} from aromatic hydrocarbons, thus eliminating the need for fractionation on silica gel prior to GC analysis.

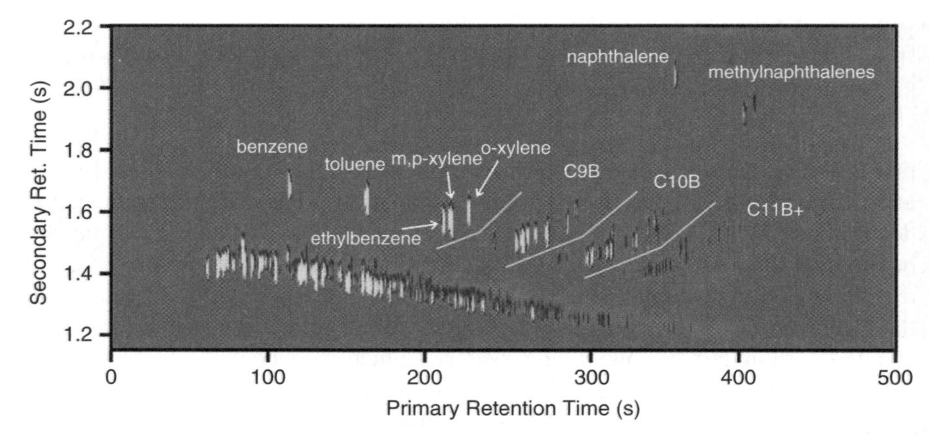

Figure 5.7 Two-dimensional chromatogram of finished gasoline generated with a Deans switch modulator. [From Seeley et al. (2007a), with permission. Copyright © 2007 by The American Chemical Society.]

The main weakness of all single-stage flow modulation strategies is that a low-modulation duty cycle is required for high resolution along the secondary dimension. This is because the temporal width of the pulses injected into the secondary column equals the amount of time the modulator samples primary effluent. Thus, narrow input pulses require that little primary column effluent is sampled, and the sensitivity of the technique decreases as the widths of the modulated pulses are reduced. This is not a major detriment when samples are fairly concentrated and plentiful (e.g., fuels); however, it does decrease the utility of single-stage modulation for trace analysis situations. The low duty cycle of single-stage flow modulation also makes it imperative that modulation ratios greater than 3.0 are employed to maintain high quantitative precision.

5.5 TWO-STAGE FLOW MODULATION

Two-stage flow modulation was employed by James Jorgenson's group in the first published account of comprehensive two-dimensional chromatography (Bushey and Jorgenson, 1990). That particular study was devoted to two-dimensional liquid chromatography. Two-stage flow modulation was subsequently shown to be effective for GC $\times$ GC separations by Seeley et al. (2000). Two-stage modulators have many features in common with single-stage modulators. For example, they can be assembled from common chromatography components and do not require additional consumables. However, unlike single-stage modulators, two-stage modulators temporarily store primary column effluent in a sample loop prior to transfer to the secondary column. This process is demonstrated in Figure 5.8 for a modulator constructed from a six-port valve. The upper portion of Figure 5.8 shows the modulator in the "collect" state, where the primary column effluent flows into the sample loop. At the desired time within the modulation cycle, the valve is switched to the "inject" state, where the auxiliary flow flushes the contents of the sample loop into the secondary column.

Two-stage modulators have been used in two different modes. The first mode attempts to minimize pulse widths and produce analyses with maximum resolving power. The second mode maximizes the modulator duty cycle and results in analyses with optimal sensitivity and quantitative precision.

5.5.1 Low-Duty-Cycle Two-Stage Modulation

The first mode of two-stage flow modulation employs a small sample loop and a small modulation duty cycle. This leads to narrow pulses and high resolution along the secondary dimension, but also decreases the sensitivity of the analysis. The main advantage that this mode has over single-stage modulation is that the flow in the primary column is decoupled from the flow in the secondary column. Thus, the flows in both columns can be optimized independently to maximize resolution.

Robert Synovec's group was first to employ this approach (Sinha et al., 2003a). They used a high-speed six-port diaphragm valve fitted with a low-volume sample

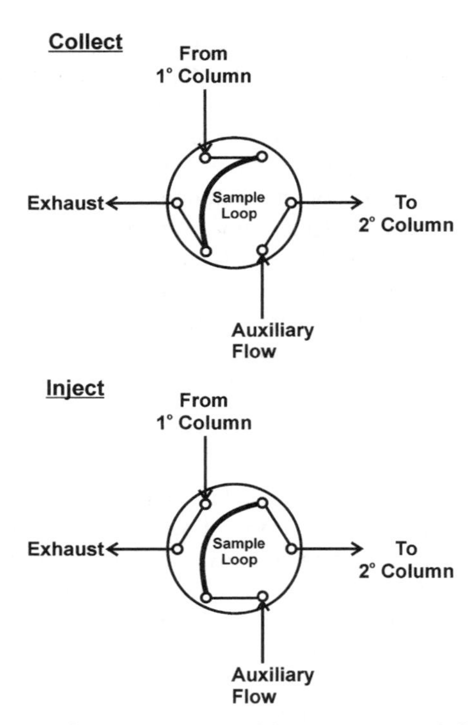

Figure 5.8 Two states of a two-stage modulator constructed from a six-port valve. Primary effluent is collected in the sample loop during the "collect" state (see the top of the figure). When the valve is placed in the "inject" state (see the lower portion of the figure) the auxiliary flow flushes the contents of the loop into the secondary column.

loop. Modulation periods of 1.0 s combined with duty cycles of 0.02 resulted in 20-ms-wide output pulses. The two-stage modulation approach allowed secondary columns with diameters as small as 0.10 mm to be employed. The peak widths at half-maximum of components exiting the secondary column ranged from 30 to 60 ms. The Synovec group also demonstrated that the diaphragm valves could be used for GC × GC separations of semivolatile compounds. By mounting the valves so that only the wetted portions extended into the oven, temperatures as high as 265°C could be used without damaging the valves. Synovec et al. used their modulators to perform GC × GC analysis of mixtures of volatile organic compounds, gasoline, and diesel fuel.

Recently, this approach has also been used in a comprehensive three-dimensional gas chromatography (GC^3) system (Watson et al., 2007). One six-port valve was used to transfer primary column effluent to a secondary column with a 5-s modulation period, and a second six-port valve was used to transfer secondary column effluent to a tertiary column with a 0.20-s modulation period. This approach provides independent retention information on each of the three stationary phases and generates a theoretical peak capacity of 3500 for a 50-min analysis.

In 2003, the Synovec group also used a two-stage flow modulator to produce the first example of flow modulation GC × GC–MS (Sinha et al., 2003b). The end of the secondary column was inserted directly into a time-of-flight mass spectrometer operated at a 50-Hz scan rate. This system was found to provide high-resolution separations with high sensitivity. A variety of studies were performed, including quantifying the levels of organophosphorous compounds, evaluation of deconvolution algorithms, and the examination of plant metabolites (Sinha et al., 2004).

Hamilton et al. (2003) also developed a low-duty-cycle two-stage modulation approach. A rotary valve fitted with a 2-μL sample loop was used to couple a 25 m × 0.32 mm primary column to a 2.5 m × 0.10 mm secondary column. The primary flow was maintained at 6 mL/min, while the auxiliary flow was 2.5 mL/min. The modulation period was set at 3.8 s, with the valve in the flush state for 300 ms each modulation cycle. Peak widths along the secondary dimension ranged from 50 to 100 ms. Hamilton et al. performed a side-by-side analysis with cryogenic modulation GC × GC and found that the valve system generated better resolution along the primary and secondary axes. A comparison of the two chromatograms is shown in Figure 5.9. However, it is important to note that this high resolution was generated by operating at a very small duty cycle. In this particular mode of two-stage modulation, the sample loop is overfilled during the collection period; thus, the duty cycle is given by the ratio of the loop volume divided by the volume of gas leaving the primary column per modulation cycle. In this case, the duty cycle was around 0.005. Thus, the thermal modulation separation produced far greater sensitivity for trace components because of its duty cycle, which was greater by a factor of 200.

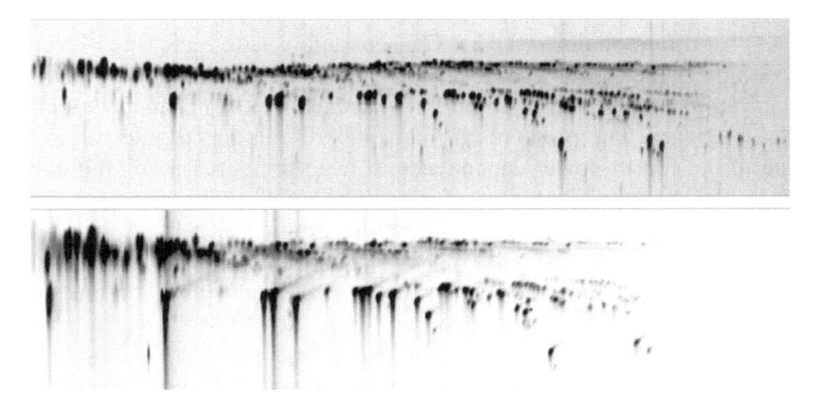

Figure 5.9 Analysis of gasoline with flow modulation GC × GC (top chromatogram) and cryogenic modulation GC × GC (bottom chromatogram). Flow modulation produces sharper peaks along both axes, but thermal modulation transfers a much larger portion of the hydrocarbons from the primary column to the secondary column. [From Hamilton et al. (2003), with permission. Copyright © 2002 by Wiley-VCH Verlag GmbH & Co. KGaA.]

5.5.2 High-Duty-Cycle Two-Stage Modulation

The second mode of two-stage modulation is known as *differential flow modulation*. It is distinguished by the use of a large sample loop and an auxiliary flow that is substantially greater than the primary column flow. In this mode, the modulator is held in the "collect" state for the majority of the modulation period. The sample loop volume is large enough not to overfill during the collection stage. When the modulator is switched to the "inject" state, the sample loop is flushed in a fraction of the collection time because the auxiliary flow rate is substantially greater than the primary flow rate. For example, if the auxiliary flow is kept 20 times greater than the primary flow, the sample loop is flushed in 5% of the fill time. Thus, differential flow modulators produce fairly narrow pulses while maintaining a high duty cycle. It is important to note that unlike thermal modulation, differential flow modulation does not increase the mole fraction of the modulated components. Instead, it drastically increases the velocity of the material leaving the modulator, and hence increases component flux. The narrow pulses broaden as they pass through the secondary column, but the flux of material leaving the secondary column is still substantially larger than the flux of material that entered the modulator. Since many chromatography detectors are mass sensitive and respond to component flux (e.g., the flame ionization detector), differential flow modulation often leads to a substantial increase in detector response.

The high auxiliary flow required for the generation of narrow peaks places a limit on the effectiveness of the secondary separation. It is well known that plate height often increases as the carrier gas velocity is increased (Golay, 1958). Fortunately, the peak capacity of the secondary separation is not determined exclusively by the plate height but, instead, by the total number of theoretical plates and the range of secondary retention times (Shen and Lee, 1998). The standard approach for attaining a reasonable peak capacity in the secondary dimension for differential flow GC × GC separations is to increase the secondary column length as the flow is increased. It should be noted that high flows and long secondary columns are incompatible with secondary column diameters of much less than 0.25 mm. The pressures in narrow-bore columns would be extremely high due to the fourth-power dependence of head pressure on column diameter. A work by Seeley et al. (2000) was the first GC × GC study to employ differential flow modulation. They used a six-port diaphragm valve fitted with a 20-μL sample loop to couple a 15 m × 0.25 mm primary column to a 5 m × 0.25 m secondary column. The primary flow was set at 0.75 mL/min, and the auxiliary flow was 15 mL/min. The modulator was held in the "collect" state 80% of the 1.0-s modulation period, resulting in a duty cycle of 0.8. Modulated peaks with widths ranging from 50 to 80 ms were observed. This efficacy of differential flow GC × GC was demonstrated by separating several different classes of volatile organic compounds (VOCs).

The negative aspects of high secondary column flow can be reduced by splitting some of the auxiliary flow downstream of the modulator before it enters the secondary column. This leads to a loss in sensitivity as a smaller fraction of material is transferred from the primary column to the secondary column.

Fortunately, a high duty cycle is maintained because the flow is split after modulation. One particularly constructive way to implement postmodulation splitting is to partition the modulated flow into two even streams that are passed into two equal-length secondary columns having different stationary phases. Dual secondary column GC × GC (often denoted GC × 2GC) generates a pair of two-dimensional chromatograms for each chromatographic analysis. The primary retention times are identical in each chromatogram, but the secondary retention times are different and often complementary. This approach improves peak identification and often leads to a larger number of fully resolved components. Seeley et al. demonstrated this approach in 2001 and then used it to perform a comprehensive analysis of the VOCs in a wide variety of gaseous samples, including outdoor air, indoor air, and exhaled breath (Seeley et al., 2002). A typical chromatogram of breath is shown in Figure 5.10. The increased resolving power of

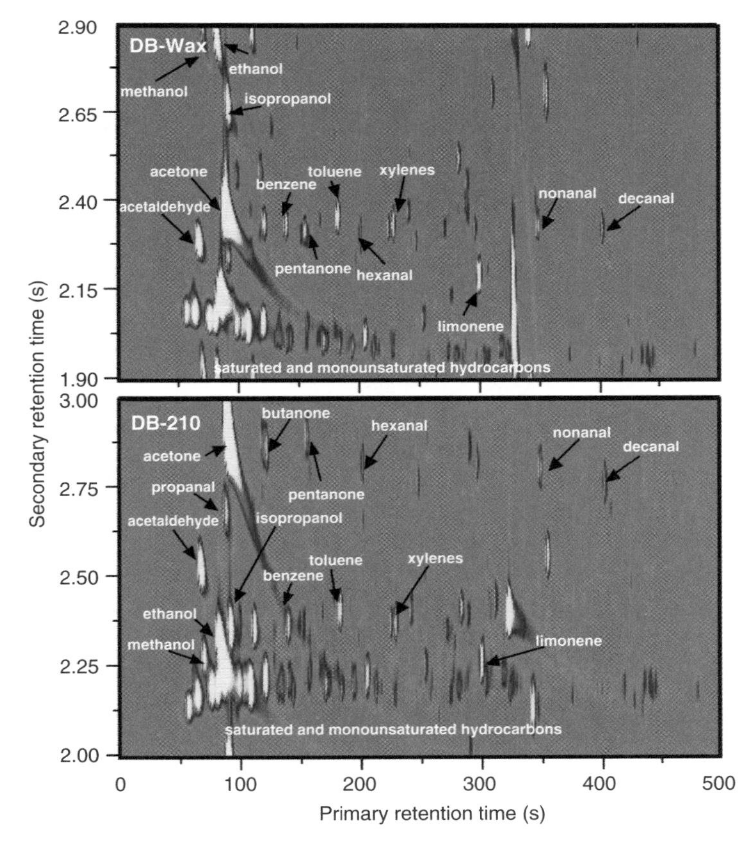

Figure 5.10 Pair of GC × GC chromatograms of exhaled breath generated with a GC × 2GC instrument fitted with a six-port diaphragm valve differential flow modulator. [From Seeley et al. (2002), with permission. Copyright © 2002 by Wiley-VCH Verlag GmbH & Co. KGaA.]

GC × 2GC allowed over 100 VOCs in breath to be fully resolved in less than 10 min.

Several types of differential flow modulators have been developed since the original six-port valve design. All of these new designs have employed full primary effluent sampling (i.e., a duty cycle of 1.0). The main benefit of modulating with a duty cycle of 1.0 is that quantitative precision is not reduced when operating at small modulation ratios. It is important to note that undersampling still results in degraded primary resolution, regardless of the duty cycle.

In 2005, Diehl and Di Sanzo demonstrated the use of an eight-port rotary valve fitted with two sample loops, as shown in Figure 5.11. Their modulator used the same configuration as the original flow modulator developed by Bushey and Jorgenson (1990). The dual loops allow one sample loop to be filled while the other sample loop is flushed. At the end of each modulation period, the valve is actuated so that the opposite loops are filled and flushed. Diehl and Di Sanzo

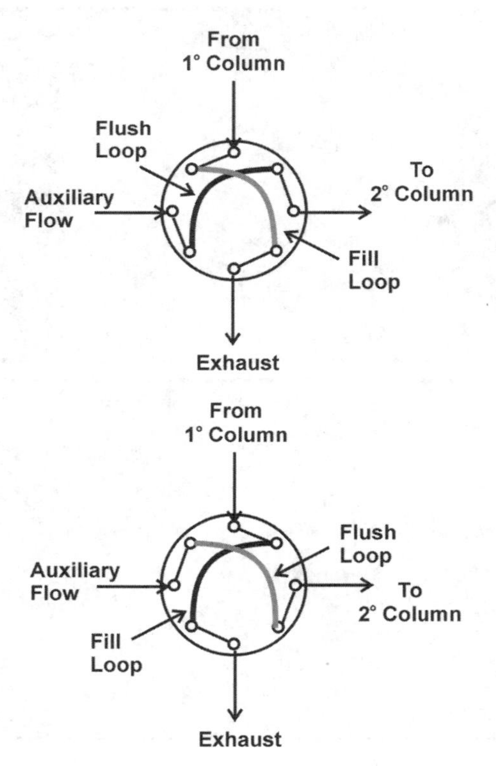

Figure 5.11 Two states of a two-stage modulator constructed from an eight-port valve. Primary effluent is collected in one sample loop while the auxiliary flow flushes the other sample loop onto the secondary column. The valve is switched at the end of each modulation period. The state at the top of the figure shows the filling of the gray loop and the flushing of the black loop, while the bottom of the figure shows the flushing of the gray loop and the filling of the black loop.

used this approach with long, wide-bore columns, an eight-port rotary valve, and with a very large modulation period of 1.0 min. The eight-port rotary valve could be operated at temperatures as high as 325°C. They were able to accurately characterize the aromatic composition of gasoline with excellent precision (e.g., RSDs were near 1%). However, the 90-min analysis time was somewhat excessive compared to similar flow modulation GC × GC analyses. Diehl and Di Sanzo also coupled their differential flow GC × GC instrument to a quadrupole mass spectrometer. The 50-mL/min auxiliary flow was too large for full transfer into the vacuum chamber, so an open-split interface was used. The slow modulation conditions allowed standard MS scan speeds to be employed. The main drawback of an eight-port valve modulator is that the sample loops are not flushed in a symmetric manner: One loop is flushed in the same direction in which it was filled, while the other loop is flushed in the opposite direction. This leads to peaks that are split along the secondary dimension if the sample loops were only partially filled before injection.

In 2006, the Synovec group (Mohler et al., 2006) developed a modulator that sampled all of the primary column effluent (i.e., had a duty cycle of 1.0) using a six-port diaphragm valve and a sample loop. Their design was very similar to the six-port valve modulator developed by Seeley et al. (2000) except that the valve port normally used as the primary column exhaust vent (see Figure 5.8) is plugged. This causes the flow in the primary column to slow as the loop is filled during the collection phase. When the valve is placed in the "inject" state, the loop contents are flushed onto the secondary column and the primary column flow is essentially halted. Despite the high primary column pressures and extreme changes in primary column flow rate, this modulation approach produced reasonable-looking chromatograms. The peaks were especially narrow along the secondary axis, with widths at half maximum ranging from 25 to 90 ms. This modulation approach was demonstrated further by analyzing gasoline and eucalyptus oil.

In 2008, Wang described the use of two four-port rotary valves fitted with two sample loops, as shown in Figure 5.12. This approach employs the same principles as those of the eight-port valve modulator but it also solves the asymmetrical loop flushing problem. The modulator was operated with a modulation period of 8.0 s, an auxiliary flow of 100 mL/min, and a 3 m × 0.53 mm secondary column. Wang used a GC × GC system fitted with this modulator to analyze a variety of petrochemical samples, including a naphtha refinery stream and a diesel refinery stream. The system was also used to analyze mixtures of fatty acid methyl esters (FAMEs) and mixtures of polychlorinated biphenyls. The dual-loop modulator was found to be robust and capable of being operated routinely at oven temperatures as high as 300°C. It remains to be seen if this method will work with the shorter modulation period and narrower secondary columns required for higher-speed analyses.

Differential flow modulators based on fluidic devices have also been developed. In 2004, Bueno and Seeley developed a differential flow modulator that

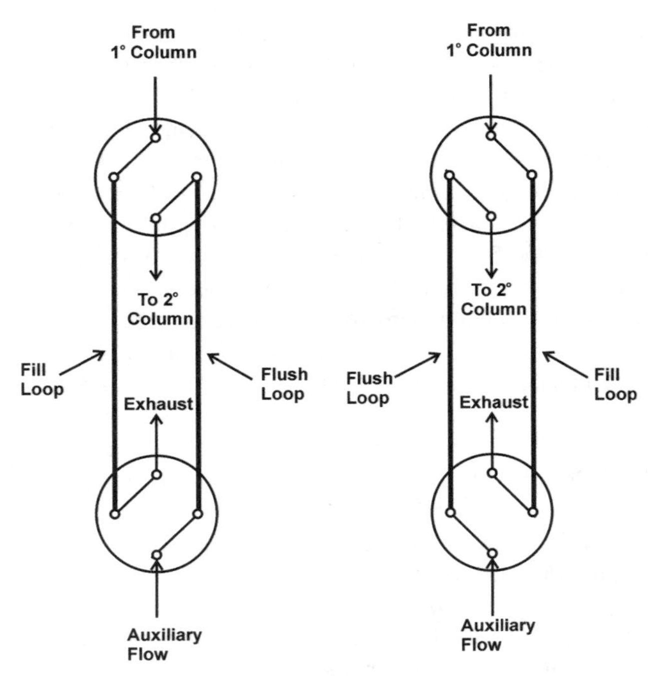

Figure 5.12 Two states of a two-stage modulator constructed from two four-port valves. Primary effluent is collected one sample loop while the auxiliary flow flushes the other sample loop onto the secondary column. The valve is switched at the end of each modulation period. The state on the left side of the figure shows the filling of the left loop and the flushing of the right loop, while the right side of the figure shows the flushing of the left loop and the filling of the right loop.

resembles a Deans switch fitted with two sample loops. Their device is constructed from four T-unions, pieces of deactivated fused-silica capillaries, and a three-port two-way solenoid valve. The dimensions of the fused-silica capillaries are selected to cause the primary effluent to fill one of the sample loops while the auxiliary flow flushes out the other sample loop. The solenoid valve is switched at the end of each modulation period to begin filling and flushing the opposite loops. The conduits and unions in the sample path contain no moving parts, so they can easily be constructed from robust materials that can withstand a wide range of temperatures. The device provides full sampling of the primary effluent and generates narrow pulses with minimal disturbances to the flow. A thorough performance analysis of this device has been published (LaClair et al., 2004). The modulator was used to analyze a variety mixtures, including gasoline, diesel fuel, and organic compounds collected in rural atmospheres. Quantitative results with high accuracy and high precision were also obtained for the aromatic content of gasoline (Micyus et al., 2005). The main drawback of this modulator is that it can be difficult to construct and works only over a narrow range of flow conditions.

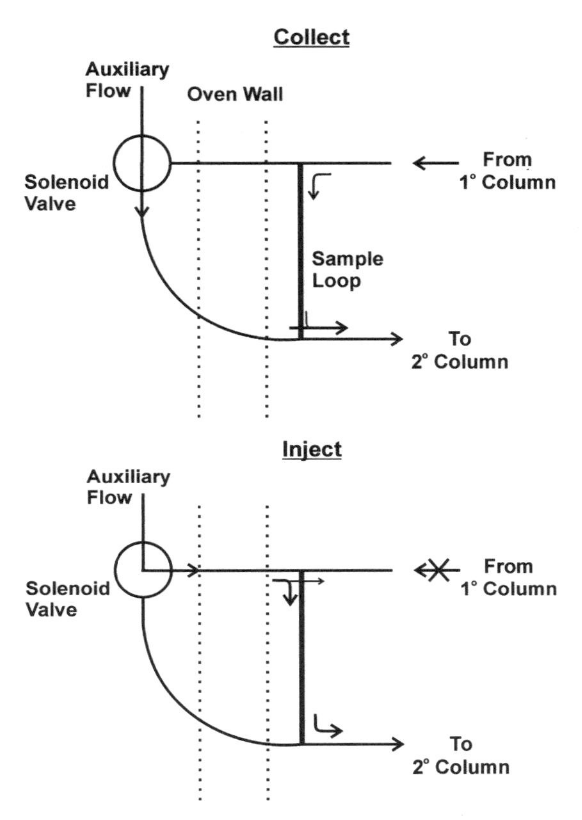

Figure 5.13 Two states of the in-line fluidic modulator. The device is normally kept in the "collect" state shown at the top of the figure, where the primary effluent flows into the sample loop while the auxiliary flow is directed to the secondary column through the lower T-junction. Pulses are generated by placing the device briefly in the "inject" state by directing the auxiliary flow to the upper T-junction. This simultaneously halts the primary column flow and flushes the content of the sample loop onto the head of the secondary column.

The complexity of the device described previously motivated the search for a simpler fluidic device for differential flow modulation GC × GC. In 2005, Seeley et al. (2005, 2006) described a fluidic device built from two T-junctions, a single sample loop, and a three-port two-way solenoid valve. A schematic of this in-line fluidic modulator is shown in Figure 5.13. One of the T-junctions is connected to the outlet of the primary column, and the other T-junction is connected to the inlet of the secondary column. As is always the case for differential flow modulators, an auxiliary flow that is substantially larger than the primary column flow is used to generate component pulses. The auxiliary flow enters the common port of the solenoid valve. The output ports of the solenoid valve are connected to the two T-junctions. The T-junctions are connected to one another with a single piece of deactivated fused silica tubing that serves as the sample loop.

The state of the device is determined by the state of the solenoid valve. The device is normally in the "collect" state shown at the top of Figure 5.13, where the solenoid valve directs the auxiliary flow to the T-junction connected to the secondary column. In this state, the flow from the primary column enters the opposite T-junction (shown as the upper union in Figure 5.13) and then moves through the sample loop. The device is switched to the "inject" state before the sample loop is completely filled with effluent from the primary column. In the "inject" state, the auxiliary flow is directed to the upper T-junction, where it quickly flushes the contents of the sample loop into the secondary column. After the sample loop is flushed completely, the solenoid valve is switched back and the device returns to the "collect" state. The large difference between the auxiliary and primary flow rates allows the device to stay in the "collect" state for approximately 90% of the modulation period.

It is desirable to halt the ingress of primary column effluent during the "inject" state to eliminate baseline changes associated with the mixing of primary effluent and auxiliary flow. An easy way to stop the primary flow momentarily at the entrance of the modulator is to generate a small pressure pulse when the modulator is switched from "collect" to "inject." This can be accomplished by making the flow resistance of the tubing connecting the solenoid to the downstream T-junction (the lower T-junction in Figure 5.13) greater than the flow resistance of the tubing connecting the solenoid valve to the upstream T-junction. This causes the pressure in the solenoid valve to be higher than the pressure in the upstream T-junction. Thus, when the device is switched to the "inject" state, a short pressure pulse occurs in the upstream T-junction. This temporarily reverses the primary column flow near the entrance to the device. This approach has been shown to generate peaks with minimal tailing (Seeley et al., 2006).

The in-line fluidic modulator has been applied to several analyses. Seeley et al. (2006) have shown that it can be used to characterize the aromatic composition of gasoline. A 10-min separation produced highly accurate and precise concentrations (i.e., most RSDs were less than 1%) for all of the aromatic compound classes in gasoline. A 1.5-s modulation period was used to produce secondary peaks, with widths at half-maximum ranging from 65 to 80 ms. It should be noted that these widths are essentially the minimum that can be expected for a 1.5-s modulation period with a 1 : 20 primary/auxiliary flow ratio. A similar study also demonstrated that GC × GC analysis with an in-line fluidic modulator can accurately characterize the aliphatic alcohols found in high-ethanol-containing motor fuels such as E85 (Seeley et al., 2008). The modulator was also used to develop a GC × GC method for determining the fatty acid methyl ester (FAME) content of biodiesel blends (Seeley et al., 2007b). A typical chromatogram of a biodiesel blend is shown in Figure 5.14.

Research groups outside the Seeley group have also verified the efficacy of this approach. Agilent Technologies has recently introduced a modulator based on this design. Their monolithic device integrates the sample loop, T-junctions, and valve transfer lines into a single unit, and increases the ease at which a differential flow modulator can be employed for GC × GC analysis.

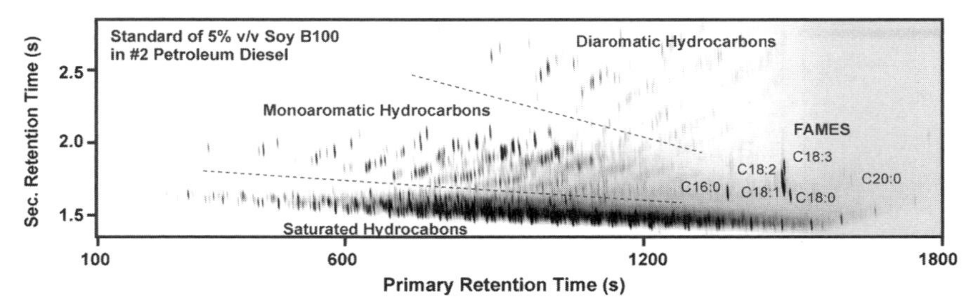

Figure 5.14 Two-dimensional chromatogram of a 5% blend of soy biodiesel in petroleum diesel.

Amirav et al. have developed a similar modulator (Poliak et al., 2008b). They have shown that a large auxiliary flow can produce pulse widths as small as 20 ms. Their GC × GC instrument is coupled to a quadrupole mass spectrometer fitted with a supersonic molecular beam interface. The differentially pumped interface of the mass spectrometer can easily handle all of the high auxiliary flow. They tailor their primary and auxiliary flows to generate secondary column peaks that are compatible with their spectrometer scan speeds. Their instrument was demonstrated by analyzing diesel fuel and pesticides in agricultural matrices (Kochman et al., 2006; Poliak et al., 2008a).

5.6 SUMMARY OF FLOW MODULATORS

Eleven different flow modulators have been described in this chapter. A summary of their operating characteristics is shown in Table 5.1. Low-duty-cycle modulators have evolved into two effective designs: six-port high-speed valves fitted with low-volume-sample loops and the Deans switch fluidic modulator. The six-port valve approach has the advantage of decoupling the flow in the primary and secondary columns. Thus, both flows can be optimized independently. The six-port diaphragm valve has proven to be especially effective for high-speed separations. The Deans switch benefits from an upper temperature limit that exceeds that of stationary phases; thus, a wide range of separations can be performed. Both modulators are robust and can be used for years without servicing.

High-duty-cycle modulators have also evolved into multiport valve and fluidic designs. The key advantage of multiport valves is that they provide direct flow control and are more immune to tailing due to the dead volume. Wang's twin four-port valve approach (Wang, 2008) appears to be especially robust and straightforward, but it remains to be seen if it is effective at smaller modulation periods and lower auxiliary flows. The in-line fluidic modulator (Seeley et al., 2006) is a simple and effective way to modulate with a high duty cycle using a moderate auxiliary flow. Since this design has only one moving part (the poppet in the solenoid valve), it can operate for years without major service or replacement. It is possible that high-duty-cycle sampling of the in-line modulator could

TABLE 5.1 Characteristics of the Published Flow Modulators

Modulator	Number of Stages	Duty Cycle	Max. T(°C)	Typical Auxiliary Flow (mL/min)	Typical Modulation Period (s)	Date of First Publication
4-Port diaphragm valve	1	≤ 0.1	175	15	0.75	1998
Partial modulation with flow pulses	1	≤ 0.1	330	2	2.0	2004
Deans switch	1	0.07	340	9	1.0	2007
6-Port diaphragm valve with small-volume sample loop	2	0.02	265	3	1.0	2002
Rotary valve with small-volume sample loop	2	0.005	N.A.[a]	2.5	3.8	2003
6-Port diaphragm valve with large-volume sample loop	2	0.8	200	15	1.0	2000
8-Port valve with dual sample loops	2	1.0	325	50	60	2005
Two 4-port valves with dual sample loops	2	1.0	300	100	8	2008
6-Port diaphragm valve with stopped flow	2	1.0	200	30	1.5	2006
Dual-loop fluidic modulator	2	1.0	250	20	1.5	2004
In-line fluidic modulator	2	1.0	260	15	1.5	2005

[a] N.A., not applicable.

be combined with post-modulation splitting to generate especially high resolution GC × GC separations that have high precision and only a small decrease in sensitivity.

5.7 BRIEF COMPARISON TO THERMAL MODULATION

Thermal modulation is the *ultimate* modulation technique, as it produces extremely narrow concentrated pulses, has a duty cycle of 1.0, works well at flows near 1 mL/min, has reasonable modulation speed, and works well over the range of temperatures commonly employed in GC separations. Numerous studies have shown that thermal modulation can drastically increase both the sensitivity and the resolution of GC × GC analysis (see Chapter 4). The downside of thermal modulation is that it normally requires a large amount of cryogenic fluid to operate properly. This represents a large increase in the cost of operation over that of conventional single-column GC. Thus, thermal modulation is the best choice when performance is the paramount concern and cost is not a determining factor. On the other hand, thermal modulation might not be the best choice when low cost or portability is required.

Flow modulation is a feasible alternative for analysts seeking a simple, low-cost GC × GC approach. It is also interesting to note that the performance gap between thermal modulation and flow modulation decreases as the speed of the analysis increases. This is because increasing the separation speed often results in narrower primary peaks; thus, smaller modulation periods are required. Flow modulators have pulse widths that decrease with decreasing modulation period (assuming a constant duty cycle). In contrast, thermal modulators generate pulse widths that are essentially independent of the modulation period. Thus, the decrease in peak capacity along the secondary dimension that occurs as the modulation period is reduced is not as significant for flow modulators. This is probably the reason that the majority of flow modulation GC × GC analyses have been performed with faster temperature ramps and smaller modulation periods than those of thermal modulation GC × GC analyses.

5.8 CONCLUDING REMARKS

It has been 12 years since the first flow modulation GC × GC article was published. Since that time, numerous hardware improvements have made it easy to apply flow modulation GC × GC to a wide range of analyses with excellent results. Yet with the exception of the laboratories of the hardware developers, few groups have adopted flow modulation GC × GC. The cause of this reluctance is unknown. Perhaps it is because neither thermal nor flow modulation GC × GC has yet to be widely adopted for routine analyses, or perhaps it is because commercially produced flow modulation GC × GC instruments have only been available in the last two years. The next few years will probably determine if flow modulation evolves into a widely adopted practical form of GC × GC or if flow modulation is just a temporary steppingstone to more elaborate forms of GC × GC modulation.

REFERENCES

Bruckner CA, Prazen BJ, Synovec RE. *Anal. Chem*. 1998; 70:2796–2804.

Bueno PA, Seeley JV. *J. Chromatogr. A* 2004; 1027:3–10.

Bushey MM, Jorgenson JW. *Anal. Chem*. 1990; 62:161–167.

Cai HM, Stearns SD. *Anal. Chem*. 2004; 76:6064–6076.

Deans DR. *Chromatographia* 1968; 1:18–22.

Diehl JW, Di Sanzo FP. *J. Chromatogr. A* 2005; 1080:157–165.

Fraga CG, Prazen BJ, Synovec RE. *Anal. Chem*. 2000; 72:4154–4162.

Golay MJE. In: Desty DH, Ed., *Gas Chromatography 1958*. London: Butterworths, 1958, p. 36.

Hamilton JF, Lewis AC, Bartle KD. *J. Sep. Sci*. 2003; 26:578–584.

Khummueng W, Harynuk J, Marriott PJ. *Anal. Chem*. 2006; 78:4578–4587.

Kochman M, Gordin A, Alon T, Amirav A. *J. Chromatogr. A* 2006; 1129:95–104.

LaClair RW, Bueno PA, Seeley JV. *J. Sep. Sci*. 2004; 27:389–396.

Liu ZY, Phillips JB. *J. Chromatogr. Sci*. 1991; 29:227–231.

Micyus NJ, McCurry JD, Seeley JV. *J. Chromatogr. A* 2005; 1086:115–121.

Mohler RE, Prazen BJ, Synovec RE. *Anal. Chim. Acta* 2006; 555:68–74.

Murphy RE, Schure MR, Foley JP. *Anal. Chem*. 1998; 70:1585–1594.

Poliak M, Fialkov AB, Amirav A. *J. Chromatogr. A* 2008a; 1210:108–114.

Poliak M, Kochman M, Arnirav A. *J. Chromatogr. A* 2008b; 1186:189–195.

Prazen BJ, Johnson KJ, Weber A, Synovec RE. *Anal. Chem*. 2001; 73:5677–5682.

Seeley JV. *J. Chromatogr. A* 2002; 962:21–27.

Seeley JV, Kramp F, Hicks CJ. *Anal. Chem*. 2000; 72:4346–4352.

Seeley JV, Kramp FJ, Sharpe KS. *J. Sep. Sci*. 2001; 24:444–450.

Seeley JV, Kramp FJ, Sharpe KS, Seeley SK. *J. Sep. Sci*. 2002; 25:53–59.

Seeley JV, Seeley SK, Primeau NJ. *Abstr. Pap. Am. Chem. Soc*. 2005; 230:338–ANYL.

Seeley JV, Micyus NJ, McCurry JD, Seeley SK. *Am. Lab*. 2006; 38:24–26.

Seeley JV, Micyus NJ, Bandurski SV, Seeley SK, McCurry JD. *Anal. Chem*. 2007a; 79:1840–1847.

Seeley JV, Seeley SK, Libby EK, McCurry JD. *J. Chromatogr. Sci*. 2007b; 45:650–656.

Seeley SK, Bandurski SV, Brown RG, McCurry JD, Seeley JV. *J. Chromatogr. Sci*. 2007c; 45:657–663.

Seeley JV, Libby EM, Seeley SK, McCurry JD. *J. Sep. Sci*. 2008; 31:3337–3346.

Shen YF, Lee ML. *Anal. Chem*. 1998; 70:3853–3856.

Sinha AE, Johnson KJ, Prazen BJ, Lucas SV, Fraga CG, Synovec RE. *J. Chromatogr. A* 2003a; 983:195–204.

Sinha AE, Prazen BJ, Fraga CG, Synovec RE. *J. Chromatogr. A* 2003b; 1019:79–87.

Sinha AE, Hope JL, Prazen BJ, Fraga CG, Nilsson EJ, Synovec RE. *J. Chromatogr. A* 2004; 1056:145–154.

Wang FCY. *J. Chromatogr. A* 2008; 1188:274–280.

Watson NE, Siegler WC, Hoggard JC, Synovec RE. *Anal. Chem*. 2007; 79:8270–8280.

6

COMPREHENSIVE TWO-DIMENSIONAL GAS CHROMATOGRAPHY COMBINED WITH MASS SPECTROMETRY

PETER Q. TRANCHIDA AND LUIGI MONDELLO
University of Messina, Messina, Italy
SAMUEL D.H. POYNTER AND ROBERT A. SHELLIE
University of Tasmania, Hobart, Tasmania, Australia

Ever since the inception of comprehensive two-dimensional gas chromatography (GC × GC) there have been comparisons made between GC × GC analysis and GC–mass spectrometric (MS) analysis. Initially, these comparisons were drawn to illustrate the multidimensionality of the GC × GC approach. For example, Figure 6.1 shows the point of agreement between GC × GC and GC–MS. GC × GC uses a separation column rather than a mass analyzer in the second analytical dimension, but this figure clearly illustrates that in both cases a device for chemical measurement is used to analyze the portions of the sample being eluted from the first-dimension GC column. The second-dimension GC column in the GC × GC instrument produces a series of chromatograms from the portions of the first-dimension effluent, while a series of mass spectra are produced from the equivalent sample portions in the GC–MS instrument. Most GC–MS analyses utilize electron-induced ionization (EI), so the spectra of coeluted compounds exist as overlapping envelopes of fragment ions from each of the coeluted compounds. For this reason, GC–MS with EI is referred to as a two-dimensional analysis, not a two-dimensional separation, although it is

Comprehensive Chromatography in Combination with Mass Spectrometry, First Edition.
Edited by Luigi Mondello.

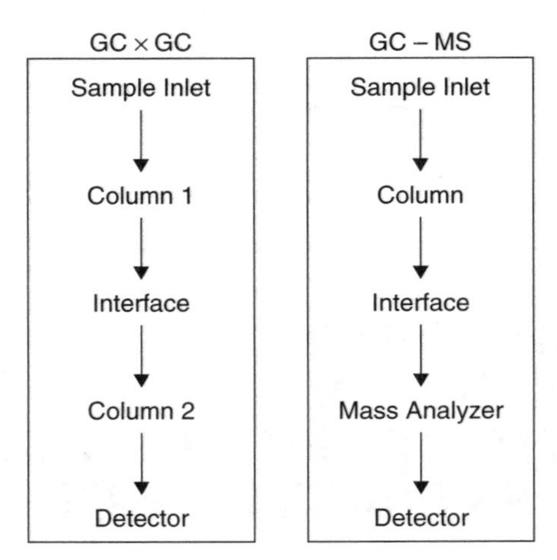

Figure 6.1 Comparison of two multidimensional approaches: GC × GC versus GC–MS. [Adapted from Phillips and Beens (1999).]

noteworthy that corresponding parts of the overlapping spectra can be apportioned very successfully to individual compounds by using MS deconvolution software such as the freely available Automated Mass Spectral Deconvolution and Identification System (AMDIS, http://chemdata.nist.gov/mass-spc/amdis/). GC × GC very clearly produces a two-dimensional separation, with compounds being chromatographically resolved in each of the first and second dimensions, respectively. It is also possible to perform a two-dimensional separation with a GC–MS instrument. This experiment relies on soft ionization to produce a single ion for each compound, and when this criterion is met, compounds are separated according to their mass/charge ratio in the mass analyzer. If the entire sample is subjected to these two separation dimensions, these separation dimensions fit the definition (Schoenmakers et al., 2003) of being comprehensively coupled, and GC × MS is denoted using the multiplex sign. This GC × MS concept is discussed in more detail later in the chapter.

The obvious congruity between GC–MS and GC × GC has also led several commentators to propose that GC × GC should be fit to perform analyses that are typically made by GC–MS, without recourse to a mass analyzer. However, this promise is largely unfulfilled. GC × GC chromatograms often reveal literally thousands of peaks in truly complex samples. Under these circumstances, reliance on pattern formation or injection of reference materials for the purpose of qualitative peak assignment will fail. Even though GC × GC is a powerful technique for the physical separation of volatile mixtures, it can realistically only give a broad indication of the nature or chemical class of many of the resolved components. The characterization of individual components and their structural elucidation is the realm of selective and spectroscopic detectors. The complexity of many samples further dictates that mass spectrometry be employed for confirmation of peak assignment and peak purity. The obvious extension of the technique involves coupling GC × GC with MS, giving rise to an undisputedly powerful analysis approach. Having three independent dimensions of analysis, in a GC × GC–MS experiment, there is unquestionably more qualitative information available to the analyst than in approaches offering less dimensionality. Harnessing this information in a meaningful way can be a challenge, but users generally find that by distilling the information down to smaller manageable bundles makes interpretation no more challenging than GC–MS data interpretation. Assuming that the aim of an experiment is to identify unknown components by way of GC × GC–MS, with electron-induced ionization to give library-searchable spectra, it is generally acceptable to zoom a single second-dimension modulation slice using GC–MS data analysis software and perform the library search routine according to standard practice for GC–MS.

Specialized GC × GC–MS software is sparingly available, but regular automated functionality such as MS database searching can also be performed, with minimal user input using dedicated commercial GC × GC–MS software packages. Compared to GC–MS, the advantage of using GC × GC–MS is the ease with which positive assignment of peak identity can be made by way of database matching. There is unequivocal evidence that a carefully optimized GC × GC–MS separation makes peak assignment more facile than equivalent GC–MS data. The high-quality spectra result from two of the general benefits of GC × GC. The improved resolution offered by secondary chromatographic separation allows the introduction of a higher-purity analyte flux to the MS. Furthermore, zone compression arising from the modulation process refocuses analyte bands and eliminates band broadening from the first-dimension separation, which is often operated outside optimum separation parameters. Combined, these enhancements greatly assist qualitative analysis of complex samples. By utilizing selected ion monitoring, or by careful use of extracted ion chromatograms, it is possible to reveal the presence or absence of key target compounds or compound classes. Married to the highly praised "structured chromatogram" that is typical of a well-designed GC × GC separation, the combination of group patterns and MS information often provides undisputable confirmation of peak identity. Similarly, quantitative analysis is enhanced by combination of three analytical dimensions but has proven difficult to implement. Proposed methods are generally tedious and may require considerable development by the analyst for the application at hand. For this reason, only a limited number of quantitative applications of GC × GC–MS have been published, of which many focus on a limited number of target components, generally fewer than 20. The details of these applications are discussed in more detail toward the end of the chapter.

6.1 INSTRUMENT REQUIREMENTS FOR GC × GC–MS

As described in Chapter 4, to provide effective sampling of the first-dimension effluent, the second-dimension separation speed in GC × GC has to be very fast. Should this not be the case, insufficient sampling of each component is made and the phenomenon of analyte wraparound may be observed. This would lead to deviation from the system optimum, and the analyst runs into difficulty with correct assignment of the peaks. Consequently, the MS detector acquisition frequency needs to be sufficiently high to interpret the separated components correctly and thus permit GC × GC–MS analysis. Frysinger and Gaines (1999) were first to demonstrate GC × GC–MS. Noting experimental challenges in their initial investigation, which were related primarily to the slow acquisition rate of the quadrupole MS used in the experiments, the first approach for implementing the GC × GC–MS system was realized by slowing the typical analysis speed of the first-dimension separation by a factor of about 7, with concomitant reduction of second-dimension throughput to the order of five times. The first GC × GC–MS chromatogram reported depicting the separation of a marine diesel sample is shown in Figure 6.2. The separation was achieved using a 13 m × 100 μm i.d. first-dimension column with a thick-film (3.5-μm) 100% dimethyl polysiloxane stationary phase, coupled to a 2 m × 100 μm i.d. 14% cyanopropylmethyl polysiloxane stationary-phase column installed in a thermally modulated GC × GC system. The experimental difficulties that Frysinger and Gaines described clearly portray the importance of high-speed detectors for GC × GC–MS. Notwithstanding the valuable illustration of GC × GC–MS, the detection rate of 2.43 full-scan spectra per second meant that the nearly 1-s-wide second-dimension peaks were severely undersampled. In the example given here,

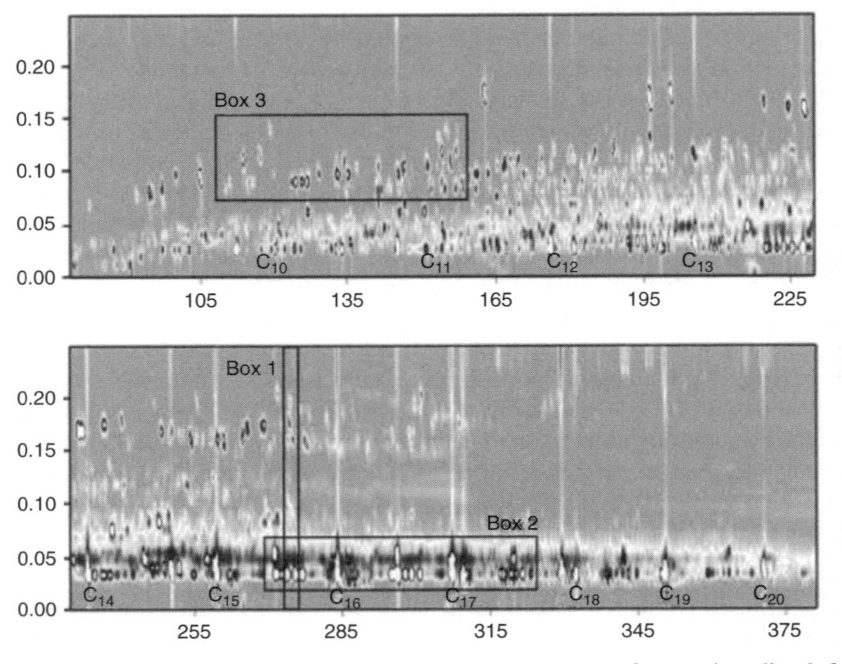

Figure 6.2 GC × GC–MS total ion current chromatogram of a marine diesel fuel sample. [From Frysinger and Gaines (1999), with permission. Copyright © 1999 by Wiley-VCH Verlag GmbH & Co. KGaA.]

an average of only one full-scan mass spectrum was recorded for each second-dimension peak. Although the reconstructed two-dimensional (2D) separation space provides a representation of the GC × GC separation, not enough data points are recorded to faithfully portray the chromatographic result and permit measurement of important chromatographic metrics, such as peak width, peak asymmetry, and resolution.

A list of accepted peak widths associated with different GC separation speeds is presented in Table 6.1. The importance of data density being compatible with the chromatographic time scale is illustrated in Figure 6.3, where Gaussian peaks have been generated using the retention times and peak widths shown in Table 6.1. Three data acquisition rates are provided with the data points superimposed on each peak to illustrate the capabilities of a range of instrument types. First, 4 Hz represents the full-scan mass spectrum acquisition that is easily achieved by most quadrupole mass analyzers. This data acquisition rate is suitable for conventional GC, but it is hardly sufficient for fast GC, and very few full-scan spectra per

TABLE 6.1 Characteristic Figures of Merit for Standard, Fast, Very Fast, and Ultrafast GC Analysis

Separation Speed	Column i.d. (mm)	Column Length (m)	Theoretical Plates	Retention Time (s)	Peak Width 2.354σ (s)
Standard	0.32	25	75,000	100	0.7
Fast	0.05	10	260,000	60	0.2
Very fast[a]	0.05	1	24,000	2	0.03
Ultrafast	0.05	0.3	6,500	0.3	0.01

Source: van Deursen et al. (2000a), with permission.

[a]The second dimension separation speed for GC × GC–MS typically fits the very fast category.

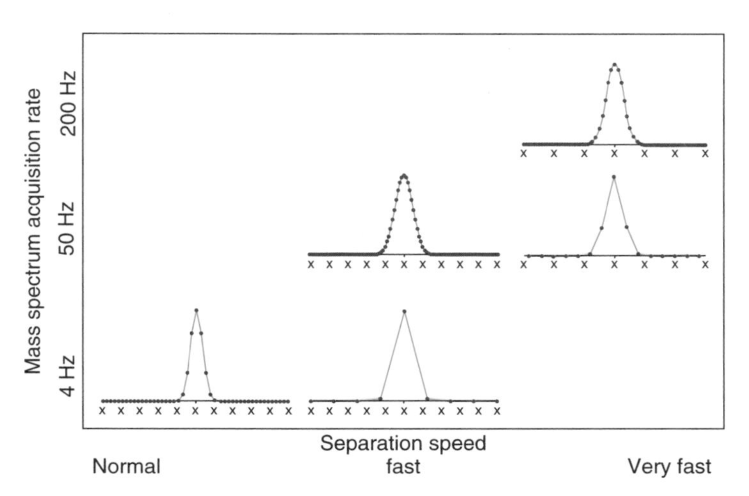

Figure 6.3 Detection requirements for GC × GC–MS shown by drawing ideal peaks and superimposing possible data points at various detection speeds.

peak will be collected using this instrument configuration with fast GC separation (Figure 6.3). The physical laws governing mass selection processes limit the ability of quadrupole mass analyzers to achieve very fast data acquisition rates. A quadrupole MS passes ions of various m/z (one m/z at a time) to the detector by simultaneously ramping direct-current (dc) and radio-frequency (RF) voltages carried by the quadrupole rods. Contemporary quadrupole MS instruments are capable of mass scanning rates of around 10,000 amu/s. Thus, for a full-scan mass range of 50 to 350 amu, the maximum spectral acquisition rate would be expected to be less than 30 spectra/s (including an interscan period). Although these figures have improved considerably over the past decade, there is a limit to the scanning speed because transmission of the ions selected is affected if the RF and dc voltages change during the transit time of an ion through the quadrupole (Holland et al., 1983). Very fast GC is best mated with nonscanning time-of-flight mass spectrometry, which is capable of very fast full-spectrum data acquisition rates. Figure 6.3 shows that acquisition of 50 spectra/s will suffice for fast GC, but ultrafast separations demand even faster data acquisition. Although there are some examples of ultrafast second-dimension GC × GC separations in the literature, these are outside the norm, and 50 to 100 spectra/s is generally applicable for GC × GC–MS analysis.

6.1.1 Quadrupole MS

Following Frysinger and Gaines' introduction of GC × GC–MS, there were no reports in the periodical literature regarding hyphenation of GC × GC with quadrupole instrumentation until 2002. This apparent lack of activity can be ascribed almost entirely to the speed limitations of MS detection systems. However, there is alternative approach for coupling quadrupole MS instrumentation with GC × GC that does not depend on slowing the GC × GC analysis excessively, but this is achieved at the expense of sacrificing some MS information to overcome the inherent spectral acquisition rate limitations. In 2002, Shellie and Marriott described an enantioselective–GC × GC experiment using a quadrupole mass analyzer in single-ion selected ion monitoring mode, with a nominal data acquisition rate of 20 spectra/s. Although faster, this detection regime lacks information richness, due to the loss of almost all mass spectral information. The experiments that Shellie and Marriott performed were designed to utilize the vacuum outlet condition of the MS detector to capitalize upon the theoretical efficiency benefits of low-pressure GC (Cramers and Leclercq, 1988). The dynamic diffusion coefficient of an analyte in the carrier gas and the optimum carrier gas linear velocity are inversely proportional to the column pressure, so under certain conditions, vacuum-outlet operation can have speed and efficiency benefits over atmospheric-outlet operation. However, with the GC × GC equipment installed on their GC–MS instrument, Shellie and co-workers decided serendipitously to attempt full-scan detection. With the initial experiments aimed at the characterization of volatile plant extracts comprising terpenoid compounds, a reduced mass-scan range from 41 to 229 amu was deemed appropriate, since this permits detection of molecular ions of mono- and sesquiterpenes as well as of the oxygenated monoterpenes. The speed advantage attainable by reducing the mass range can be estimated by considering the quadrupole duty cycle divided by the mass range. The absolute data acquisition rate will be less than this because the interscan delay must also be considered. Thus, by using a mass analyzer with a maximum scan speed of 4000 amu/s and a mass scan range of 188 amu as used by Shellie and Marriott, a data acquisition rate of approximately 20 scans/s is attainable (Shellie and Marriott, 2003a). Although the data acquisition rate was relatively slow compared to conventional GC × GC applications with fast detection, giving only about four data points for each narrow second-dimension peak, the quality of the spectra for the separated components was reportedly very high (Shellie and Marriott, 2003a; Shellie et al., 2003).

Interestingly, by speeding up the detection rate as described above, the total GC × GC–MS analysis times reported by Shellie and co-workers were only about 30% longer than the programmed solvent delay of the GC × GC–MS system reported by Frysinger and Gaines (1999) using similar equipment. The GC × GC–MS chromatogram of a *Pelargonium graveolens* essential oil analyzed using a 30 m × 250 μm i.d. nonpolar [5% phenyl/95% poly(dimethylsiloxane)] first-dimension column coupled to a 0.5 m × 320 μm i.d. polar [poly(ethylene glycol)] second-dimension column is shown in Figure 6.4. Thus, these early studies, combined with the continual speed improvements being made to contemporary quadrupole instruments themselves, have provided the impetus for further application of GC × GC–MS using quadrupole MS instrumentation (Adahchour et al., 2005a; Debonneville and Chaintreau, 2004; Kallio et al., 2003; Ochiai et al., 2007; Ryan et al., 2004). The key results from these investigations are expanded later in the chapter.

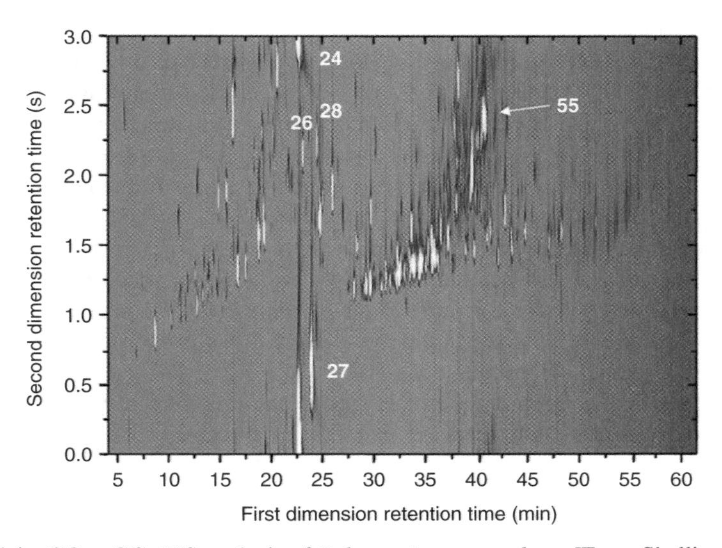

Figure 6.4 GC × GC–MS analysis of *Pelargonium graveolens*. [From Shellie and Marriott (2003a), with permission of The Royal Society of Chemistry.]

6.1.2 Time-of-Flight MS

The major advantage that time-of-flight MS (TOF MS) has over other MS technologies is that a complete spectrum is produced at the detector for every pulse of ions from the ion source. In addition, the spectral acquisition rate is determined essentially by the transit time of the heaviest ion in the mass spectrum, so very fast data acquisition is possible. Considering an ion of 800 amu with an applied acceleration potential of 2000 eV, the ion will take approximately 90 μs to travel from the source to the detector in a 2-m flight tube (Holland et al., 1983). In this example, the instrument could acquire a full spectrum from 0 to 800 amu every 90 μs, which equates to a spectral acquisition rate in excess of 11,000 spectra/s.

The first illustration of GC × GC–TOF MS confirmed the suitability of this detection system for GC × GC analysis (van Deursen et al., 2000b). Having already investigated the feasibility of the approach by generating ultrafast one-dimensional separations (van Deursen et al., 2000a), the researchers coupled a 10 m × 250 μm nonpolar column to a 0.7 m × 100 μm polar column via a 0.07 m × 100 μm thick-film transfer capillary to facilitate the operation of a thermal sweeper modulation interface. A total ion current (TIC) chromatogram from the separation of a kerosene sample from this study is shown in Figure 6.5: the total run time was 73 min, with a 7.5-s modulation period. This plot was produced using software developed in-house. A second-dimension chromatogram from the GC × GC–TOF MS separation of kerosene is also shown in Figure 6.5, illustrating the class separation of alkanes (1), mononaphthenes (2), and dinaphthenes (3, 4) from monoaromatics (5–8). Of particular interest here is the residual peak overlap in the second-dimension column, because the synergistic relationship between chromatographic and mathematical separation is highlighted. Even the best GC × GC separation will not separate the most complex sample; similarly, the best deconvolution algorithm will fail if there is no chromatographic separation (the caveat being that we are speaking of low-resolution MS). Fortunately, in this example, the carrier gas velocity in the second-dimension column was close to optimum, and the column efficiency was calculated to be approximately 4000 plates. This is significantly lower than the theoretical plate number of such a column, but it was thought that this resulted from insufficient zone focusing of the reinjected analyte plug on the second column, delivered by the thermal sweeper modulator. However, armed with highly reproducible nonskewed mass spectra across the narrow second-dimension peaks, mass spectral deconvolution reliably finds the response for these eight partially resolved peaks.

Shellie and co-workers built upon these developments and published the separation of an essential oil sample using GC × GC–TOF MS in 2001. They utilized the comparatively new technique of cryogenic modulation to generate the comprehensive separation and reported excellent applicability of the technique to their sample. Although the sensitivity was exceptional, 38 resolved components were elucidated in a lavender oil sample using TOF MS detection, as opposed to over 150 using a similar approach with a flame ionization detector (FID).

Enhancements and developments of the technique have led to the almost routine application of GC × GC–TOF MS for a suite of analysis situations. The key

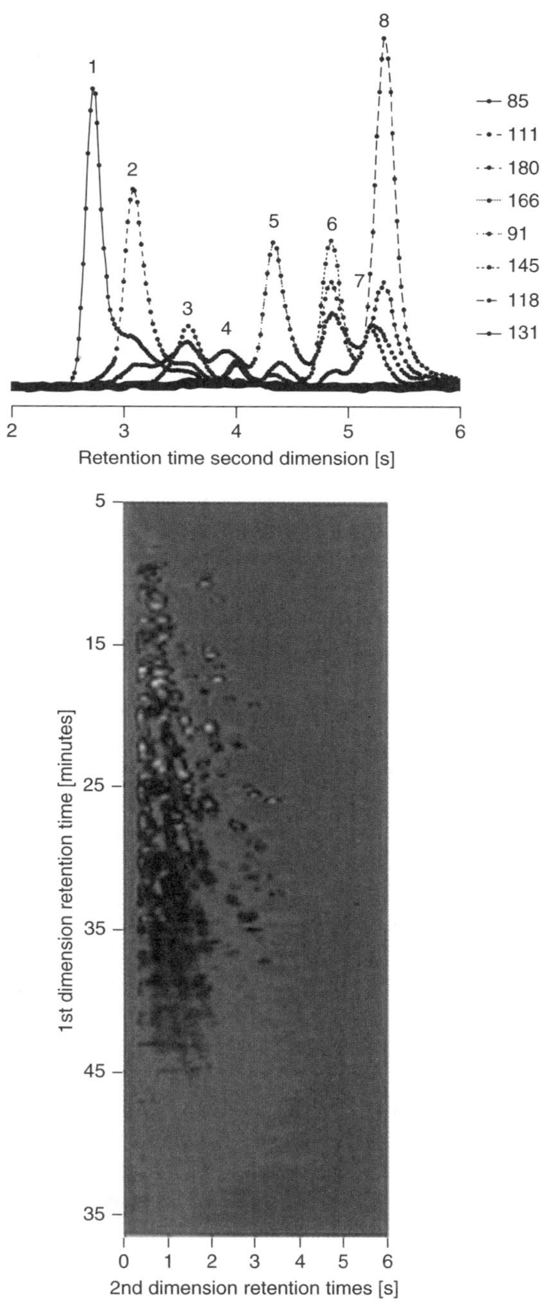

Figure 6.5 Example of a second-dimension chromatogram from the analysis of kerosene using GC × GC–MS (upper trace), with the unique ion traces of each compound shown separately. Full-scan GC × GC–TOF MS chromatogram (lower trace) of kerosene. [From van Deursen et al. (2000b), with permission. Copyright © 2000 by Wiley-VCH Verlag GmbH & Co. KGaA.]

results from investigations employing GC × GC–TOF MS are highlighted later in the chapter.

6.2 DATA PROCESSING OF GC x GC–TOF MS RESULTS

Following Shellie's successful application of GC × GC–TOF MS, Dallüge and co-workers saw an opportunity to optimize and characterize the technique and reported this a year later (Dallüge et al., 2002a). Although it is somewhat outside the scope of this chapter, Dallüge et al. covered the aspects of temperature programming, column selection, modulation temperature, time and frequency, and reported analytical performance data. A key part of their findings incorporated the difficulties imposed by the quantity of data generated from separations and how it is handled by the analyst. Early GC × GC–TOF MS studies were plagued by troubles associated with large data files. A 73-min GC × GC–TOF MS analysis was reported to have collected 2.1×10^5 mass spectra (over 100 Mb of data), where a data acquisition rate of 50 spectra/s was used (van Deursen et al., 2000b). At the time of publication, no automated data handling was available, and the screening of analysis results was laborious.

One of the setbacks encountered in peak assignment in GC × GC–TOF MS stems from the production of a two-dimensional chromatogram utilizing a single detector collecting a one-dimensional signal array. As such, the time of primary elution and secondary injection are not recorded, making correct assignment of retention time significantly more difficult than in a one-dimensional separation. Should a component be strongly retained on the second-dimension column, its secondary retention time may be greater than the modulation frequency. This effect, known as *wraparound* (see Chapter 4), can introduce ambiguity in the correct assignment of peaks. Although prediction of retention times in two-dimensional separations has been successful (Beens et al., 1998; Vendeuvre et al., 2005), Micyus and co-workers reported the noteworthy development of an algorithm to determine the absolute retention time of wrapped-around analytes in 2005. Despite not being specifically geared for GC × GC–TOF MS data, it could be described as an intermediate step toward fully automated data analysis for this technique. Using the algorithm, the retention time of a compound undergoing six wraparound cycles could be determined accurately by repeating the injection with a 14% increase in modulation time.

Hoggard et al. (2008) have investigated an automated method for applying parallel factor analysis (PARAFAC) to complete GC × GC–TOF MS chromatograms for peak assignment and resolution. The use of chemometrics techniques in comprehensive multidimensional GC is covered elsewhere, but in brief, the group could correctly identify and assign both fully and partially resolved peaks without resorting to a labor-intensive manual approach. This greatly reduces the time required for comprehensive analysis and processing of large data files and is a step toward routine application of the technique. These once-troublesome large amounts of data reflect the power of GC × GC–TOF MS to unravel previously unreachable sample information.

6.3 METHOD TRANSLATION IN GC × GC–MS

Method translation is used widely by GC and GC–MS practitioners to permit transfer of a chromatographic method between different GC systems, which may have, for example, different column length, carrier gas, or detector while maintaining selectivity for all peaks in a given temperature-programmed separation (Klee and Blumberg, 2002). Retention-time locking is an extension of the method translation approach, which leads to exact retention matching (Klee and Blumberg, 2002). These features are also desirable in GC × GC separations. However, by comparing GC × GC–MS and GC × GC–FID chromatograms of ginseng extracts, in a set of experiments that used the same nominal average linear velocity throughout the entire GC × GC column set, with the same column set moved between the MS- and FID-equipped instruments, Shellie et al. (2003) reported differences in retention times. The impediment to achieving matched retention times comes about via the dissimilar column outlet pressures of the two types of detector. Different outlet pressure is easily accounted for in one-dimensional separations, but changes in outlet pressure affect the pressure at the confluence of the first-and second-dimension columns and make it more difficult to retention-time-lock GC × GC chromatograms. This challenge is exacerbated further when dissimilar carrier gases are used to perform the GC × GC–FID and GC × GC–MS separations, and retention time locking requires careful adjustment of the pressure drop across each section of a GC × GC-coupled column set (Shellie et al., 2004a). In practice, this can be achieved by using a T-union at the end of the second-dimension column and providing electronic pressure control of both the inlet and outlet of the GC × GC column set. Under this arrangement it is possible to obtain nearly matching retention in GC × GC–FID and GC × GC–MS (Shellie et al., 2004a).

6.4 GC × MS

The use of a nonfragmentation ionization method for MS detection allows the analyst to perform two-dimensional separation using conventional GC–MS instrumentation. As mentioned previously, to use the "separation" tag to define an analytical procedure, the detection mechanism must apply a distinguishable label to resolved components from an original mixture. Using conventional hard-ionization techniques such as electron impact results in a slew of fragments for a single compound which may be conserved across the chemical class from which the compound originated. A typical example of this is the *n*-alkanes, which all show similar fragmentation patterns. The use of a soft-ionization technique only produces parent ions for each compound reaching the detector, the mass of which is not conserved between isomers or members of an analogous series or compound class. This is reflected as a comparable separation step to a nonpolar second-dimension column in a comprehensive two-dimensional system, as the higher mass analytes in a complex mixture generally have higher boiling points (Venkatramani et al., 1996).

Obviously, several critical parameters must be met to allow such a separation, most notably the amenability of the sample to soft ionization. The most common soft- and selective-ionization methods for gaseous compounds are (i) chemical ionization, where the analyte molecules are ionized by ion–molecule chemical reactions; (2) field ionization, wherein very high electric fields in close proximity of an emitter needle ionizes gas-phase molecules; and (3) photoionization, where ionization is achieved by means of absorption of ultraviolet photons. Zimmermann and co-workers recently published an interesting review on the use of photoionization mass spectrometry as a detection method for single-dimension and comprehensive gas chromatography (Zimmermann et al., 2008).

Wang and co-workers presented the first example of this technique in 2005, based on the groundwork provided by Qian's coupling of GC and soft-ionization MS in 2002 (Qian and Dechert, 2002). Wang utilized a TOF MS instrument equipped with field ionization coupled with a GC instrument for the separation of a diesel sample. Naturally, the use of a modulation device is not required in this technique, as there is only a single chromatographic separation step. By using a polar GC column in the GC × MS experiment, compounds are separated in a temperature program according to their boiling point and their specific polarity-based interaction with the stationary phase. This has a negative impact on the orthogonality of the separation but provides no adversity to the validity of the data collected.

For a GC × MS chromatogram to look like the two-dimensional GC × GC chromatogram and provide similar compound class separation information, one needs to manipulate the display. First the results are plotted according to time versus mass. The transformation is done according to the approach illustrated in Figure 6.6. In essence, the retention-time axis is transformed such that the *n*-alkanes present in the sample are forced to lie along a straight line. Following the transformation, the homologous series within the sample all align and the separation closely resembles a typical GC × GC separation obtained by analysis using an apolar–polar columns set. The conversion of GC × MS data to a three-dimensional plot allows the visualization of the compound classes in a way that is not possible with conventional GC–MS techniques, thus permitting a simpler fingerprinting method for class identification to essentially the same degree as a GC × GC separation. Wang reported that the separation of individual compound groups within classes is easier using GC × MS than GC × GC, due to the similarity in relative polarity provided by the latter technique. Furthermore, because the use of the soft-ionization technique allows the collection of parent ion masses, individual chromatograms of sulfur- or nitrogen-containing compounds can be extracted in a manner similar to the single-ion monitoring mode used in GC–MS. This feature would require the use of an element-specific detector in conventional GC × GC–MS as established by Ochiai and colleagues in 2007.

Other soft-ionization methods have been examined for GC × MS. Welthagen et al. (2007) used laser photo-ionization coupled with TOF MS to perform a separation similar to that of Wang but followed up with the obvious extension of the GC × MS approach by coupling the photoionization to a GC × GC system to

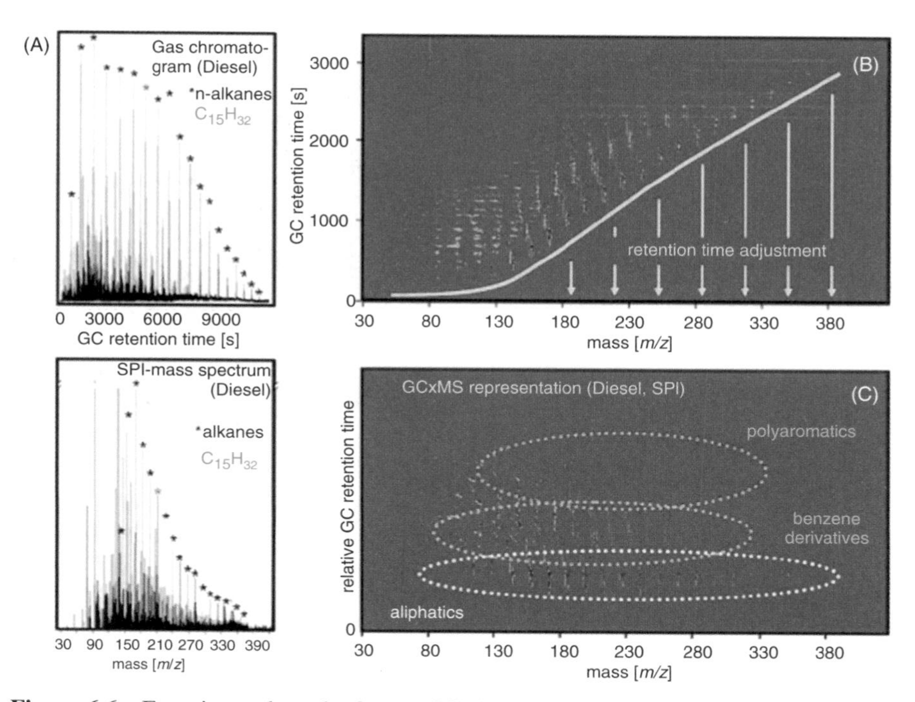

Figure 6.6 Experimental results from a GC–laser SPI–TOF MS coupling. (A) Demonstration of the separation similarity of a gas chromatogram using a nonpolar column (TIC of diesel GC–MS chromatogram with EI ionization, top) and a soft ionization mass spectrum (SPI mass spectrum of diesel obtained by summing up all mass spectra of a GC–SPI–MS run, bottom): the homolog row of the alkenes is indicated by asterisks, respectively. (B) Two-dimensional retention time/molecular mass representation of a GC–SPI–TOF MS run. The course of the *n*-alkanes is indicated by the continuous line. By transformation of the representation according to the arrows indicated, one can obtain the GC × GC plot shown in (C). (C) Comprehensive two-dimensional GC × MS representation generated from (B) by "linearization" of the *n*-alkane row, exhibiting "separation" properties similar to those of classical GC × GC. [From Zimmermann et al. (2008), with permission. Copyright © 2008 by Elsevier.]

give a true three-dimensional GC × GC × MS separation. Figure 6.7 provides a schematic of the instrument setup for GC × GC × MS, with a typical result from the separation of a diesel sample. Three largely orthogonal separations can be achieved by combining a carbowax first-dimension column for separation according to "polarity," a 50% phenyl/50% methyl polysiloxane second-dimension column for separation according to "polarizability," and soft-ionization MS mimicking a "volatility" separation (Welthagen et al., 2007). The primary limitation observed with the technique was the frequency of the ionization laser, which limited spectral acquisition to 5 Hz and was not sufficient to generate five data points per peak for the early-eluting compounds. As such, optimization of the technique was hampered and the chromatographic steps were operated outside

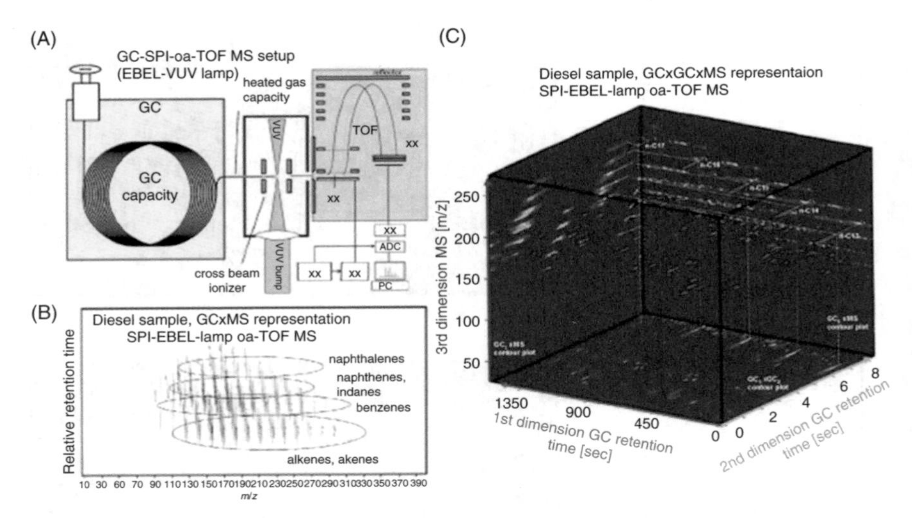

Figure 6.7 (A) Experimental setup of the gas chromatography SPI–oa–TOF MS instrument. SPI is performed in a separate ionization chamber with 126-nm photons from an Ar-filled EBEL VUV lamp. (B) Comprehensive two-dimensional GC × MS representation obtained from a gas chromatographic analysis of a diesel sample with an EBEL VUV lamp for SPI. (C) If GC × GC is combined with soft SPI–TOF MS, a three-dimensional comprehensive separation can be realised with the separation axes first and second retention time and molecular mass (GC × GC × SPI–MS). The figure depicts a section of a GC × GC × SPI–MS separation of a diesel sample using the EBEL VUV lamp technology (Ar, 126 nm) for SPI. The position of the *n*-alkanes is indicated. [From Zimmermann et al. (2008), with permission. Copyright © 2008 by Elsevier.]

the most desirable parameters, but the feasibility of such an application was implemented successfully. It was proposed that the use of a vacuum ultraviolet photoionization method would enable continuous data acquisition and eliminate the problems encountered in the published study. Although the publication of further applications of this technique have been limited compared to the abundance of traditional comprehensive GC techniques, research is continuing, especially in the area of novel applications for the technology (Hejazi et al., 2009).

6.5 CONVENTIONAL AND ALTERNATIVE MODULATION TECHNIQUES FOR GC × GC–MS

Traditionally, the use of mass spectrometric detection has dictated the coupling of chromatographic separation dimensions with a thermal modulator. As described in Chapter 4, thermal modulation interfaces can be broken down into cryogenic or heating techniques. In brief, the former rely on the application of a cryogen to rapidly cool first-dimension column eluent below that of the GC oven, thus retarding its thoroughfare in a transfer line between the analytical dimensions.

Reheating the transfer line by cessation of cryogen flow or movement of the modulator causes rapid remobilization and zone focusing of trapped analytes onto the second-dimension column. Alternatively, the use of a heated sweeping device may be used to rapidly advance the progress of packets of analytes, slowed by interaction with a thick-film stationary phase applied to a transfer line between the two dimensions. This technique predates the use of cryogenic modulation. A good comparison of the two techniques was presented by Marriott and co-workers in 2000.

Over the last few years, increasing interest has been lavished on the development of low-cost alternatives for GC × GC modulation interfaces. Development of heated modulators often yielded results with insufficient robustness. Although cryogenic modulation is a robust technique, the cost of consumables may be inappropriate for some applications. The use of differential flow and valve-based modulators (as described in Chapter 5) has provided an avenue for research. Although John Seeley could arguably be denoted the father of differential flow modulation based on his seminal publication in 2000 (Seeley et al., 2000), it took another three years for Sinha and co-workers to present the first application of a valve-based modulator for GC × GC–MS, using novel valve placement allowing the working temperature range of the instrument to be increased to 250°C (Sinha et al., 2003). It was necessary to utilize a mere 10% of the primary column effluent for second-dimension analysis, which naturally had an adverse effect on the sensitivity available. Detection limits were still acceptable, however. Diehl and Di Sanzo (2005) used a similar open-split arrangement in their valve-based modulator. The use of such an arrangement came as a result of the relatively high flow rates necessary for the appropriate operation of the modulation interface, which is generally outside the optimum range for most mass spectrometers.

Another low-cost GC × GC alternative to thermal modulation—pulsed-flow modulation—has received an appreciable amount of interest in the past few years. The operation of pulsed-flow modulation GC × GC is described in Chapter 5, and the reader should refer to this chapter for clarification of its operational principles. As with valve-based modulation, the direct coupling of pulsed-flow GC × GC with MS is problematic, due to the necessary use of very high second-dimension column carrier gas flow rates, which are generally on the order of 10 to 20 mL/min (Seeley et al., 2006). In 2008, Shellie described GC × GC–MS results that were acquired by direct coupling of a pulsed-flow GC × GC modulation system with a quadrupole MS of high pumping capacity (Shellie, 2008). The author was forced to use an extremely high flow rate to produce an adequate flow compression ratio (ca. 10 mL/min He), which was well outside the optimum parameters for the separation. This was duly noted, and the applicability of the published technique for complex samples was critically examined.

To circumvent the low carrier flux dictated by conventional mass spectrometers, Amirav and co-workers examined the performance capabilities of a unique kind of GC × GC–MS, based on direct coupling with a supersonic molecular beam (SMB) interface (Kochman et al., 2006). Unlike most GC–MS interfaces, SMB–MS requires a high helium flow (typically 90 mL/min) for proper operation

(Amirav et al., 2008). Under this general interface configuration, the coupling of GC × GC with tandem mass spectrometry was later introduced with practical demonstration of the analysis of pesticides in vegetable matrices (Poliak et al., 2008). Although the application of this technique was a resounding success, it was duly noted that it was primarily a target method which is not suitable for all samples, and the technique was incompatible with library detection. In Figure 6.8

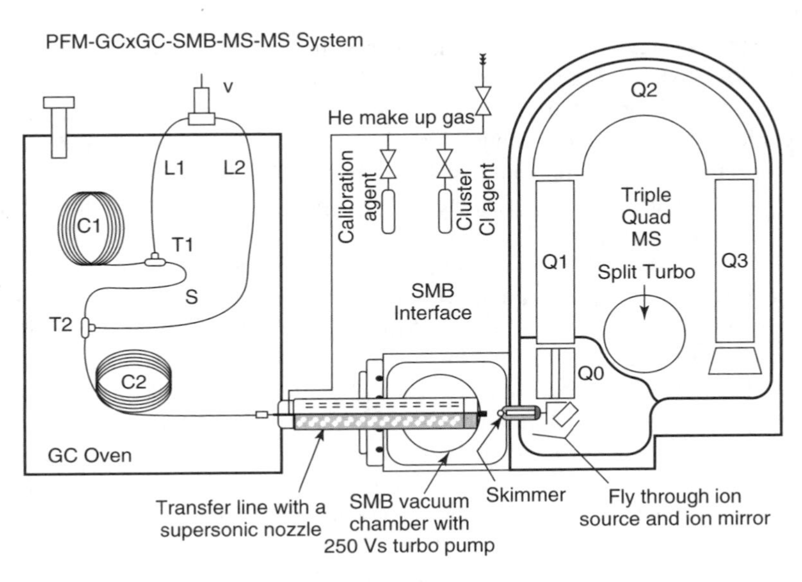

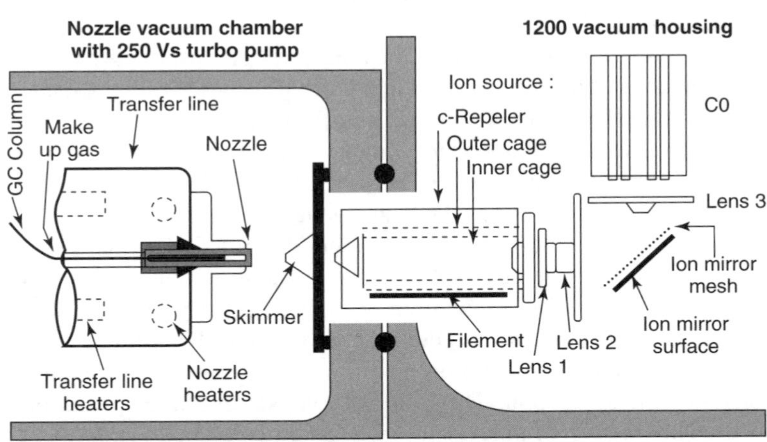

Figure 6.8 Pulsed-flow modulation setup for GC × GC–MS (and GC × GC–MS–MS) with supersonic molecular beams inside a Varian 3800 GC. The bottom schematic diagram shows the SMB interface and its fly-through ion source at the entrance to the 1200 triple-quadrupole MS system. [From Poliak et al. (2008), with permission. Copyright © 2008 by Elsevier.]

a pulsed-flow GC × GC modulator is shown inside the oven of a Varian 3800 GC, which is hyphenated with a Varian 1200-I triple quadrupole-based MS with SMB.

To further extend the development of differential flow-based modulation for GC × GC–MS, it seems prudent to explore the minimization of separation column dimensions. This apparently unexplored option would yield a concomitant reduction in the total carrier gas flow rates in both separation dimensions while retaining a sufficiently high flow ratio between them. Evidently, this approach is not as easy as it sounds, but if the operator is mindful of the constraints imposed by the operation method, this approach could deliver a system capable of direct coupling with conventional quadrupole or TOF mass spectrometers.

6.6 GC x GC–MS APPLICATIONS

It is a common habit of many GC × GC authors to describe their 2D chromatograms as being "characterized by thousands of peaks." If such a challenging analytical situation occurs, it is unthinkable to rely only on the aforementioned characteristics of GC × GC for reliable peak assignment. Hence, it is unquestionable that mass spectrometry is an obliged choice if structural elucidation at the molecular level is desired. The combination of a third MS analytical dimension to a GC × GC instrument generates the most powerful analytical tool available today for the elucidation of complex mixtures of volatiles and semivolatiles (Mondello et al., 2008).

A wide series of GC × GC–MS experiments reported in the last 11 years is listed in the reference section. To the best of our knowledge, all the GC × GC–MS papers from 1999 to present (June 2010) are reported at the end of the chapter. Review papers containing GC × GC–MS data are also reported. The only review focused specifically on the subject of the present chapter was published in 2008 (Mondello et al., 2008). We apologize in advance if any contributions are missing. In the 1999–2002 period, only nine works were published, whereas in the 2003–2005 period, over 60 were reported, with an almost equal interyear distribution. Instead of publishing a long table with brief details of every GC × GC–MS experiment, extensive descriptions of some of the most significant contributions to this field are provided herein. Characterizing analytical trends and developments related to the evolution of the technique over the 1999–2010 time range will be linked to specific periods. The treatment of GC × GC–MS data using data interpretation methodologies is discussed extensively in Chapter 12. An entire book would be necessary to report essential information related to all published GC × GC–MS work, and thus it is obvious that many interesting experiments cannot be described here. A graph of the GC × GC–MS applications (grouped on the basis of MS type), over the 1999–2010 (June) time range is shown in Figure 6.9.

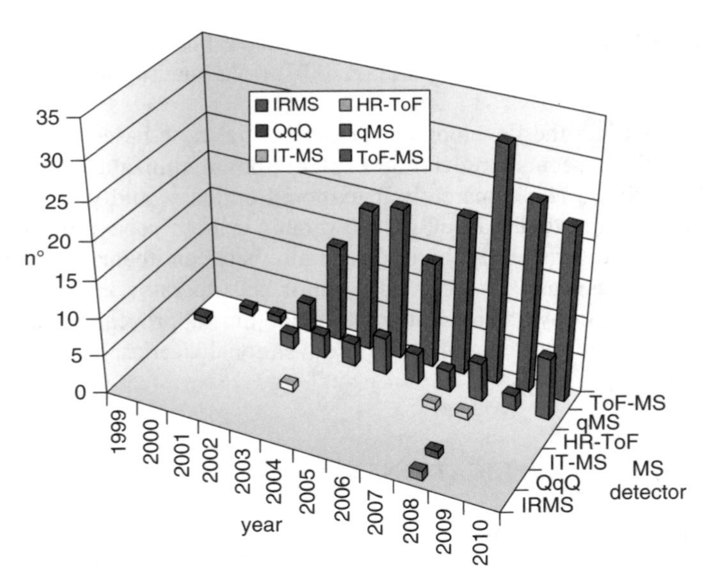

Figure 6.9 GC × GC–MS applications, grouped on a year and MS-type basis, throughout the 1999–2010 (June) period.

As mentioned previously, the first GC × GC–MS attempt was reported in 1999 (Frysinger and Gaines, 1999), but the quadrupole MS system used was not suitable for "normal" GC × GC experimental conditions. It is noteworthy that the authors stated, with hindsight, that the best solution would be the use of a TOF MS instrument. The first GC × GC–TOF MS experiments came soon after, in 2000 (Dimandja et al., 2000; van Deursen et al., 2000b). In these three-dimensional experiments, there was no compromise between the GC × GC and MS operational conditions. Ten years have gone by since those primordial GC × GC–MS applications, and there has been quite a considerable degree of research and progress in this fascinating separation science field.

6.6.1 GC x GC–MS Early Years: 2001–2002

During 2001 and 2002, further GC × GC–TOF MS works appeared (Adahchour et al., 2002; Dallüge et al., 2002a–c; Shellie et al., 2001).

In particular, Dallüge et al. (2002b) described the GC × GC–TOF MS analysis of a very complex sample (i.e., mainstream cigarette smoke), emphasizing both the problems encountered during partially automated data processing and the necessity of dedicated software. In the study, the authors used GC–TOF MS software to derive information from the raw GC × GC–TOF MS chromatograms.

Smoke volatiles were first collected in a sample tube containing three sorbents, then thermally desorbed and reconcentrated onto a cold trap. The sample was launched onto the first nonpolar dimension by heating the trap rapidly to 250°C. A second polar dimension collected bands from the primary column every 6 s and (ideally) directed entirely resolved smoke constituents to the ion source of a Pegasus II TOF MS (LECO) which generated 100 spectra/s. The use of a thick film (1 μm) caused rather high first-dimension elution temperatures, limiting the application range to the more volatile constituents, because of the low upper temperature limit of the second-dimension poly(ethylene glycol) phase. In general, whenever such a phase is employed in single-oven GC × GC experiments, the volatile elution range is always restricted. With regard to the GC × GC–TOF MS operational parameters, these were set using the conventional GC–TOF MS software. The latter, as is obvious, did not foresee the employment of a different i.d., twin-column combination and hence could not calculate the correct gas flow.

One can appreciate the high degree of complexity of cigarette smoke by observing the baseline-corrected TIC GC × GC–TOF MS chromatogram (C_7 to C_{12} elution range) illustrated in Figure 6.10. The entire chromatogram is shown only in the inset, because of the presence of a series of highly overloaded compounds. A few points can be made after a quick observation of the chromatogram. First, many unresolved volatiles (saturated and unsaturated hydrocarbons), located mainly in the nonpolar zone of the bidimensional chromatogram (between 0 to 1 s), are clearly evident. Second, peaks are spread rather homogeneously across the chromatogram, suggesting that the operational conditions have been nicely tuned. Most certainly, many polar constituents undergo wraparound, which is quite acceptable. In general, wraparound can be considered more an advantage than a disadvantage, because it can enable a more extensive occupation of the 2D chromatography space (i.e., the dead-time space can be exploited).

From Figure 6.10 it is clear that the separation power of GC × GC fails to match the overwhelming number of analytes contained in cigarette smoke and that the MS deconvolution used by the authors was indeed necessary. In fact, the software deconvolution algorithm generated decent-quality mass spectra for many unresolved compounds, as shown in Figure 6.11. An expansion of the chromatogram illustrated in Figure 6.10, with a vertical line to indicate a single second-dimension analysis, is shown in (A); the related raw chromatogram, illustrated in (B), shows an overlapping cluster, composed mainly of alkanes, alkenes, dienes, trienes, and benzene; in this 2-s second-dimension elution zone, 18 peaks, indicated by horizontal lines in (B), were unraveled by the GC–MS software. Nine compounds were tentatively identified after setting a minimum spectral similarity value. As an example of mass spectral quality, the deconvoluted spectrum of 2-methyl-1,5-hexadiene and its best library match are illustrated in (C) and (D), respectively.

The enormous amount of data generated in comprehensive GC–TOF MS experiments and the consequential data-processing difficulties immediately became evident in the experiment. At the end of the analysis, bidimensional chromatograms were generated by using two external programs. Further data

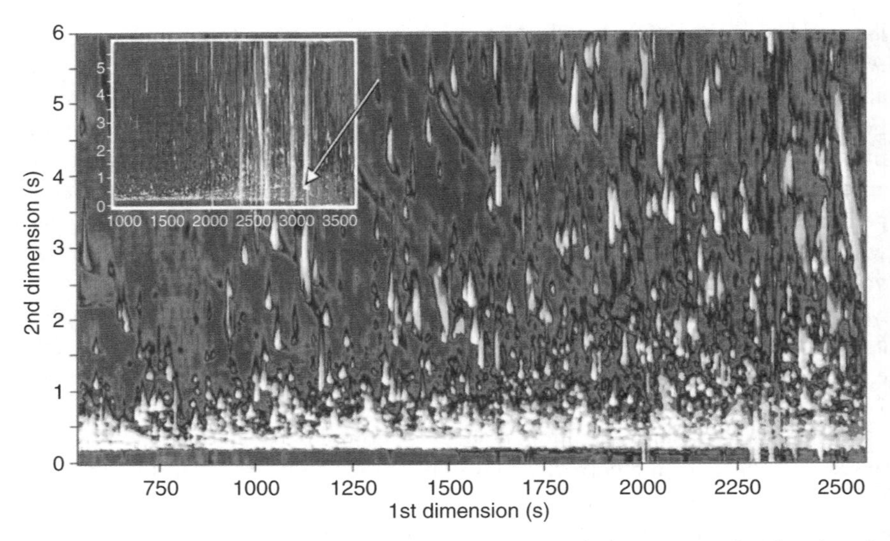

Figure 6.10 TIC GC × GC–TOF MS chromatogram of cigarette smoke showing the first-dimension range of 500 to 2600 s. [From Dallüge et al. (2002b), with permission. Copyright © 2002 by Elsevier.]

processing was carried out using the GC–TOF MS software; all signals in the raw GC × GC chromatogram, with a signal/noise ratio (S/N) >30, were found and then mass spectra for all peaks were derived by means of deconvolution. The latter process was carried out on 500-s chromatogram segments, because the maximum number of compounds that could be included in the software peak table was "only" 9999! An MS library search was then performed, and the results were combined in a single table. The three procedures described—peak finding, deconvolution, and library matching—were carried out in an automated fashion, and required 7 h. Although the final table contained 30,000 peaks, the number of different component names was much less, 7500, because peak modulation produces a series of sequential pulses that contain the same analyte. Each component defined by a compound name, CAS number, similarity [this factor expresses the similarity (range 0 to 999) between the experimental and library spectra, considering all masses], reverse [this factor expresses the similarity (range 0 to 999) between the experimental and library spectra, considering only the library masses], and probability [this factor expresses the uniqueness (range 0 to 9999) of a spectrum, compared with all other library spectra].

The peak table was exported to a spreadsheet program for further processing that required manual interaction. To filter out mass spectra with low match factors, the authors chose minimum values of 800 and 900, for similarity and reverse, respectively. With the additional support of literature-derived linear retention index (I) values, only 152 compounds from a group of 2500 volatiles with good

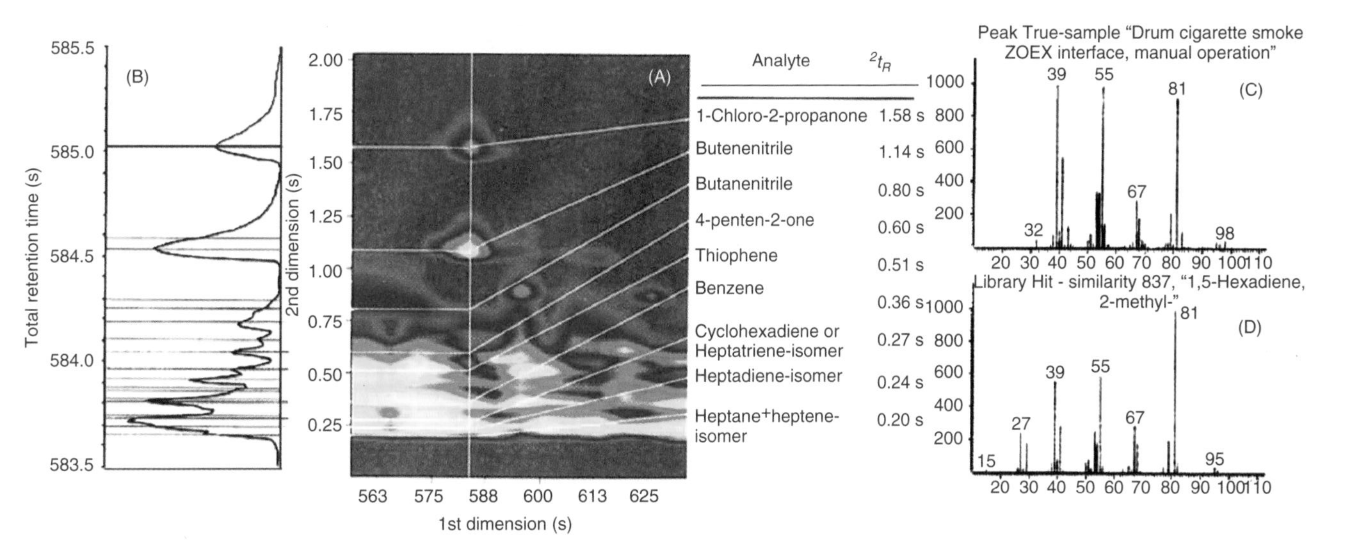

Figure 6.11 (A) Detail of the GC × GC chromatogram of Figure 6.10. The vertical line at 583 s indicates the second-dimension chromatogram, which is shown separately in (B). In (B) the horizontal lines indicate the positions where peaks were found by the deconvolution algorithm of the GC–TOF MS software. Provisional identifications are summarized on the right-hand side of the figure. (C) Deconvoluted mass spectra of the peak at 0.24 s. (D) Corresponding library spectrum. [From Dallüge et al. (2002b), with permission. Copyright © 2002 by Elsevier.]

and acceptable match factors were identified (reference I values were found for only 238 compounds).

In a further GC × GC–TOF MS investigation, carried out by the same research group (Dallüge et al., 2002a), apart from software issues, various aspects related to method optimization (column characteristics, modulation period and temperature, gas velocities, temperature program) and validation (precision, sensitivity, linearity) were discussed. A nonpolar 15 m × 0.25 mm i.d. and a polar 0.8 m × 0.10 mm i.d. column were employed in the first and second dimensions, respectively. A primary column He velocity of 30 cm/s was applied by measuring the dead time of the column set; such a gas velocity can be considered as ideal. With regards to the secondary column He velocity, a value of 230 cm/s was calculated on the basis of a 1.3-mL/min flow; such a gas velocity is certainly far from the optimum value, which is about 100 cm/s. The authors studied the influence of flow on the GC × GC separation: a pesticide mixture was analyzed twice using flow conditions of 1.3 and 3.7 mL/min (ca. 50 cm/s in the first dimension). The higher primary column gas velocity led to reduced retention times, to lower elution temperatures, and to more intense analyte/stationary phase interactions in the second dimension. The two competing effects, the lower elution temperatures and the higher linear velocity, counterbalanced one another; thus, no substiantial resolution differences were observed in the two applications. The same sort of outcome would have been in attained if a lower flow was used (e.g., 0.8 mL/min): in this case, the lower primary gas velocity would have led to increased retention times, to higher elution temperatures, and to less intense analyte/stationary phase interactions in the second dimension. Consequently, hardly any benefits would be derived from a closer-to-ideal linear velocity in the second dimension. Such a dead-end situation is easily circumvented if a double-oven configuration is employed.

Considering the modulation process that was achieved with an LMCS device, the authors emphasized that the separation achieved on the first column should not be destroyed by the modulation process. A good example is illustrated in Figure 6.12, where two pesticides, metalaxyl and prometryn, are partially resolved

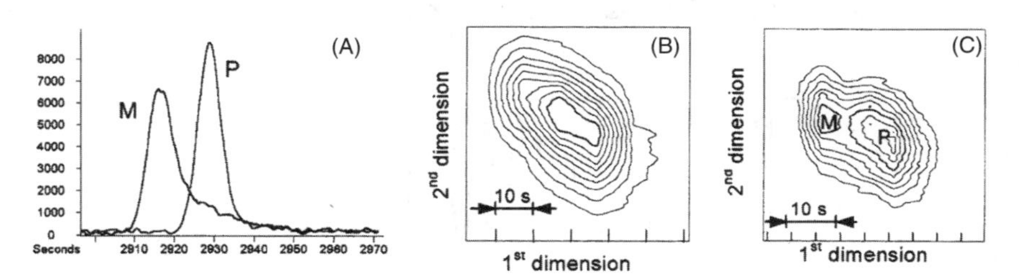

Figure 6.12 Influence of modulation time on the separation of metalaxyl (M) and prometryn (P). (A) First-dimension chromatogram with partial separation; (B) GC × GC contour plot, 9.9 s modulation time, loss of first-dimension resolution; (C) GC × GC contour plot, 5 s modulation time, first-dimension separation maintained. [From Dallüge et al. (2002a), with permission. Copyright © 2002 by Wiley-VCH Verlag GmbH & Co. KGaA.]

in the first dimension, with peak widths of about 20 s (A); using a modulation period of 9.9 s, each peak was modulated three times. As can be seen in (B), the separation achieved in the first dimension was lost because the two analytes were remixed in the modulator and there was no degree of separation in the second dimension. When the modulation period was reduced to 5 s, leading to four or five modulations per peak, the first-dimension separation was preserved (C). It must be added that four modulations would probably have been sufficient. In general, excessive sampling of a peak is not useful because modulation periods become short, thus reducing the overall separation space; furthermore, sensitivity decreases as the number of modulations tends to increase. An advantage of a short modulation period is the less common occurrence of secondary column overloading.

Limits of detection (LODs), calculated at the $S/N = 3$ level, were in the range 5 to 23 pg for six pesticides. Sensitivity was improved considerably (by a factor of 5 to 7) by using the highest multichannel plate voltage (i.e., 2000 V). However, as affirmed by the authors, such an approach reduces the detector lifespan. Figure 6.13 illustrates the influence of voltage on the S/N value for a single pesticide (fenitrothion). LODs were compared between GC × GC and single-column GC experiments for a series of pure standard pesticides; comprehensive GC–TOF MS LODs were two to five times lower than those observed in GC–TOF MS. The authors justified such a limited sensitivity increase by the fact that peaks were modulated two or three times rather than once.

In these initial works it became clear immediately that powerful PCs and fully integrated software were necessary for instrumental control, to locate and identify peaks in automatically generated contour plots and, in general, to process the enormous amount of data generated by GC × GC–TOF MS experiments. Other GC × GC–MS experiments, with a quadrupole mass spectrometer (qMS), were described three years after the first paper (Frysinger et al., 2002; Shellie and Marriott, 2002). Frysinger et al. (2002) used the same qMS instrument as that described in their previous GC × GC–MS experiment (Frysinger and Gaines,

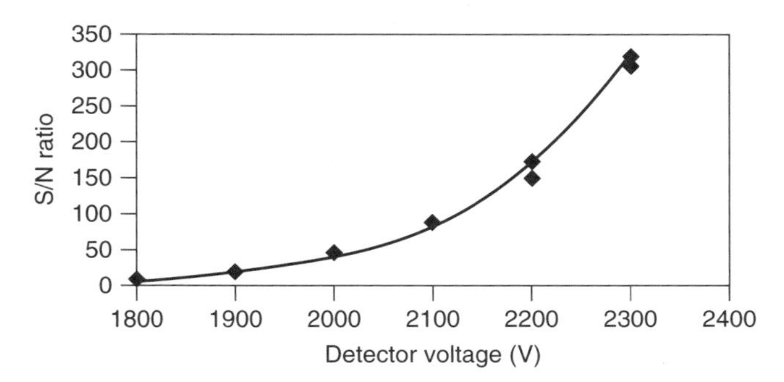

Figure 6.13 Influence of voltage on the signal-to-noise ratio for fenitrothion. [From Dallüge et al. (2002a), with permission. Copyright © 2002 by Wiley-VCH Verlag GmbH & Co. KGaA.]

1999) in a diesel fuel application and hence encountered the same type of analytical problems. Shellie and Marriott used a microbore first-dimension apolar column (10 m × 0.1 mm i.d.) and a second-dimension wider-bore chiral column (1 m × 0.25 mm i.d.), to achieve the first-ever boiling-point GC × enantio-GC separation (Shellie and Marriott, 2002). The main obstacle, and novelty, of the study consisted of the separation of enantiomer pairs on a short secondary column. This rather difficult task was achieved through proper selection of column i.d.s, to exploit the low-pressure outlet conditions.

In the field of GC, it is well known that higher optimum gas velocity values can be attained by increasing the diffusion coefficient of the solute in the mobile phase (D_{AB}). Apart from using a "lighter" gas, an increase in D_{AB}(A = analyte; B = gas) can be attained through a reduction of the intracolumn pressure; the second option can be obtained easily by locating the outlet of a GC column in the ion source of an MS instrument. Subatmospheric pressure conditions can extend across the entire column length if a short wide-bore capillary is used (requisite I), while if a microbore column is employed, the advantages of low pressure are exploited only in a short-end segment. Generally, a restrictor is connected to the head of the wide-bore (requisite II) column to prevent an excessively low pressure at the injection point and to reduce the gas flow (which can exceed the MS pumping capacity). Additional benefits are represented by the high sample capacity and by the generation of greater peak widths; hence, very high mass-spectral acquisition rates are not required, suiting the low acquisition rates of qMS systems. By using the aforementioned column combination, requisites I and II were met. The primary microbore column was exploited both for separation and as a restrictor; the secondary wide-bore column was suited for low-pressure rapid chiral separations. GC × enantio-GC–qMS applications were first carried out on a mixture of standards and then on a bergamot essential oil. Figure 6.14 illustrates a nice separation achieved on a test mixture containing chiral compounds: although wraparound is rather evident, this suited the slow acquisition rate of the qMS (8.33 Hz) because the widths of bands along the *y*-axis were rather wide. Moreover, the authors' main objectives were fully reached: namely, the achievement of rapid second-dimension enantiomer separations.

6.6.2 GC × GC–MS Landmark Year: 2003

The year 2003, during which 17 GC × GC papers were published [13 (ca. 76%) were focused on the use of TOF MS], can be considered as a landmark in the GC × GC–MS field. One of the main reasons relates to the introduction of a twin-oven comprehensive GC–TOF MS instrument (Pegasus 4D, LECO) with a cryogenic two-stage quad-jet modulator (Dimandja, 2003). The system was equipped with fully integrated software (ChromaTOF) for instrument control and entirely automated data processing. During the two previous years, GC–TOF MS instrumental software had proven to be inadequate in dealing with GC × GC–TOF MS data.

A GC × GC–TOF MS experiment was carried out on a sample of diesel using an apolar–polar column setup. Different visualization options (2D, 3D,

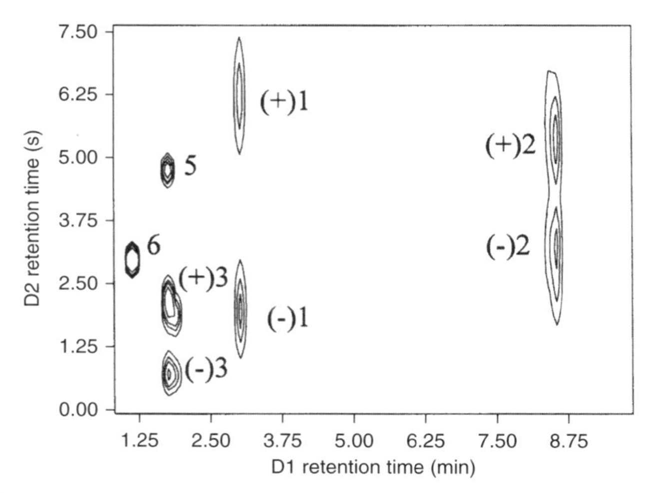

Figure 6.14 Two-dimensional chromatogram for the GC × enantio-GC–qMS analysis of a standard mixture. Identity of compounds: 1, linalool; 2, linalyl acetate; 3, limonene; 5, 1,8-cineole; 6, (±)-α-pinene. [From Shellie and Marriott (2002), with permission. Copyright © 2002 by The American Chemical Society.]

reconstructed 1D chromatogram) could be employed during the GC run time. Moreover, as with standard GC–MS programs, a specific ion or a selection of different ions could be monitored during the analysis. GC × GC–MS quantitative profiling was possible only at the end of the application. The TOF MS produced 100 spectra/s; the experiment lasted 52 min, thus leading to the generation of 312,000 spectra per/analysis. It was reported that the file sizes for each experiment were 331.4 MB. In general, acquisition rates should be chosen carefully, because excessively high rates lead to the production of unnecessarily large data files and to a decrease in sensitivity. For most GC × GC–MS applications, a 50-Hz sampling frequency provides a satisfactory result.

An additional interesting GC × GC–TOF MS method was developed and used for the characterization of semivolatile organic compounds (SVOCs) in German airborne particulate matter with a diameter of up to 2.5 μm (Welthagen et al., 2003). As demonstrated by many epidemiological studies, ambient particles have a highly negative impact on human health, caused by physical and chemical properties of the substances inhaled. The particle samples were extracted from the city air by using quartz fiber filters on a daily basis. The volatiles were desorbed from the filters inside the GC injector (direct thermal desorption). Initially, the SVOCs were analyzed through GC–TOF MS, the chromatogram was characterized by a typical kerosene-style hump of unresolved carbonaceous matter (UCM) that contained a large amount of SVOCs. The authors estimated that only 15% of the latter were quantified, and affirmed that probably many toxic compounds were contained in the UCM band.

GC × GC–TOF MS was exploited to increase knowledge on the consitituents and chemical classes present in aerosol samples. An additional objective was the reduction of the amount of data attained to a suitable dimension for epidemiological studies. The employment of a second polar dimension was successful in distributing SVOCs along the y-axis on the basis of increasing polarity; the ChromaTOF software found about 15,000 peaks. Chromatographic peak data (first- and second-dimension elution times, peak areas, MS peak list) were exported and were subjected to analysis with external programs (e.g., EXCEL 2000 software). The 2D chromatograms were reconstructed as bubble plots, where each chromatographic peak was visualized as a circle whose position and diameter were related to retention data and areas, respectively. The authors classified the huge number of resolved peaks and set selection rules based on second-dimension analyte retention data (group-type patterns) and well-known MS fragmentation information. Seven groups of components were defined in relation to a specific y-axis retention-time range [e.g., alkanes were in the range 1 to 1.5 s, and polar benzenes (with or without alkyl groups) presented elution times of over 2s] and to mass fragmentation criteria (e.g., alkanes present an m/z 57 or 71 base peak, with the second largest peak at m/z 71 or 57, whereas polar benzenes present peaks at m/z 77 above 25% relative intensity). Each group was defined by a color in the bubble plot to facilitate data interpretation. The approach was of interest, although the employment of additional software was again necessary. The fact that GC × GC–TOF MS methods were limited by the lack of a universal data analysis tool was highlighted by the authors in the concluding comments.

In the field of food analysis, an effective analytical method must not only enable the qualitative and/quantitative determination of major food constituents but also be sufficiently sensitive to allow the analysis of trace-amount contaminants. Multiresidue GC methods in food analysis are usually based on the use of selective detectors, such as the electron capture and/or nitrogen–phosphorus detector. The results attained are then generally confirmed by using GC–MS in the selected ion monitoring (SIM) mode. However, the GC–MS analysis of pesticides can be very challenging because the target analytes are usually smothered by large amounts of coextracted matrix components. Zrostlíková et al. (2003) evaluated the employment of GC × GC–TOF MS in the analysis of fruit pesticides. The main objective of the authors was to reliably identify pesticides at levels equal or lower than 10 ppb (μg/kg), which is the maximum concentration allowed for pesticide residues in baby foods. A peach sample was spiked with 20 pesticides; in almost all cases, the limits of detection ($S/N = 5$) were well below 10 ppb, while a certain variability was observed when sensitivity was compared to a GC–TOF MS experiment. Strangely, very little or no sensitivity enhancement at all was observed for some contaminants.

Chapter 5 is devoted to a very interesting topic: pneumatically modulated GC × GC. The first description of a valve-based GC × GC system, combined with a TOF MS instrument (Figure 6.15), appeared in 2003 (Sinha et al., 2003): A low-polarity 60 m × 0.25 i.d. first dimension was connected to a six-port micro-diaphragm valve; the latter was bridged by a 5-μL sample loop and was positioned on top of the GC, with its wetted portions inside the oven and the temperature-sensitive O-rings outside. The second dimension, a high-polarity

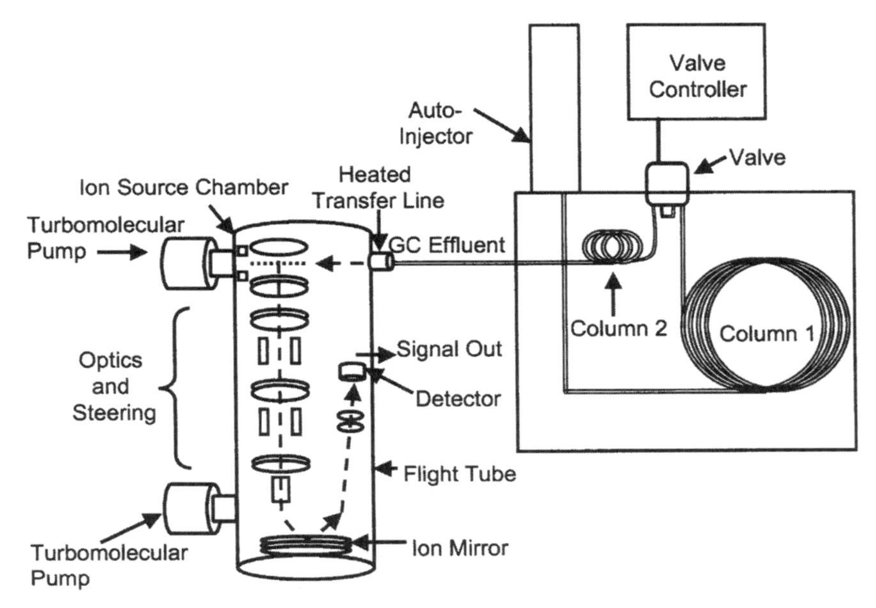

Figure 6.15 Valve-based GC × GC–TOF MS instrument schematic. [From Sinha et al. (2003), with permission. Copyright © 2003 by Elsevier.]

3 m × 0.18 i.d. capillary, bridged the valve and the TOF MS. An He flow of 1 mL/min (ca. 17 μL/s) was used in column 1, while column 2 was operated using an auxiliary gas source with a constant pressure of 138 kPa. On the basis of simple flow calculations, it can be derived that the initial flow (40°C) in the second dimension corresponded to nearly 7 mL/min (ca. 300 cm/s) and that the end flow (230°C) corresponded to about 3 mL/min (ca. 210 cm/s). Attention was paid by the authors not to exceed the maximum flow allowed for the TOF MS, 10 mL/min. The valve was operated in the accumulation mode for 0.25 s (an effluent volume of ca. 4 μL) and in the reinjection mode for 2.25 s; during the reinjection period, analytes were directed to waste. Under the proposed experimental conditions, it was estimated that only 10% of the sample was analyzed in the second dimension. However, the experiment described was a GC × GC experiment because each peak was sampled every 2.5 s, causing no loss of separation information attained in the first dimension.

Throughout the GC × GC history, quadrupole and time-of-flight systems have dominated the mass spectrometry scene almost entirely. As will be seen, though, other MS instruments have been employed every now and then. For example, Wahl et al. (2003) tried to combine GC × GC with an ion-trap (IT) MS system using a portable multidimensional GC system capable of either heart-cutting or comprehensive GC analyses. A nonpolar 12 m × 0.32 i.d. first dimension was linked, via a 10-port switching valve, to two equal polar secondary columns (6 m × 0.25 mm i.d.), which analyzed fractions from the primary capillary in an alternate mode. Temperature programming was applied in both dimensions.

The second-dimension bandwidths were necessarily wide (a value of 6.5 s was reported) because the MS acquisition speed was very low (1 spectra per/1.5 s); the modulation period corresponded to 120 s. On the basis of the information provided, it does appear that each peak was sampled an insufficient number of times to justify the term *comprehensive*, and that the experiment described was somewhere between heart-cutting MDGC and GC × GC.

During the 1999–2002 period, it became rather clear that the most suitable MS instrument for GC × GC analysis was the TOF mass spectrometer; a negative aspect, also considered widely, was the high cost of such instrumentation. Quadrupole MS systems are less expensive but are characterized by much lower spectra production rates, because the mass analyzer scans individual ion groups on a m/z basis. It is obvious that in almost all academic and industrial organizations, benefit/cost ratios must be maintained at the highest possible levels. Consequently, the first appearances of the use of qMS systems, operated under relatively fast-scanning conditions, in a series of GC × GC experiments, were met with great interest. Three GC × GC papers, focused on the use of qMS, were published during 2003 (Kallio et al., 2003; Shellie and Marriott, 2003a; Shellie et al., 2003). In particular, Shellie et al. (2003) applied GC × GC–qMS to the analysis of ginseng volatiles. The use of a rapid scanning system (5973 mass selective detector, Agilent) enabled the application of common GC × GC operating conditions: a 30 m × 0.25 mm i.d. low-polarity column was used for the first-dimension separation, while a 1.2 m × 0.10 mm i.d. polar column was used to resolve analytes, prior to their release into the MS. The head pressure (He) applied was 297 kPa, corresponding to initial linear velocities of about 20 and 215 cm/s, in the first and second dimensions, respectively. Mass spectra were acquired at 20 Hz, in a reduced range of 41 to 228.5 m/z, to enable the detection of molecular ions up into the oxygenated sesquiterpene region. The authors reported that reliable quantitative data could not be derived with only three or four data points per peak, which was sufficient, on the other hand, for identification. Although not observed in the experiment, insufficient peak sampling can also cause inconsistent second-dimension retention times for pulsed peaks that belong to the same compound. Mass spectral skewing was also investigated, with only slight spectral variations observed across a single peak. The main restrictions of the method employed, emphasized by the authors, were related to the impossibility to derive quantitative data and to the nonidentification of higher-molecular-weight compounds due to the limited mass range.

6.6.3 GC × GC–MS Foothold Years: 2004–2006

During the 2004–2006 period, GC × GC–MS methodologies began to gain gradual favor in both academic and industrial fields; moreover, the contrast between TOF and rapid-scanning quadrupole MS instrumentation continued (especially in the 2004–2005 period). During 2004, 20 GC × GC–MS papers were published; in that year, Ryan et al. used a rapid-scanning qMS (MS-QP2010, Shimadzu) and a TOF MS (Pegasus III) system in LMCS GC × GC experiments for the analysis of coffee volatiles isolated by using solid-phase microextraction (SPME).

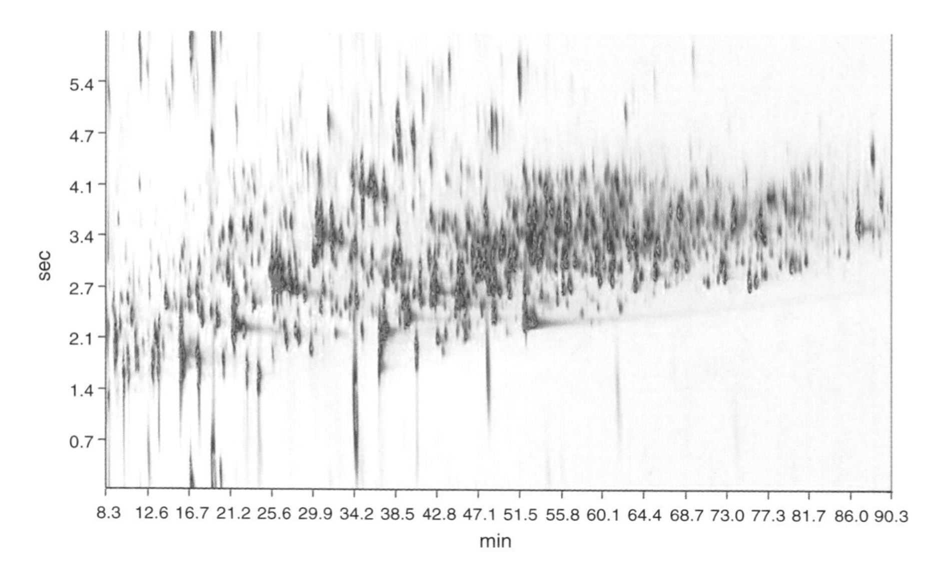

Figure 6.16 TIC GC × GC–qMS result relative to SPME-extracted *Arabica* coffee bean volatiles. [From Mondello et al. (2008), with permission. Copyright © 2008 by Wiley-VCH Verlag GmbH & Co. KGaA.]

The analytes were separated on a polar–apolar column set, which provided a more satisfactory result than the orthogonal setup [as reported previously in GC × GC–FID research (Mondello et al., 2004)]. The high complexity of an *Arabica* coffee sample was revealed by the 1000+ number of peaks scattered across the 2D plane, as can be seen in the TIC GC × GC–qMS chromatogram illustrated in Figure 6.16. The rapid-scanning qMS instrument enabled the application of a normal mass range (40 to 400 amu) at a scanning rate of 20 spectra/s, which was sufficient for reliable peak identification but not for correct peak reconstruction.

In the middle of the chromatography chaos illustrated in Figure 6.16, some ordered chemical-class structures were present; for example, as shown in Figure 6.17, pyrazines with the same degree of carbon substitution (i.e., dimethylpyrazines and ethylpyrazine) were aligned along distinct horizontal bands. Although characteristic and high-quality pyrazine mass spectra were generated, these were characterized by very similar fragmentation patterns. Peak identification was achieved by combining MS information with that derived from the specific pyrazine location and one-dimensional LRIs. The amount of information derived from a GC–qMS application on the same sample was very far from that generated in the GC × GC–qMS experiment.

Considering the GC × GC–TOF MS experiment, the mass spectrometer was operated at an acquisition rate of 100 Hz, over a mass range 41 to 415 amu. TIC chromatograms were processed automatically with ChromTOF software; the maximum number of processed peaks was restricted to 1000 ($S/N > 100$),

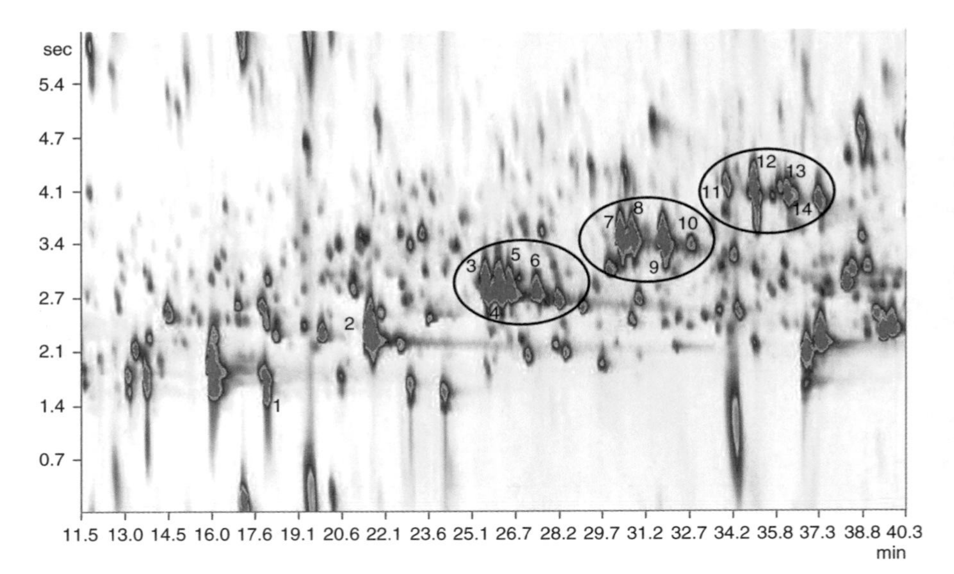

Figure 6.17 Pyrazine zone relative to the chromatogram illustrated in Figure 6.16. Peak identification: 1, pyrazine; 2, 2-methylpyrazine; 3, 2,5-dimethylpyrazine; 4, 2,6-dimethylpyrazine; 5, 2-ethylpyrazine; 6, 2,3-dimethylpyrazine; 7, 2-ethyl-6-methylpyrazine; 8, 2-ethyl-5-methylpyrazine; 9, 2,3,5-trimethylpyrazine; 10, 2-ethyl-3-methylpyrazine; 11, 2,6-diethylpyrazine; 12, 2-ethyl-3,5-dimethylpyrazine; 13, 2,3-diethylpyrazine; 14, 2-ethyl-3,6-dimethylpyrazine. [From Mondello et al. (2008), with permission. Copyright © 2008 by Wiley-VCH Verlag GmbH & Co. KGaA.]

a choice related to the extensive time (8 h) required for data processing (the generation of large data files was stressed by the authors).

In full-scan quadrupole mass spectrometry, the spectra production frequency is dependent on two factors: the time necessary to scan a specific mass range and the interval between one scan and the next (interscan dead time). The scan time can easily be derived by dividing the mass range by the specific scan speed of the instrument. Consequently, a rapid-scanning qMS system with a scan speed of 10,000 amu/s, operated at a mass range 40 to 360 amu, will be characterized by a scan time of 32 ms. If that same mass spectrometer requires 10 ms for the scan interval, approximately 24 spectra/sec (1000 ms/42 ms) will be generated. In SIM quadrupole mass spectrometry, the data acquisition frequency is dependent on the number of ions selected. A specific time is necessary to monitor each specific mass, while between one m/z value and the next there is an interchannel deadtime.

Since the introduction of rapid-scanning quadrupole mass spectrometers, quite a number of GC × GC–qMS papers have been published. The authors of these contributions have all stated more or less the same thing: that qMS systems are fine for qualitative purposes but fall short of quantitative requirements (Mondello et al., 2008). A few authors have reported the employment of a qMS system for quantification purposes. Although there are several differents opinions on "how many data points are necessary to correctly reconstruct a chromatography peak," it is generally accepted that 10 points are enough (Poole, 2003). The first example of quantification in a comprehensive GC application, using a qMS system (5972, Agilent), was reported by Debonneville and Chaintreau (2004): known analytes (fragrance allergens) were analyzed by monitoring selected ions in predefined retention-time windows. A detection frequency of 30.7 Hz was reported and affirmed to be sufficient for peak quantitation: comprehensive GC–qMS data were in good agreement with data derived from a single-column GC–qMS experiment. Such a data acquisition frequency still appears to be insufficient for the proper construction of the narrower GC × GC analyte bands (e.g., 100 to 200 ms). Furthermore, the use of the SIM mode is obviously applicable only to known target analytes.

During 2005, 23 GC × GC–MS papers were published, 19 of which were focused on the use of TOF systems, while the remaining experiments were based on the use of a qMS instrument. In particular, Korytár et al. (2005a) were the first to use a rapid-scanning qMS (Perkin-Elmer Clarus 500), with chemical ionization [electron-capture negative ion mode (ECNI)], in a GC × GC experiment. As aforementioned, mass spectrometry can certainly be considered as a separation dimension, because the mass analyzer can resolve analyte molecular ions (and fragments) on the basis of their mass-to-charge ratios; additionally, even if light wavelengths are not involved in the production of a mass spectrum, MS can also be compared to a spectroscopic method (i.e., UV absorbance), because the fragmentation pattern acts as a molecule-specific fingerprint. However, MS experiments are generally carried out by using electron ionization, in which the fingerprint aspect overshadows the separation aspect. As emphasized previously, GC × GC and GC–MS are characterized by several points in common. The similarity increases between the two methodologies if soft ionization techniques such as chemical ionization are employed.

Among the variety of applications reported by Korytár et al., that regarding polychlorinated *n*-alkanes (PCA) is described here. PCA mixtures are characterized by high complexity, because single constituents present different degrees of chlorination and chain lengths (short, medium, or long). Conventional GC–MS, employed widely in PCA analysis, generates typical "humped" chromatograms, which indicate the occurrence of extensive overlapping. Moreover, electron ionization is not advisable, because the extensive fragmentation hinders identification and thus the ECNI approach is preferable.

In the initial parts of the research, the authors studied the performance of the MS instrument, finding it capable of producing 23 spectra/s at a 300-amu mass range, which can be considered as the minimum range extension required for reliable GC–MS spectral library searching. The qMS was found capable of generating 63 spectra/s at a highly restricted 100-amu mass range. Furthermore, a 90-Hz acquisition frequency was reported when operating in the SIM mode, using a single ion [nearly three times that reported by Debonneville and Chaintreau (2004)]. After testing the MS performance, the influence on band

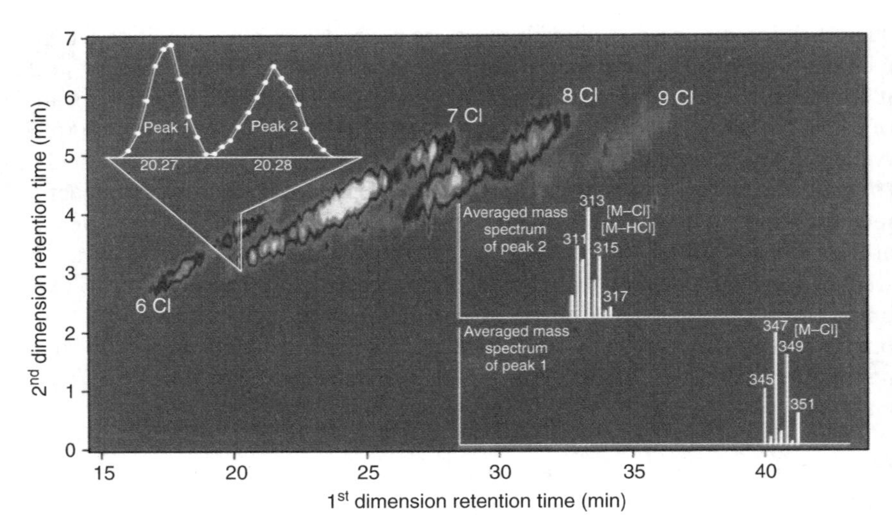

Figure 6.18 Full-scan (m/z 210 to 490 Da) GC × GC–ECNI qMS chromatogram of a mixture of polychlorinated decanes with average chlorine content of 65 wt%. Left-hand-side inset: separation of 6- and 7-Cl decanes in the second dimension and number of data points per peak. Right-hand-side inset: part of averaged mass spectra of peaks 1 and 2. [From Korytár et al. (2005a), with permission. Copyright © 2005 by Elsevier.]

broadening of the higher pressure present in the ion source (due to methane, the reagent gas), was studied; compared to EI applications, no differences were observed. Figure 6.18 illustrates a full-scan GC × GC–ECNI qMS chromatogram of a mixture of polychlorinated (PC) decanes with an average Cl content of 65 wt%. The group-type patterns are visible, with PC decanes characterized by the same number of Cl atoms (six to nine) aligned along distinct bands; within each group, retention increases when the substituents are distributed over the entire chain length. The $C_{10}Cl_6$–$C_{10}Cl_9$ clusters tend to widen as the degree of chlorination increases, due to the higher number of congeners and diastereoisomers. On the basis of the soft-ionization mass spectra attained and of the formation of group-type patterns (clearly visible in Figure 6.18), the presence of decanes with six to nine 9 Cl atoms was confirmed. The low degree of MS fragmentation, which simplified analyte differentiation, can be observed in the two averaged mass spectra reported in Figure 6.18; the hexa-Cl and hepta-Cl decanes are characterized by totally different spectral profiles. The untransformed peaks, relative to the two decanes and illustrated in the same figure, are reconstructed with 10 and 14 data points (sampling frequency: 23 Hz). Although both peaks appear to be adequately reconstructed, it must be added that their widths are rather wide (ca. 500 to 600 ms); both of the highly polar compounds are most certainly characterized by high k values on the secondary stationary phase (65% phenyl).

As mentioned previously, although the use of a limited mass range in GC × GC–qMS experiments can maximize the sampling frequency, it can also be the cause of poor mass spectral library results. Song et al. (2004a) came up with an idea to overcome such a disadvantage. A drug MS library was created by analyzing standard compounds, using a reduced mass range (42 to 235 amu). Although the creation of an ex novo MS library is certainly not very attractive (in working terms), the solution proposed was an interesting one. Furthermore, the authors highlighted the potential occurrence of peak skewing, emphasizing that attention must be paid when choosing a mass spectrum for library searching. Figure 6.19I shows three spectra (A–C) taken across different points of a GC × GC–qMS peak, identified as clorpheniramine. The continuous changing of analyte concentrations in the ion source, during each scan caused considerable variations in ion abundances (see ions 58 and 203 m/z), even though the peak illustrated in Figure 6.19I is a rather wide (GC × GC) one (ca. 500 ms), and the sampling frequency (19.36 Hz) should have generated approximately 10 spectra. An easy way to circumvent spectral inconsistency is to use an averaged spectrum for library matching. The same GC × GC peak was identified using a TOF MS system [Figure 6.19II (D–F)]; the experiment was carried out using a mass range of 40 to 900 amu and a 50-Hz sampling frequency. In TOF MS, the ions

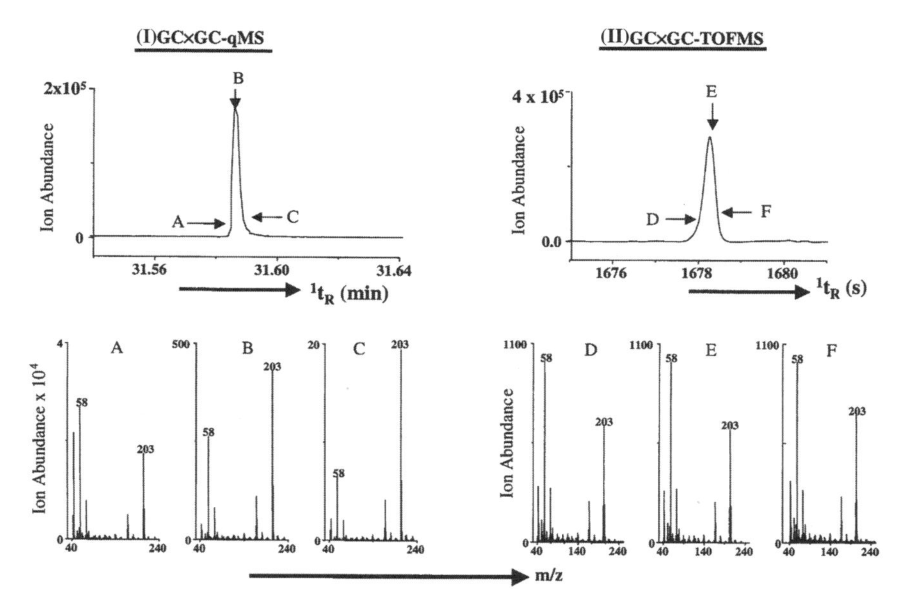

Figure 6.19 Comparison of ion abundances of chlorpheniramine at different regions of the chromatographic peak obtained by GC × GC–qMS and GC × GC–TOF MS. (I) TIC GC × GC–qMS chromatogram, using a range of 42 to 235 amu and acquired at 19.36 Hz. (A) Mass spectrum taken at the front of the 2D peak. (B) Mass spectrum taken at the apex of the 2D peak. (C) Mass spectrum taken at the back of the 2D peak. Note the variation of the relative ion abundances of 58 and 203 amu in A to C. (II) TIC GC × GC–TOF MS chromatogram, using a range of 40 to 900 amu and acquired at 50 Hz. (D) Mass spectrum taken at the front of the 2D peak. (E) Mass spectrum taken at the apex of the 2D peak. (F) Mass spectrum taken at the back of the 2D peak. [From Song et al. (2004a), with permission Copyright © 2004 by Elsevier.]

produced in the ion source are pulsed at very brief intervals into the flight tube. This process guarantees a high degree of spectra consistency, as can be seen in three spectra (D–F) reported in Figure 6.19II.

Adahchour et al. (2005a) reported a nice study on the principles, practicability, and potential of rapid-scanning qMS instrumentation in comprehensive 2D GC. The performance of an ultimate generation qMS system (Shimadzu QP2010), which neared GC × GC requirements, was studied for qualitative and quantitative purposes. The detector was characterized by a maximum scan speed of 10,000 amu/s and could reach the ultimate (GC × GC) goal of 50 spectra/s, at an excessively restricted mass range (95 amu). In the experiment it was emphasized that the MS sampling rate should be sufficiently high to reconstruct correctly the chromatogram and to avoid any peak skewing. First, the influence of the acquisition rate on peak-area precision was determined through the analysis of standard solutions, carried out at 20, 25, 33, and 50 Hz. The last two acquisition rates were generated by restricting the MS scan ranges to 195 and 95 amu, respectively. It was found that a 33-Hz sampling rate produced at least seven data points (only above-baseline points were counted) for peaks with a base width of 200 ms or higher. This number of experimental data points was found to be the minimum necessary for reliable quantitation. If a wider mass range was required, a 20- or 25-Hz sampling rate needed to be selected, but peak reconstruction was much less reliable. For peaks of <200 ms, a rate of 50 Hz was necessary, and limited the mass range even more. Peak skewing was evaluated by plotting the ratios of abundant ions in mass spectra, relative to a series of compounds, at acquisition rates of 33 and 20 Hz. It was found that considerable ion-abundance variation occurred at the lower rate, whereas ratios were essentially constant at the higher frequency. From the study, it appeared that a 30-Hz rate could be fast enough for GC × GC–MS applications if the operational conditions were tuned properly. During the 2004–2006 period, it appeared that the gap separating rapid-scanning qMS systems from the correct quantification of GC × GC peaks was a small one and was not dependent only on the qMS duty cycle. An optimized GC × GC–MS method does not automatically mean that 100 to 200-ms peaks will be introduced into the ion source; on the other hand, the generation of 400 to 500-ms peak widths can be attained if the series of variables that characterize a GC × GC–MS system are optimized in a proper fashion and with no costs in terms of end performance.

Mondello et al. (2005) continued the qMS trend, reporting a GC × GC–qMS application (a Shimadzu QP2010 was used) on a very complex commercial perfume; the authors used a mass range of 40 to 400 amu (scan frequency: 20 Hz). Pure standard compounds, MS library matching, and one-dimensional LRI data (LRIs were used as a filter during library searches, with the elimination of matches outside a predefined LRI window) were used for positive peak assignment. An apolar 30 m × 0.25 mm i.d. column and a polar 1 m × 0.25 mm i.d. capillary were employed in the first and second dimensions, respectively. It was found that the influence of the secondary column, in terms of LRI variation with respect to reference MS library values, was negligible for the apolar analytes; on the contrary, the more-polar components, subject to more intense interactions, were characterized by greater LRI variability. A total of 866 peaks were counted on the 2D space plane, whereas only 186 were detected in an unmodulated experiment; although such a comparison was not entirely fair (the uncoupled first dimension would probably have provided better GC–MS performance), the extent of that difference revealed the degree of peak overlapping. The shortcomings of GC–MS were fully demonstrated by the authors, showing a peak, unreliably identified as estragole and with a 72% spectral similarity. In the comprehensive GC–qMS experiment, the same peak was fully resolved into eight compounds, four of which were reliably identified (a 98% similarity was attained for estragole).

The use of a wider-bore second dimension, under vacuum outlet conditions, had beneficial effects on analyte separation, as reported previously by Shellie and Marriott (2002). Furthermore, band broadening was increased, and hence, excessively high sampling rates were not required. In fact, the authors reported the attainment of at least four data points per/peak (for the narrowest bands), sufficient for identification purposes.

If quadMS applications were not lacking in the 2004–2006 period, the same is true for TOF applications. Shellie et al. (2004a) started off the TOF series (considering the year) by successfully seeking an approach to match GC × GC–FID and GC × GC–TOF MS chromatograms. Additionally, emphasis was devoted to the fact that GC × GC experiments are carried out under suboptimum flow conditions. The research was worthy and deserves an extensive description. In GC, the matching of GC–FID and GC–MS chromatogram profiles is often sought for, is a rather well-known procedure, and is useful for qualitative (MS) and quantitative (FID) analysis. In gas chromatography, retention time locking is achieved when the dead times are the same for two specific applications; such an event can be attained by proper adjustment of the column head pressure (Blumberg and Klee, 1998). Shellie et al. demonstrated that things were much more complicated in the comprehensive GC field, because the column setup consists of the combination of two capillaries with differing internal diameters. Consequently, although FID and MS experiments might be carried out at the same nominal average linear velocity, different retention times will be noted in both dimensions; this factor is due to the different pressure profiles in each column, under atmospheric and vacuum conditions. To calculate the linear velocity in each GC × GC dimension, it is necessary to determine the equivalent dimensions of the twin column set, through the evaluation of flow relationships between columns of a different length and internal diameter. The outlet column volumetric flow [$F_{o(c)}$] can be derived by using the Poiseulle equation:

$$F_{o(c)} = \frac{60\pi r^4}{16\eta L} \frac{p_i^2 - p_o^2}{p_o} \frac{T\text{ref}}{T} \tag{6.1}$$

where r stands for the capillary radius; η is the dynamic viscosity of the mobile phase (at the operating temperature); p_i and p_o are the absolute inlet and outlet pressures, respectively; L is the column length; T_{ref} is the reference temperature, typically 25°C (298 K); and T is the oven temperature. Shellie et al. reported the use of a primary 30 m × 0.25 mm i.d. column and a secondary 0.5 m × 0.10 mm i.d. capillary. The equivalent 0.25 mm i.d. capillary relative to the second dimension, in terms of flow resistance, can be derived by using Eq. (6.1) as follows:

$$F_{o(c)(L\times 0.25\ \text{mm i.d.})} = F_{o(c)(L\times 0.10\ \text{mm i.d.})} \tag{6.2}$$

Resolving Eq. (6.2), it can easily be derived that a 0.5 m × 0.10 mm i.d. column is equivalent to a 19.53 m × 0.25 mm i.d. capillary with respect to flow resistance. As a consequence, the 30 m × 0.25 mm i.d. + 0.5 m × 0.10 mm i.d. column set is "equivalent" to a 49.53 m × 0.25 mm i.d. capillary. The most immediate advantage of such a calculation is that the correct column flow (and injector split flow) can be calculated by any conventional GC–FID or GC–MS software. The second benefit is that if the equivalent column dimension is known, the pressure at any point along the column (p_z), can easily be calculated using the following relationship:

$$P_z = \sqrt{P^2 - (z/L)(P^2 - 1)} \tag{6.3}$$

where P is equal to the ratio p_i/p_o, P_z is equal to the ratio p_z/p_o, and z is equal to the column point (e.g., 30 m if the p_z value at the head of the secondary column is desired). Once the head pressure, the pressure at the head of the secondary column, and the outlet pressure are known, it is rather simple to derive good estimates for the first- and second-dimension outlet linear velocities by using

$$u_o = \frac{r^2}{16\eta L}\frac{p_i^2 - p_o^2}{p_o} \tag{6.4}$$

As mentioned earlier, in one-dimensional GC analyses, alignment of void times enables the correlation of retention times. Table 6.2 shows that things are totally different in the GC × GC field; for example, with GC × GC–FID method A, an applied head pressure of 28.7 psia will lead to an overall average H_2 linear velocity of 38.4 cm/s; the linear velocities in the first and second dimensions will be 34.1 and 295 cm/s, respectively. If the same column set is employed in a GC × GC–MS experiment using helium as carrier gas (method B), and the head pressure is adjusted to equal the dead time of method A, linear velocities in the first and second dimensions will be 32.6 and 382 cm/s, respectively. As noted, the different pressure drops in methods A and B across each dimension will generate different retention times. Consequently, a modification of the GC × GC–MS outlet pressure was carried out by Shellie et al. by employing an additional gas supply, which was directed to a T-union between the secondary column outlet and the MS interface. Initially, a GC × GC–MS experiment was achieved using helium as the carrier gas under atmospheric outlet conditions (method C). If, again, the head pressure is adjusted to equal the dead time of method A,

TABLE 6.2 Absolute Inlet Pressure (psia) and Average Linear Velocity ($\bar{u}$; cm/sec) Across Each Dimension of a GC x GC Column Set[a]

Method	Gas	p_i	p_2	p_o	p_i/p_o	$\bar{u}$	p_i/p_2	$^1\bar{u}$	p_i/p_o	$^2\bar{u}$
A	H_2	28.7	21.3	14.7	1.95	38.4	1.34	34.1	1.45	295
B	He	40.1	25.2	≈ 0	large	38.3	1.59	32.6	large	382
C	He	47.6	32.0	14.7	3.23	38.4	1.49	33.4	2.18	334
D	He	63.8	47.5	32.7	1.95	38.4	1.34	34.1	1.45	295

Source: Shellie et al. (2004a), with permission. Copyright © 2004 by Wiley-VCH Verlag GmbH & Co. KGaA.

[a]Temperature = 333 K. The dimensions of the equivalent column are 49.53 m × 0.25 mm i.d. The pressure at the head of the secondary column is given by p_2. A, GC × GC–FID analysis using hydrogen; B, equivalent analysis under vacuum outlet conditions, using helium; C, analysis under atmospheric pressure outlet conditions, using helium; D, conditions for absolute retention matching with method A, using helium, with the outlet pressure elevated as required.

linear velocities in the first and second dimensions will be 33.4 and 334 cm/s, respectively. It is clear that the chromatograms derived from methods A and C will still not match perfectly. The different viscosities of the two gases [$H_2 \eta$ = 95.9 μP, He η = 213 μP (333 K)] leads to differing pressure drops across each dimension, even when the outlet pressure is the same. The condition for matching chromatograms can be met by adjusting both the head and the outlet pressures: the He–H_2 viscosity ratio equals 2.22, and equivalent linear velocities, as in method A, can be achieved (method D) by using a head and outlet pressure increased by a factor of 2.2 with respect to method A. The authors analyzed a mixture containing 18 compounds of various polarities and found that average absolute retention time differences were 3.7 s in the first dimension and 42 ms in the second dimension. A further advantage of the approach described was that column changing was a much easier task. At the end of the article the authors emphasized the far-from-ideal linear velocities generated in the second dimension, in all applications.

Over the past decades, persistent organic pollutants (POPs) have been widely subjected to conventional GC analysis, either with selective detection and/or MS detection. These sample types are highly complex, so it is not surprising that GC × GC–MS has been widely employed in this analytical field. Among the different classes of POPs, the 209 polychlorinated biphenyl (PCB) congeners are very challenging: PCBs can be classified in 10 homolog groups on the basis of their Cl content, and due to their similar chemical–physical properties, the analytical difficulties encountered are considerable.

An experiment focused on the separation of the highest possible number of PCBs using GC × GC–TOF MS was described by Focant et al. (2004a). Four thermally stable column sets were tested, with the best set consisting of a slightly polar first dimension [(8% phenyl)-polycarborane-siloxane] and a polar second dimension [(50% phenyl) polysilphenylene-siloxane]. Obviously, the use of thermally stable columns was related to the high boiling point (and, therefore, high elution temperature) of many PCBs. Separation of a PCB standard

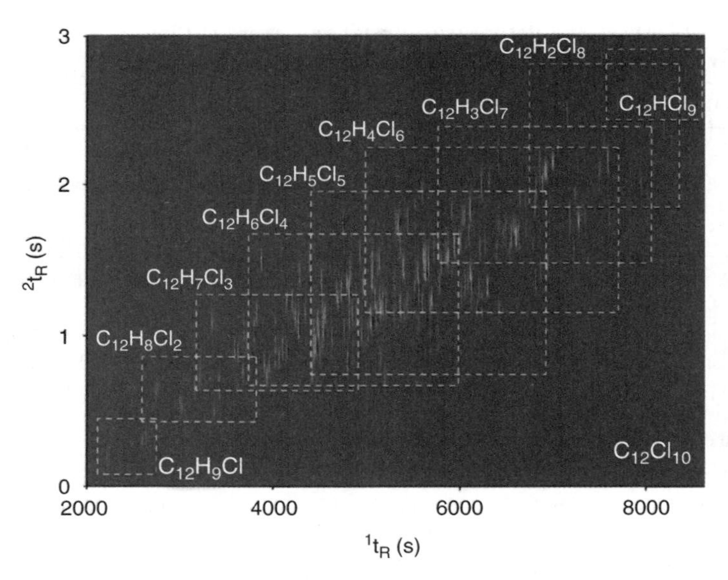

Figure 6.20 GC × GC–TOF MS chromatogram of the 209 PCB congeners using an HT-8/BPX-50 column set. [From Focant et al. (2004a), with permission. Copyright © 2004 by Elsevier.]

solution is illustrated in Figure 6.20, with the analytes distributed in a diagonal band across the 2D plane. The great amount of unoccupied 2D space was dependent on the similar chemical–physical properties of the PCBs. The organized structure of the 2D chromatogram was of great help in terms of peak identification: homologs are located in distinct zones, even though vicinal families tend to overlap. Moreover, depending on the number of Cl substituents, up to five subseries were separated within the same group. In fact, the first-dimension carborane phase enabled the separation of PCBs on the basis of the degree of *ortho*-substitution, from non-*ortho*-CBs to tetra-*ortho*-PCBs. The method developed allowed the chromatographic separation of 188 congeners, an additional four required deconvolution, and the remaining 17 analytes were distributed in eight complete coelutions. All the dioxin-like World Health Organization PCBs, as well as the European Union marker PCBs, were well resolved.

In another investigation, Focant et al. (2004b) used ^{13}C isotope-dilution TOF MS, combined with comprehensive 2D GC, for the simultaneous analysis of 59 target POPs in human serum and milk: 38 PCBs, 11 organochlorine pesticides (OCPs), and 10 brominated flame retardants [polybrominated diphenyl ethers (PBDEs)]. All of these POPs are highly lipophilic and tend to accumulate in body lipid compartments. The authors (from the U.S. Centers for Disease Control and Prevention) affirmed that no previously reported method (usually, GC–IDHRMS) allowed the simultaneous determination of these compounds in human fluids. The GC × GC column combination consisted of a completely apolar first dimension and of a slightly polar second dimension. The separation of the 59 target POP standards is shown in Figure 6.21A; the rather closely eluting analytes are located in a diagonal band which would appear to show a lack of orthogonality of the column setup (a substantial part of the chromatogram is unoccupied). In truth, the (sufficient) second-dimension analyte scattering is due to the similar chemical properties of the POPs. The polar column was also responsible for the isolation of an enlargened band of matrix-related interferences located along the x-axis (Figure 6.21B,C). These results highlight a further favorable aspect of GC × GC, the reduced requirements of tedious sample cleanup processes (obviously, this is also dependent on the number of interfering analytes). The instrumental detection limits, which were determined by considering the lowest quantity of a solute that generated a S/N ratio >3, ranged between 0.5 and 10 pg/μL. Method detection limits, determined by spiking bovine serum, were between 1 and 15 pg/μL. The newly developed approach was tested against a validated GC–IDHRMS procedure, which consisted of three separate applications with three different temperature programs. Precision was nearly as good as in the single-column technique, and the POP levels measured were very similar.

As seen previously, Korytár et al. (2005a) used a rapid-scanning qMS in the ECNI mode in the GC × GC–MS analysis of PCAs (2005a). In the same year, Korytár et al. (2005b) used a newly introduced TOF MS system operated in the ECNI mode (ThermoElectron). The 100% methylpolysiloxane column used in the first dimension and the thermally stable 65% phenyl-methylpolysiloxane stationary phase employed in the second dimension, generated highly structured chromatograms. With regard to ECNI–TOF MS detection, methane was used as a reagent gas, the mass range was 50 to 700 Da, and the acquisition rate was 40 Hz. The chromatographic and ECNI mass spectral behavior of PC decanes with a Cl content of 55% (a standard mixture was subjected to analysis) is shown in Figure 6.22; the GC × GC–TOF MS m/z 70 to 73 extracted ion chromatogram (EIC), which corresponds to the nonspecific $[Cl_2]^{\bullet-}$ and $[HCl_2]^-$ ions, is shown in Figure 6.22A. The same peak distribution as in Korytár et al. (2005a) was observed. PC decanes characterized by four to seven Cl atoms are aligned along distinct bands; within each chemical group, polarity increases when the substituents are scattered over the entire chain length (e.g., 2,5,6,9-$C_{10}Cl_4$ is more polar than 1,1,1,3-$C_{10}Cl_4$).

The general PCA distribution behavior was confirmed through the visualization of EICs for m/z 243 to 245 ($C_{10}Cl_4$ cluster), 276 to 283 ($C_{10}Cl_5$ cluster), 311 to 319 ($C_{10}Cl_6$ cluster), and 345 to 355 ($C_{10}Cl_7$ cluster), which correspond to the $[M - Cl]^-$ and $[M - HCl]^{\bullet-}$ ion clusters. However, for penta-, hexa-, and hepta-Cl PCAs, more than a single group was present on the 2D plane. If Figure 6.22B is observed, the presence of additional small groups, located below and above the $C_{10}Cl_6$ band, is evident. The upper compounds belong to the $C_{10}Cl_5$ class and are characterized by the spectral presence of the $[M]^{\bullet-}$ ion; the lower components belong to the $C_{10}Cl_7$ class, and are characterized by the $[M - 2HCl]^{\bullet-}$ ion (mass spectra inserts are shown in Figure 6.22B).

In order to verify the location of specific congeners in the various bands, eight individual PC decanes were added to the standard mixture (their positions are

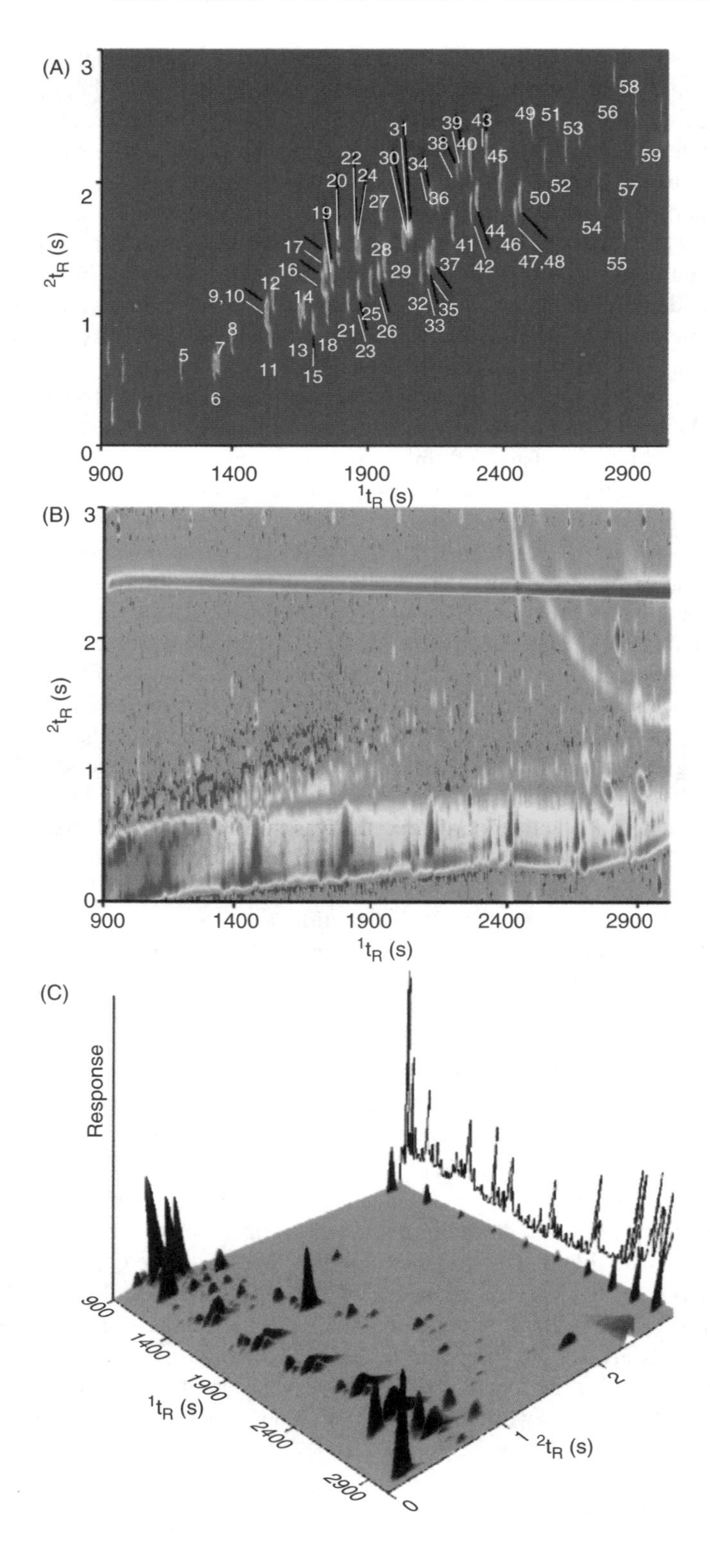

(A)
^2t_R (s)
^1t_R (s)
(B)
^2t_R (s)
^1t_R (s)
(C)
Response
^1t_R (s)
^2t_R (s)

indicated by black circles). Although many PC decanes followed the expected 2D chromatographic behavior, a series of congeners appeared to "slip" a chemical-class band; for example, 1,1,1,3,9,10-$C_{10}Cl_6$ is situated at the rear of the $C_{10}Cl_7$ group and, on the basis of its location, could be identified as a $C_{10}Cl_7$ compound. The authors affirmed, though, that for technical mixtures the content of these "outliers" can be considered as negligible. PC decanes with a Cl content of 65% were also subjected to study (Figure 6.22C), showing a distribution altogether similar to that reported by Korytár et al., (2005a).

In the 2004–2005 period, there was significant interest in the analysis of metabolites using GC × GC–TOF MS (Hope et al., 2005b; Shellie et al., 2005; Sinha et al., 2004b,c; Welthagen et al., 2005). The analytical method appeared to be particularly well suited, because the determination of volatile metabolites in a specific biological situation is a cumbersome task.

Welthagen et al. (2005) applied GC and GC × GC, both combined with TOF MS, to the analysis of spleen extracts of obese (NZO strain) and lean (C57BL/6 strain) mice. A 10-fold higher sample amount was injected onto the single GC column to compensate for the lower sensitivity. GC–TOF MS enabled the detection of 538 peaks, whereas >1220 were reported for the comprehensive GC experiment. The difference between the number of peaks detected would have been considerably more if equal sample amounts had been injected. Despite the number of compounds detected, the authors reported that only 10% were positively identified through library matching and the use of LRIs (a 500-compound MS library was used). A further aspect, worthy of note, was the analytical "purity," which refers to the combination of chromatographic and mass spectrometric capacities to isolate compounds. A compound is considered "pure" if either one or both of these separative dimensions fully resolves it from the rest of the sample constituents. Purity approaches zero in the ideal case and (theoretically) infinitum in problematic situations. In the GC × GC–TOF MS experiment, the number of peaks with acceptable purity (<1 values were considered as sufficient) was increased by a factor of 7. The effectiveness of the three-dimensional method for differential metabolomic biomarker determination can be observed in Figure 6.23; four obese mice were compared to five lean controls. The first point that is immediately evident is the apparently good stability of the separation pattern. Moreover, metabolite variability between the samples becomes observable if 2D peak intensities are considered; for example, two sugar alcohols (indicated by the light circles) are present in lower amounts in the obese mice spleen tissues. Student's t-test was used to evaluate differences between the two sample types;

←

Figure 6.21 TIC GC × GC–TOF MS chromatogram of (A) a 100-pg/μL native compound multianalytes calibration solution and (B) a real human serum sample. The shaded surface plot and the reconstructed one-dimensional trace in (C) are based on specific extracted ion current for the same human serum sample as in (B). The 2D scale was shifted by 1.5 s. [From Focant et al. (2004b), with permission. Copyright © 2004 by The American Chemical Society.]

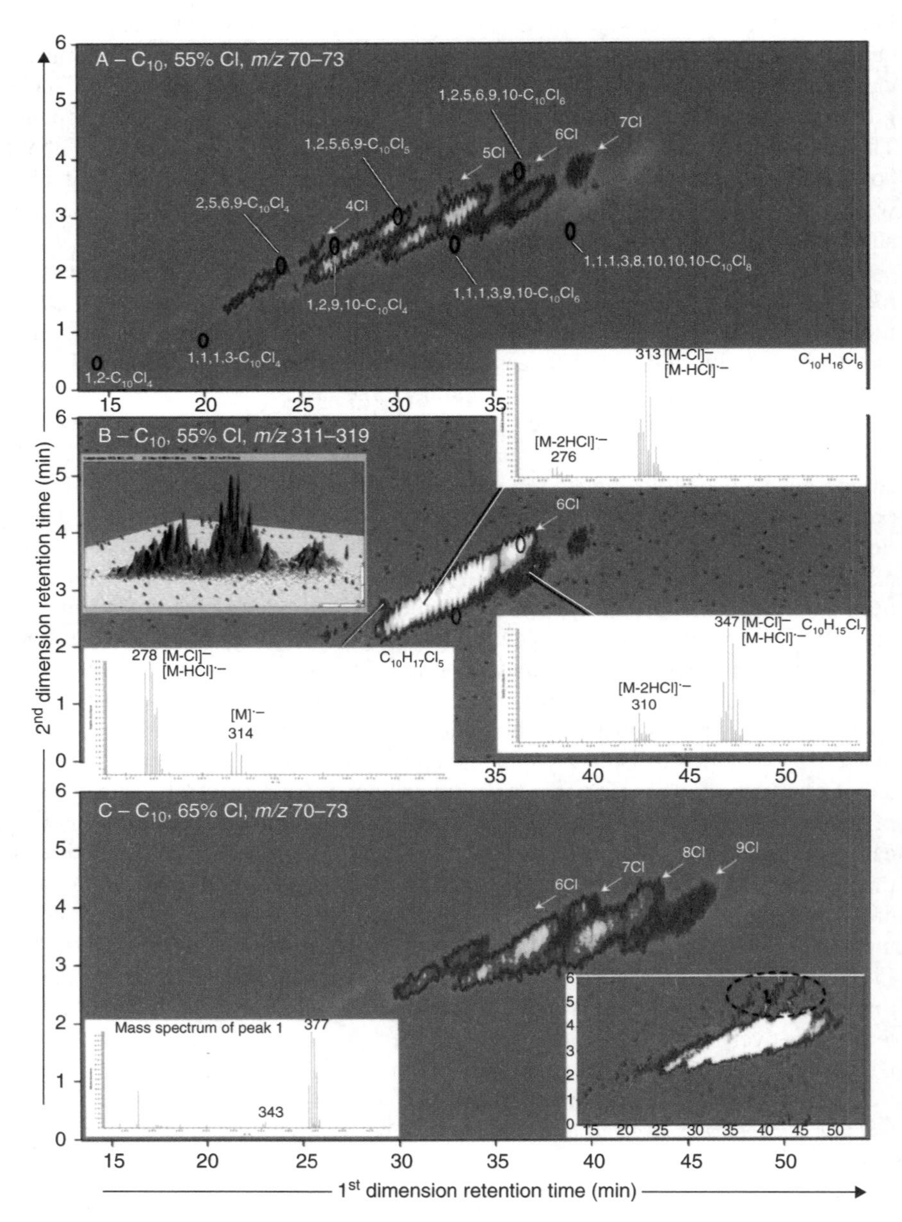

Figure 6.22 GC × GC–ECNI–TOF MS extracted-ion chromatograms of polychlorinated decanes: (A) 55% (w/w) m/z 70 to 73; (B) 55% (w/w) m/z 311 to 319; (C) 65% (w/w) m/z 70 to 73. Insets of (B) show its 3D presentation and averaged mass spectra of selected peaks. Insets of (C) show its zoom-out visualization and mass spectrum of selected peak. [From Korytár et al. (2005b), with permission. Copyright © 2005 by Elsevier.]

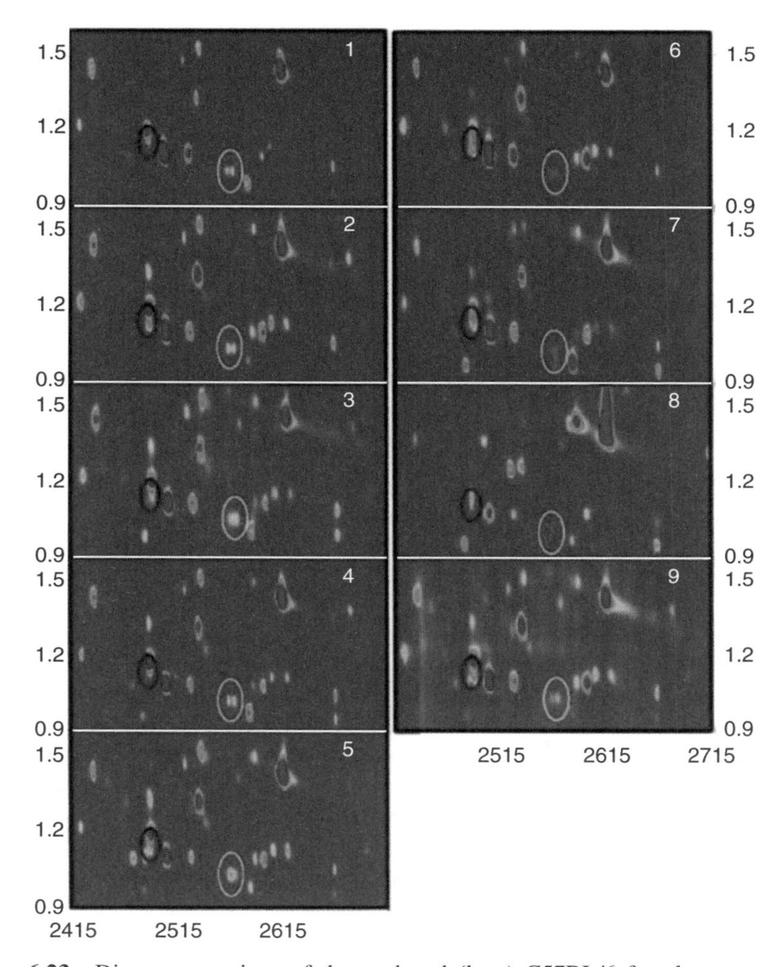

Figure 6.23 Direct comparison of the analyzed (lean) C57BL/6 female mouse spleen samples (left panel, samples 1 to 5) with the (obese) NZO female mouse samples (right panel, samples 6 to 9). The circled compounds were used for exemplary statistical evaluation of biomarker efficiency. [From Welthagen et al. (2005), with permission of Springer.]

although the number of replicates were too limited to achieve actual statistical significance levels, the results indicated that the content of a series of compounds could have indeed been different between the two groups.

The GC analysis of volatile organic compounds (VOCs) in breath samples is an interesting noninvasive method that can replace or complete blood analysis. Most breath VOCs are present in very low amounts, hence, a preconcentration step is required prior to GC analysis. Sanchez and Sacks (2006) used GC × GC–TOF MS, combined with a novel online concentration device (Figure 6.24), for the analysis of breath samples. On the basis of the results provided by the authors, it would appear that the high method sensitivity rather than the separation power

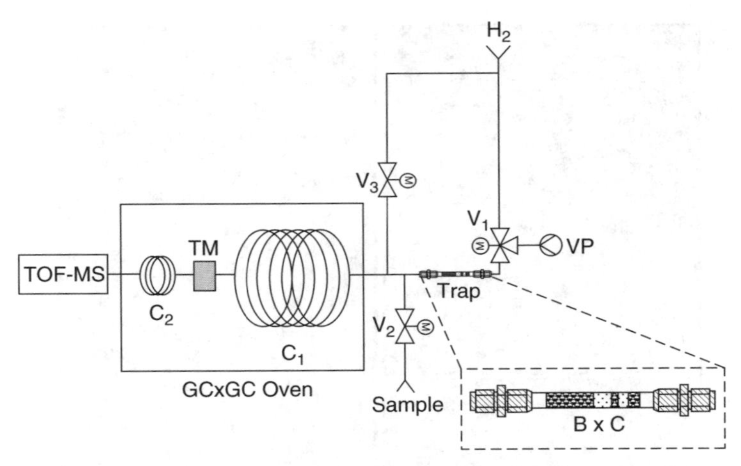

Figure 6.24 Multibed sorption trap/GC × GC–TOF MS system. Valves V_1, V_2, and V_3 are used to control the gas flow direction through the trap tube. TM, thermal modulator; VP, vacuum pump; B, Carbopack B; X, Carbopack X; C, Carboxen 1000. Expansion shows details of the three-bed trap design. [From Sanchez and Sacks (2006), with permission. Copyright © 2006 by The American Chemical Society.]

was exploited. A primary apolar 30 m × 0.25 mm i.d. column was linked to a secondary poly(ethyleneglycol) 2 m × 0.10 mm i.d. capillary; the latter ended in the ion source of a LECO TOF MS. Breath volatiles were entrapped on three different C-based adsorbents, separated by quartz-wool plugs, inside an 8 cm × 1.35 mm i.d. metal tube. A vacuum pump (VP) was employed to pull the analytes from the left to the right end of the metal tube (from the weakest to the strongest adsorbent). A pair of two-way valves (V_2, V_3) and a single three-way valve (V_1) were used to control the gas flow direction within the sampling device. For analyte entrapment, V_2 and V_3 were open, while V_1 was set to the VP direction. At the end of analyte entrapment, V_2 was closed for 10 s, to empty the lines of sample gas. V_3 was then closed and V_1 was set to the trap direction, hence, carrier gas flowed through the metal tube (Figure 6.24). During sample collection it was necessary to employ V_3 to avoid suction of part of the sample toward the TOF MS. The breath volatiles were released from the adsorbents through resistive heating of the trap tube (ca. 250°C). A very nice TIC GC × GC–TOF MS chromatogram relative to a 54-VOC test mixture is illustrated in Figure 6.25. Two pairs of isomers, *m*-xylene +*p*-xylene (29 and 30) and 2-chlorotoluene + 4-chlorotoluene (39 and 40), were not separated on either column. The concentrations of the VOCs contained in the sampling (Tedlar) bag ranged from 19 to 63 ppb. In the study, calibration data (for 33 standard compounds) were collected by taking samples for different time periods between 20 and 600 s. The concentration relative to each calibration point was extrapolated by considering the sampling flow rate, time, and the analyte concentration in the Tedlar bag. The amount collected in the trap was assumed to be derived from a volume equal to 560 mL. The first calibration points showed values down in

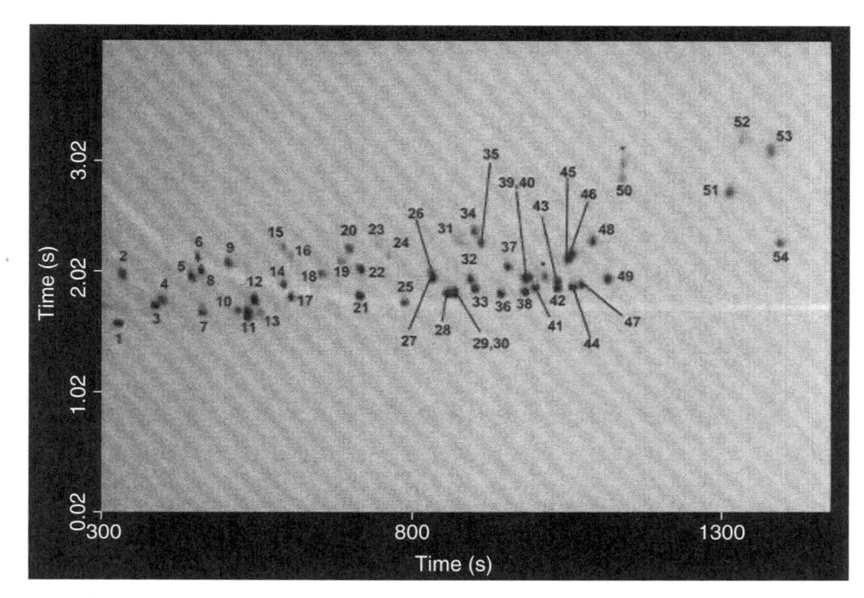

Figure 6.25 Two-dimensional chromatogram of a 54-compound test mixture. For peak identification, refer to Sanchez and Sacks (2006). The symbol * corresponds to *R*-methylstyrene and (*R*, *R*)-dimethylbenzenemethanol, which were released from the trap itself. [From Sanchez and Sacks (2006), with permission. Copyright © 2006 by The American Chemical Society.]

the ppt range. The authors affirmed that the method detection limits were below the values of the first calibration point and that sensitivity could reach sub-ppt levels by increasing the sampling time. Real breath samples were collected in 1-L Tedlar bags using a sampling time of 14 min and a flow of 40 mL/min (560 mL). Figure 6.26 shows the TIC GC × GC–TOF MS chromatogram relative to the breath of a person working in a heavily contaminated laboratory with a methylene chloride concentration of about 50 ppb.

6.6.4 GC x GC–MS Recent Years: 2007–2009

During the 2007–2009 period, although the appearence of rapid-scanning qMS instrumentation continued, GC × GC–TOF MS became increasingly well established, gaining a firm position as prime GC × GC–MS approach. A considerable amount of GC × GC–TOF MS applicational work was published, whereas limited article space was occupied by true hardware innovation. Even though the TOF mass spectrometer dominated the GC × GC scene (ca. 87%, 82%, and 93% of the GC × GC–MS published work during 2007, 2008, and 2009, respectively), the use of other MS instruments was reported.

Time-of-flight MS approaches can be divided essentially into two categories. The first is related to instruments characterized by a high data acquisition frequency (i.e., 100 to 500 Hz) and low resolution, generally in the range 300 to

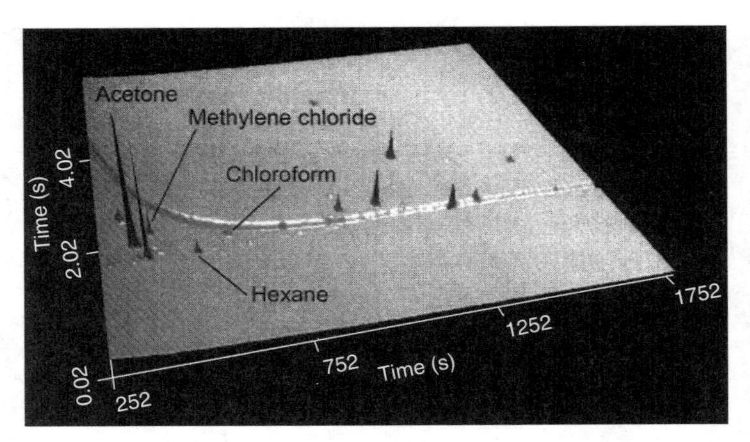

Figure 6.26 GC × GC–TOF MS TIC chromatogram relative to the analysis of a breath sample from a person working in a chemistry laboratory in a period of contamination. [From Sanchez and Sacks (2006), with permission. Copyright © 2006 by The American Chemical Society.]

1500. The other TOF MS category is characterized by systems with a high mass accuracy (i.e., 5 to 10 ppm) and a relatively low data acquisition frequency (Dalluge et al., 2002d). Ochiai et al. (2007) employed a high-resolution (HR) TOF MS (Micromass) to identify peaks eluting from an apolar–polar column combination. The HR TOF MS was operated using electron-based ionization, and provided full-scan spectra in the mass range 45 to 500 m/z, at a 25-Hz data acquisition frequency. 2,4,6-Tris-trifluoromethyl-[1,3,5]triazine was used as an internal standard for mass calibration. The sample analyzed consisted of atmospheric nanoparticles ($d_p < 50$ nm) collected from a Japanese road characterized by intense traffic. Atmospheric nanoparticles have a proven negative effect on human health; thus, their chemical characterization is important. The authors subjected the samples to thermal desorption and entrapped the released volatiles at $-100°$C in a PTV injector. At the end of the thermal desorption process, the volatiles were directed onto the primary GC column by rapidly heating the PTV injector. The authors used extracted ion chromatograms with 0.1- and 0.05-Da wide windows for identification purposes. Figure 6.27 shows two extracted ion TD–GC × GC–TOF MS chromatograms relative to a sample containing nanoparticles 29 to 58 nm in diameter. The highly selective (0.05-Da-wide window) chromatogram was derived by using five ions, specific to oxygenated polyaromatic hydrocarbons (PAH); m/z values of 180.0575 [9*H*-fluorene-9-one and 1*H*-phenalene-1-one ($C_{13}H_8O$)], 194.0732 [9(10*H*)-anthracenone ($C_{14}H_{10}O$)], 198.0317 [naphto(1,2-*c*)furan-1,3-dione ($C_{12}H_6O_3$)], 230.0732 [7*H*-benz[de]anthracene-7-one and 11*H*-benzo[*a*]fluorine-11-one ($C_{17}H_{10}O$)], and 258.0681(naphthacene-5,12-dione and benz[*a*]anthrace-7,12-dione) were employed. The oxygenated PAHs found in the sample were 9*H*-fluorene-9-one (−0.0007 Da mass error: 4 ppm), 1*H*-phenalene-1-one (0.0046 Da mass error: 26 ppm), 9(10*H*)-anthracenone

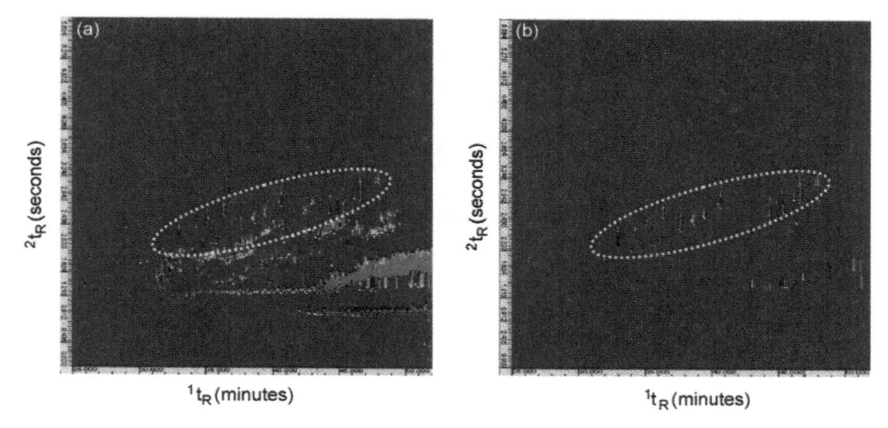

Figure 6.27 Extracted ion TD–GC × GC–TOF MS chromatogram (sum of five selected ions for oxy-PAHs; m/z180.0575, 194.0732, 198.0317, 230.0732, and 258.0681), with (a) 0.1-Da- and (b) 0.05-Da- wide windows relative to a sample containing nanoparticles of diameter 29 to 58 nm. [From Ochiai et al. (2007), with permission. Copyright © 2007 by Elsevier.]

(0.0005 Da mass error: 3 ppm), naphto(1,2-*c*)furan-1,3-dione (0.0031 Da mass error: 16 ppm), 7*H*-benz[de]anthracene-7-one (0.0016 Da mass error: 7 ppm), and 11*H*-benzo[*a*]fluorine-11-one (0.0039 Da mass error: 17 ppm).

Polychlorinated dibenzo-*p*-dioxins (PCDDs) and dibenzofurans (PCDFs) are structurally related, highly toxic environmental and food contaminants that tend to accumulate in fatty tissues. Gas chromatography, combined with isotope dilution-sector HR MS, is commonly used for the determination of these dangerous substances in purified extracts. The use of GC-sector HR MS is a costly issue requiring high levels of expertise.

A GC × GC–HR TOF MS experiment, focused on the analysis of PCDDs/Fs, was reported by Shunji et al. (2008). A low-polarity primary column (60 m × 0.25 mm i.d. × 0.1 μm d_f) was connected to a polar secondary capillary (1.5 m × 0.075 mm i.d. × 0.1 μm d_f) whose outlet ended in the ion source of a HR TOF MS (JEOL). The latter was operated using electron-based ionization and provided full-scan spectra in the mass range 35 to 550 m/z, at a 25-Hz data acquisition frequency; mass resolution was 5000 (full width half-maximum). A good example of the HR TOF MS potential for highly selective mass discrimination is shown in Figure 6.28. The spectrum, derived at 40.28 min and relative to a GC × GC application on an incinerator ash extract, is characterized by a 337.86378 m/z ion [M^+], which corresponds to 2,3,4,7,8-pentachlorinated dibenzofuran (PeCDF). Also present are the 339.85974 m/z $[M+2]^+$, 341.85751 m/z $[M+4]^+$, and 343.85615 m/z $[M+6]^+$ ions, due to the presence of one, two, and three ^{37}Cl atoms, respectively. Other ions, such as 337.38438 m/z and 338.88269 m/z, probably derived from interferences.

The highly selective nature of the three separation dimensions (boiling point × polarity × high-resolution mass differentiation) proposed by Ochiai et al. and

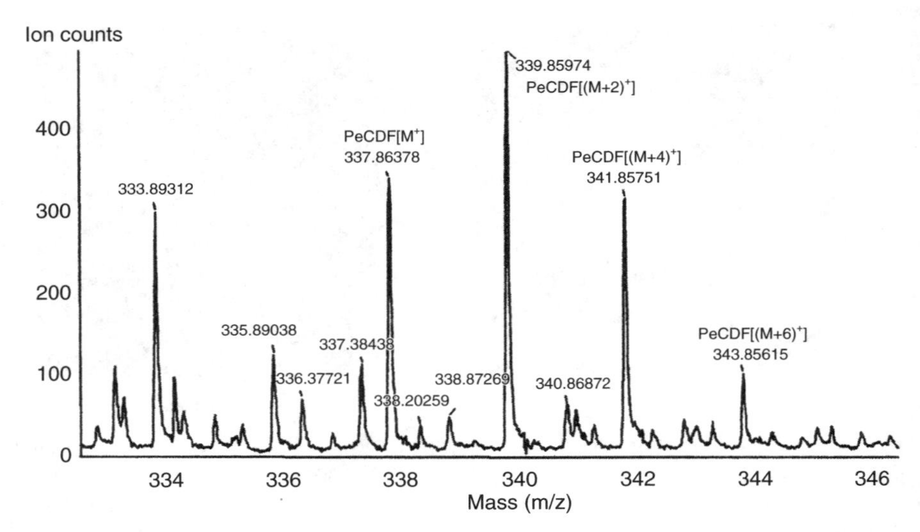

Figure 6.28 Mass profile from 335 to 346 (m/z) (at 40.28 min) of a crude fly ash extract, analyzed using GC × GC–HR–TOF MS. $m/z = 337.86378$, 339.85974, 341.85751, and 343.85615 are the mass peaks of a PeCDF. [From Shunji et al. (2008), with permission. Copyright © 2008 by Elsevier.]

Shunji et al. was certainly interesting. However, the very high cost of the GC × GC–HR TOF MS instrument must also be accounted for when evaluating the advantages/disadvantages of the method. Both experiments were directed to the analysis of target analytes; perhaps the employment of less costly, optimized GC × GC–qMS in the SIM mode could have produced the same or similar results.

During the 1999–2007 period, the MS systems employed in the GC × GC field, classified on the basis of the number of applications, were (1) low-resolution TOF, (2) qMS, (3) HR TOF, and (4) ion trap. A fifth and sixth MS system—an isotope ratio (Tobias et al., 2008) and a triple quadrupole mass spectrometer (Poliak et al., 2008)—were added to this list in 2008.

Most elements of interest (C, H, O, N, Cl, etc.) have two or more stable isotopes, with the lightest present in greater amounts than the heavier ones. Among stable isotopes, a great deal of analytical work has been devoted to C and N isotopes, because these two elements are found in the earth, air, and in all living things. ^{13}C and ^{15}N are characterized by a natural abundance of about 1% or less, with ^{12}C and ^{14}N making up almost all of the difference.

High-precision isotope ratio mass spectrometry (IRMS) can be defined as the technique that deals with the measurement of deviations of isotope abundance ratios from an accepted standard by only a few parts per thousand; nowadays, IRMS is considered as a valuable tool in disciplines such as food authentication, biomedicine, geochemistry, archaeology, and forensic science. Each element must be transformed from its chemical form into a gas (e.g., CO_2, N_2), and purified prior to its introduction into the IRMS ion source (Brenna, 1994).

As for all MS techniques, the analysis of a mixture of volatiles is easier if the mass spectrometric step is preceded by a GC step. The online combination of gas GC and IRMS enables compound-specific isotope analysis. In particular, the measurement of $^{13}C/^{12}C$ ratios through GC–IRMS is well established. Compared to the other MS systems, IRMS instrumentation does not generate fragmentation profiles, only the measurement of isotopic abundances. The development of a GC × GC–IRMS instrument is no easy task; all GC–IRMS systems are characterized by a combustion interface, because the carbon contained in each compound must be converted to CO_2. Conventional combustion chambers cause a certain degree of band broadening (ca. 1 s), while the IRMS response is in the same time order; such instrumental characteristics are not compatible with compounds eluting rapidly from the second dimension of a GC × GC system. Consequently, it is clear that instrumental modification was necessary; the main components subjected to consideration were the solvent elimination process, combustion reactor, transfer lines, water trap, and the open split. A schematic of the GC × GCC–IRMS instrument developed and used in steroid analysis is illustrated in Figure 6.29; a GC was combined with an IRMS (Thermo Finnigan). An LMCS device used for thermal modulation (4-s period) was installed between the most typical GC × GC column set, namely a 30 m × 0.25 mm i.d. apolar capillary and a 1 m × 0.10 mm i.d. polar column. A PTV injector was used to prevent the introduction of solvent onto the first dimension. A low-dead-volume lab-constructed combustion interface was employed; the combustion reactor was composed of a 0.45 m × 0.25 mm i.d. × 0.36 mm o.d. deactivated fused-silica capillary containing two 0.19 mm × 0.10 mm wires, one composed of Cu/Mn/Ni

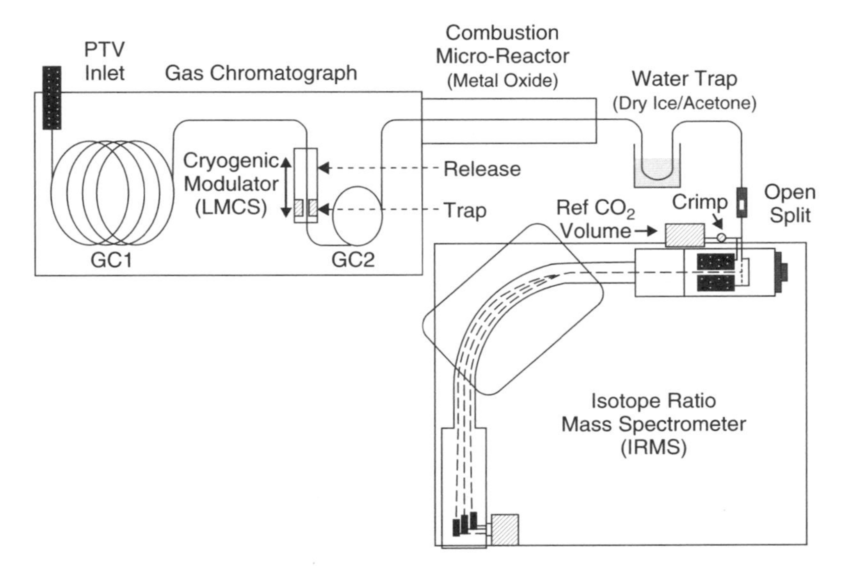

Figure 6.29 GC × GCC–IRMS instrument. [From Tobias et al. (2008), with permission. Copyright © 2008 by The American Chemical Society.]

(84 : 12 : 4%) and the other of Pt (>99%). In conventional GCC–IRMS applications, ceramic or quartz tubes with i.d.s$\geq$0.5 mm are normally employed. A 1 m $\times$ 0.1 mm i.d. deactivated fused-silica column was employed as a transfer line between the reactor and the open split. The latter enables coupling and decoupling the column gas flow to the IRMS. Water vapor generated from the combustion process was removed (avoiding its entrance to the MS) by immersing 10 cm of the transfer capillary into a water trap held at -78°C. Connections between capillaries were achieved through press-tight fittings. The dry effluent was directed to the IRMS by placement of the upper end of a 1 m $\times$ 0.075 mm i.d. IRMS sampling capillary at the end of the transfer column within the press-tight fitting. The IRMS data acquisition frequency was 25 Hz. Figure 6.30A shows a nonmodulated m/z 44 peak formed of 5β-androstan-3α-ol-11,17-dione acetate (11k-AC) and 5β-pregnan-3α-20α-diol (5βP). GC $\times$ GC modulation enabled the baseline separation of m/z 11k-AC and 5βP, producing rather broadened peaks (Figure 6.30B). Sensitivity increased considerably, passing from the unmodulated to the modulated application. Peak intensities are approximately the same even if 10-fold-lower amounts were used in the GC $\times$ GC experiment. The unmodulated m/z 45/44 trace, for 11k-AC and 5βP, is illustrated in Figure 6.30C; as expected, there was a rise and fall in the isotope ratio indicative of the arrival of m/z 45 CO_2, prior to m/z 44 CO_2, at the detectors. The modulated m/z 45/44 trace, for 11k-AC and 5βP, is shown in Figure 6.30D. Apart from baseline resolution, the six consecutive peak modulations obviously follow the rise and fall in the isotope ratio across the first-dimension peak and are characterized by decreasing $^{13}C/^{12}C$ values. As can be seen, 11k-AC is more ^{13}C-enriched than is 5βP. GC $\times$ GCC–IRMS $^{13}C/^{12}C$ values were calculated in part manually and were characterized by acceptable statistics.

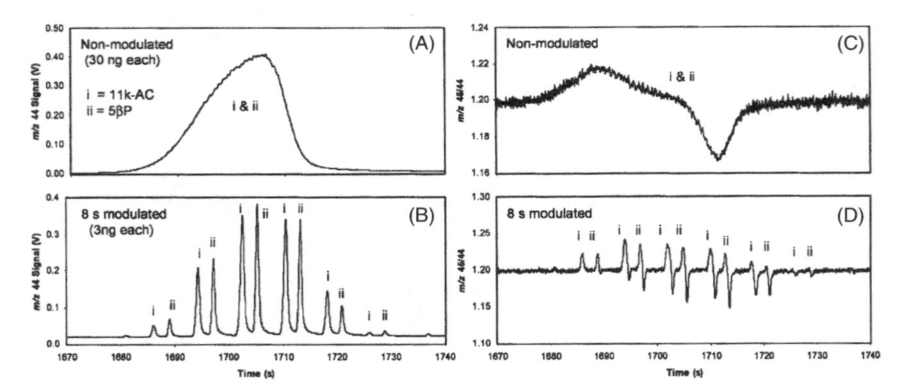

Figure 6.30 GC $\times$ GCC–IRMS analysis of 5β-androstan-3α-ol-11,17-dione acetate (11k-AC) and 5β-pregnan-3α-20α-diol (5βP): (A) unmodulated m/z 44, (B) 8-s modulated m/z 44, (C) unmodulated m/z 45/44, and (D) 8-s modulated m/z 45/44 chromatograms. [From Tobias et al. (2008), with permission. Copyright © 2008 by The American Chemical Society.]

Triple-quadrupole mass spectrometers (QqQs) are highly selective instruments, used widely in combination with a GC pre-separation step. When operating a QqQ MS in the tandem MS mode (MS–MS), the first quadrupole (Q_1) acts as the primary mass-selective dimension (MS_1), while the second quadrupole (q_2) serves either as a field-free region in which metastable dissociations occur, or (more often) as a cell for collision-induced dissociation (CID) experiments. The third quadrupole (Q_3) is employed to separate the ions that exit q_2 (Gross, 2004). Apart from the common MS–MS procedures, QqQ MS systems can be operated in the full-scan or SIM modes (Gross, 2004). As mentioned previously, pneumatically modulated GC × GC–MS–MS experiments, achieved using a triple-quadrupole MS (Varian), were reported in 2008 (Poliak et al., 2008). MS ionization was achieved using supersonic molecular beam (SMB) EI; the latter, defined as "cold EI," generates intense molecular ion peaks (Amirav et al., 2008). Figure 6.31 illustrates GC–SMB–MS, GC × GC–SMB–MS and GC × GC–SMB–MS–MS chromatogram expansions relative to the analysis of diazinon in coriander (100 ppb). The GC–SMB–MS chromatogram is characterized by evident extensive coelution, and provided a poor-quality, fragment-rich spectrum for diazinon. On the contrary, the diazinon spectrum attained using GC × GC–SMB–MS was of much higher quality; a hit of about 94% was attained during MS library matching. Both full-scan experiments were carried out using a mass range of 50 to 400 amu and a 6.25 data acquisition frequency (scan speed: 2100 amu/s). For the GC × GC–SMB–MS–MS experiment, Q_1 isolated a parent ion with an m/z of 304; the CID process in q_2 generated a daughter ion with an m/z of 179, which was isolated in Q_3. As can be seen in Figure 6.31, the employment of MS–MS enabled the complete elimination of the other sample interferences. However, the authors showed no GC–SMB–MS–MS chromatogram and did not report information on the time necessary to carry out a single MS–MS analysis. Although the experiment described by Poliak et al. was certainly interesting, it must be noted that (1) tandem MS is commonly employed for target analyte analysis, (2) a conventional GC pre-separation would appear to be sufficient in most cases, considering the extremely high MS^2 selectivity, and (3) quantification is straightforward using GC–MS–MS. Considering very complex samples, it is obvious that GC × GC–MS is the prime choice for the analysis of unknowns, whereas both GC × GC–MS and GC–MS–MS are powerful tools for target analyte analysis. The question that must be asked is if the use of GC × GC prior to MS–MS is really necessary or even suitable, considering quantification issues.

6.6.5 GC x GC–MS Current-Year Work: January–June 2010

During the first six months of 2010, the GC × GC–TOF MS : qMS ratio was a little lower than that of previous years (23 : 8). A series of GC × GC–qMS investigations were reported that held promise for the future employment of qMS systems, for quantification purposes. For example, GC × GC–qMS research was reported (Tranchida et al., 2010a), focused on the operation of a 50-μm i.d. secondary column, under optimized gas-flow conditions. The experiment

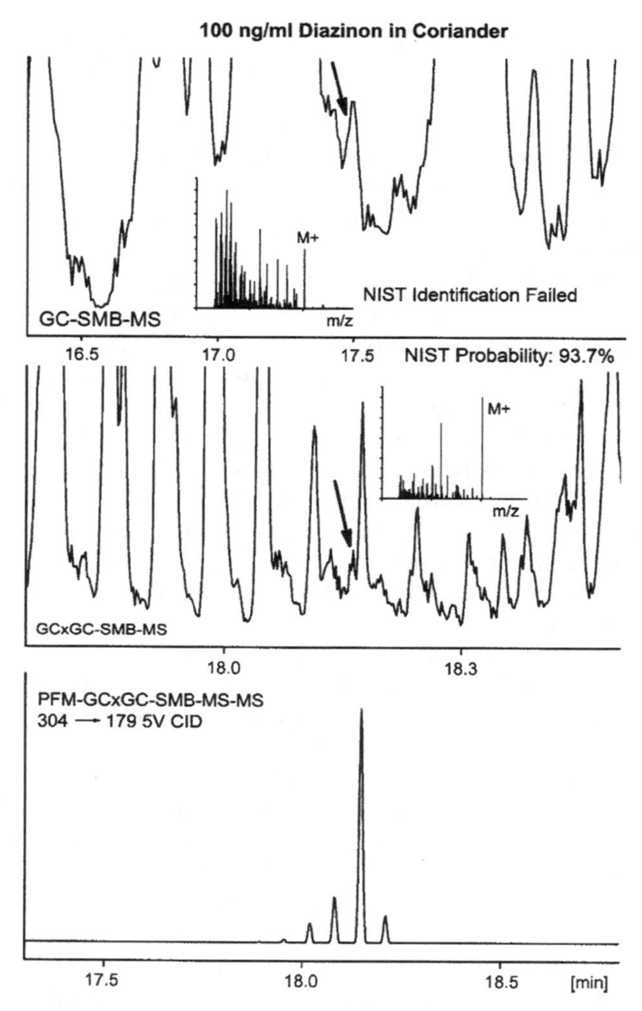

Figure 6.31 Diazinon analysis in coriander using GC–SMB–MS (upper chromatogram), by PFM GC × GC–SMB–MS (middle chromatogram) and by PFM GC × GC–SMB–MS–MS (bottom chromatogram) (4 s modulation period). The insets at the upper and middle traces show the mass spectra of diazinon obtained and the results of the NIST library search. [From Poliak et al. (2008), with permission. Copyright © 2008 by Elsevier.]

described was directly related to previous work carried out by the same research group using GC × GC–FID (Tranchida et al., 2007a, 2009a,b).

If the GC × GC–MS literature is consulted, it can be observed that the 0.25 mm i.d. + 0.1 mm i.d. column combination has been widely exploited (Mondello et al., 2008). Such a capillary configuration satisfies the requirements of wide first-dimension chromatography bands, and fast high-resolution separations on the second column. If a volatile compound is analyzed on a 1 m × 0.1 mm i.d. column (with a high β value) under optimum conditions of temperature ($k = 5$ to 10) and helium velocity (ca. 100 cm/s), approximately 10,000 theoretical plates ($H_{min} = 0.1$ mm) should be generated. In fact, from fundamental GC theory it is well known that for columns with a high β value, the H_{min} value approaches the column i.d. (David et al., 1999). The following question arises spontaneously from previous considerations: if a 1 m × 0.05 mm i.d. column can potentially generate 20,000 plates, why not benefit from such an analytical option in GC × GC–MS? The authors carried out a series of applications by using a twin-oven split-flow GC × GC–qMS system, with a cryogenic loop modulator (Figure 6.32). The stationary-phase combination employed consisted of an apolar (silphenylene polymer) 30 m × 0.25 mm i.d. column, linked by means of a T-union, to an MS-connected 1 m × 0.05 mm i.d. polar column [poly(ethylene glycol)], and to a 0.20 m × 0.05 mm i.d. uncoated capillary segment; the latter was connected to a manually operated split valve.

Initially, a commercial perfume was analyzed, under conventional GC × GC–qMS conditions (the split valve was closed). Helium velocities were very slow and fast in the first and second dimensions: about 5.5 and 194 cm/s. The "best" result, attained using an initial 10-min temperature hold time in GC2 (practically, a −30°C offset), is shown in Figure 6.33. As can be observed, only a limited amount of the 2D space available was exploited. The peak assignment process generated a peak table containing 30 compounds, only six of which were characterized by an MS similarity >95% (Table 6.3). On the basis of the results obtained, the reasons that it would not be advisable to employ a 50-μm i.d. second dimension were evident. The authors concluded that both the primary and secondary capillaries provided a poor performance, and showed that it would be much more preferable to use an optimized GC–qMS method (Table 6.3).

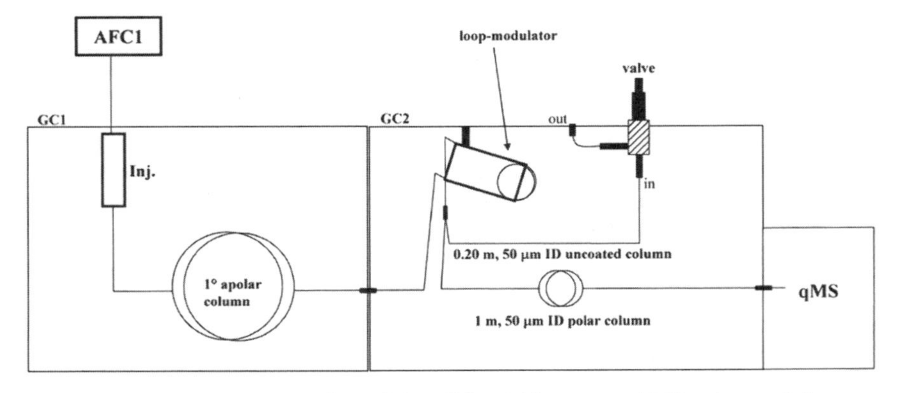

Figure 6.32 Twin-oven, split-flow GC × GC–qMS system. AFC, advanced flow controller; Inj., injector. [From Tranchida et al. (2010a), with permission. Copyright © 2010 by Elsevier.]

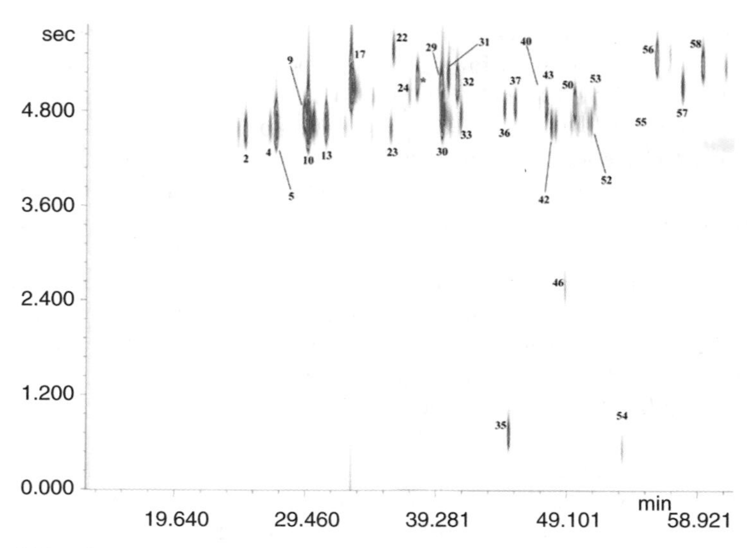

Figure 6.33 Conventional GC × GC–qMS analysis of a commercial perfume using a 50-μm i.d. secondary column. For peak identification, refer to Table 6.3. [From Tranchida et al. (2010a), with permission. Copyright © 2010 by Elsevier.]

Various split-flow GC × GC–qMS experiments were carried out by regulating the split valve manually at different stages. The best operational conditions were attained with the split-valve opened completely; gas velocities of about 17 and 120 cm/s were calculated in the first and second dimensions, respectively. When using a 30 m × 0.25 mm i.d. first dimension, a gas velocity near the 20-cm/s mark can be considered as ideal because (1) the column efficiency is satisfactory, and (2) peak widths are wide enough for a sufficient number of cuts. With regard to a 1 m × 0.05 mm i.d. capillary, a helium velocity in the range 100 to 130 cm/s is desirable because (1) the column plate number is high; (2) the second-dimension separations are still sufficiently rapid; and (3) the reduced velocity generates increased peak widths and hence more compatible conditions for a quadrupole mass spectrometer. Approximately 83% of the primary column flow was lost, and thus the injector split ratio was adjusted to 20 : 1 to avoid a sensitivity decrease. The optimized split-flow GC × GC–qMS experiment, achieved with no temperature offset in GC2, is illustrated in Figure 6.34. A total number of 58 compounds were identified (Table 6.3), with 23 of these characterized by an MS similarity of greater than 95%. As can be observed, the GC × GC outcome was greatly improved, both in terms of amount of 2D space occupied and number of separated solutes. For example, considering the primary column separation, compounds 5 and 6 are well resolved, while they coelute in the conventional GC × GC analysis, forming peak 5. With regard to the secondary column separation, a direct comparison between the conventional and split-flow applications is shown in Figure 6.35. A single "raw" chromatogram, derived from the conventional GC × GC trace and characterized by a single peak, is illustrated in

TABLE 6.3 Peak Identification, Library LRI Values, GC–qMS, Conventional GC x GC–qMS, Split-Flow GC x GC–qMS % Similarities, and LRI Values (in Parentheses)

Compound	LRI Library	GC–qMS %Similarities (LRI)	Conventional GC × GC–qMS %Similarities (LRI)	Split-Flow GC × GC–qMS %Similarities (LRI)
1. α-Thujene	927	97 (924)	n.d.[a]	93 (929)
2. α-Pinene	933	97 (932)	94 (919)	96 (937)
3. α-Fenchene	948	n.d.	n.d.	90 (954)
4. Sabinene	972	96 (971)	90 (956)	95 (976)
5. β-Pinene	978	98 (977)	96 (967)	97 (983)
6. Myrcene	991	97 (988)	n.d.	95 (993)
7. 3-Octanone	986	91 (984)	n.d.	95 (991)
8. Hexyl acetate	1012	92 (1017)	n.d	93 (1017)
9. *p*-Cymene	1025	98 (1023)	95 (1010)	97 (1030)
10. Limonene	1030	98 (1029)	98 (1018)	97 (1034)
11. (*Z*)-β-Ocimene	1035	90 (1034)	n.d.	93 (1047)
12. (*E*)-β-Ocimene	1046	96 (1045)	n.d.	94 (1049)
13. γ-Terpinene	1058	97 (1057)	96 (1043)	96 (1062)
14. Sabinene hydrate	1069	n.d.	n.d.	90 (1079)
15. Terpinolene	1086	95 (1085)	n.d.	94 (1089)
16. *trans*-Linalool oxide	1086	92 (1087)	n.d.	91 (1086)
17. Linalool	1101	98 (1100)	98 (1084)	96 (1109)
18. 3-Acetoxyoctene	1109	90 (1107)	n.d.	96 (1111)
19. Dihydrolinalool	1136	96 (1134)	n.d.	93 (1141)
20. Camphor	1149	n.d.	n.d.	90 (1158)
21. *trans*-β-Terpineol	1149	n.d.	n.d.	90 (1158)
22. Benzyl acetate	1167	96 (1162)	90 (1147)	97 (1169)
23. Linalool ethyl ether	1166	94 (1166)	92 (1145)	96 (1170)
24. Terpinen-4-ol	1180	90 (1181)	91 (1175)	93 (1190)
25. Butanoic acid, hexyl ester	1195	n.d.	n.d.	93 (1196)
26. Estragole	1201	93 (1197)	n.d.	98 (1205)
27. α-Terpineol	1195	94 (1195)	n.d.	93 (1205)
28. Acetic acid, octyl ester	1214	n.d.	n.d.	94 (1215)
29. Neral	1238	97 (1238)	94 (1225)	98 (1246)
30. Linalyl acetate	1250	97 (1249)	90 (1225)	96 (1255)
31. Carvone	1246	95 (1244)	93 (1241)	97 (1253)
32. Geranial	1268	97 (1268)	97 (1257)	97 (1277)
33. Lavandulyl acetate	1284	91 (1283)	91 (1259)	97 (1287)
34. α-Terpinyl acetate	1349	n.d.	n.d.	90 (1355)
35. Eugenol	1357	92 (1351)	92 (1345)	96 (1364)

(*continued overleaf*)

TABLE 6.3 ***(Continued)***

Compound	LRI Library	GC–qMS %Similarities (LRI)	Conventional GC × GC–qMS %Similarities (LRI)	Split-Flow GC × GC–qMS %Similarities (LRI)
36. Neryl acetate	1361	95 (1358)	94 (1337)	97 (1364)
37. Geranyl acetate	1380	98 (1377)	93 (1356)	97 (1382)
38. β-Patchoulene	1383	n.d.	n.d.	90 (1393)
39. β-Elemene	1390	n.d.	n.d.	93 (1396)
40. Jessemal	1414	n.d.	92 (1400)	93 (1422)
41. α-Santalene	1418	n.d.	n.d.	91 (1424)
42. (*E*)-Caryophyllene	1424	95 (1419)	92 (1424)	96 (1429)
43. Nopyl acetate	1423	96 (1421)	95 (1413)	96 (1429)
44. *trans*-α-Bergamotene	1432	n.d.	n.d.	94 (1438)
45. α-Guaiene	1438	n.d.	n.d.	92 (1443)
46. Coumarin	1438	n.d.	94 (1452)	94 (1454)
47. (*E*)-β-Farnesene	1452	93 (1452)	n.d.	95 (1455)
48. Seychellene	1445	92 (1449)	n.d.	92 (1460)
49. α-Patchoulene	1459	91 (1461)	n.d.	95 (1472)
50. α-Isomethylionone	1473	94 (1472)	94 (1471)	97 (1481)
51. α-Amorphene	1482	n.d.	n.d.	90 (1486)
52. α-Bulnesene	1505	94 (1502)	90 (1503)	94 (1508)
53. α-Methylionone	1512	92 (1514)	93 (1509)	97 (1521)
54. Tropional	1566	93 (1563)	92 (1568)	94 (1576)
55. Caryophyllene oxide	1587	n.d.	92 (1597)	92 (1595)
56. Hedione	1642	95 (1648)	96 (1648)	95 (1658)
57. Patchouli alcohol	1668	93 (1671)	95 (1700)	94 (1687)
58. α-Hexylcinnamaldehyde	1746	94 (1745)	95 (1750)	97 (1755)

Source: Tranchida et al. (2010a), with permission. Copyright © 2010 by Elsevier.
[a]n.d., not determined.

Figure 6.35 (the peak, defined by the symbol*, can also be seen in Figure 6.33). The library search did not produce a match, and thus the minimum degree of acceptable spectral similarity, a library-matching parameter, was lowered from 90% to 80%. A match for estragole, with 86% MS similarity, was found by the GC–MS software. If the same 2D chromatogram zone in the split-flow analysis is considered, the presence of two partially resolved peaks can be observed (Figure 6.35b): peak 26 was identified as estragole with a 98% spectral similarity, while peak 27 was assigned as α-terpineol, with a 93% value.

The rapid-scanning qMS employed by Tranchida et al. (2010a) generated a sufficient number of spectra for qualitative purposes, and nearing the requirements for correct peak re-construction. The slower second-dimension chromatography,

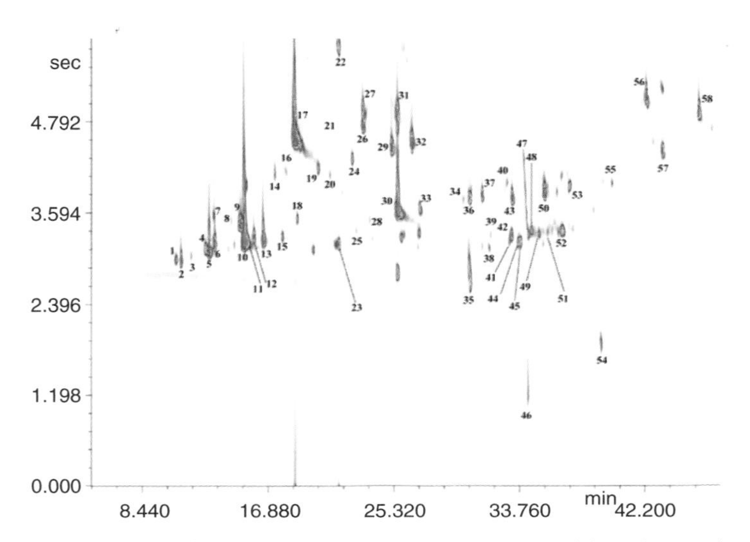

Figure 6.34 Split-flow GC × GC–qMS analysis of a commercial perfume using a 50 – μm i.d. secondary column. For peak identification, refer to Table 6.3. [From Tranchida et al. (2010a), with permission. Copyright © 2010 by Elsevier.]

apart from optimizing the separation, allowed the acquisition of a higher number of spectra: for example, considering α-thujene (1) in Figure 6.34, it was modulated twice, with both peaks reconstructed with six data points; considering α-hexylcinnamaldehyde (58), it was modulated four times, with all peaks formed by at least seven data points.

A further GC × GC–qMS study, carried out by Tranchida et al., (2010b), was directly related to the previous GC × GC–qMS work. The main aims of the GC × GC–qMS experiment were (1) to generate second-dimension chromatographic bands that could be accurately reconstructed (minimum 10 data points per/peak); (2) to operate both columns under optimum gas flow conditions so that no analytical compromise was required to achieve objective (1); and (3) to avoid the use of a narrow mass range, to increase the MS acquisition frequency; hence, no MS compromise must be employed to achieve objective (1). The method developed, applied to the analysis of a commercial perfume, used a primary apolar 30 m × 0.25 mm i.d. capillary connected to a secondary polar 1.0 m × 0.05 mm i.d. capillary and to a 0.10 m × 0.05 mm i.d. uncoated segment. The latter directed most of the primary column effluent to waste. The best GC × GC–qMS result (about 10% of the flow reached the MS) was achieved with He linear velocities of about 20 and 80 cm/s in the first and second dimensions, respectively, using a +20°C temperature offset in GC2. The 2D peak widths (6σ) of 10 low-amount analytes, belonging to different chemical groups (monoterpene/sesquiterpene hydrocarbons, ketones, an aldehyde, and an alcohol) and spanning the entire chromatogram, were measured ($n = 3$). Peaks with $S/N < 10$ and thus not subject to quantification were not taken into consideration. As can be seen from the data reported in Table 6.4, peak widths were

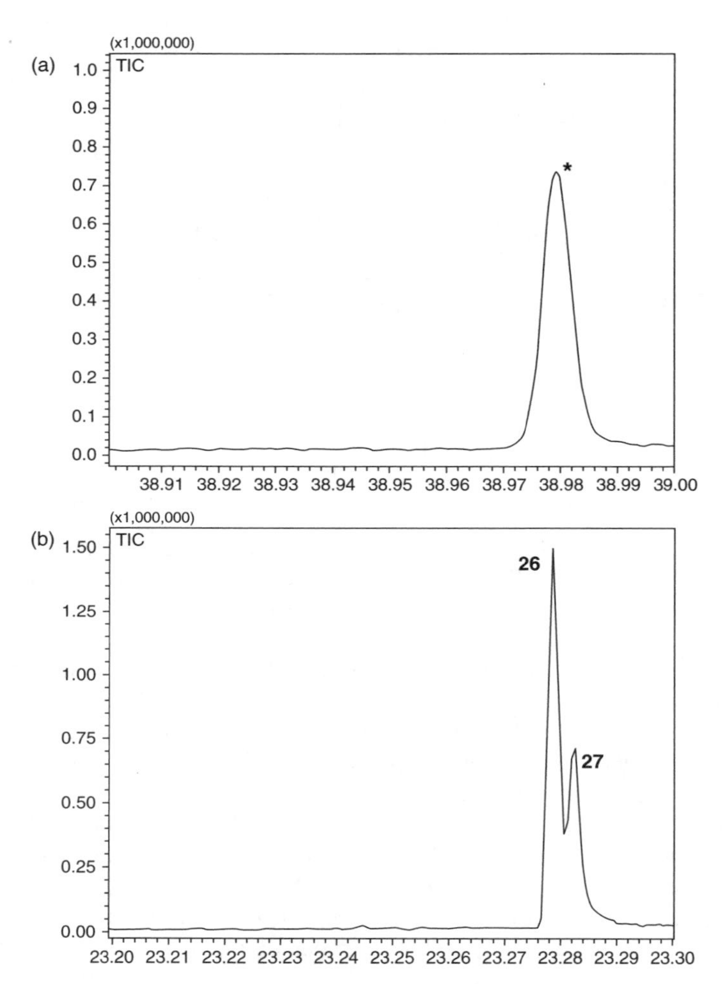

Figure 6.35 Single untransformed second-dimension separations relative to the same 2D chromatogram zone in (a) conventional and (b) split-flow GC × GC–qMS experiments. For peak identification, see Table 6.3. [From Tranchida et al. (2010a), with permission. Copyright © 2010 by Elsevier.]

in the range 360 to 576 ms. The qMS was operated at a 25-Hz scan frequency (a "normal" range of 40 to 360 amu was employed) and generated 10 and 16 spectra for the 360- and 576-ms bands, respectively. The spectra numbers, listed in Table 6.4, comprise the two points at peak baseline; CV% values, regarding all peaks (widths), were acceptable (Table 6.4), typical of cryogenic GC × GC. A chromatography band narrower than that of bicyclogermacrene ($S/N \sim 13$), with $S/N > 10$, was not found. The bicyclogermacrene peak, with each data point indicated, is shown in Figure 6.36. The two additional data points reported in the figure have not been numbered because they are situated in the peak tail.

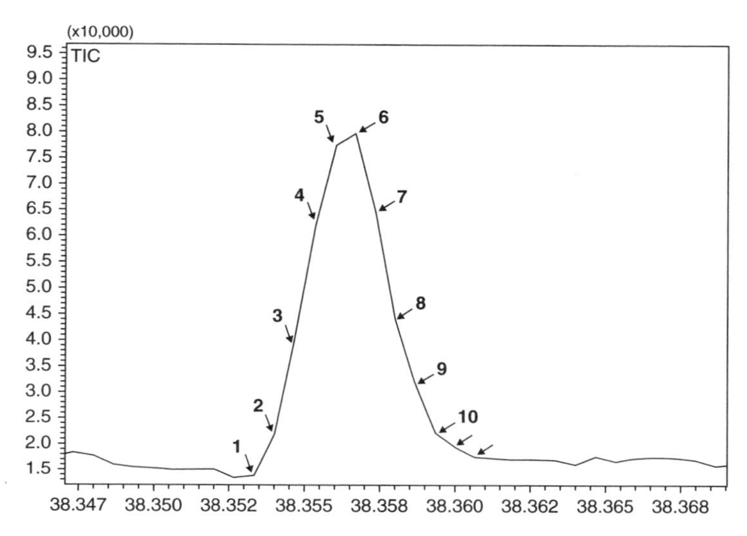

Figure 6.36 Bicyclogermacrene peak with the data points indicated. [From Tranchida et al. (2010b), with permission. Copyright © 2010 by Wiley-VCH Verlag GmbH & Co. KGaA.]

TABLE 6.4 Data Relative to Peak Identity, Peak Base Widths (6σ), Average Widths, CV Values, and Number of Spectra/Peak in Three Consecutive GC × GC–qMS Analyses

Peak/Compound	6σ Width (ms)	Average Width/CV%	Number of Spectra per Peak
1. α-Thujene	432/468/444	448/4.1	11/13/12
2. α-Pinene	444/444/456	448/1.5	12/12/13
3. Camphor	576/552/540	556/3.3	16/15/15
4. Geranyl acetate	444/456/456	452/1.5	12/12/12
5. (*E*)-Caryophyllene	540/552/552	548/1.3	15/15/15
6. α-Isomethylionone	540/552/540	544/1.3	15/15/15
7. Bicyclogermacrene	372/366/360	366/1.6	10/10/10
8. [1-Methyl-, 4-(4-methyl-3-pentenyl)] cyclohex-3-ene-1-carboxaldehyde	480/492/516	496/3.7	12/13/15
9. Unknown	504/504/480	496/2.8	13/14/13
10. 4-(Diethoxymethyl)- α, α-dimethyl-1-cyclohexene-1-butanol	456/480/492	476/3.8	12/13/13

Source: Tranchida et al. (2010b), with permission. Copyright © 2010 by Wiley-VCH Verlag GmbH & Co. KGaA.

TABLE 6.5 Average Mass Spectral Ion Ratios (Indicated by the Ion Ratio) at Sequential Scan Points for Six Compounds[a]

α-Pinene (96%)			Camphor (94%)		
Scan Points	*m/z 79/77*	*m/z 93/91*	*Scan Points*	*m/z 95/81*	*m/z 110/108*
7	0.92 (2.8)	1.89 (5.0)	9	1.02 (4.4)	0.32 (10.0)
Geranyl Acetate (96%)			**(E)-Caryophyllene (94%)**		
Scan Points	*m/z 69/67*	*m/z 93/53*	*Scan Points*	*m/z 91/65*	*m/z 133/95*
8	4.61 (8.9)	2.82 (9.2)	9	4.75 (11.8)	3.65 (12.1)
Bicyclogermacrene (91%)			**4-(Diethoxymethyl)-α,α-Dimethyl-1-cyclohexene-1-butanol (92%)**		
Scan Points	*m/z 107/79*	*m/z 121/93*	*Scan Points*	*m/z 103/75*	*m/z 164/135*
6	0.99 (16.9)	1.25 (6.9)	7	0.91 (4.5)	0.53 (10.6)

Source: Tranchida et al. (2010b), with permission. Copyright © 2010 by Wiley-VCH Verlag GmbH & Co. KGaA.

[a] Values in parentheses indicate the library spectral similarity, beside the compound names; and the CV% value, beside the average mass spectral ion ratio. Copyright © 2010 by Wiley-VCH Verlag GmbH & Co. KGaA.

The extent of skewing was evaluated by calculating two ion ratios, at sequential data points, for six compounds: CV values were in the range 2.8 to 16.9%, with an average CV value of 8.6% (Table 6.5). In a previous study by Adahchour et al. (2005a) devoted to the evaluation of the performance of the same mass spectrometer (Shimadzu QPMS-2010), CV values for mass ratios, derived from consecutive spectra, were always less than 7% when a 33-Hz frequency was used; however, when the scan frequency was decreased to 20 Hz, CV values as high as 32% were observed. As is to be expected, the skewing results attained by Tranchida et al. (2010b) were similar to those described by Adahchour et al.: better than the 20-Hz result but worse than the 33-Hz result (Adahchour et al., 2005a; Tranchida et al., 2010b). In any case, mass spectral skewing can easily be overcome by using averaged spectra for MS library searching; in fact, satisfactory spectral similarity results, always over 90%, were attained (Table 6.5).

During 2010, a research was described that focused on the evaluation of a novel rapid-scanning qMS detector characterized by a 20,000-amu/s scan speed and a 50-Hz scan frequency, using a "normal" mass range (40 to 330 m/z) (Purcaro et al., 2010). The quadrupole mass spectrometer (Shimadzu QP2010-Ultra) was essentially an upgraded instrument of the qMS system tested by Adahchour et al. (2005a). The performance of the qMS system was evaluated by analyzing mixtures of 24 allergens, as well as a perfume sample, through GC × GC–qMS. The extent of peak reconstruction was assessed by counting the number of data points per/peak above the baseline and over the half-height of the peak, comparing three different acquisition frequencies (50, 33, and 25 Hz), and maintaining the same mass range. For such a purpose, four standard compounds were considered: linalool, eugenol, lilial, and benzyl salicylate. The results are shown in Table 6.6

TABLE 6.6 Target Ion (T) and Qualifier Ions (Q1 and Q2), Peak Width at Baseline and at Half-Height, and Number of Data Points Acquired, for the Main Modulated Peak of Four Compounds

					50 Hz				33 Hz				25 Hz			
Compound	T	Q1	Q2	LM	Peak Width (ms)	Points	Half Peak (ms)	Points	Peak Width (ms)	Points	Half Peak (ms)	Points	Peak Width (ms)	Points	Half Peak (ms)	Points
Linalool	93	71	121	55	360	18	180	9	360	12	180	6	360	9	180	4
Eugenol	164	103	149	55	480	24	240	12	480	16	240	8	480	12	240	6
Lilial	189	147	204	57	420	21	190	10	420	14	180	6	360	9	180	5
Benzyl salicylate	91	228	65	—	480	24	240	12	480	16	240	8	480	12	240	6

Source: Purcaro et al. (2010), with permission from the American Chemical Society.

and are related to the main modulated peak of each compound. It was seen that the 10-data-point-per-peak requirement (Poole, 2003) can be satisfied by using an acquisition frequency of 25 Hz only if the peak width is above 400 ms, while in such a case, more than 20 data points per peak were obtained through the use of a 50-Hz frequency. The mass spectral quality relative to the four compounds was assessed at the three acquisition frequencies tested by (1) evaluating the similarity match (MS%), with MS library-contained compounds at each data point; (2) calculating the intensity ratio between the target ion (T) and the qualifiers ions (Q1 and Q2) to evaluate the consistency of mass spectra profiles. A further evaluation of peak skewing was achieved by considering the ion intensity ratio between the highest mass (HM) and lowest mass (LM) ion present, with a significant abundance in the mass spectrum (an ion value above 50 m/z was chosen to reduce the noise contribution to its relative abundance). With regard to the spectral quality observed for the four compounds considered, MS% values were nearly always constant at each data point (all above 90%), T/Q ratios were comparable under all the conditions, tested while the coefficients of variation relative to the HM/LM ion ratios were between 5 and 10% when using 33 and 50 Hz and were in the range 10 to 15% when using 25 Hz. The GC × GC–qMS method developed was validated and proved to be suitable for the analysis of perfume allergens, according to the requirements of Directive 2003/15/EC. The research was the first report of true full-scan quantification in GC × GC–qMS (extracted ion chromatograms were used) using a standard comprehensive 2D GC setup. It was reported that more than 15 data points per peak were attained, fully meeting the requirements for reliable peak reconstruction. The authors concluded the paper by affirming that over the 1999–2010 period it has been affirmed, time and time again, that a qMS is fine for identification purposes, but a time-of-flight MS is necessary for accurate peak reconstruction. It is obvious that notwithstanding the validity of TOF MS instrumention, the possibility of using a quadrupole MS for quantification in GC × GC analysis is certainly a desirable additional option.

6.7 CONCLUDING REMARKS

Comprehensive two-dimensional gas chromatography in combination with mass spectrometry fulfills the requirements of many, if not practically all GC–MS analysts: unprecedented selectivity (three separation dimensions, related to volatility, polarity, and mass), high sensitivity (through band compression), enhanced separation power, structured chromatograms, and increased speed (comparable to very fast GC experiments if the number of peaks resolved per unit of time is considered). It must be added that every picture is not so nice because whereas GC × GC–MS is continuously gaining popularity and favor, it is still far from being well established, with the main hinderances probably deriving from a natural scepticism toward new methodologies as well as the high costs per analysis (in the case of cryogenic modulation), total instrumental costs, and the scarcity of

GC × GC software packages. It may be anticipated that if further improvement occurs, wherever necessary, GC × GC–MS will undergo a gradual and constant expansion in the next decade.

Summarizing the data reported here, during the 1999–2010 (June) period, just over 200 papers have been published in the GC × GC–MS field, with over 80% of these based on the use of TOF MS and about 16% on the employment of a qMS. With regard to other MS devices exploited in the GC × GC field, at the moment they have had little or no impact. Hopefully, in this chapter we have provided the reader with a comprehensive perspective on the evolution of GC × GC–MS and on its analytical potential in various research fields. We would also like to think that the information provided could act as a stimulant, and hence contribute to the expansion of this powerful 3D methodology.

A final note is devoted to the inventor of comprehensive two-dimensional gas chromatography, certainly one of the most revolutionary discoveries in gas chromatography, and to the many other scientists who have achieved fundamental advances in the GC × GC field. The wealth of information and intuitions contained in many primordial papers can still be exploited to bring progress to this wonderful area of research.

REFERENCES AND FURTHER READING

Adahchour M, Beens J, Vreuls RJJ, Batenburg AM, Rosing EAE, Brinkman UAT. *Chromatographia* 2002; 55:361–367.

Adahchour M, van Stee LLP, Beens J, Vreuls RJJ, Batenburg AM, Brinkman UAT. *J. Chromatogr. A* 2003; 1019:157–172.

Adahchour M, Beens J, Vreuls RJJ, Batenburg AM, Brinkman UAT. *J. Chromatogr. A* 2004; 1054:47–55.

Adahchour M, Brandt M, Baier H-U, Vreuls RJJ, Batenburg AM, Brinkman UAT. *J. Chromatogr. A* 2005a; 1067:245–254.

Adahchour M, Wiewel J, Verdel R, Vreuls RJJ, Udo AT. Brinkman UAT. *J. Chromatogr. A* 2005b; 1086:99–106.

Adahchour M, Beens J, Vreuls RJJ, Brinkman UAT. *Trends Anal. Chem*. 2006a; 25:438–454.

Adahchour M, Beens J, Vreuls RJJ, Brinkman UAT. *Trends Anal. Chem*. 2006b; 25:540–553.

Adahchour M, Beens J, Vreuls RJJ, Brinkman UAT. *Trends Anal. Chem*. 2006c; 25:726–741.

Adahchour M, Beens J, Vreuls RJJ, Brinkman UAT. *Trends Anal. Chem*. 2006d; 25:821–840.

Adahchour M, Beens J, Brinkman UAT. *J. Chromatogr. A* 2008; 1186:67–108.

Adam F, Bertoncini F, Brodusch N, Durand E, Thiébaut D, Espinat D, Hennion M-C. *J. Chromatogr. A* 2007; 1148:55–64.

Adam F, Bertoncini F, Coupard V, Charon N, Thiébaut D, Espinat D, Hennion M-C. *J. Chromatogr. A* 2008; 1186:236–244.

Akoto L, Frans S, Irth H, Vreuls RJJ, Pel R. *J. Chromatogr. A* 2008; 1186:254–261.

Almstetter MF, Appel IJ, Gruber MA, Lottaz C, Timischl B, Spang R, Dettmer K, Oefner PJ. *Anal. Chem.* 2009; 81:5731–5739.

Amador-Mu noz O, Villalobos-Pietrini R, Aragón-Pi na A, Tran TC, Morrison P, Marriott PJ. *J. Chromatogr. A* 2008; 1201:161–168.

Amirav A, Gordin A, Poliak M, Fialkov AB. *J. Mass Spectrom.* 2008; 43:141–163.

Araújo RCS, Pasa VMD, Marriott PJ, Cardeal ZL. *J. Anal. Appl. Pyrol.* 2010; 88:91–97.

Ávila BMF, Aguiar A, Gomes AO, Azevedo DA. *Org. Geochem.* doi:10.1016/j.orggeochem.2010.03.008

Banerjee K, Patil SH, Dasgupta S, Oulkar DP, Patil SB, Savant R, Adsule PG. *J. Chromatogr. A* 2008; 1190:350–357.

Beens J, Tijssen R, Blomberg J. *J. Chromatogr. A* 1998; 822:233–251.

Bianchi F, Careri M, Conti C, Musci M, Vreuls R. *J. Sep. Sci.* 2007; 30:527–533.

Blumberg LM, Klee MS. *Anal. Chem.* 1998; 70:3828–3839.

Booth AM, Sutton PA, Lewis CA, Lewis AC, Scarlett A, Chau W, Widdows J, Rowland SJ. *Environ. Sci. Technol.* 2007; 41:457–464.

Breme K, Tournayre P, Fernandez X, Meienhenrich UJ, Brevard H, Joulain D, Berdagué JL. *J. Agric. Food Chem.* 2010; 58:473–480.

Brenna JT. *Acc. Chem. Res.* 1994; 27:340–346.

Čajka T, Hajšlová J, Cochran J, Holadová K, Klimánková E. *J. Sep. Sci.* 2007; 30:534–546.

Čajka T, Hajšlová J, Pudil F, Riddellova K. *J. Chromatogr. A* 2009; 1216:1458–1462.

Čajka T, Riddellova K, Klimánková E, Cerna M, Pudil F, Hajšlová J. *Food Chem.* 2010; 121:282–289.

Cardeal ZL, Marriott PJ. *Food Chem.* 2009; 112:747–755.

Cardeal ZL, Gomes da Silva MDR, Marriott PJ. *Rapid Commun. Mass Spectrom.* 2006; 20:2823–2836.

Cardeal ZL, de Souza PP, Gomes da Silva MDR, Marriott PJ. *Talanta* 2008; 74:793–799.

Chin S-T, Che Man YB, Tan CP, Hashim DM. *JAOCS* 2009; 86:949–958.

Cochran J. *J. Chromatogr. A* 2008; 1186:202–210.

Cordero C, Rubiolo P, Sgorbini B, Galli M, Bicchi C. *J. Chromatogr. A* 2006; 1132:268–279.

Cordero C, Bicchi C, Joulain D, Rubiolo P. *J. Chromatogr. A* 2007; 1150:37–49.

Cordero C, Bicchi C, Galli M, Galli S, Rubiolo P. *J. Sep. Sci.* 2008; 31:3437–3450.

Cordero C, Liberto E, Bicchi C, Rubiolo P, Reichenbach SE, Tian X, Tao Q. *J. Chromatogr. Sci.* 2010; 48:251–261.

Cortes HJ, Winniford B, Luong J, Pursch M. *J. Sep. Sci.* 2009; 32:883–904.

Cramers CA, Leclercq PA. *CRC Crit. Rev. Anal. Chem.* 1988; 20:117–147.

d'Acampora Zellner B, Casilli A, Dugo P, Dugo G, Mondello L. *J. Chromatogr. A* 2007; 1141:279–286.

Dallüge J, Vreuls RJJ, Beens J, Brinkman UAT. *J. Sep. Sci.* 2002a; 25:201–214.

Dallüge J, van Stee LLP, Xu X, Williams J, Beens J, Vreuls RJJ, Brinkman UAT. *J. Chromatogr. A* 2002b; 974:169–184.

Dallüge J, van Rijn M, Beens J, Vreuls RJJ, Brinkman UAT. *J. Chromatogr A* 2002c; 965:207–217.

Dallüge J, Roose P, Brinkman UAT. *J. Chromatogr. A* 2002d; 970:213–223.

Dallüge J, Beens J, Brinkman UAT. *J. Chromatogr. A* 2003; 1000:69–108.

Dasgupta S, Banerjee K, Patil SH, Ghaste M, Dhumal KN, Adsule PG. *J. Chromatogr. A* 2010; 1217:3881–3889.

David F, Gere DR, Scanlan F, Sandra P. *J. Chromatogr. A* 1999; 842:309–319.

de Souza PP, Cardeal ZdL, Augusti R, Morrison P, Marriott PJ. *J. Chromatogr. A* 2009; 1216:2881–2890.

Debonneville C, Chaintreau A. *J. Chromatogr. A* 2004; 1027:109–115.

Diehl JW, Di Sanzo FP. *J. Chromatogr. A* 2005; 1080:157–165.

Dimandja JMD. *Am. Lab*. 2003; 35:42–53.

Dimandja JMD. *Anal. Chem*. 2004; 76: 167A–174A.

Dimandja JMD, Grainger J, Patterson DG Jr. Pittsburgh Conference, New Orleans, LA, 2000. Abstract 267.

Dimandja JMD, Clouden GC, Colón I, Focant J-F, Cabey WV, Parry RC. *J. Chromatogr. A* 2003; 1019:261–272.

Dück R, Wulf V, Geißler M, Baier H-U, Wirtz M, Kling HW, Gäb S, Schmitz OJ. *Anal. Bioanal. Chem*. 2010; 396:2273–2283.

Eganhouse RP, Pontolillo J, Gaines RB, Frysinger GS, Gabriel FLP, Kohler H-PE, Giger W, Barber LB. *Environ. Sci. Technol*. 2009; 43:9306–9313.

Eyres G, Dufour J-P, Hallifax G, Sotheeswaran S, Marriott PJ. *J. Sep. Sci*. 2005; 28:1061–1074.

Eyres G, Marriott PJ, Dufour J-P. *J. Chromatogr. A* 2007a; 1150:70–77.

Eyres G, Marriott PJ, Dufour J-P. *J. Agric. Food Chem*. 2007b; 55:6252–6261.

Flego C, Gigantiello N, Parker WO Jr, Calemma V. *J. Chromatogr. A* 2009; 1216:2891–2899.

Focant J-F, Sjödin A, Patterson DG Jr. *J. Chromatogr. A* 2003; 1019:143–156.

Focant J-F, Sjödin A, Patterson DG Jr *J. Chromatogr. A* 2004a; 1040:227–238.

Focant J-F, Sjödin A, Turner WE, Patterson DG Jr. *Anal. Chem*. 2004b; 76:6313–6320.

Focant J-F, Reiner EJ, MacPherson K, Kolic T, Sjödin A, Patterson DG Jr, Reese SL, Dorman FL, Cochran J. *Talanta* 2004c; 63:1231–1240.

Focant J-F, Eppe G, Scippo M-L, Massart A-C, Pirard C, Maghuin-Rogister G, De Pauw E. *J. Chromatogr. A* 2005; 1086:45–60.

Freitas LS, Von Mühlen C, Bortoluzzi JH, Zini CA, Fortuny M, Dariva C, Coutinho RCC, Santos AF, Caram ao EB. *J. Chromatogr. A* 2009; 1216:2860–2865.

Frysinger GS, Gaines RB. *J. High Resolut. Chromatogr*. 1999; 22:251–255.

Frysinger GS, Gaines RB, Reddy CM. *Environ. Forens*. 2002; 3:27–34.

Göğü F, Özel MZ, Lewis AC. *J. Sep. Sci*. 2006; 29:1217–1222.

Goldstein AH, Worton DR, Williams BJ, Hering SV, Kreisberg NM, Panić O, Górecki T. *J. Chromatogr. A* 2008; 1186:340–347.

Górecki T, Harynuk N, Panic O. *J. Sep. Sci*. 2004; 27:359–379.

Górecki T, Panic O, Oldridge N. *J. Liq. Chromatogr*. 2006; 29(7–8):1077–1104.

Gröger Th, Schäffer M, Pütz M, Ahrens B, Drew K, Eschner M, Zimmermann R. *J. Chromatogr. A* 2008a; 1200:8–16.

Gröger T, Welthagen W, Mitschke S, Schäffer M, Zimmermann R. *J. Sep. Sci*. 2008b; 31:3366–3374.

Gross JH. *Mass Spectrometry: A Textbook*. Heidelberg, Germany: Springer-Verlag, 2004.

Guthery B, Bassindale A, Pillinger CT, Morgan GH. *Rapid Commun. Mass Spectrom*. 2009; 23:340–348.

Hajšlová J, Pulkrabová J, Poustka J, Čajka T, Randák T. *Chemosphere* 2007; 69:1195–1203.

Hamilton JF. *J. Chromatogr. Sci*. 2010; 48:274–282.

Hamilton JF, Webb PJ, Lewis AC, Hopkins JR, Smith S, Davy P. *Atmos. Chem. Phys*. 2004; 4:1279–1290.

Hamilton JF, Webb PJ, Lewis AC, Reviejo MM. *Atmos. Environ*. 2005; 39:7263–7275.

Hamilton JF, Lewis AC, Millan M, Bartle KD, Herod AA, Kandiyoti R. *Energy Fuels* 2007; 21:286–294.

Hao C, Headley JV, Peru KM, Frank R, Yang P, Solomon KR. *J. Chromatogr. A* 2005; 1067:277–284.

Harju M, Bergman A, Olsson M, Roos A, Haglund P. *J. Chromatogr. A* 2003; 1019:127–142.

Hayward DG, Pisano TS, Wong JW, Scudder RJ. *J. Agric. Food Chem*. 2010; 58:5248–5256.

Hejazi L, Ebrahimi D, Guilhaus M, Hibbert DB. *Anal. Chem*. 2009; 81:1450–1458.

Hoggard JC, Synovec RE. *Anal. Chem*. 2007; 79:1611–1619.

Hoggard JC, Siegler WC, Synovec RE. *J. Chemometr*. 2008; 23:421–431

Hoggard JC, Wahl JH, Synovec RE, Mong GM, Fraga CG. *Anal. Chem*. 2010; 82:689–698.

Hoh E, Mastovska K, Lehotay SJ. *J. Chromatogr. A* 2007; 1145:210–221.

Hoh E, Lehotay SJ, Mastovska K, Huwe JK. *J. Chromatogr. A* 2008; 1201:69–77.

Hoh E, Lehotay SJ, Mastovska K, Ngo HL, Vetter W, Pangallo KC, Reddy CM. *Environ. Sci. Technol*. 2009a; 43:3240–3247.

Hoh E, Lehotay SJ, Pangallo KC, Mastovska K, Ngo HL, Reddy CM, Vetter W. *J. Agric. Food Chem*. 2009b; 57:2653–2660.

Holland JF, Enke CG, Allison J, Stults JT, Pinkston JD, Newcome B, Watson JT. *Anal. Chem*. 1983; 55: 997A–1012A.

Hope JL, Sinha AE, Prazen BJ, Synovec RE. *J. Chromatogr. A* 2005a; 1086:185–192.

Hope JL, Prazen BJ, Nilsson EJ, Lidstrom ME, Synovec RE. *Talanta* 2005b; 65:380–388.

Houtman CJ, Booij P, Jover E, Pascual del Rio D, Swart K, van Velzen V, Vreuls R, Legler J, Brouwer A, Lamoree MH. *Environ. Chem*. 2006; 65:2244–2252.

Huang X, Regnier FE. *Anal. Chem*. 2008; 80:107–114.

Humston EM, Dombek KM, Hoggard JC, Young ET, Synovec RE. *Anal. Chem*. 2008; 80:8002–8011.

Humston EM, Zhang Y, Brabeck GF, McShea A, Synovec RE. *J. Sep. Sci*. 2009; 32:2289–2295.

Humston EM, Knowles JD, McShea A, Synovec RE. *J. Chromatogr. A* 2010; 1217:1963–1970.

Ieda T, Horii Y, Petrick G, Yamashita N, Ochiai N, Kannan K. *Environ. Sci. Technol*. 2005; 39:7202–7207.

Jeleń HH, W1sowicz E. *J. Chromatogr. A* 2008; 1215:203–207.

Jover E, Adahchour M, Bayona JM, Vreuls RJJ, Brinkman UAT. *J. Chromatogr. A* 2005; 1086:2–11.

Jover E, Matamoros V, Bayona JM. *J. Chromatogr. A* 2009; 1216:4013–4019.

Kaal E, de Koning S, Brudin S, Janssen H-G. *J. Chromatogr. A* 2008; 1201:169–175.

Kalinová B, Kindl J, Jiroš P, Žážcek P, Vašicková S, Budčšinský M, Valterová I. *J. Nat. Prod*. 2009; 72:8–13.

Kallio M, Hyötyläinen T, Lehtonen M, Jussila M, Hartonen K, Shimmo M, Riekkola M-L. *J. Chromatogr. A* 2003; 1019:251–260.

Kallio M, Jussila M, Rissanen T, Anttila P, Hartonen K, Reissell A, Vreuls R, Adahchour M, Hyötyläinen T. *J. Chromatogr. A* 2006; 1125:234–243.

Khummueng W, Trenerry C, Rose G, Marriott PJ. *J. Chromatogr. A* 2006; 1131:203–214.

Klee MS, Blumberg LM. *J. Chromatogr. Sci*. 2002; 40:234–247.

Klimánková E, Holadová K, Hajšlová J, Eajka T, Poustka J, Koudela M. *Food Chem*. 2008; 107:464–472.

Kochman M, Gordin A, Alon T, Amirav A. *J. Chromatogr. A* 2006; 1129:95–104.

Koek MM, Muilwijk B, van Stee LLP, Hankemeier T. *J. Chromatogr. A* 2008; 1186:420–429.

Kohl A, Cochran J, Cropek DM. *J. Chromatogr. A* 2010; 1217:550–557.

Korytár P, van Stee LLP, Leonards PEG, de Boer J, Brinkman UAT. *J. Chromatogr. A* 2003; 994:179–189.

Korytár P, Parera J, Leonards PEG, de Boer J, Brinkman UAT. *J. Chromatogr. A* 2005a; 1067:255–264.

Korytár P, Parera J, Leonards PEG, Santos FJ, de Boer J, Brinkman UAT. *J. Chromatogr. A* 2005b; 1086:71–82.

Korytár P, Covaci A, Leonards PEG, de Boer J, Brinkman UAT. *J. Chromatogr. A* 2005c; 1100:200–207.

Kortyár P, Haglund P, de Boer J, Brinkman UAT. *Trends Anal. Chem*. 2006; 25:373–396.

Kouremenos KA, Pitt J, Marriott PJ. *J. Chromatogr. A* 2010a; 1217:104–111.

Kouremenos KA, Harynuk JJ, Winniford WL, Morrison PD, Marriott PJ. *J. Chromatogr. B* 2010b; 878:1761–1770.

Kuhn F, Natsch A. *J. R. Soc. Interface* 2009; 6:377–392.

Kusano M, Fukushima A, Kobayashi M, Hayashi N, Jonsson P, Moritz T, Ebana K, Saito K. *J. Chromatogr. B* 2007; 855:71–79.

Laitinen T, Martin SH, Parshintsev J, Hyötyläinen T, Hartonen K, Riekkola M-L, Kulmala M, Pavón JLP. *J. Chromatogr. A* 2010; 1217:151–159.

Li X, Xu Z, Lu X, Yang X, Yin P, Kong H, Yu Y, Xu G. *Anal. Chim. Acta* 2009; 633:257–262.

Lojzova L, Riddellova K, Hajšlová J, Zrostlíková J, Schurek J, Čajka T. *Anal. Chim. Acta* 2009; 641:101–109.

Lu X, Cai J, Kong H, Wu M, Hua R, Zhao M, Liu J, Xu G. *Anal. Chem*. 2003; 75:4441–4451.

Lu X, Zhao M, Kong H, Cai J, Wu J, Wu M, Hua R, Liu J, Xu G. *J. Chromatogr. A* 2004a; 1043:265–273.

Lu X, Zhao M, Kong H, Cai J, Wu J, Wu M, Hua R, Liu J, Xu G. *J. Sep. Sci*. 2004b; 27:101–109.

Ma C, Wang H, Lu X, Li H, Liu B, Xu G. *J. Chromatogr. A* 2007; 1150: 50–53.

Ma C, Wang H, Lu X, Wang H, Xu G, Liu B. *Metabolomics* 2009; 5: 497–506.

Mao D, Lookman R, Van De Weghe H, Weltens R, Vanermen G, De Brucker N, Diels L. *Environ. Sci. Technol*. 2009; 43:7651–7657.

Marriott PJ, Shellie R. *Trends Anal. Chem*. 2002; 21:573–583.

Marriott PJ, Kinghorn RM, Ong R, Morrison PJ, Haglund P, Harju M. *High Resol. Chromatogr*. 2000; 23:253–258.

Matamoros V, Jover E, Bayona JM. *Anal. Chem*. 2010; 82:699–706.

Mateus E, Barata RC, Zrostlíková J, Gomes da Silva MDR, Paiva MR. *J. Chromatogr. A* 2010a; 1217:1845–1855.

Mateus EP, Zrostlíková J, Gomes da Silva MDR, Ribeiro AB, Marriott P. *J. Appl. Electrochem*. 2010b; 40:1183–1193.

Mayadunne R, Nguyen T-T, Marriott PJ. *Anal. Bioanal. Chem*. 2005; 382:836–847.

McGuigan M, Waite JH, Imanaka H, Sacks RD. *J. Chromatogr. A* 2006; 1132:280–288.

Melbye AG, Brakstad OG, Hokstad JN, Gregersen IK, Hansen BH, Booth AM, Rowland SJ, Tollesfen KE. *Environ. Toxicol. Chem*. 2009; 28:1815–1824.

Micyus NJ, Seeley SK, Seeley JV. *J. Chromatogr. A* 2005; 1086:171–174.

Mitrevski BS, Brenna JT, Zhang, Y Marriott PJ. *J. Chromatogr. A* 2008; 1214:134–142.

Mitrevski BS, Wilairat P, Marriott PJ. *Anal. Bioanal. Chem*. 2010a; 396:2503–2511.

Mitrevski BS, Wilairat P, Marriott PJ. *J. Chromatogr. A* 2010b; 1217:127–135.

Mitschke S, Welthagen W, Zimmermann R. *Anal. Chem*. 2006; 78:6364–6375.

Mleth M, Schubert JK, Gröger T, Sabei B, Kischkel S, Fuchs P, Hein D, Zimmermann R, Miekisch W. *Anal. Chem*. 2010; 82:2541–2551.

Moeder M, Martin C, Schlosser D, Harynuk J, Górecki T. *J. Chromatogr. A* 2006a; 1107:233–239.

Mohler RE, Dombek KM, Hoggard JC, Young ET, Synovec RE. *Anal. Chem*. 2006b; 78:2700–2709.

Mohler RE, Tu BP, Dombek KM, Hoggard JC, Young ET, Synovec RE. *J. Chromatogr. A* 2008; 1186:401–411.

Mondello L, Casilli A, Tranchida PQ, Dugo P, Costa R, Festa S, Dugo G. *J. Sep. Sci*. 2004; 27:442–450.

Mondello L, Casilli A, Tranchida PQ, Dugo G, Dugo P. *J. Chromatogr. A* 2005; 1067:235–243.

Mondello L, Casilli A, Tranchida PQ, Lo Presti M, Dugo P, Dugo G. *Anal. Bioanal. Chem*. 2007; 389:1755–1763.

Mondello L, Tranchida PQ, Dugo P, Dugo G. *Mass Spectrom. Rev*. 2008; 27:101–124.

Morales-Mu ñoz S, Vreuls RJJ, Luque de Castro MD. *J. Chromatogr. A* 2005; 1086:122–127.

Mullins OC, Ventura GT, Nelson RK, Betancourt SS, Raghuraman B, Reddy CM. *Energy Fuels* 2008; 22:496–503.

Ochiai N, Ieda T, Sasamoto K, Fushimi A, Hasegawa S, Tanabe K, Kobayashi S. *J. Chromatogr. A* 2007; 1150:13–20.

O'Hagen S, Dunn WB, Knowles JD, Broadhurst D, Williams R, Ashworth JJ, Cameron M, Kell DB. *Anal. Chem*. 2007; 79:464–476.

Özel MZ, Göğü F, Lewis AC. *Food Chem*. 2003; 82:381–386.

Özel MZ, Göğü F, Hamilton JF, Lewis AC. *Chromatographia* 2004; 60:79–83.

Özel MZ, Göğü F, Hamilton JF, Lewis AC. *Anal. Bioanal. Chem*. 2005; 382:115–119.

Özel MZ, Göğü F, Lewis AC. *Anal. Chim. Acta* 2006a; 566:172–177.

Özel MZ, Göğü F, Lewis AC. *J. Chromatogr. A* 2006b; 1114:164–169.

Özel MZ, Ward MW, Hamilton JF, Lewis AC, Raventos-Duran T, Harrison RM. *Aerosol Sci. Tech*. 2010; 44:109–116.

Pangallo K, Nelson RK, Teuten EL, Pedler BE, Reddy CM. *Chemosphere* 2008; 71:1557–1565.

Parsi Z, Górecki T, Poerschmann J. *LC-GC Eur*. 2005; 18:582–587.

Peacock EE, Arey JS, DeMello JA, McNichol AP, Nelson RK, Reddy CM. *Energy Fuels* 2010; 24:1037–1042.

Perestrelo R, Petronilho S, Câmara JS, Rocha SM. *J. Chromatogr. A* 2010; 1217:3441–3445.

Peters R, Tonoli D, van Duin M, Mommers J, Mengerink Y, Wilbers ATM, van Benthem R, de Koster Ch, Schoenmakers PJ, van der Wal Sj. *J. Chromatogr. A* 2008a; 1201:141–150.

Peters R, van Duin M, Tonoli D, Kwakkenbos G, Mengerink Y, van Benthem RATM, de Koster CG, Schoenmakers PJ, van der Wal SJ. *J. Chromatogr. A* 2008b; 1201:151–160.

Phillips JB, Beens J. *J. Chromatogr. A* 1999; 856:331–347.

Pierce KM, Hope JL, Hoggard JC, Synovec RE. *Talanta* 2006a; 70:797–804.

Pierce KM, Hoggard JC, Hope JL, Rainey PM, Hoofnagle AN, Jack RM, Wright BW, Synovec RE. *Anal. Chem*. 2006b; 78:5068–5075.

Poliak M, Fialkov AB, Amirav A. *J. Chromatogr. A* 2008; 1210:108–114.

Poole CF. *The Essence of Chromatography*. Amsterdam: Elsevier, 2003, pp. 66–67.

Pripdeevech P, Wongpornchai S, Marriott PJ. *Phytochem. Anal*. 2010; 21:163–173.

Purcaro G, Morrison P, Moret S, Conte LS, Marriott PJ. *J. Chromatogr. A* 2007; 1161:284–291.

Purcaro G, Tranchida PQ, Ragonese C, Conte L, Dugo P, Dugo G, Mondello L. *Anal. Chem*. 2010; 82:8583–8590.

Qian K, Dechert GJ. *Anal. Chem*. 2002; 74:3977–3983.

Qiu Y, Lu X, Pang T, Zhu S, Kong H, Xu G. *J. Pharm. Biomed. Anal*. 2007; 43:1721–1727.

Qiu Y, Lu X, Pang T, Ma C, Li X, Xu G. *J. Sep. Sci*. 2008; 31:3451–3457.

Ralston-Hooper K, Hopf A, Oh C, Zhang X, Adamec J, Sepúlveda MS. *Aquat. Toxicol*. 2008; 88:48–52.

Ratel J, Engel E. *J. Chromatogr. A* 2009; 1216:7889–7898.

Reichenbach SE, Kottapalli V, Mingtian N, Visvanathan A. *J. Chromatogr. A* 2005; 1071:263–269.

Roberts MT, Dufour J-P, Lewis AC. *J. Sep. Sci*. 2004; 27:473–478.

Rocha SM, Coelho E, Zrostlíková J, Delgadillo I, Coimbra MA. *J. Chromatogr. A* 2007; 1161:292–299.

Rochat S, de Saint Laumer J-Y, Chaintreau A. *J. Chromatogr. A* 2007; 1147:85–94.

Rochat S, Egger J, Chaintreau A. *J. Chromatogr. A* 2009; 1216:6424–6432.

Ryan D, Shellie R, Tranchida P, Casilli A, Mondello L, Marriott P. *J. Chromatogr. A* 2004; 1054: 57–65.

Ryan D, Watkins P, Smith J, Allen M, Marriott P. *J. Sep. Sci*. 2005; 28:1075–1082.

Sanchez JM, Sacks RD. *Anal. Chem*. 2006; 78:3046–3054.

Schmarr H-G, Bernhardt J. *J. Chromatogr. A* 2010; 1217:565–574.

Schnelle-Kreis J, Welthagen W, Sklorz M, Zimmermann R. *J. Sep. Sci*. 2005; 28:1648–1657.

Schoenmakers PJ, Marriott PJ, Beens J. *LC-GC Eur*. 2003; 16:1–4.

Schurek J, Portolés T, Hajslova J, Riddellova, Hernández F. *Anal. Chim. Acta* 2008; 611:163–172.

Sciarrone D, Tranchida PQ, Costa R, Donato P, Ragonese C, Dugo P, Dugo G, Mondello L. *J. Sep. Sci*. 2008; 31:3329–3336.

Seeley JV, Kramp F, Hicks CJ. *Anal. Chem*. 2000; 72:4346–4352.

Seeley JV, Micyus NJ, McCurry JD, Seeley SK. *Am. Lab*. 2006; 38:24–26.

Shellie R. *LC-GC Eur*. 2008; 21:572–578.

Shellie R, Marriott PJ. *Anal. Chem*. 2002; 74:5426–5430.

Shellie RA, Marriott PJ. *Analyst* 2003a; 128:879–883.

Shellie RA, Marriott PJ. *Flavour Fragr. J*. 2003b; 18:179–191.

Shellie R, Marriott P, Morrison P. *Anal. Chem*. 2001; 73:1336–1344.

Shellie RA, Marriott PJ, Huie CW. *J. Sep. Sci*. 2003; 26:1185–1192.

Shellie R, Marriott P, Morrison P, Mondello L. *J. Sep. Sci*. 2004a; 27:503–512.

Shellie R, Marriott P, Morrison P. *J. Chromatogr. Sci*. 2004b; 42:417–422.

Shellie RA, Welthagen W, Zrostliková J, Spranger J, Ristow M, Fiehn O, Zimmermann R. *J. Chromatogr. A* 2005; 1086:83–90.

Shunji H, Yoshikatsu T, Akihiro F, Hiroyasu I, Kiyoshi T, Yasuyuki S, Masa-aki U, Akihiko K, Kazuo T, Hideyuki O, Katsunori A. *J. Chromatogr. A* 2008; 1178:187–198.

Silva AI Jr, Pereira HMG, Casilli A, Conceiç ao FC, Aquino Neto FR. *J. Chromatogr. A* 2009; 1216:2913–2922.

Sinha AE, Prazen BJ, Fraga CG, Synovec RE. *J. Chromatogr. A* 2003; 1019:79–87.

Sinha AE, Fraga CG, Prazen BJ, Synovec RE. *J. Chromatogr. A* 2004a; 1027:269–277.

Sinha AE, Hope JL, Prazen BJ, Nilsson EJ, Jack RM, Synovec RE. *J. Chromatogr. A* 2004b; 1058:209–215.

Sinha AE, Hope JL, Prazen BJ, Fraga CG, Nilsson EJ, Synovec RE. *J. Chromatogr. A* 2004c; 1056:145–154.

Skoczycska E, Korytár P, de Boer J. *Environ. Sci. Technol*. 2008; 42:6611–6618.

Song SM, Marriott P, Wynne P. *J. Chromatogr. A* 2004a; 1058:223–232.

Song SM, Marriott P, Kotsos A, Drummer OH, Wynne P. *Forens. Sci. Int*. 2004b; 143:87–101.

Stanimirova I, Üstün B, Čajka T, Riddelova K, Hajšlová J, Buydens LMC, Walczak B. *Food Chem*. 2010; 118:171–176.

Tobias HJ, Sacks GL, Zhang Y, Brenna JT. *Anal. Chem*. 2008; 80:8613–8621.

Tran TC, Marriott PJ. *Atmos. Environ*. 2008; 42:7360–7372.

Tranchida PQ, Dugo P, Dugo G, Mondello L. *J. Chromatogr. A* 2004; 1054:3–16.

Tranchida PQ, Casilli A, Dugo P, Dugo G, Mondello L. *Anal. Chem*. 2007a; 79:2266–2275.

Tranchida PQ, Donato P, Dugo G, Mondello L, Dugo P. *Trends Anal. Chem*. 2007b; 26:191–205.

Tranchida PQ, Purcaro G, Conte L, Dugo P, Dugo G, Mondello L. *J. Chromatogr. A* 2009a; 1216:7301–7306.

Tranchida PQ, Purcaro G, Conte L, Dugo P, Dugo G, Mondello L. *Anal. Chem*. 2009b; 81:8529–8537.

Tranchida PQ, Purcaro G, Fanali C, Dugo P, Dugo G, Mondello L. *J. Chromatogr. A* 2010a; 1217:4160–4166.

Tranchida PQ, Purcaro G, Sciarrone D, Dugo P, Dugo G, Mondello L. *J. Sep. Sci*. 2010b; 33:2791–2795.

Tranchida PQ, Shellie RA, Purcaro G, Conte LS, Dugo P, Dugo G, Mondello L. *J. Chromatogr. Sci*. 2010c; 48:262–266.

van der Lee MK, van der Weg G, Traag WA, Mol HGJ. *J. Chromatogr. A* 2008; 1186:325–339.

van Deursen MM, Beens J, Janssen H-G, Leclercq PA, Cramers CA. *J. Chromatogr. A* 2000a; 878:205–213.

van Deursen M, Beens J, Reijenga J, Lipman P, Cramers C, Blomberg J. *J. High Resolut. Chromatogr*. 2000b; 23:507–510.

van Stee LLP, Beens J, Vreuls RJJ, Brinkman UAT. *J. Chromatogr. A* 2003; 1019:89–99.

Vaz-Freire LT, da Silva MDRG, Freitas AMC. *Anal. Chim. Acta* 2009; 633:263–270.

Vendeuvre C, Bertoncini F, Thiebaut D, Martin M, Hennion M-C. *J. Sep. Sci*. 2005; 28:1129–1136.

Venkatramani CJ, Xu J, Phillips JB. *Anal. Chem*. 1996; 68:1486–1492.

Ventura GT, Kenig F, Reddy CM, Frysinger GS, Nelson RK, Van Mooy B, Gaines RB. *Org. Geochem*. 2008; 39:846–867.

Veriotti T, Hilton D. *Am. Lab*. 2006; 38:10–12.

Vial J, Noçairi H, Sassiat P, Mallipatu S, Cognon G, Thiébaut D, Teillet B, Rutledge DN. *J. Chromatogr. A* 2009; 1216:2866–2872.

Vogt L, Gröger T, Zimmermann R. *J. Chromatogr. A* 2007; 1150:2–12.

von Mühlen C, Alcaraz Zini C, Bastos Caram ao E, Marriott PJ. *J. Chromatogr. A* 2006; 1105:39–50.

von Mühlen C, Alcaraz Zini C, Bastos Caram ao E, Marriott PJ. *J. Chromatogr. A* 2008; 1200:34–42.

Wahl JH, Riechers DM, Vucelick ME, Wright BW. *J. Sep. Sci*. 2003; 26:1083–1090.

Wang FC-Y, Qian K, Green LA. *Anal. Chem*. 2005; 77:2777–2785.

Webb PJ, Hamilton JF, Lewis AC, Wirtz K. *Polycycl. Aromat. Compounds* 2006; 26:237–252.

Welthagen W, Schnelle-Kreis J, Zimmermann R. *J. Chromatogr. A* 2003; 1019:233–249.

Welthagen W, Schnelle-Kreis J, Zimmermann R. *J. Aerosp. Sci*. 2004; 35:17–28.

Welthagen W, Shellie RA, Spranger J, Ristow M, Zimmermann R, Fiehn O. *Metabolomics* 2005; 1:65–73.

Welthagen W, Mitschke S, Mühlberger F, Zimmermann R. *J. Chromatogr. A* 2007; 1150:54–61.

Windt M, Meier D, Marsman JH, Heeres HJ, de Koning S. *J. Anal. Appl. Pyrol*. 2009; 85:38–46.

Wu J, Lu X, Tang W, Kong H, Zhou S, Xu G. *J. Chromatogr. A* 2004; 1034:199–205.

Wulf V, Wienand N, Wirtz M, Kling H-W, Gäb S, Schmitz OJ. *J. Chromatogr. A* 2010; 1217:749–754.

Xu X, van Stee LLP, Williams J, Beens J, Adahchour M, Vreuls RJJ, Brinkman UA. *Atmos. Chem. Phys*. 2003; 3:665–682.

Yang S, Sadilek M, Synovec RE, Lidstrom ME. *J. Chromatogr. A* 2009; 1216:3280–3289.

Zhang D, Huang X, Regnier FE, Zhang M. *Anal. Chem*. 2008; 80:2664–2671.

Zhu S, Lu X, Dong L, Xing J, Su X, Kong H, Xu G, Wu C. *J. Chromatogr. A* 2005a; 1086:107–114.

Zhu S, Lu X, Xing J, Zhang S, Kong H, Xu G, Wu C. *Anal. Chim. Acta* 2005b; 545:224–231.

Zhu S, Lu X, Ji K, Guo K, Li Y, Wu C, Xu G. *Anal. Chim. Acta* 2007a; 597:340–348.

Zhu S, Lu X, Qiu Y, Pang T, Kong H, Wu C, Xu G. *J. Chromatogr. A* 2007b; 1150:28–36.

Zimmermann R, Welthagen W, Gröger T. *J. Chromatogr. A* 2008; 1184:296–308.

Zrostlíková J, Hajšlová J, Čajka T. *J. Chromatogr. A* 2003; 1019:173–186.

7

DETECTOR TECHNOLOGIES AND APPLICATIONS IN COMPREHENSIVE TWO-DIMENSIONAL GAS CHROMATOGRAPHY

PHILIP J. MARRIOTT
Monash University, Clayton, Victoria, Australia

The detector plays an integral part in any chromatographic system. It provides the means by which chromatographic separations are "visualized." In the process and mechanism of chromatographic separation, results should ideally be unaffected by the detector. This means that what happens "in the column" is independent of the detection step, which only provides the sensing capacity required at the column outlet. A detector converts a chemical species into a measurable electronic signal capable of being registered by a suitable device such as a chart recorder or computer. In this sense we might refer to this as a transducer device, where a chemical compound undergoes a chemical or physical process that can be transformed to an electrical signal. Just as many chemical and physical transducers are applicable to gas chromatographic (GC) detection, so the relatively new technique of comprehensive two-dimensional gas chromatography (GC × GC)—the topic of this review—can also be hyphenated to many detection systems, although progress in this direction is slow.

Although GC × GC increases the separation power of GC manyfold, the suitability or applicability of existing GC systems and detectors may not be adequate to accommodate the requirements of GC × GC. One of the primary concerns of

Comprehensive Chromatography in Combination with Mass Spectrometry, First Edition.
Edited by Luigi Mondello.

researchers is the need for precise measurement of very fast peaks that enter the detector. This can be as rapid (narrow) as 50 ms peak width at baseline (i.e., peak standard deviation of 10 to 15 ms) in some instances, so a detector acquisition rate of at least 100 Hz would be considered necessary for quantification purposes. The ability of the detector to provide a detailed and accurate report on a rapidly changing chromatographic peak profile depends on a number of factors: the detector response mechanism and chemical–physical process of the response; the design of flow within the detector body; detector gas flows (carrier, makeup, flame gases where appropriate); and electronic processing of the signal generated by the detector then acquired by the recording device. The various detectors that have been reported for GC × GC applications are discussed here, first with a brief introduction to the detector mechanism, then giving the technical demands of the use of such detectors in GC × GC, then presenting a number of applications of their use for different samples. The use of mass spectrometry detectors is covered elsewhere and is only touched on here as a means to provide an overview comparison with other detector techniques. The present chapter expands on a work published in 2006 which reviewed detectors used for GC × GC up to that date (von Mühlen et al., 2006). In updating that work, a similar style is retained for consistency, and a number of newly reported detectors for GC × GC have been included. Brief commentary on mass spectrometry is included, as are recent studies on the use of dual detection methods in GC × GC.

7.1 DETECTION IN GC x GC

Comprehensive two-dimensional (2D) gas chromatography has firmly established its foothold on the GC scene, having its genesis in the early 1990s (Górecki et al., 2004; Marriott and Shellie, 2002; Phillips and Beens, 1999). The manner in which GC × GC has significantly altered the way in which the GC experiment is implemented means that it is necessary to reevaluate many aspects of GC technology (Figure 7.1); the new nomenclature required of this multidimensional separation technique has been considered (Schoenmakers et al., 2003). Thus, GC × GC requires consideration of many new concepts in GC: modulation processes for the migrating chromatographic band (Adahchour et al., 2006), to which the term *modulation ratio* has been added (Khummueng et al., 2006a); experimental setup, orthogonality, interpretation and optimization (Harynuk et al., 2002; Ryan et al., 2005a); data presentation methods (Reichenbach et al., 2003, 2004); and quantitative considerations (Beens et al., 1998). Analysts who undertake GC × GC are presented with a conceptual change (and challenge) in their implementation of classical GC to accomplish GC × GC. Basing the total experiment on coupled column systems, routinely performing very fast and ultrafast separations, a new format for interpretation of results, and experimental optimization are all new to the routine user. These are—or should be—based on well-established and logical principles of GC (Harynuk et al., 2002), but do represent a departure from classical one-dimensional (1D) GC.

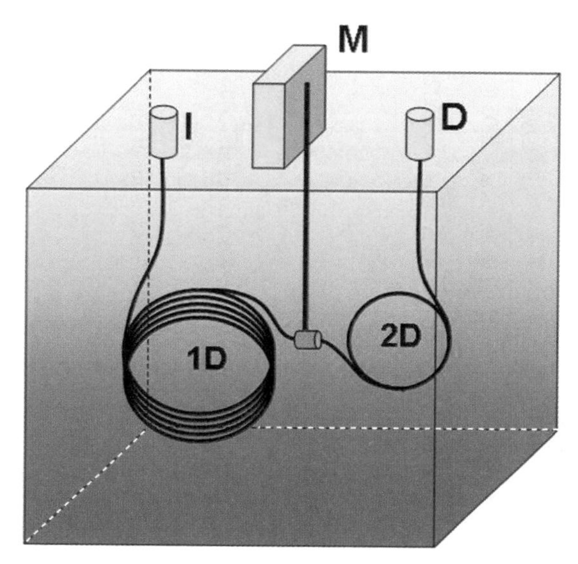

Figure 7.1 GC × GC instrument, showing the use of a short, fast elution second-dimension column which produces very narrow peaks at the detector. I, injector; M, modulator; D, detector.

One important component of the GC × GC experiment is the detector (Adahchour et al., 2006), although it appears that considerably less attention has been focused on the detector at the expense of reporting new applications and attempting to understand and interpret aspects of the separation phenomenon. Given that most detectors used in GC—apart from the universal flame ionization detector (FID)—are used to provide selectivity of detection, it can be argued that the extra separation performance of GC × GC automatically reduces the need for a selective detector; one of the purposes of using selective detection in conventional GC analysis is that the selective (specific) detector can compensate for nonresponding matrix interferences which might coelute with classical universal detection. This is in addition to the often improved sensitivity of most selective detectors.

The reliance on the FID workhorse detector, which operationally is an extremely easy detector to use, meets the specifications on speed of data acquisition, and the focus of researchers on presenting new applications studies can mean that perceived operational difficulties with most selective detectors when applied to GC × GC reduce the literature on alternative detectors used with GC × GC. Various selected applications where element-selective detectors have been used with GC × GC in the literature are summarized in Table 7.1.

Adapting a different detector for use with GC × GC requires primarily that information on its acquisition speed be assessed. Detector speed relates to the analog-to-digital conversion output of the transducer, which allows electronic processing of the input signal to a computerized data system, and one should

TABLE 7.1 Applications of Comprehensive Two-Dimensional Gas Chromatography with Element-Selective Detectors (Non-mass Spectrometric)

Detector	Analyte	Column Sets; ^{1}D then ^{2}D listed	Detector Temperature (°C)	Acquisition Rate (Hz)	Reference
NPD	Methoxypyrazines in wine	BPX5 (30 m × 0.25 mm × 0.25 μm) BP20 (1 m × 0.1 mm × 0.1 μm)	300	100	Ryan et al., 2005a
μ-ECD	Polychlorinated biphenyls, polychlorinated dibenzodioxins, polychlorinated dibenzofurans	1. HP-1 (30 m × 0.25 mm × 0.25 μm) HT-8 (1 m × 0.1 mm × 0.1 μm) 2. HP-1 (30 m × 0.25 mm × 0.25 μm) SupelcoWax-10 (1 m × 0.1 mm × 0.1 μm)	300	50	Korytár et al., 2002b
	Toxaphene	HP-1 (30 m × 0.25 mm × 0.25 μm) HT-8 (1 m × 0.1 mm × 0.1 μm)	300	50	Korytár et al., 2003
	Polychlorinated dibenzodioxins, polychlorinated dibenzofurans, polychlorinated biphenyls	DB-XLB (30 m × 0.25 mm × 0.25 μm) LC-50 (0.9 m × 0.18 mm × 0.1 μm)	300	50	Korytár et al., 2004
	Chiral PCBs	BGB-176SE (30 m × 0.25 mm × 0.25 μm) SupelcoWax-10 (2 m × 0.18 mm × 0.1 μm)	300	50	Bordajandi et al., 2005
	Polychlorinated dibenzodioxins, polychlorinated dibenzofurans, polychlorinated biphenyls	DB-XLB (30 m × 0.25 mm × 0.25 μm) LC-50 (0.9 m × 0.18 mm × 0.15 μm)	300	50	Danielsson et al., 2005

	Atropisomeric polychlorinated biphenyls	Chirasil-Dex CB (10 m × 0.1 mm × 0.1 μm) LC-50 (1.0 m × 0.1 mm × 0.15 μm)	260	50	Harju et al., 2001
SCD	S-containing compounds	Five different column sets	800	100	Hua et al., 2003
	S-containing compounds in light catalytically cracked cycle oil–heavy gas oil mixture	DB-1 (10 m × 0.25 mm × 0.25 μm) BPX-50 (17.5 m × 0.1 mm × 0.05 μm)	800	50	Blomberg et al., 2004
	S-containing compounds in crude oils	VB-5 (6 m × 0.18 mm × 3.5 μm) 007-17 (2 m × 0.1 mm × 0.1 μm)	800	50	Hua et al., 2004
NCD	N-containing compounds in diesel fuel	SPB-5 (30 m × 0.25 mm × 0.1 μm) BPX-50 (3 m × 0.25 mm × 0.25 μm)		100	Wang et al., 2004
AED	S-containing compounds in crude oil	DB1 (15 m × 0.25 mm × 0.25 μm) BPX50 (0.6 m × 0.1 mm × 0.1 μm)	Plasma discharge	10	van Stee et al., 2003
FPD	P and S pesticides; S compounds in diesel	Rxi-5 ms (30 m × 0.25 mm × 0.1 μm) Stabilwax (1 m × 0.1 mm × 0.1 μm)	250	100	Chin et al., 2010
NPD	N compounds in heavy gas oil	BPX5 (30 m × 0.25 mm × 0.25/0.5 μm) BPX50 (0.5–1 m × 0.25–0.5 mm × 0.1–0.15 μm)	325	50 or 100	von Mühlen et al., 2007
NPD–ECD	Dual detection; fungicides	BPX5 (18 m × 0.25 mm × 0.25/0.5 μm)	Both detectors	NPD 100	Khummueng et al., 2008
		BPX50 (0.75 m × 0.15 mm × 0.15 μm)	320	ECD 50	

expect this to be very fast in a modern GC instrument. To most users, this is probably not likely to be a parameter of importance, assuming that the detector speed will be suitable for the application. Slow signal processing would impose an unacceptable delay, lead to broadening, and reduce the number of data points measured across a peak. The second factor concerns the chemical–physical considerations of the detector, particularly the detector response to the input chemical signal (i.e., the chemical compound as it enters the detection region), in order to produce an electronic measure of the instantaneous amount of chemical. This effectiveness of the detector as a chemical transducer must consider factors such as (1) the detector geometry; (2) the detector gases and makeup gas flow rates; (3) any unswept or excessive volumes within the detector cell; (4) chemical reactions in flames, plasmas, or at surfaces; and (5) the sensing elements recording the signal changes. These same factors play critical roles whenever a fundamental change (improvement) in GC performance is implemented.

The chromatographic peak profile generated at the detector is a sum of the contributions to band broadening from all sources in the system. Generally, this is summarized as injector (i)-, chromatographic column (c)-, and detector (d)-based variances:

$$\sigma_{\mathrm{tot}}^2 = \sigma_{\mathrm{i}}^2 + \sigma_{\mathrm{c}}^2 + \sigma_{\mathrm{d}}^2$$

For best chromatographic performance, it is a general goal that injector- and detector-based broadening be reduced to negligible proportions compared with column broadening, so that the best possible resolution and response are obtained. The narrower the chromatographic peak, the more important therefore will be the imposition that any detector-sourced broadening may have on the recorded signal (and, of course, injector-sourced noninstantaneous injection).

There have been essentially three significant developmental stages in the history of GC: when packed columns were supplanted by the general acceptance of capillary GC, and now when very fast GC (and GC × GC) performance is demanded. Each stage necessitates the chromatographic manufacturer to redesign equipment to cater for the new demands placed on each of the critical steps of injection and detection systems, along with improved oven control, with operation at carrier flows perhaps 10 to 50 times smaller in the capillary system. The technical challenge from normal capillary GC to GC × GC technology may require similar performance improvements, and these are under way as the interest in very fast one-dimensional (1D) GC (and GC × GC) expands. We should ask whether the present suite of detectors will all be (or need to be) retooled to cater to the demands of the new requirements of speed and instantaneous response. Detectors that suit such developments might correspond to a "third generation" of technology.

The normal GC × GC experiment produces peaks at the end of the second-dimension column, which may be 100 ms wide at the base (or less) (Adahchour et al., 2006; Harynuk et al., 2002), and very fast GC × GC peaks with narrow-bore 50-μm i.d. second-dimensional columns may be on the order of 25 ms wide (Adahchour et al., 2003; Junge et al., 2007). These are perhaps 10 to 50 times narrower than classical 1D GC peaks. If we require 10 detection data points (about the minimum commensurate with acceptable quantification) per peak width at baseline ($w_b = 4\sigma$), this is about three data points per peak standard deviation. For a peak of 100 ms w_b, this implies an acquisition rate of 3/25 ms = 120 Hz.

The success with which detectors reported for GC × GC accomplish the required acquisition speed and achieve chemical–physical transduction for the sensing reaction is discussed in this chapter. A summary of applications of GC × GC employing different detectors, insight into the information provided by these detectors, and how detector performance affects the quality of GC separation are all discussed.

7.2 COMMENTS ON GC × GC WITH MASS SPECTROMETRY

Although mass spectrometry (MS) is dealt with elsewhere, a few pertinent comments need to be made to place MS in context compared with other detectors used in GC × GC. GC × GC is an information-rich separation method, and separation of components in 2D space allows facile identification of components. Identification may often be achieved even by using methods such as simple coinjection with a reference standard, because peak position (retention) in 2D space is a more reliable identification measure than retention in 1D GC. However, the very complexity of most samples that have been analyzed by using GC × GC, and the sheer number of compounds, make coinjection an impractical approach to use for multicomponent identification, especially with unknown components present. This strategy is tedious, so the only option is to resort to MS for component identification. This does pose an interesting question: If MS is used in 1D GC methods to permit unique identification of overlapping compounds, and if GC × GC has the ability to separate these overlapping components, in what circumstances is MS required for fully resolved peaks? This is a conceptual difference in the role of MS for this instance.

Further interesting questions may be asked in respect to the comparative roles of improved resolution in the separation domain (GC vs. GC × GC or MDGC) compared with resolution in the mass domain (quadrupole or low-resolution MS vs. high-resolution MS or MS–MS methods). In all cases, though, improved separation should permit improved performance of MS selectivity, and this will also be the case for other selective detectors (see below). The suitability of use of MS with GC × GC is concerned largely with the speed of detection that can be accomplished with MS. Not surprisingly, the MS of choice is the time-of-flight (TOF) system, due to its fast acquisition speed. GC × GC–TOF MS was described by Shellie et al. (2001). Notwithstanding the slower scan rate of quadrupole MS (qMS), the need to provide some identification capacity with GC × GC meant that even the qMS system was soon tested with GC × GC. For this purpose, and acknowledging the limitation of qMS, it was apparent that qMS was able to achieve the identification so dearly sought by GC × GC users. Newer qMS systems are capable of scanning in excess of 10,000 amu/s, so is borderline for use with GC × GC. Early studies with qMS include studies for geranium essential oil (Shellie and Marriott, 2003a) and ginseng analysis (Shellie et al., 2003b), where the scan range was reduced as much as possible to increase scan speed. Since then, an increasing number of studies have appeared. More recently, the isotope ratio MS method was used with GC × GC for steroids and fatty acids (Tobias et al., 2008), following from an earlier attempt to provide faster IRMS sensing for the CO_2 isotopes by minimizing transfer line dimensions and ensuring that the speed of the combustion process was suitably fast (Sacks et al., 2007). A case study of separation of overlapping compounds in 1D GC that made IRMS measurements somewhat problematic revealed that GC × GC apparently generated useful data for the components resolved. The efforts made by some researchers to improve the applicability of qMS to GC × GC by measures such as reduced mass scan range (Shellie et al., 2003a) proved to be acceptable for low-mass-range components, but were less suitable for higher-molar-mass compounds such as drugs, making it difficult to select a suitable restricted mass range (e.g., over a reduced mass range of about 200 units) to either a low- or high-mass region (Song et al., 2004). In summary, MS detection has been a defining study in developing the best MS approaches for GC × GC, or in trying to adapt available methods to the GC × GC experiment, with the single goal in mind of providing identification capabilities for the GC × GC separation.

7.3 FLAME IONIZATION DETECTION IN GC x GC

Invented by McWilliam and Dewar (1958), not surprisingly flame ionization detection (FID) is the most commonly employed detection method in GC (Amirav, 2001; Holm, 1999). The origin of this important detection system was reviewed some years ago (McWilliam, 1983). The ubiquity of the FID means that it was also the first detector chosen for application to GC × GC (Liu and Phillips, 1991); however, this can be suited to GC × GC only if it is capable of the data acquisition speed required in GC × GC. Fortunately, the FID has always been the one detector capable of fast electronic transduction, so most manufacturers offer FID acquisition rates up to 500 Hz, fast enough for adequate measurement of the fast GC × GC peak at the end of the second-dimension column (Shimadzu Corporation, 2010; ThermoElectron Corporation, 2010; Varian, Inc., 2010). The high gas flows of the detector aid the quick expulsion of compounds from the detector, with fast—essentially instantaneous—response.

A compound containing carbon will respond in the FID by producing ions in a hydrogen–air flame—hence the name of the detector is definitive for the mechanism of response. A schematic diagram of a FID detector is presented in Figure 7.2. The simplified response for the FID can be given as follows:

$$\text{C-containing compound} \xrightarrow{H_2\text{-air flame}} CHO^+ \rightarrow\rightarrow \xrightarrow{+\,H_2O} [H_3O(H_2O)_n]^+$$

The study of organic compounds in flames typically employed in FID by using mass spectrometry showed that an important intermediate ion was the formylium

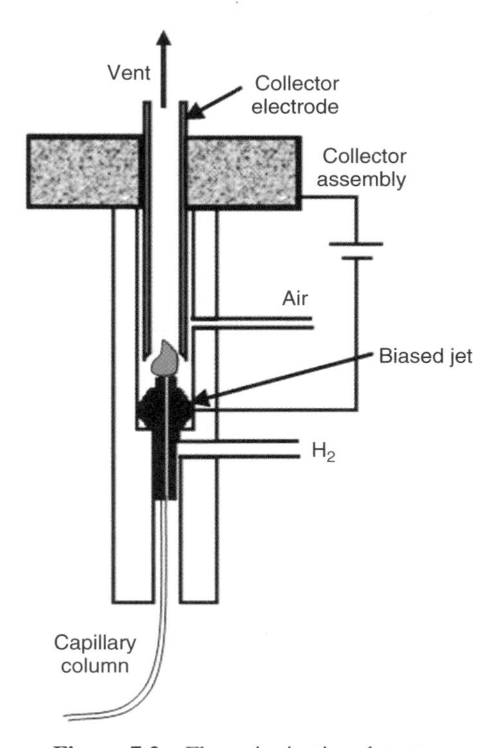

Figure 7.2 Flame ionization detector.

ion CHO^+ (Holm, 1999), so the presence of this ion is confirmed. For most chromatographers, the physical process involved is not so crucial, just the fact that it works as a superb detector for GC.

The ion current produced in the FID cell will be proportional to the instantaneous amount of carbon compound in the FID, so on the chromatographic time scale, ions are generated instantaneously. The final charge-carrying species is believed to be the hydrated hydronium ion, but it must be related proportionally to the C-compound and the intermediates in the flame.

While the FID response is approximately proportional to the carbon content of a compound, so that the FID is effectively a "carbon counting" device, compounds of an homologous series produce an ionization response in proportion to the number of carbon atoms (Holm, 1999). Thus, x moles of butane give about half the response of x moles of octane, and about x nanograms of butane give about the same response as x nanograms of octane. The carbon-counting process for FID response is not a precise metric. The chemical nature of the sample molecule affects the flame combustion effectiveness to produce the species responsible for the ionization response of the detector. Carbon counting is therefore not quantitative, and the presence of hetroatoms alters the ionization quantum (David, 1974). Thus, variable response factors are found for different compounds, and tables are available for various response factors for different classes of compound.

In compounds containing no carbon (and for this we might modify the statement to say no C—H bond), the response factor is diminishingly small. Thus, CO_2, NO_2, CO, CCl_4, N_2, and similar compounds have vanishingly small FID response. For the purposes of the present chapter, the FID response mechanism will not be discussed further.

The universal FID detector has well served the development of GC × GC, with its suitability for fast recording of peak widths at half-height as narrow as 25 ms (Junge et al., 2007) being faithfully recorded. This has allowed the respective improvements in sensitivity and resolution over the 1D GC method to be demonstrated (Lee et al., 2001) and method development to be discussed. The unique quantification approaches and generally better capabilities of GC × GC in the presence of matrix compared with 1D GC have been described. Negligible internal volume (Dallüge et al., 2003) of the FID, combined with fast transduction of the chemical signal, ensures that the FID imposes minimal detector-based broadening on the chromatographic signal. As a result, the FID produces narrow peaks, and these are usually narrower than the narrowest peaks possible with other detectors for operation under equivalent conditions. The short second-dimension column in GC × GC places a premium on detector response, and this is met by the FID. Not only is the FID response fast, but favorable detection limits in the low $pg \cdot s^{-1}$ of carbon range, and a large linear response range of about 10^6 or 10^7 for most compounds with good FID response are obtained (refer to manufacturers Web sites for general and comparative detector performance in this respect). Perhaps the other characteristics of the FID that makes it so popular with users are its reliability, ease of operation, stability, robust and reproducible response, and low electronic and flame-based noise, which all contribute to its set-and-forget performance. However, this makes specific information relating to the nature of the FID response per se relatively sparse. Most authors of GC × GC literature do not discuss precise details about the FID used in their studies, and this makes discussions of specific detection conditions in the GC × GC–FID literature rather mundane.

That said, however, the FID has been instrumental in presenting and evaluating the differences between conventional GC and GC × GC: not only the aforementioned sensitivity and resolution enhancement afforded by GC × GC operation, but the inherent interpretation of compounds in 2D separation space in GC × GC, which represents a significant departure from the informational content in conventional GC analysis. Thus, while the FID is employed for quantification (Fraga et al., 2000) and may be integrated with GC × GC/TOF MS data (Zhu et al., 2005), there has been widespread recognition of the inherent value of the bidimensional retention of a component in 2D space that can be employed for enhanced identification. Diagnostic interpretation based on FID data for peak identity is understandably a key feature of any gas chromatographic system, since compound retention is a constant for constant conditions. For a given target compound, its 1D GC retention must be the same as that of an authentic compound under the same conditions.

However, in many applications it is common to require retention data on two columns to provide the requisite identity, precisely because matrix interferences prevent adequate identification of the target peaks. The GC × GC experiment provides these two sets of data in one experiment, and the probability of matrix overlap is reduced considerably. Western and Marriott (2002, 2003) and later Bieri and Marriott (2006, 2008) developed concepts of retention index estimation in each dimension of the GC × GC experiment. The latter studies developed a method for sampling alkanes through a second injector, to deliver the homologous alkane series directly to the start of the second column for improved second-dimension retention index calculation. The FID has been employed for environmental analysis (Frysinger et al., 2003; Reddy et al., 2002; Slater et al., 2005), for industrial processes associated with petrochemicals (Vendeuvre et al., 2004, 2005), and for general analytical problems in areas such as essential oils and aromas (strawberry aroma, pepper volatiles), which traditionally required use of a GC–MS prior to the advent of GC × GC methodologies (Cardeal et al., 2006; Williams et al., 2005). However, the use of MS to provide greater certainly in identification is still important. Mitrevski et al. (2010) employed GC × GC–FID and GC × GC–TOF MS for steroid analysis in doping control, using the World Anti-doping Agency key anabolic agents to develop a general strategy for GC × GC analysis as compared with analogous 1D GC analysis.

Beens et al. (2000) used GC × GC–FID to characterize nonaromatic solvents. By grouping like compound classes into different bands, alkenes were separated from mono- and dinaphthenes, but within the alkane band, overlap of alkanes with different carbon numbers still resulted. This application represents the PIONA (paraffins, isoparaffins, olefins, naphthenes, aromatics) analysis application; however, the classical PIONA analyzer has a limited analysis range, being restricted to naphthenes and alkanes with carbon numbers below about 11. The complexity and time-consuming nature of the PIONA analysis cannot be underestimated. By comparison, with GC × GC the foregoing limits to mass range do not apply, and the mass can be extended to up to 20 or more carbons. Thus, PIONA analysis merely requires a suitable metric to be developed to allow automated data extraction from the GC × GC experiment.

Heavy gas oil quantification is also a complex analytical application. A similar group-type approach using hyphenation of LC with GC (LC–GC) and GC × GC–FID (Beens et al., 1998) allowed excellent quantitative analyses to be performed. Although the individual compound identification may not be secure, the identification of isomer clusters and better detection limits ensure that the GC × GC technique has a unique role to fulfill for this and similar analytical challenges. FID response factors of related compounds such as hydrocarbon isomers tend to be well characterized. FID response of different classes of compounds can also be predicted with some accuracy. Response factors depend on the mass of carbon in a molecule; however, a C— H bond is also required. Response factors are also amended according to the presence of heteroatoms in the molecule. Thus, analyte quantification can be performed using a surrogate molecule to represent the total isomer suite in the absence of individual standards for the estimation of precise response factors. The structured retention patterns within the GC × GC 2D space allow related compounds (homologs; structural isomers) to be relatively easily recognized, so the FID response can be approximated for such a cluster of compounds. This application depends on the unique structured separation of compounds in 2D space and so has no real analogy in 1D GC for a multiple-class overlapping suite of different compounds.

The role of mass spectrometry can be reassessed in some instances for GC analysis in light of the improved resolution provided by GC × GC. In many cases for overlapping peaks, precise measurement of these peaks cannot be assured in the absence of spectroscopic detection (i.e., mass spectrometry). In this case, MS compensates for information shortfall by allowing individual unique ions to be selected for both confirmation and quantification. But if peaks are quantitatively separated, does the need for MS detection still exist in this case? The answer must be no! We must still be sure that a peak has been identified correctly, but once this is confirmed, each analysis can be conducted by GC × GC–FID instead of GC × GC–MS (or GC–MS). This allows us to revert our concept of analysis back to the pure notion of separation being the key requirement in GC analysis.

7.4 ELECTRON CAPTURE DETECTION IN GC × GC

The electron capture detector (ECD) had its genesis in the late 1960s, when Lovelock (1974, 1975) proposed improved selectivity and sensitivity for detection arising from this ionization detector. The ECD uses a radioactive beta emitter (e.g., generated by a radioactive source such as ^{63}Ni) to ionize the carrier gas, producing a standing current between biased electrodes. Organic molecules containing electronegative functional groups (halogens, phosphorus and nitrogen groups) capture some of the electron flux, reduce the standing current, and this is reflected in a change in response between the electrodes. The outcome is a traditional GC detector that is probably the most sensitive and selective of the detectors commonly available; its application space is predominantly in the detection and analysis of high-electron-affinity compounds—hence, largely halogen-containing compounds (Chen and Chen, 2004). Therefore, the ECD detector has long been recognized as the detector of choice for trace analysis of organochlorine pesticides, herbicides, pollutants, and halogenated hydrocarbon in a range of samples. For multiresidue analysis of such compounds using GC × GC, ECD has been a natural choice of detection system, although FID has also been used either as a detector in its own right, or as a comparative system to contrast performance of ECD versus FID. Mechanistically, carrier gas molecules are ionized by ejecting thermal electrons (e^-; low-energy electrons), which migrate to an anode and thereby generate a current signal. In the presence of a sample in the detection region containing compounds with high e^--affinity substituents, capture of the thermal e^- results and the subsequent reduction in baseline current provide a signal response related to the thermal e^- captured by the compound. Either a constant potential applied across the cell—the *DC mode*—or with a pulsed potential across the cell—the *pulsed mode*—can be used to monitor the ECD result. The process described above can be represented as follows:

$$^{63}Ni \rightarrow \beta \text{ particles}$$

$$N_2 \text{ (carrier gas)} + \beta \rightarrow N_2{}^+ + e^- \text{ (thermal electron)}$$

$$\text{halogen X (or other high-electron-affinity atom)} + \text{thermal } e^- \rightarrow X^-$$

$$\text{reduction of } e^- \text{ flux} \rightarrow \text{response of detector}$$

In the dc mode, H_2 or N_2 carrier gas can be used with a small potential applied across the cell. This provides a small standing current to the cell. Using the example of a halogen-containing molecule that enters the cell, electrons are captured by the molecule, the molecule becomes charged, and the standing current is reduced. In the pulsed mode of operation, it is common to use a mixture of 10% methane in argon, to change the electron-capturing environment. Electrons generated by the radioactive source assume a lower, thermal energy. In the absence of collecting potential, thermal e^- exists at the source surface about 2 mm deep at room temperature and about 4 mm deep at 400°C. Application of a short-period square-wave pulse applied to the electrode collects the e^- and produces a baseline current. The standing current, in the presence of 10% methane in argon, is about 10^{-8} A with a noise level of about 5×10^{-12} A. Modification of the basic ECD design continues, especially with respect to the design of detectors that better suit the advent of fast and very fast GC analysis, as outlined below.

7.4.1 Micro-Electron Capture Detector

A significant recent development in the improvement of ECD design has resulted in miniaturization of the detector, making it more responsive to and compatible with lower columns flows, which are more usually employed with narrower-bore capillary columns and therefore respond more rapidly to changing peak fluxes that enter the detector. The micro-ECD (μ-ECD) from Agilent Technologies is an example of such a modified design (Figure 7.3). The design elements of this ECD make it more appropriate to miniaturized flows and reduced internal cell volume. Thus interpretation of the key features driving the detection to faster and more sensitive applications are instructive in terms of how manufacturers have responded to increased demands placed on the detection step in narrow-bore fast GC.

The μ-ECD design has a reduced internal volume of 150 μL, which was claimed to be a factor of 10 smaller than that of the prior macro-ECD design. Since it is critical to interpret detector performance with respect to greater sensitivity and acquisition speed, which hopefully makes it appropriate for use with GC × GC, it is useful to analyze the manufacturer's claims for the stated performance attributes of this new detector in light of the future of very fast GC. The validation of μ-ECD performance was reported by Klee et al. (1999), and

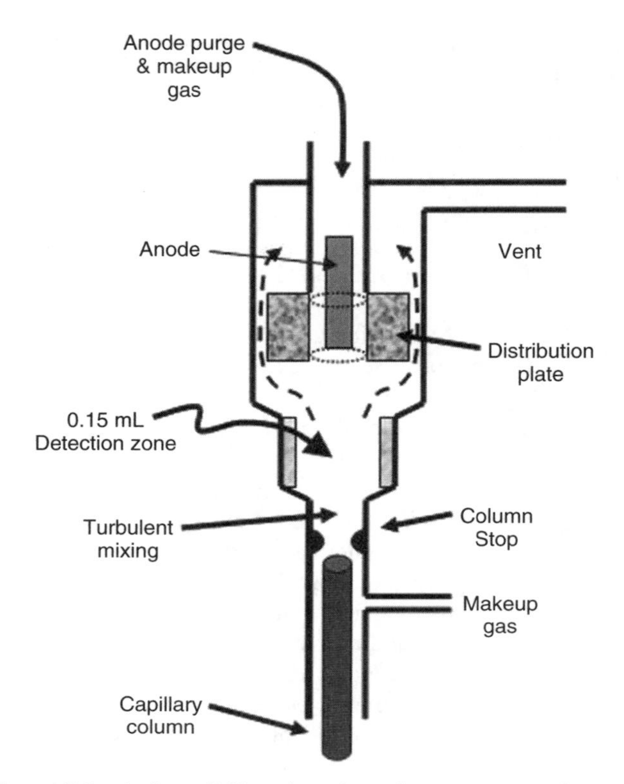

Figure 7.3 Agilent 6890 series micro-electron capture detector.

these are reported in Table 7.2, for the purpose of relating the manufacturer's reasons for introducing the μ-ECD. It is interesting to consider the upper limit of 50 Hz for the chromatographic signal sampling rate of the μ-ECD. This may not be as fast as some applications in GC × GC may demand (consider the 500-Hz rate of the FID), but it is not clear whether this is a design constraint or a practical constraint. The suitability of the μ-ECD for 1D GC should be straightforward; however, it will still attract relevant studies to assess its performance. Any claimed improvement toward trace-level analysis of pesticide residues such as organochlorines (OCs), organophosphorus compounds (OPs), and polychlorinated biphenyls (PCBs) in various samples (Aybar-Munoz et al., 2005; Conka et al., 2005; Gomara et al., 2002; Kusakabe et al., 2003; Quan et al., 2002; (Zrostlíková et al., 2002) will be relevant to decreasing the detection limits of such analyses. Assessment of long-term stability, robustness, response reliability at the most sensitive levels detected, and other requirements of the residue analyst are important to acceptance of the new detector. Thus, while the small internal volume and maximum acquisition rate of 50 Hz address the compatibility of the μ-ECD with fast 1D gas chromatography (allowing for peak widths at baseline of about 1 s or less), to couple the μ-ECD with GC × GC, the performance demands are more critical given peak widths at baseline of 10^{-1} s. There are

TABLE 7.2 Characteristics of the Agilent μ-ECD Detector

Parameter	Comparative Performance Reported for the μ-ECD Detector
Sensitivity	Improved sensitivity compared with conventional ECDs; detection sensitivity of less than 8 fg/s of lindane
Internal volume	A detection zone volume 10 times smaller than other ECDs with
Linear dynamic range	A broad dynamic linear range from femto- to nanograms per liter; for example, a 5×10^5 linear dynamic range for lindane
Flow velocity through the detector	Higher linear velocities through the detection zone, reduced analyte residence time, decreased cell contamination, and improved operational uptime
Anode arrangement design	Isolated anode (isolated from the direct carrier flow), which minimizes the chance for contamination of the anode
Sensing electric field	An optimized electric field that minimizes contamination effects on cell performance
Liner and column alignment	A replaceable liner that serves as a physical stop for the column, ensuring reproducible column installation and decreasing column contamination of the cell
Sampling rate	A variable sampling rate, from 5 to 50 Hz, suitable for fast chromatography

[a]As reported by Klee et al. (1999).

two primary considerations here: (1) Is there any observable skewing of the rapid peak flux through the detector, that may impose on the detector response some asymmetry compared with the instantaneous recorded peak profile?; and (2) is the detector acquisition rate fast enough to record peak changes in the detector? It has been reported that it is necessary to operate the μ-ECD with a sufficiently high makeup gas flow (greater than that recommended by the manufacturer) in order to increase the linear flow velocity through the detector (Korytár et al., 2002b). This has the less desirable outcome of decreasing the response magnitude (height, and therefore area) due to dilution of the compound. The need to sharpen up the peaks is very important in GC × GC, due to the need to separate peaks in the fast second dimension rather than allowing peaks to overlap "in the detector." Thus, while the data acquisition rate (50 Hz) should be adequate for fast GC peaks, it will possibly not be sufficiently fast for peaks that result from GC × GC operation. Needless to say, much attention has been focused on an analysis of PCBs using GC × GC both to observe the structured patterns possible in this sort of sample and to evaluate if complete separation of all the congeners is possible. Figure 7.4 is an example of PCB analysis using GC × GC. Classical multidimensional GC has been applied to this problem, with the most common approach being analysis of the target PCB congeners (Kinghorn et al., 1996).

In the GC × GC domain, Korytár et al. optimized the temperature program and column sets (first and second dimensions) for GC × GC with μ-ECD. A focus on 90 PCB congeners was used for this study using a HP-1/HT-8 column

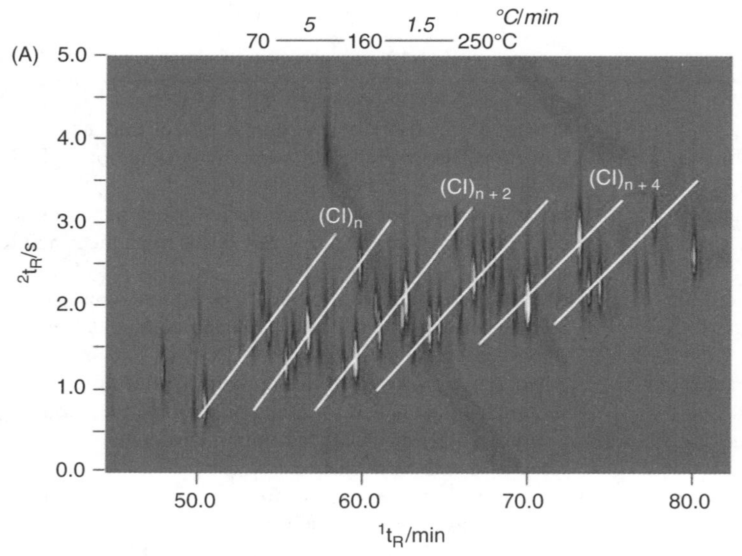

Column set: **BP X 5** 30 m x 0.25 mm x 0.5 μm, **BP X 50** 1.0 m x 0.15 mm x 0.15 μm

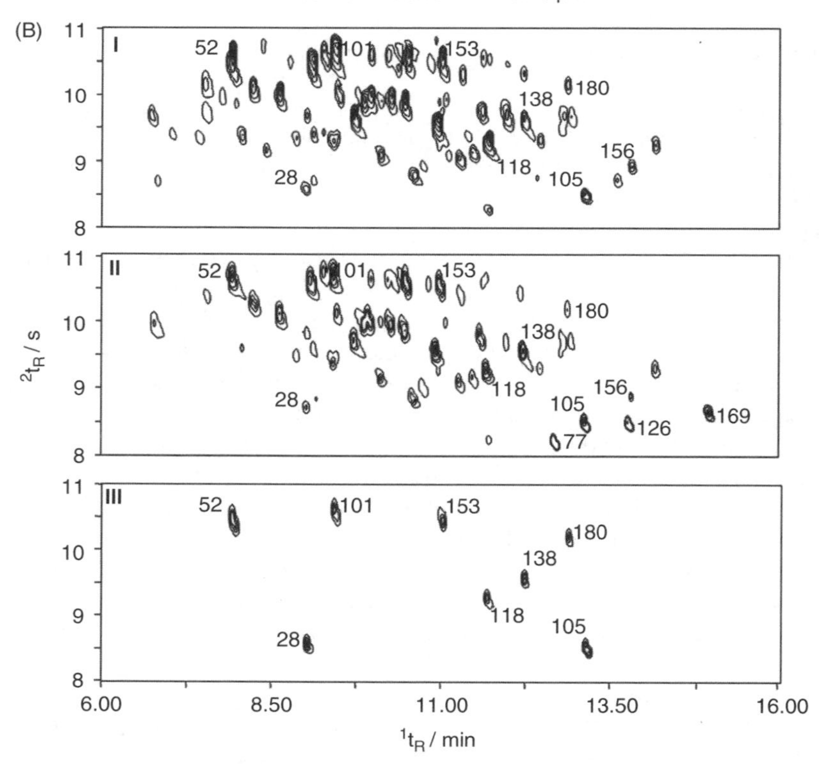

Figure 7.4 Analysis of PCB samples: (A) Arochlor 1260 sample, GC × GC–ECD (50 Hz) on a column set and conditions shown; (B) Clophen A50 sample I. Clophen A50; II I spiked with PCBs 77, 105, 126, and 169; III I spiked with PCBs 28, 52, 101, 105, 118, 138, 153, and 180.

set, with the μ-ECD detector operated at 300°C, with N_2 used as a makeup gas at a flow rate of 60 mL/min and data acquisition of 50 Hz (Korytár et al., 2002b). Limits of detection with a μ-ECD of 10 fg can generally be expected for tetra- or higher substituted chlorobiphenyl congeners in GC × GC. Kristensson et al. (2003) studied six different modulator systems and ECDs from three different suppliers for the analysis of high-boiling halogenated OCs using a test mixture of PCBs, fluorinated polycyclic aromatic hydrocarbons (F-PAHs), and two chlorinated alkanes. The focus of the ECD comparison was with respect to detector-band broadening, and for suitability in fast GC and/or GC × GC analysis. The different detectors and operational conditions employed in this study are reported in Table 7.3. The study concluded that the internal volume of the detector is not the only factor that is important when selecting an ECD for use in GC × GC. Although the μ-ECD from Agilent Technologies had the smallest internal detector volume (150 μL), it still suffered additional band broadening up to threefold greater than that for FID, and exhibited second-dimension tailing. This study also confirmed that optimization of the makeup flow in order to reduce peak widths can have a dilution effect on the solute flux and so reduce signal response. As shown in Table 7.3, these makeup flows can be far greater than those recommended for routine operation. The contrasting need to obtain sensitivity while minimizing peak widths can result in a compromise between these two conflicting requirements. Note that in 1D GC, band broadening in the detector will be relatively negligible (1D peaks are already broad), so operating at a high makeup flow is not needed; the sensitivity will then be the key optimization parameter.

In a study of fungicide residues using GC × GC with separate detection steps of NPD and μ-ECD (Khummueng et al., 2006b) and with PTV injection, it was noted that peak widths for GC × GC operation were generally greater than those with FID, that matrix components in vegetable samples that would normally have interfered with 1D GC analysis were now well resolved in the 2D space, and that an unusual instance of degradation of iprodione in the injector and on the column could be recognized in the 2D GC × GC result. Further study of dual parallel NPD/μ-ECD detection (Khummueng et al., 2008) for a range of pesticides and related compounds is reported in Section 7.11. In summary, introduction of the

TABLE 7.3 Characteristics of ECD Designs and Operational Conditions for GC x GC[a]

Manufacturer	ECD Internal Volume (μL)	Temperature (°C)	Makeup Flow (mL/min)	Acquisition Frequency
Agilent	150	320	60–450[b]	50
Shimadzu	1500	100–340	15–200	250
Thermo Finnegan	480	300	2050[c]	?

[a]As reported by Kristensson et al. (2003).
[b]450 mL/min flow achieved by using an additional gas supply with a T-piece.
[c]Maximum makeup flow is 375 mL/min, but the flow regulator was replaced with wider tubing and a needle valve.

μ-ECD detector is an important and valuable step forward in the selective and sensitive analysis of organohalogens in GC × GC and the μ-ECD will be the detector of choice in many such studies. The rationale for use of the ECD is very much the same as for 1D GC: additional sensitivity and the selectivity of detection response in the 2D space, which allows for facile detection and identification of the target compounds, notwithstanding the inherent improved separation power of GC × GC.

7.5 SULFUR CHEMILUMINESCENCE DETECTION IN GC × GC

For many reasons, sulfur offers a number of challenges for GC analysis. The chemical reactivity of many sulfur-containing compounds can make both analysis of the compound at high temperature of concern, but also relating the compound in a sample to the level that is measured at the detector somewhat problematic. Thus, precise isolation of the sulfur-compound from a matrix may be difficult to assure. That sulfur compounds offer unique properties to many foods and beverages (Sarrazin et al., 2007) in terms of aroma and off-odors is a matter of popular knowledge. This makes sulfur an important element, but may also be a major heteroelement in coal and petroleum components, diesel fuel, and gasoline, which must be removed in order to reduce the impact of emissions from fossil fuels. Often, sulfur compounds are present at trace levels, with detection and quantification being complicated by high levels of interfering components, as found in the headspace analysis of roast beef aroma (Rochat et al., 2007).

The detection of organic sulfur compounds may be accomplished by using a number of analytical techniques: for example, oxidation to sulfate followed by ion chromatography. However, GC with a variety of detectors such as FID, flame photometric detection (FPD), atomic emission detection (AED), sulfur chemiluminescence detection (SCD), and various mass spectrometry methods have been used. The use of GC with SCD combines specificity of detection of trace sulfur-containing compounds with suppression of matrix response, and low-level detection that may be critical for some samples. With respect to highlighting the breadth of applications to which the SCD with GC has been employed, descriptions of GC–SCD for analysis of breath components (Paetznick et al., 2010), petroleum and petrochemical products (Jaycox and Olsen, 2000; Ng et al., 2000; Pham et al., 1995), natural gases (Chawla and Di Sanzo, 1992), foods and beverages (Benn and Peppard, 1996; Mestres et al., 2000; Rauhut et al., 1998; Rochat et al., 2007; Sarrazin et al., 2007), and environmental samples (Gaines et al., 1990) have been reported. The role of sulfur-selective detection combined with sensorial evaluation is often highlighted (Chaintreau et al., 2006; Vermeulen et al., 2001). The chemiluminescent process for detection of sulfur compounds in air (Benner and Stedman, 1989) was recognized as a suitable mechanism for sulfur-based chemiluminescence in chromatography (Hutte et al. 1990). The SCD possesses a linear response to the amount of sulfur compound injected, with a response that is largely equimolar for different classes of sulfur compounds and an absence of quenching (both of these properties are compromised in the FPD system; see Section 7.9). Its sensitivity (<0.5 pg S/s) and high selectivity (S : C $> 10^8$) are both attractive attributes for a GC detector (Shearer and Meyer, 1999). An overview of sulfur and nitrogen chemiluminescence detection with GC—mechanisms, operating principles, selectivity, sensitivity, linearity and applications—is available (Yan, 2002).

The process of chemiluminescent radiation generated from sulfur compounds requires initial oxidation to SO_2 followed by high-temperature combustive reduction in a H_2 environment to produce sulfur monoxide (SO). SO is believed to react with ozone to form excited-state sulfur dioxide (SO_2^*), which generates light when it returns back to the ground state with emission of chemiluminescent radiation. In general terms, the sulfur gases are combusted in a hydrogen-rich chamber, yielding sulfur monoxide, which should be linear with the input distribution of the chromatographic peak; reaction with ozone produces sulfur dioxide, oxygen, and light. The light produced is detected by a photomultiplier tube. The net effect is that the response is proportional to the amount of compound in the sample. The process can be summarized as follows:

1. Conversion of organic sulfur compound to sulfur chemiluminescent species:

$$\text{Oxidation}: \quad R{-}S + O_2 \rightarrow CO_2 + H_2O + SO_2$$

$$(R = \text{hydrocarbon group})$$

$$\text{Combustion}: \quad SO_2 + H_2 \rightarrow H_2O + SO$$

2. Detection of sulfur chemiluminescence:

$$\text{Chemiluminescence}: \quad SO + O_3 \rightarrow SO_2^* + O_2 \rightarrow SO_2 + h\nu \text{ (blue; UV)}$$

where $h\nu$ is light energy in the blue region of the spectrum.

The dual plasma burner of the Sievers 355 (SCD) is shown in Figure 7.5. The SCD mechanism offers the most sensitive and selective chromatographic detector available for analysis of sulfur compounds. Components of the SCD include a dual plasma burner for high-temperature combustion of sulfur-containing compounds for stepwise formation of SO_2 and then the monoxide (SO). A photomultiplier tube (PMT) records light produced by the reaction of SO with ozone. The SCD dual plasma burner enhances production of the sulfur monoxide (SO) intermediate; the lower flame is oxygen-rich and the upper is hydrogen-rich. The conversion of sulfur-containing compounds to sulfur monoxide (SO) occurs within the ceramic reaction chamber housed in the burner assembly. The key features and benefits of the SCD include:

- Sulfur-specific detection for gas chromatography
- Picogram detection limits
- Hydrocarbon quenching not observed

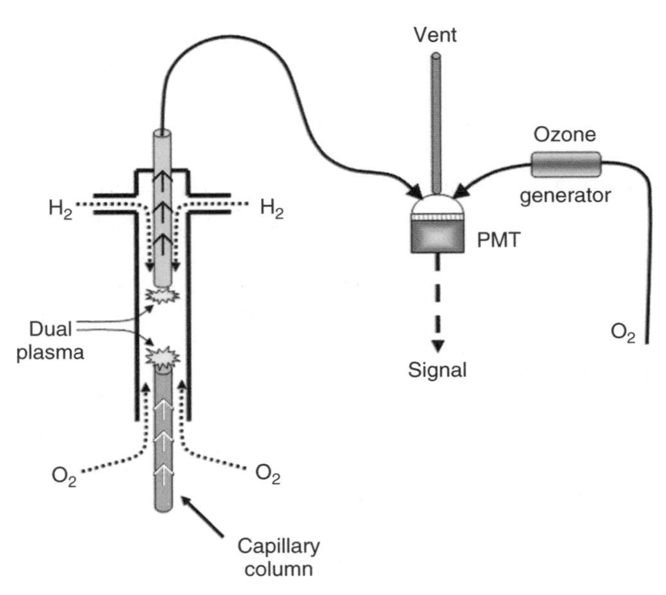

Figure 7.5 Dual plasma SCD detector. Note that this detector can also incorporate an intermediate FID response through combustion at the first flame, prior to entering the H_2-rich reducing environment and passage of effluent to the chemiluminescent zone.

- Linear response with equimolar response to (most) sulfur compounds

In a further modification to the detector, the first "plasma" in the detector may be arranged to provide a FID response, so as to permit dual "universal" and selective detection. Not unsurprisingly, GC × GC users will focus primarily on the response speed of the detector reaction processes (and electronics). To maximize the throughput of a sulfur-compound as it traverses the detector cell, any unswept or excessive volume that may serve to broaden peaks or dilute the solute should be minimized. Clearly, flow rates of ancillary gases into the detector (O_2, H_2, and O_3) should be important considerations.

Hua et al. (2003, 2004) used an SCD detector with GC × GC for determination of sulfur-containing compounds in diesel oils; optimization of the detector and choice of column set were considered. The SCD detector used in this study was operated at 800°C with radiation detected at a wavelength of 260 to 480 nm and a data acquisition rate of 100 Hz. Clearly, this speed is adequate to follow peak changes, and provided that detector-based peak broadening is negligible, the profile should represent that which enters the detector. This can be gleaned from the general peak width measures. Results indicated peak widths in the GC × GC–SCD system to be 0.8 to 1.2 s, which is much wider than that generated by an equivalent GC × GC–FID system (0.2 to 0.3 s). A bigger SCD cell volume of 480 μL may account for this, although it would be instructive to attempt to isolate the source of peak broadening. Even though a 100 Hz data acquisition rate would appear promising, this is not compatible with the apparently slow travel of the narrow GC × GC peak through the SCD. One would therefore assume that the SCD arrangement is not optimum for GC × GC performance. Based on this study, GC × GC–SCD gave reported relative standard deviations (RSDs) for ${}^1t_R$ and ${}^2t_R$ of 0.93 and 4.25%, respectively, with the value for ${}^2t_R$ apparently rather large. By contrast, using GC × GC–FID, Shellie et al. (2003c) and Mitrevski et al. (2010) quoted values on the order of 1% for ${}^2t_R$. The GC × GC–SCD separation required 135 min to speciate the various sulfur compound groups and target sulfur compounds in diesel oil fractions. Various compound classes—thiols, sulfides, disulfides, and thiophenes (TPs), benzothiophenes (BTs), dibenzothionphenes (DBTs), and benzonaphthothiophenes (BNTs)—were detected, and some classes formed contour horizons in the 2D plot almost parallel to one another. Recently, the FPD system reported similar conclusions for sulfur compounds in aged diesel and kerosene samples (Chin et al., 2010). Coupling GC × GC with SCD detection will have advantages such as ease of operation and reliable performance when applied to group separation, identification, and quantification of sulfur-containing compounds in diesel oil fractions. Suppression of the complex hydrocarbon matrix due to the selective SCD response aids the identification and quantification process significantly, and this has major advantages over conventional GC–FID, GC–SCD, and GC × GC–FID analysis. The ease of profiling samples during such industrial applications as diesel oil processing for desulfurization, and trace sulfur compound identification in downstream products should be useful process-monitoring tools.

Blomberg et al. (2004) reported that the speed of the electronics used for classical SCD operation accounted for the apparent "lack of speed" of the SCD and so concluded that it does not arise from detector physical dimensions, flow cell designs, or any inherent chemical reaction sources within the plasma or during the chemiluminescent step. This supports the contention that manufacturers need to construct detectors with all factors optimized for maximum performance. In their experiment, Blomberg found that initially, the SCD did gave rise to additional band broadening but that once the slow response was identified and attributed to being caused by system electronics rather than chemiluminescence chamber dead volumes, transfer line effects between the burner and reaction chamber gas flow rate, or slow PMT response, improved electronics could overcome the limitations of the commercial system. The final outcome was a design that allowed successful coupling of GC × GC with the SCD, and this makes a useful case study of a detection technology that could be reengineered to deliver the good performance demanded of GC × GC.

7.6 NITROGEN CHEMILUMINESCENCE DETECTION IN GC × GC

The nitrogen chemiluminescence detector (NCD) is useful for the analysis of nitrogen-containing compounds in different types of samples. Advantages of the

NCD over other nitrogen-based detectors include equimolar response, higher selectivity, lack of quenching due to matrix effects, and reliability of operation. Deciding the application portfolio of the NCD is as simple as recognizing the types of samples and compounds that contain nitrogen-containing compounds that can be chromatographed, thus, petroleum and petrochemical (Chawla, 1997; Wing et al., 1984), food and beverage flavor (Gautschi et al., 1997; Yan, 2006), and environmental (Kashihira et al., 1982) products contain such compounds and an NCD can or may be used for these analyses. The NCD can be used as a single detector or in a dual-detection arrangement coupled with a FID, allowing simultaneous acquisition of selective and universal detection chromatographic traces from both detectors (refer to Section 7.11 for such applications in GC × GC).

Operationally, the NCD is similar to the SCD mode of response and mechanism. The effluent from the GC column enters the ceramic combustion tube in a stainless-steel burner. The H_2/O_2 plasma converts nitrogen compounds to nitric oxide (NO; $T > 1800^\circ C$). Nitric oxide reacts with ozone to form electronically excited nitrogen dioxide (NO_2^*), with NO_2^* emitting light in the red–infrared region of the spectrum when it relaxes to the ground state. The intensity of light emitted, I_{em}, is proportional to the amount of nitrogen in the detector, and also in the sample; I_{em} is related directly to nitrogen concentration. A typical detection limit is 10 μg/L. Diatomic nitrogen (N_2) is generally not detected; however, a slight response to N_2 implies that air leaks can be a problem. The detector requires precise control of the hydrogen and airflow rates in order to maintain good reproducible response, and a vacuum system. The chemiluminescence process mimics that of the SCD:

$$\text{nitrogen} - \text{compounds} + O_2 \rightarrow NO + \text{other products}$$

$$NO + O_3 \rightarrow NO_2^* + O_2 \rightarrow NO_2 + h\nu \text{ (NIR)}$$

The Sievers 255 NCD employs a dual plasma burner for high-temperature combustion of nitrogen compounds, to form NO; a photomultiplier tube detects emission produced by the subsequent chemiluminescent reaction of NO with ozone. Little or no interference arises for complex sample matrices, due to the specificity of the reaction, and so makes such analyses feasible. Complete conversion of the matrix to products such as carbon dioxide and water in the dual plasma burner only results in response to the NO_2 species. Burner gas flow dynamics of this detector will not be illustrated separately, due to the similarity with those shown for the SCD.

Wang and co-workers (2004) speciated nitrogen-containing compounds in diesel fuel with detection by GC × GC–NCD. This typical refinery stream analysis in the fraction boiling between 150 and 430°C corresponds to carbon numbers from approximately C_8 to C_{28}. The capability of GC × GC–NCD for separation of individual compounds and classes of nitrogen-containing compounds such as indoles and carbazoles with an SPB-5/BPX50 column set combination documents well the role of this detector for such applications. The NCD was operated at a data acquisition rate of 100 Hz, and the analysis proved the general applicability of the NCD for similar petroleum products. For component identification, major peaks of nitrogen-containing compounds were identified by retention times matching those of pure compounds, or peaks were identified by using alternative identification approaches. The bands of nitrogen-containing compounds comprising carbazoles and indoles were generally in agreement with later studies using GC × GC–NPD (von Mühlen et al., 2007) and GC × GC–TOF MS (von Mühlen et al., 2010). The capabilities for fingerprint profiling of nitrogen-containing compounds in the sample were described, and complex reactions and processes monitoring in diesel production were compared with the difficulty of achieving this goal in the absence of GC × GC technology.

7.7 ATOMIC EMISSION DETECTION IN GC x GC

The multielement potential of the AED detector arises from the basic property of elements that possess atomic emission lines in the region of the spectrum (ultraviolet, visible, and infrared) that allows for spectral monitoring. Depending on the instrumental arrangement, more than 23 different elements may be monitored. Applicable elements for the AED, including N, S, C (universal), H (universal), O, Pb, Mn, F, Te, Se, and Si, can all be detected (Quimby and Giarrocco, 1997). Note that some of these can be considered universal modes. Compounds eluted from the capillary column are introduced into a microwave helium plasma, usually coupled to a wavelength-dispersive diode array optical emission spectrometer (Skoog et al., 1998). The plasma atomizes relevant elements of each compound, exciting their characteristic atomic emission spectra, as shown in Figure 7.6. Emitted light is separated via the diffraction grating into individual lines, then each line or energy is monitored by a photodiode array (PDA); the PDA recording diodes are located at the characteristic wavelength of each element according to those that are available in the system. Each individual signal channel records emission intensity versus time, to produce chromatograms comprising peaks for each compound whose wavelength is measured. Clearly, it is important to know which elements are present, or which elements can be monitored by the experimental setup of the AED. The AED system includes (1) an interface for the GC capillary column designed to pass the effluent into the microwave-induced plasma chamber; (2) the microwave chamber and all necessary control gases and microwave discharge; (3) a cooling system for the chamber; (4) a dispersive optical system comprising mirror(s), diffraction grating, and optics to focus, then disperse, the atomic lines; and (5) a position-adjustable photodiode array with computer control and intensity measurement.

Simultaneous multichannel detection using GC–AED allows multiple monitoring of different elements at the same time. Excellent limits of detection (LODs) for most important elements (LODs: 1 to 3 pg/s), response linearity often over three to five orders of magnitude, and good element-to-carbon selectivity of about four or five orders of magnitude generally makes the AED useful for empirical

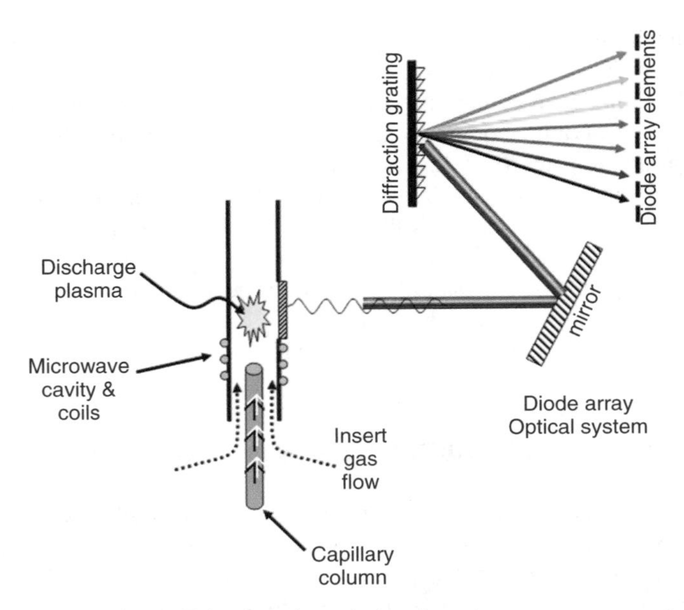

Figure 7.6 Atomic emission detection system.

formula estimation in 1D GC applications (van Stee et al., 2002). However, a maximum acquisition rate of the AED of 10 scans/s is rather slow for adequate quantitative sampling of the GC × GC peaks. Peak broadening by use of a transfer line between the 2D column of the GC × GC system and the AED allowed Van Stee et al. to increase the number of sample points monitored for each compound, but this must necessarily accompany worse resolution and loss of performance of the GC × GC method (van Stee et al., 2004). This strategy led to peak width at baseline of about 500 ms, with some tailing. The resulting six data points per sampled peak makes this system just suitable for GC × GC qualitative purposes; that is, it should allow identification of some peaks (provided that they are resolved), but quantitative capability will be compromised.

Conceptually, there is no reason why the AED cannot provide fast data acquisition: the microwave source produces fast response, as does light dispersion and recording; the diode array elements can be sampled fast. Thus, development of a suitable AED system for GC × GC need only await modification of the AED to provide fast signal generation. GC × GC–AED was used to analyze sulfur-containing compounds in crude oil and fluid catalytic cracking (FCC) products, with correlation of GC × GC–TOF MS data permitting additional identification of peaks. Comparison of GC × GC–AED plots prior to, and after, catalytic cracking confirms the shortening of carbon chains of sulfur compounds. Although compounds present in crude oil samples comprise primarily alkylated benzothiophenes (BTs), dibenzothiophenes (DBTs), and benzonaphthothiophenes (BNTs) in the FCC product, high-boiling sulfur compounds were absent, and moderately alkylated (C_1 to C_6) aromatic sulfur compounds dominate. Correlation of AED and TOF MS detection permits further study of unknown sulfur compounds in a sample, although direct alignment of GC × GC–AED and GC × GC–TOF MS 2D plots was not possible in this system due to different retention times in both dimensions. The information derived from both systems could be used to reduce the number of possible molecular formulas for unknowns and to aid identification of the unknowns. Shellie et al. (2004) proposed an approach to obtain better correlation of FID and MS data, accounting for the vacuum system in the latter technique. Compared with other sulfur-selective detectors, the only advantage of the AED for sulfur detection must be its empirical formula prediction, since its slow response will generally make it a less suitable detector for sulfur detection.

7.8 THERMIONIC DETECTION IN GC x GC

The term *thermionic detector* (TID) refers to the nature of the ionization process; sample molecules are converted to negative ions in the detector by extracting electrons emitted from a hot solid surface (Carlsson et al., 2001). The TID is often also referred to as a nitrogen–phosphorus detector (NPD), but strictly, this should be used only when the TID is used in the specific mode of detecting nitrogen- and phosphorus-containing compounds (Patterson, 1986). As shown in Figure 7.7, the basic difference between a FID and a TID is the bead, an

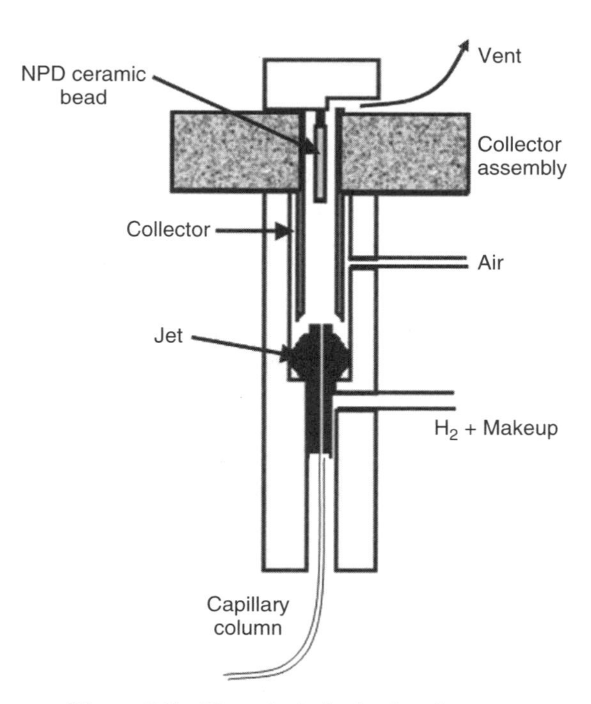

Figure 7.7 Thermionic ionization detector.

electrically heated thermionic source consisting of an alkali salt in an inorganic ceramic cement matrix (Carlsson et al., 2001). Although the operating principles of the NPD are yet to be completely understood and a detailed mechanism for the specificity of ionization process eludes researchers (Carlsson et al., 2001; Patterson, 1986; Ryan and Marriott, 2006), an acceptable theory is that of surface catalytic ionization (Olah et al., 1979; Patterson, 1978). Accordingly, alkali atoms incorporated into the ceramic bead do not leave the bead surface but rather, catalyze electron transfer taking place on the bead surface (Patterson and Howe, 1978).

Since development of the TID, a number of different bead compositions have been developed. Patterson and Howe studied the effects of detector response on bead composition, surface temperature, voltage, and composition of the gaseous environment surrounding the bead's surface. The bead surface temperature and the composition of the gaseous environment evidently determine the gaseous products formed from the decomposition of sample compounds (Patterson, 1978) and so affect detector selectivity. Sensitivity and specificity for several compounds differs according to the gas compositional environment and would also appear to affect the peak shape for compounds (Patterson, 1986). That said, however, and one might be excused in thinking that the NPD is not likely to be a reliable detector, the NPD has been used with confidence in many applications that demand good, reproducible data. Although there was some suspicion that the detection mechanism might be considered too slow to achieve the high acquisition frequencies necessary for GC × GC (notwithstanding the fact that the NPD has a reported maximum electronic acquisition rate of 200 Hz), due to this technique producing much narrower peaks (100 ms) than those obtained with conventional fast and regular gas chromatography (<1 s) (Ryan and Marriott, 2006), studies related to GC × GC analysis of various standard compounds carefully recorded the effect of detector response time and any lack of symmetry arising from the detection step. Although recognizing this possible limitation, Ryan et al. (2005b) studied NPD detection with GC × GC analysis for methoxypyrazines in wine. With an acquisition frequency of 100 Hz, analysis by GC × GC–NPD enabled detection limits of 0.5 ng·L^{-1} for the quantification of 2-methoxy-3-(2-methylpropyl)pyrazine. Comparative detection using GC × GC coupled with TOF MS gave a detection limit of 1.95 ng·L^{-1}. The use of GC × GC–TOF MS allows stable isotope isotopic dilution of the wine sample, and this is useful when analyzing the sample by use of solid-phase microextraction. The same method of GC × GC–TOF MS with isotope dilution SPME was applied to methoxypyrazine analysis in wines from New York state by Ryona et al. (2009). Optimum performance for GC × GC–NPD required optimization of hydrogen, air, and nitrogen detector flows, with the goal of generating detector flow conditions that gave best peak magnitude and peak asymmetry. As a result of selective detection using GC × GC–NPD, the complexity of the real wine headspace was simplified. The NPD performance was correlated with peak response (Ryan and Marriott, 2006). Peak asymmetry was seen to vary from a poor value of 8.0 to an acceptable 1.6

over a range of detection gas flow settings. A flow setting that gives best peak symmetry does not necessarily deliver best response magnitude.

The Agilent NPD is capable of up to 200 Hz data acquisition and so should be suited to analyzing peaks of width down to 50 ms base width. Although fast electronics may support the notion of fast peak analysis, if the detector internal process does not preserve the peak profile as it elutes from the column, detection of the peaks will deteriorate. In this case, the "chemistry" of the detection process may be a limiting factor in preventing narrow peak shape from being obtained, imposing a delay process on the peak flux as it enters and is extracted or exhausted from the detector. Although a number of modifications for the NPD, such as an extended jet and a narrow-bore collector electrode, are available from Agilent, it was found that the extended jet is better for peak narrowness but that asymmetry deteriorates when a narrow-bore collector is used. This suggests that the rapid peaks found for GC × GC are probably beyond the design tests (e.g., tested on peaks of width 100 ms) used by manufacturers when detector improvements are made, and that future improvements should best consider GC × GC.

Good structured retentions exhibiting the roof-time effect were reported for the analysis of nitrogen-containing compounds (e.g., carbazoles, quinolines, benzocarbazoles) in heavy gas oil samples (von Mühlen et al., 2007), allowing clearly identifiable speciation of classes of compounds and good coverage of the total number of isomers for each homolog. Further study demonstrated that fungicides could be detected in the presence of a complex vegetable extract matrix by use of an NPD detector (Khummueng et al., 2006b). The study allowed easy recognition of the degradation of iprodione fungicide both on the column and arising from elevated injector temperature for a splitless injector. By contrast, programmed-temperature vaporization produced less degradation in the injector. A possible explanation for the structure of the degradation product was proposed. The exhibition of on-column degradation was recognized relatively easily by the presence of elevated baseline response in 1D GC, but is better seen in the GC × GC mode, and is not unlike the on-column molecular interconversion processes studied by this group (Marriott et al., 2001).

7.9 FLAME PHOTOMETRIC DETECTION IN GC × GC

The flame photometric detector provides a good case study for detection arrangements that may or may not be suited to GC × GC operation. The pulsed-flame photometric detector may be unsuited to GC × GC, due to the slow sampling rate of this detector, which is on the order of one measurement per second. However, an inverted flame arrangement (Hayward and Thurbide, 2006) was shown to be capable of very fast data generation for sulfur- and phosphorus-containing compounds, in an application demonstrating microcolumn fabrication technology for GC analysis (Kendler et al., 2006). A dual-flame system has been proposed to overcome some of the response dependency of different chemical structures of different compounds. This system should also be suited to fast sensing of the effluent of the GC column. In all cases, we can assume that the photomultiplier system will be fast and so will not be responsible for any rate-limiting steps in the response of the FPD. The general sensing mechanism for the FPD may be described as follows:

$$\text{S- or P-containing compound} \xrightarrow{\text{flame}} \text{S or P products,}$$

$$\text{including S and HPO}^* \text{ Then}$$

$$\text{HPO}^* \rightarrow \text{HPO} + h\nu \text{ (light at} \sim\text{525 nm)}$$

$$\text{S} + \text{S} \rightarrow \text{S}_2^* \rightarrow \text{S}_2 + h\nu \text{ (light at} \sim\text{383 nm)}$$

Since a photomultiplier, usually with a bandpass filter, monitors the emission signal, the overall response mechanism should be very fast; flame-based emission processes are fast, and photomultiplier response is fast, leading to almost instantaneous signal generation for the compound in the detector compared with the compound flux through the detector. The S-mode is less sensitive than the P-mode, by perhaps a factor of 10, and the S-mode responds according to a square law for the mass of sulfur in the detector due to the S_2*-emitting species (Marriott and Cardwell, 1981).

The first report of the application of the FPD to GC × GC was by Chin et al. in 2010. An Agilent single-flame system was employed, with both S- and P-mode dual detection demonstrated. The FPD was operated in parallel with an FID. A schematic diagram of the GC × GC–FPD system described in that work is shown in Figure 7.8. Generally, the P-mode had peak widths at half-height similar to those of the FID, but the S-mode exhibited some degree of additional peak tailing

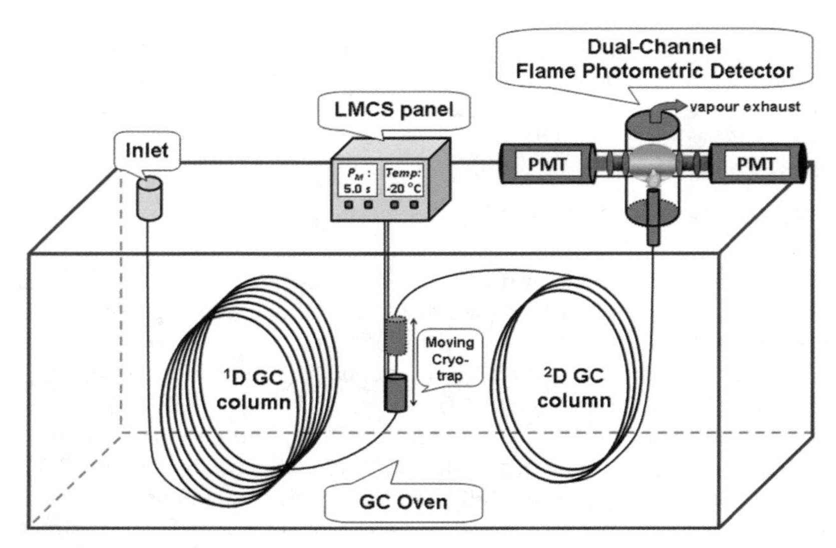

Figure 7.8 Flame photometric detector and an instrumental arrangement showing a dual-detector FID–FPD system with simultaneous S_2 and HPO detection.

(asymmetry) and broadening. There was little dependency of peak asymmetry (values about 0.8) and width with flame gas conditions for the P-mode, but more variability in the asymmetry for the S-mode, with values ranging from 0.3 to 0.66. Samples of OP-pesticides (some of which contained sulfur) and aged kerosene still retained elevated levels of sulfur-compounds. A comparative GC × GC result for a pesticide mixture with FID/FPD–S_2/FPD–HPO is presented in Figure 7.9.

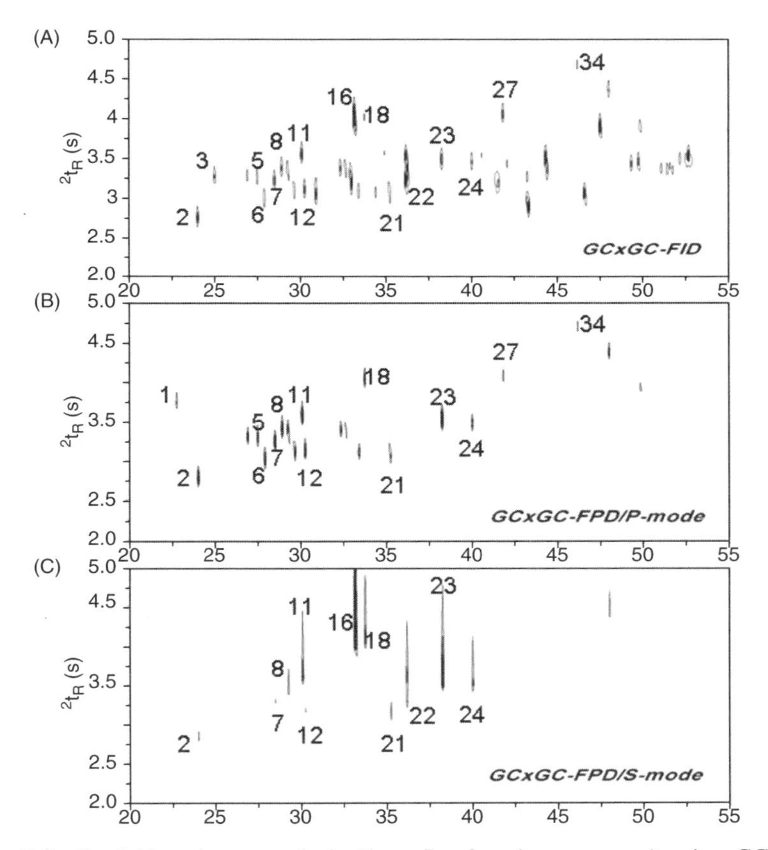

Figure 7.9 Pesticide mixture analysis (1 mg/L of each component) using GC × GC with (A) FID, (B) FPD-P mode, and (C) FPD-S mode. Selected peak identities: 1, dimethoate; 2, diazinon; 3, chlorothalonil; 5, partathion methyl; 6, fenchlorophos; 7, pirimiphosmethyl; 8, fenitrothion; 11, fenthion; 12, parathion ethyl; 16, procymidone; 18, bromophos ethyl; 21, prothiofos; 22, buprofezin; 23, ethion; 24, carbophenothion; 27, triphenyl phosphate (internal standard); 32, azinphos ethyl. Experimental conditions, 1-μL splitless injection (vent open 1 min) at 280°C; detector at 250°C; oven temperature, 50°C (1 min) to 150°C at 10°C/min, then to 280°C at 3°C/min (11 min hold). Column details: first dimension = 30 m × 0.25 mm i.d. × 0.25 μm film BPX5 (SGE); second dimension = 1 m × 0.1mm i.d. × 0.1 μm film Rxi-17 (RESTEK). Modulation, LMCS, 5 s, −20°C. FID detector gases: H_2, 30 mL/min; air, 300 mL/min; N_2, 30 mL/min. FPD detector P and S modes: H_2, 80 mL/min; air, 110 mL/min; N_2, 60 mL/min.

Evidently, none of the inverted geometry flame, pulsed flame, or dual-flame FPD arrangements of the FPD have been used with GC × GC. With only a single citation on the use of FPD with GC × GC, the literature of the application of this detector is still sparse, and extension to other studies that might prove the worth of this detector is awaited.

7.10 CASE STUDY OF GC × GC WITH SELECTIVE DETECTION

By way of an additional demonstration (see Figure 7.9) of the performance comparison of selective detection versus the FID in GC × GC, Figure 7.10A illustrates the tailing effect commonly seen with the NPD in single-column GC analysis for a selection of nitrogen- and phosphorus-containing compounds,

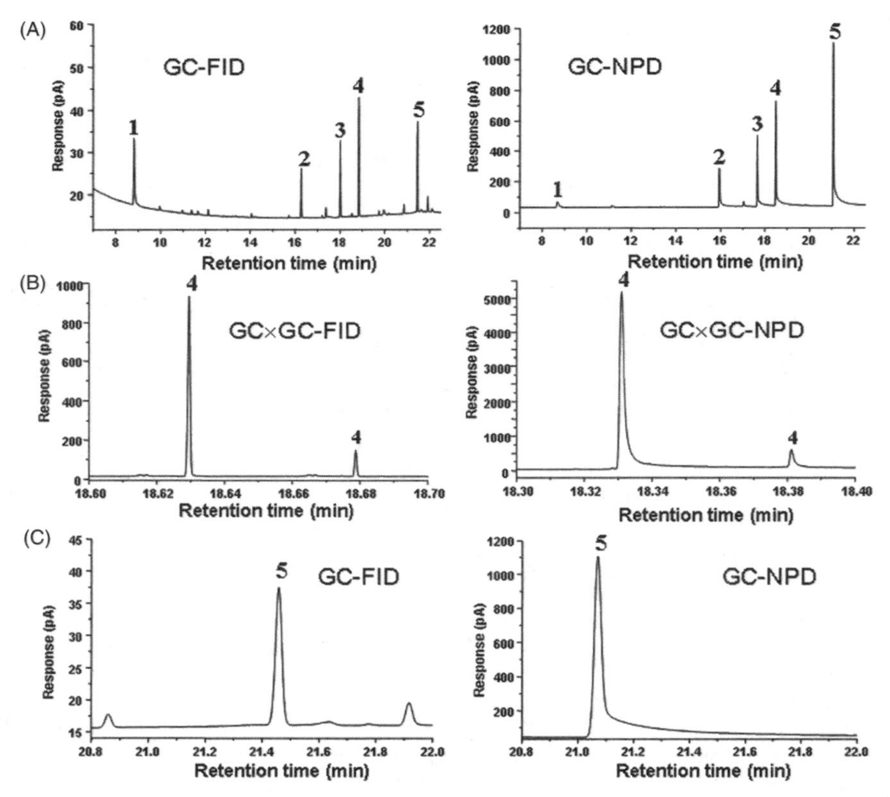

Figure 7.10 (A) GC–FID and GC–NPD chromatograms for an N- and P-containing compound mixture; (B) GC × GC–FID and GC × GC–NPD modulated peaks for fenthion, under similar conditions; (C) expanded traces showing peak shape differences for GC–FID and GC–NPD, respectively, for ethion from part (A). 1, Isobutyl methoxypyrazine; 2, dimethoate; 3, methyl parathion; 4, fenthion; 5, ethion.

with GC–FID giving a symmetric peak shape on the same column under the same conditions. The result when using cryogenic modulation to generate fast second-dimension peaks in GC × GC is illustrated in Figure 7.10B for the ethion component. In this case, if tailing arises from detector-based effects, one might suspect tailing to be worse due to the fast peak flux. Expanded GC peaks are shown in Figure 7.10C for both FID and NPD detection. The same considerations can be drawn for any other detector where performance attributes do not meet the needs of the fast response of GC × GC. Thus, the ECD would be expected to give broad peaks, and the slow data acquisition of the AED would be expected to give a truncated or nonsmooth peak profile. Rather than provide examples of each of these, only the NPD GC × GC chromatogram is included here. Figure 7.11A and B are 2D GC × GC presentations of the modulated data shown in Figure 7.10B. The advantages of selective detection are evident from the GC × GC–NPD (Figure 7.10B) plot, which now does not respond to the many impurities that occur in the GC × GC–FID result (Figure 7.10A). The tailing behavior of the NPD response is seen as elongated tails and expanded peak zones of each compound spot compared to the GC × GC–FID response. Peak position reliability is good for the two experiments.

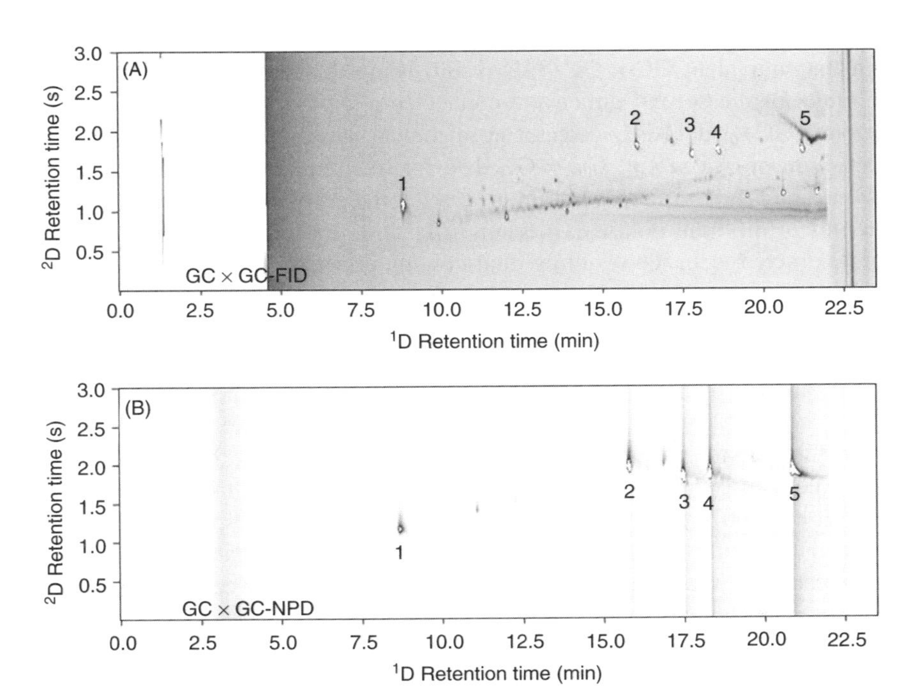

Figure 7.11 (A) GC × GC–FID and (B) GC × GC–NPD chromatograms for an N- and P-containing compound mixture under similar chromatographic conditions. Same compounds as in Figure 7.10.

7.11 DUAL DETECTION WITH GC × GC

Dual detection in GC × GC has been little exploited, although it is widely used with 1D GC. This may be due either to the reduced need for dual detection arising from the better resolution of sample components inherent in GC × GC methods, or because there may be a concern that the high efficiency demanded for the second-dimension columns might be compromised by various connections at the end of the second-dimension column. The role of a dual-detector arrangement where the effluent was split approximately equally between an FID and the two simultaneous FPD channels was mentioned earlier (Chin et al., 2010). Thus, both selective and universal detection procedures are available. It is apparently not common to employ more than one detector with GC × GC operation; however, in studying the FID–FPD dual-detector arrangement it was convenient to compare the FPD response magnitude, the peak symmetry and peak width at half-height of the selective detector with that of the FID response, where we understand the FID to be the classic fast-responding linear response detector, which provides an instantaneous profile of the peak issuing from the column. For the ECD, peak widths are also of concern, because of the internal volume of the detector, which leads to peak broadening (see Section 7.4). The FID response can confirm the relative increase in peak width of the ECD response. Similarly, Ryan and Marriott (2006) compared the peak width of the GC × GC–NPD response with the equivalent GC × GC–FID result, using the same column set but with the outlet of the second dimension connected alternately to the FID or NPD in the same GC. A dual-FID detector arrangement was employed for simultaneous acquisition of two sets of GC × GC data for retention index calculation. In this case, a single first-dimension column was interfaced with two second-dimension columns of different polarity (Bieri and Marriott, 2006). Two sets of index data were extracted from the resulting dual-column experimental arrangement.

Khummueng et al. (2008) reported use of a dual parallel detector arrangement comprising an NPD and an ECD which was used for fungicide and OP and OC pesticide analysis. Because the outlet of the second-dimension column was split equally through two short deactivated transfer lines, the peak positions in 2D space for the two detector channels were reproduced precisely. This allows direct comparison of the relative responses essentially free of matrix interference. The use of dual-detection ECD–NPD in a single-column experiment lead Jover et al. (2007) to propose a metric called the detector/response ratio (DRR), where the specific response ratio of the two detector signals provides further identification of the component. Since linear calibration curves are expected for each detector, then provided that the response is stable, the response ratio should also be characteristic of the particular compound. This is likely to be true for all compounds except those that have exactly the same response ratio in the two detectors. One might anticipate that changes in molecular structure will lead to different response magnitudes (i.e., the compounds will have different response factors). In single-column analysis, coeluting compounds that affect the response

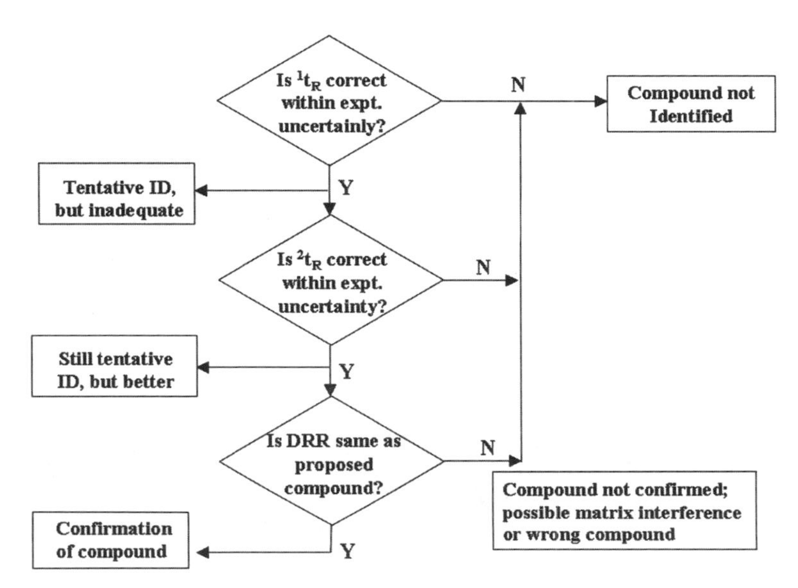

Figure 7.12 Flowchart for progressive improvement in identification of compounds based on ^1t_R, then ^2t_R, and finally, detector response ratio, DRR. [From Khummueng et al. (2008), with permission. Copyright © 2008 by Wiley-VCH Verlag GmbH & Co. KGaA.]

of either detector will lead to erroneous DRR values and hence potentially incorrect component identification (i.e., either false negatives or false positives). It was proposed that the GC × GC experiment should significantly improve the certainty of identification by the removal of coeluting compounds (Khummueng et al., 2008). The correlation of ^1t_R and ^2t_R along with the DRR should be a powerful way to provide identity to the presence of target compounds (Figure 7.12). Figure 7.13 demonstrates the simultaneous ECD–NPD detection of a standard or sample of OP pesticides and fungicides. Generally, better NPD response than ECD response is found for this sample in terms of the narrowness of the peaks.

7.12 CONCLUSIONS

There is understandably an increasing interest in selective and alternative detection systems for GC × GC. The classical need for a selective detector is to reduce the interferences from coeluting compounds in 1D GC, as dictated by the selectively of the detector toward the active element over the interfering element. A second benefit can be an improved detectability. The GC × GC experiment reduces this problem through improved separation performance; however, the other advantage of selective detection, detection sensitivity, is also a benefit. Studies of GC × GC detectors reveal that simple use of the detector in a GC × GC experiment may not be adequate to obtain maximum performance

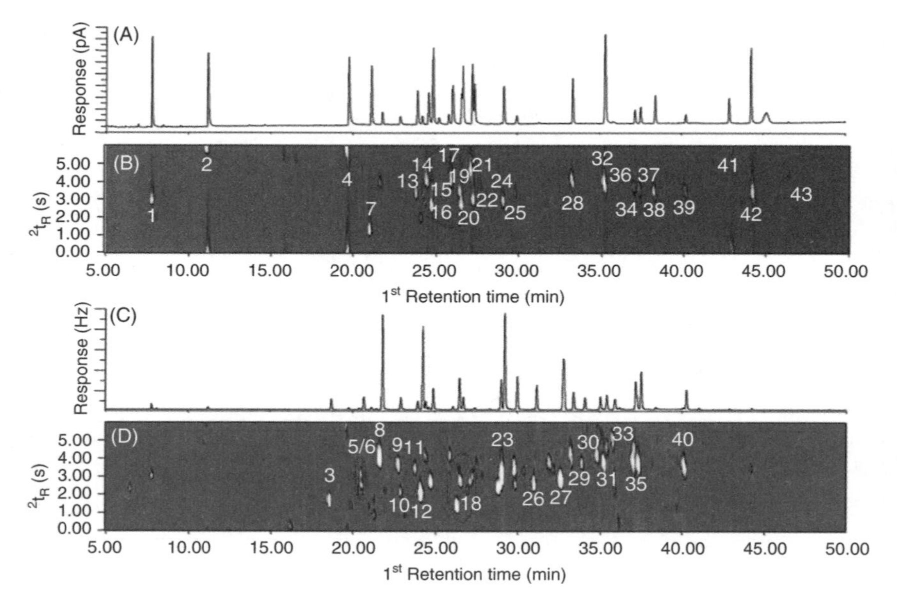

Figure 7.13 Analysis of a fungicide and pesticide mixture using (A) 1D GC–NPD, (B) GC × GC–NPD, (C) 1D GC–ECD, and (D) GC × GC–ECD. A dual simultaneous detection system was employed. [From Khummueng et al. (2008), with permission. Copyright © 2008 by Wiley-VCH Verlag GmbH & Co. KGaA.]

from the detector, or to preserve the peak shape and response fidelity of the chromatographic peak, since either the detector construction is not optimized for the very narrow peaks, or there are chemical–physical reasons (e.g., slow kinetics) associated with the response mechanism which limit the speed of the detector. Clearly, some of these factors must be addressed in a suitable range of new-generation detectors adequate for GC × GC analysis as this technology becomes more accepted in the future.

Acknowledgments

The author acknowledges the many contributions of his research group, and of our many visitors to this group over the years, to the studies that have formed the basis of some of the research studies reported here. The ongoing support of Paul Morrison in maintaining our GC systems and developing a variety of selective detection approaches is appreciated.

REFERENCES

Adahchour M, Tasöz A, Beens J, Vreuls RJJ. *J. Sep. Sci*. 2003; 26:753–760.

Adahchour M, Beens J, Vreuls RJJ, Brinkman UAT. *Trends Anal. Chem*. 2006; 25:540–553.

Amirav A. *Am. Lab*. 2001; 33:28–34.

Aybar-Munoz J, Fernandez-Gonzalez E, Garcia-Ayuso LE, Gonzalez-Casado A, Cuadros-Rodriguez L. *Chromatographia* 2005; 61:505–513.

Beens J, Boelens H, Tijssen R, Blomberg J. *J. High Resolut. Chromatogr*. 1998; 21: 47–54.

Beens J, Blomberg J, Schoenmakers PJ. *J. High Resolut. Chromatogr*. 2000; 23:182–188.

Benn SM, Peppard TL. *J. Agric. Food Chem*. 1996; 44:557–566.

Benner RL, Stedman DH. *Anal. Chem*. 1989; 61:1268–1271.

Bieri S, Marriott PJ. *Anal. Chem*. 2006; 78:8089–8097.

Bieri S, Marriott PJ. *Anal. Chem*. 2008; 80:760–768.

Blomberg J, Riemersma T, van Zuijlen M, Chaabani H. *J. Chromatogr. A* 2004; 1050:77–84.

Bordajandi LR, Ramos L, Gonzalez MJ. *J. Chromatogr. A* 2005; 1078:128–135.

Cardeal ZdL, da Silva MDRG, Marriott PJ. *Rapid Commun. Mass Spectrom*. 2006; 20:2823–2836.

Carlsson H, Robertsson G, Colmsjo A. *Anal. Chem*. 2001; 73:5698–5703.

Chaintreau A, Rochat S, De Saint Laumer J-Y. In: Bredie WLP, Petersen MA, Eds., *Flavour Science: Recent Advances and Trends*. Amsterdam: Elsevier, 2006, pp. 601–604.

Chawla B. *J. Chromatogr. Sci*. 1997; 35:97–104.

Chawla B, Di Sanzo F. *J. Chromatogr*. 1992; 589:271–279.

Chen ECM, Chen ES. *J. Chromatogr. A* 2004; 1037:83–106.

Chin ST, Wu GZ-J, Morrison PD, Marriott PJ. *Anal. Methods* 2010; 2:243–253.

Conka K, Drobna B, Kocan A, Petrik J. *J. Chromatogr. A* 2005; 1084:33–38.

Dallüge J, Beens J, Brinkman UAT. *J. Chromatogr. A* 2003; 1000:69–108.

Danielsson C, Wiberg K, Korytár P, Bergek S, Brinkman UAT, Haglund P. *J. Chromatogr. A* 2005; 1086:61–70.

David DJ. *Gas Chromatographic Detectors*. New York: Wiley, 1974.

Fraga CG, Prazen BJ, Synovec RE. *J. High Resolut. Chromatogr*. 2000; 23:215–224.

Frysinger GS, Gaines RB, Xu L, Reddy CM. *Environ. Sci. Technol*. 2003; 37:1653–1662.

Gaines KK, Chatham WH, Farwell SO. *J. High Resolut. Chromatogr*. 1990; 13:489–493.

Gautschi M, Schmid JP, Peppard TL, Ryan TP, Tuorto RM, Yang X. *J. Agric. Food Chem*. 1997; 45:3183–3189.

Gomara B, Ramos L, Gonzalez MJ. *J. Chromatogr. B* 2002; 766:279–287.

Górecki T, Harynuk J, Panić O. *J. Sep. Sci*. 2004; 27:359–379.

Harju M, Haglund P. *J. Microcol. Sep*. 2001; 13:300–305.

Harynuk J, Górecki T, Campbell C. *LC-GC N. Am*. 2002; 20:876–892.

Hayward TC, Thurbide KB. *J. Chromatogr. A* 2006; 1105:66–70.

Holm T. *J. Chromatogr. A* 1999; 842:221–227.

Hua R, Li Y, Liu W, Zheng J, Wei H, Wang J, Lu X, Kong H, Xu G. *J. Chromatogr. A* 2003; 1019:101–109.

Hua R, Wang J, Kong H, Liu J, Lu X, Xu G. *J. Sep. Sci*. 2004; 27:691–698.

Hutte RS, Johansen NG, Legier MF. *J. High Resolut. Chromatogr*. 1990; 13:421–426.

Jaycox LB, Olsen LD. *Appl. Occup. Environ. Hygine* 2000; 15:695–704.

Jover E, Gomez-Gutierrez A, Bayona JM. *Chromatographia* 2007; 66:75–79.

Junge M, Bieri S, Huegel H, Marriott PJ. *Anal. Chem*. 2007; 79:4448–4454.

Kashihira N, Makino K, Kirita K, Watanabe Y. *J. Chromatogr*. 1982; 239:617–624.

Kendler S, Reidy SM, Lambertus GR, Sacks RD. *Anal. Chem*. 2006a; 78:6765–6773.

Khummueng W, Harynuk J. Marriott PJ. *Anal. Chem*. 2006a; 78:4578–4587.

Khummueng W, Trenerry C, Rose G, Marriott PJ. *J. Chromatogr. A* 2006b; 1131:203–214.

Khummueng W, Morrison P, Marriott PJ. *J. Sep. Sci*. 2008; 31:3404–3415.

Kinghorn R, Marriott P, Cumbers M. *J. High Resolut. Chromatogr*. 1996; 19:622–626.

Klee MS, Williams MD, Chang I, Murphy J. *J. High Resolut. Chromatogr*. 1999; 22:24–28.

Korytár P, Janssen HG, Matisova E, Brinkman UAT. *Trends Anal. Chem*. 2002a; 21:558–572.

Korytár P, Leonards PEG, Boer J, Brinkman UAT. *J. Chromatogr. A* 2002b; 958:203–218.

Korytár P, Danielsson C, Leonards PEG, Haglund P, de Boer J, Brinkman UAT. *J. Chromatogr. A* 2004; 1038:189–199.

Korytár P, van Stee LLP, Leonard PEG, de Boer J, Brinkman UAT. *J. Chromatogr. A* 2003; 994:179–189.

Kristenson EM, Korytár P, Danielsson C, Kallio M, Brandt M, Mäkelä J, Vreuls RJJ, Beens J, Brinkman UAT. *J. Chromatogr. A* 2003; 1019:65–77.

Kusakabe T, Saito T, Takeichi S, Tokai J. *Exp. Clin. Med*. 2003; 28:131–138.

Lee AL, Bartle KD, Lewis AC. *Anal. Chem*. 2001; 73:1330–1335.

Liu ZY, Phillips JB. *J. Chromatogr. Sci*. 1991; 29:227–231.

Lovelock JE. *J. Chromatogr*. 1974; 99:3–12.

Lovelock JE. *J. Chromatogr*. 1975; 112:29–36.

Marriott PJ, Cardwell TJ. *Chromatographia* 1981; 14:279–284.

Marriott PJ, Shellie R. *Trends Anal. Chem*. 2002; 21:573–583.

Marriott P, Trapp O, Shellie R, Schurig V. *J. Chromatogr. A* 2001; 919:115–126.

McWilliam IG. *Chromatographia* 1983; 17:241–243.

McWilliam IG, Dewar RA. *Nature* 1958; 181: 760.

Mestres M, Busto O, Guasch J. *J. Chromatogr. A* 2000; 881:569–581.

Mitrevski BS, Wilairat P, Marriott PJ. *J. Chromatogr. A* 2010; 1217:127–135.

Ng S, Briker Y, Zhu Y, Gentzis T, Ring Z, Fairbridge C, Ding F, Yui S. *Energy Fuels* 2000; 14:945–946.

Olah K, Szoke A, Vajta Z. *J. Chromatogr. Sci*. 1979; 17:497–502.

Paetznick DJ, Reineccius GA, Peppard TL, Herkert JM, Lenton P. J. *Breath Res*. 2010; 4:1–5.

Patterson PL. *J. Chromatogr*. 1978; 167:381–397.

Patterson PL. *J. Chromatogr. Sci*. 1986; 24:41–52.

Patterson PL, Howe RL. *J. Chromatogr. Sci*. 1978; 16:275–280.

Pham TH, Janssen HGM, Cramers CA. *J. High Resolut. Chromatogr*. 1995; 18:333–342.

Phillips JB, Beens J. *J. Chromatogr. A* 1999; 856:331–347.

Quan X, Chen S, Platzer B, Chen J, Gfrerer M. *Spectrochim. Acta B* 2002; 57B:189–199.

Quimby BD, Giarrocco V. Hewlett Packard Application Note: Gas Chromatography, 1997, pp. 1–11.

Rauhut D, Kurbel H, MacNamara K, Grossmann M. *Analysis* 1998; 26:142–145.

Reddy CM, Eglinton TI, Hounshell A, White HK, Xu L, Gaines RB, Frysinger GS. *Environ. Sci. Technol*. 2002; 36:4754–4760.

Reichenbach SE, Ni M, Zhang D, Ledford EB Jr. *J. Chromatogr. A* 2003; 985:47–56.

Reichenbach SE, Ni M, Kottapalli V, Visvanathan A. *Chemometr. Intell. Lab*. 2004; 71:107–120.

Rochat S, de Saint Laumer J-Y, Chaintreau A. *J. Chromatogr. A* 2007; 1147:85–94.

Ryan D, Marriott PJ. *J. Sep. Sci*. 2006; 29:2375–2382.

Ryan D, Morrison P, Marriott PJ. *J. Chromatogr. A* 2005a; 1071:47–53.

Ryan D, Watkins P, Smith J, Allen M, Marriott PJ. *J. Sep. Sci*. 2005b; 28:1075–1082.

Ryona I, Pan BS, Sacks GL. *J. Agric. Food Chem*. 2009; 57:8250–8257.

Sacks GL, Zhang Y, Brenna JT. *Anal. Chem*. 2007; 79:6348–6358.

Sarrazin E, Shinkaruk S, Tominaga T, Bennetau B, Frérot E, Dubourdieu D. *J. Agric. Food Chem*. 2007; 55:1437–1444.

Schoenmakers P, Marriott P, Beens J. *LC-GC Eur*. 2003; 16:335–339.

Shearer RL, Meyer LM. *J. High Resolut. Chromatogr*. 1999; 22:386–390.

Shellie RA, Marriott PJ. *Analyst* 2003a; 128:879–883.

Shellie RA, Marriott PJ, Morrison PD. *Anal. Chem*. 2001; 73:1336–1344.

Shellie RA, Marriott PJ, Huie CW. *J. Sep. Sci*. 2003b; 26:1185–1192.

Shellie R, Marriott P, Leus M, Dufour J-P, Mondello L, Dugo G, Sun K, Winniford B, Griffith J, Luong J. *J. Chromatogr. A* 2003c; 1019:273–278.

Shellie R, Marriott P, Morrison P, Mondello L. *J. Sep. Sci*. 2004; 27:504–512.

Shimadzu Corporation. GC-2010 Specifications. http://www.ssi.shimadzu.com/products/product.cfm?product=gc-2010. Accessed Feb. 26, 2010.

Skoog DA, Holler FJ, Nieman TA. *Instruments for Gas–Liquid Chromatography: Atomic Emission Detectors*, 5th ed. Philadelphia: Saunders and Harcourt Brace, 1998, pp. 709–710.

Slater GF, White HK, Eglinton TI, Reddy CM. *Environ. Sci. Technol*. 2005; 39:2552–2558.

Song SM, Marriott PJ, Wynne P. *J. Chromatogr. A* 2004; 1058:223–232.

ThermoElectron Corporation. Trace GC Ultra Product Specifications. http://www.thermo.com/com/cda/product/detail/1,1055,1000001009242,00.html. Accessed Feb. 26, 2010.

Tobias HJ, Sacks GL, Zhang Y, Brenna JT. *Anal. Chem*. 2008; 80:8613–8621.

van Stee LLP, Brinkman UAT, Bagheri H. *Trends Anal. Chem*. 2002; 21:618–626.

van Stee LLP, Beens J, Vreuls RJJ, Brinkman UAT. *J. Chromatogr. A* 2003; 1019:89–99.

Varian, Inc. 450-GC FID Detector Specifications. http://www.varianinc.com/cgi-bin/nav?products/chrom/gc/gcflame&cid=LKJOJINKFIH. Accessed Feb. 26, 2010.

Vendeuvre C, Bertoncini F, Duval L, Duplan JL, Thiébaut D, Hennion MC. *J. Chromatogr. A* 2004; 1056:155–162.

Vendeuvre C, Guerrero RR, Bertoncini F, Duval L, Thiébaut D, Hennion MC. *J. Chromatogr. A* 2005; 1086:21–28.

Vermeulen C, Pellaud J, Gijs L, Collin S. *J. Agric. Food Chem*. 2001; 49:5445–5449.

von Mühlen C, Khummueng W, Zini CA, Caram ao EB, Marriott PJ. *J. Sep. Sci*. 2006; 29:1909–1921.

von Mühlen C, de Oliveira EC, Morrison PD, Zini CA, Caramão EB, Marriott PJ. *J. Sep. Sci*. 2007; 30:3223–3232.

von Mühlen C, de Oliveira EC, Zini CA, Caramão EB, Marriott PJ. *Energy Fuels* 2010; 24:3572–3580.

Wang FC-Y, Robbins WK, Greaney MA. *J. Sep. Sci*. 2004; 27:468–472.

Western RJ, Marriott PJ. *J. Sep. Sci*. 2002; 25:831–838.

Western RJ, Marriott PJ. *J. Chromatogr. A* 2003; 1019:3–14.

Williams A, Ryan D, Guasca AO, Marriott P, Pang E. *J. Chromatogr. B* 2005; 917:97–107.

Wing CY, Fine DH, Chiu KS, Biemann K. *Anal. Chem*. 1984; 56:1158–1162.

Yan X. *J. Chromatogr. A* 2002; 976:3–10.

Yan X. *J. Sep. Sci*. 2006; 29:1931–1945.

Zhu S, Lu X, Dong L, Xing J, Su X, Kong H, Xu G, Wu C. *J. Chromatogr. A* 2005; 1086:107–114.

Zrostlíková J, Lehotay SJ, Hajšlová J. *J. Sep. Sci*. 2002; 25:527–537.

8

HISTORY, EVOLUTION, AND OPTIMIZATION ASPECTS OF COMPREHENSIVE TWO-DIMENSIONAL LIQUID CHROMATOGRAPHY

Isabelle François
University of Gent, Gent, Belgium

Koen Sandra
Metablys, Research Institute for Chromatography, Kortrijk, Belgium

Pat Sandra
University of Gent, Gent, Belgium

The limited resolving power of one-dimensional liquid chromatography (1D LC) has incited the development of two-dimensional (2D LC) techniques to provide better separation of complex mixtures. In 2D LC, the sample is subjected to two different separation mechanisms and the intrinsic high selectivity (α) of LC is fully exploited. Compared to 2D GC, a variety of LC modes with distinct separation mechanisms is available and therefore the number of orthogonal combinations is theoretically higher in comprehensive 2D LC than in 2D GC.

As in all 2D approaches, the largest benefit of combining columns is the drastic augmentation of the peak capacity, which is reflected in the reduction of component overlap. As discussed in Chapter 3, the practical peak capacity theoretically approaches multiplication of the peak capacities of the individual

Comprehensive Chromatography in Combination with Mass Spectrometry, First Edition.
Edited by Luigi Mondello.

dimensions. However, similarity in separation mechanisms (low orthogonality) and the loss of first-dimension separation through insufficient fractionation of the first-dimension effluent (low sampling frequency) negatively influence the practical peak capacity. Compared to comprehensive GC separations, the peak capacities obtained by LC × LC remain limited, notwithstanding the potential high orthogonality. This is due to the fact that LC is still not able to compete with GC in terms of efficiency, despite the major efforts that have been invested lately in the development of new stationary phases and instrumentation for ultrahigh-pressure and elevated-temperature LC.

Due to the significant higher separation power of comprehensive LC compared to its 1D counterpart, the technique has experienced significant interest in diverse fields and has been the subject of various reviews (Berek, 2010; Dixon et al., 2006; Dugo et al., 2008a,b; Evans and Jorgenson, 2004; François et al., 2009; Guiochon et al., 2008; Guttman et al., 2004; Herrero et al., 2009; Jandera, 2006; Liu and Lee, 2000; Pól and Hyötyläinen, 2008; Sandra et al., 2009; Shalliker and Gray, 2006; Shellie and Haddad, 2006; Stoll et al., 2007; Stroink et al., 2005; Tranchida et al., 2007). Comprehensive LC is currently used for the separation of complex environmental and petrochemical samples, pharmaceuticals, polymers, natural products, and biological mixtures. Evidently, the technique has also attracted a great deal of interest in the proteomics arena as a valuable alternative to the widely used 2D polyacrylamide gel electrophoresis (2D-PAGE) approach, combined with mass spectrometry (MS) through electrospray (ESI) or matrix-assisted laser desorption ionization (MALDI). Given the enormous sample complexity encountered, it is no surprise that the field of proteomics can be regarded as a catalyzer of the 2D LC technique. The various LC × LC applications are described in more detail in Chapter 10.

In the present chapter we focus on a detailed description of LC × LC, including method development and bottlenecks in the instrument design. As in Chapter 4 for GC × GC separations, particular attention is devoted to instrumental hardware evolution. The present shortcomings and future developments of LC × LC are also considered. The final paragraphs of this chapter deal with the most striking achievements in this field.

8.1 METHOD DEVELOPMENT AND INSTRUMENTATION

The most general setup of a comprehensive LC system consists of two pumps, two columns, an injector, an interface, and a detector. The interface is, in general, a high-pressure switching valve, and this device is often referred to as a modulator or sampling device. Obviously, the configuration depends strongly on the sample properties and the application. As an example, in proteomics analyses, where sample quantities are very often limited, other setups needed to be developed. Next to the more general type of comprehensive LC, these configurations are discussed as well. All configurations clearly have both benefits and disadvantages, but the overall major bottleneck of comprehensive LC is the increased complexity of the system. In addition, not all separation modes are easily combined in an online system, due to incompatibility of the mobile phases, including immiscibility of mobile-phase constituents and precipitation of buffer salts. These issues are discussed later.

The possibilities of comprehensive LC were first explored by Erni and Frei (1978), who realized that the resolving power of 1D LC was not sufficient to unravel complex samples into individual solutes. They developed the idea of coupling LC columns from 2D thin-layer chromatography (TLC). An eight-port switching valve equipped with two identical sampling loops was used to couple a gel permeation (GP) column to reversed-phase (RP) LC for the separation of complex plant extracts. Their setup and separation were far from ideal: The modulation time was 75 min, resulting in the collection and reintroduction of only seven fractions of 1.5 mL over a total analysis time of 10 h. Due to the inadequate sampling of the first-dimension effluent, the comprehensiveness of this separation is often questioned. Nevertheless, the Erni and Frei article can be considered as pioneering in the field of comprehensive LC.

A more successful and practical example of interfacing in LC × LC is shown in Figure 8.1. The interface joins the two dimensions and is in this case a

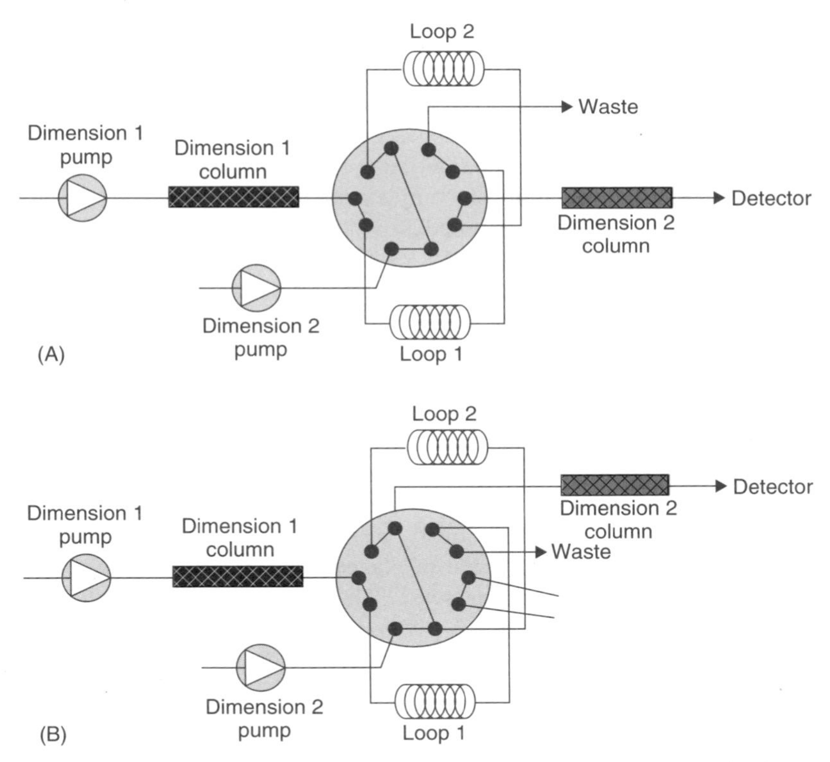

Figure 8.1 Common comprehensive setup with a two-position 10-port switching valve: (A) symmetrical arrangement; (B) asymmetrical arrangement. [From Van der Horst and Schoenmakers (2003).]

two-position 10-port switching valve equipped with two identical sampling loops, which are used alternately for the collection and reinjection of the first-dimension effluent to the secondary column. The interface is the key component in all comprehensive LC systems, since it enables continuous transfer of the primary column effluent in aliquots of predefined volumes to the second dimension in an automated manner. Due to its great importance, this part of the system is discussed in detail, together with different configurations and approaches, described in the literature.

The method development for the two dimensions in comprehensive LC is far from straightforward and deals with a lot of compromises. Prior to coupling, the methods in both dimensions should be optimized with respect to the sample characteristics, taking into account all parameters that influence the peak capacity (e.g., efficiency of the individual dimensions, orthogonality, sampling frequency, compatibility of the dimensions). To simplify the design of comprehensive LC systems for the operator, Schoenmakers et al. (2006) have suggested a theoretical protocol. They propose several guidelines, advising on optimal chromatographic parameters, including column dimensions, sampling times, and flow rates. As a practical example, they describe the RPLC × size-exclusion chromatographic (SEC) separation of a complex polymeric mixture.

A similar theoretical approach was followed by Bedani et al. (2006), giving guidance on the selection of the operational parameters and the instrumental design. A given second-dimension separation and the desired peak capacity for the entire 2D separation were taken as starting points from which optimum settings for the first dimension can be derived. They performed their calculations for the SEC × RPLC separation of complex mixtures of peptides by using an interface operated in the stop-flow mode.

Despite these guidelines, the success of comprehensive LC is largely dependent on the expertise of the operator. The appropriate mechanisms and column dimensions need to be selected, and as commercial instrumentation is only marginally available, the hands-on technical skills of the operator are of the atmost importance. In our experience, at this stage comprehensive LC is not a technique that can be implemented easily in a routine laboratory with a push-button approach; there are simply too many parameters that need optimization. However, once the instrumental configuration is defined and when robust separation mechanisms are employed, the setup does not need much more maintenance than that required for 1D LC. Obviously, problems related to the lack of commercial instrumentation and software packages will be solved in the future as the interest and number of users increase. These issues are addressed later. In the following paragraphs, details are given on the properties of the individual dimensions and aspects that need to be addressed during the development of comprehensive LC separations.

8.1.1 Orthogonality and Sampling Frequency

As discussed theoretically in Chapter 3, orthogonality is of primordial importance in LC × LC. Logically, the benefits of the 2D approach are fully exploited only by combining separation mechanisms that provide completely different elution patterns, since the possibility of coeluting sample components in both dimensions are, in this way, severely minimized. True orthogonality is technically difficult to achieve, as this feature depends not only on the separation mechanisms, but also on the properties of the solutes and the separation conditions. There is no such thing as a generic orthogonal combination, since the nature of the solutes differs with sample origin. Successful orthogonal combinations can be achieved when the appropriate stationary and mobile phases are carefully chosen with respect to the physicochemical properties of the sample constituents, including size, charge, polarity, and hydrophobicity. A diversity of stationary phases is presently available, with differences in surface chemistries, support material, carbon load, pore size, and so on, whereas the characteristics of the mobile phase can be altered by changing the modifier, pH, or temperature, or by adding ion-pair agents.

A few examples of orthogonal combinations are the normal-phase (NPLC) × RPLC separation of essential oils containing coumarins and psoralens (separation according to polarity and hydrophobicity) (Dugo et al., 2004; François et al., 2006) and the silver-ion (SI) LC × nonaqueous (NA) RPLC separation of triacylglycerols (separation according to the degree of unsaturation and hydrophobicity) (Mondello et al., 2005). Retention in orthogonal dimensions may still be weakly correlated when a series of analytes has similar physicochemical properties and behaves similarly in both dimensions. As an example, one could consider the SEC × RPLC combination as orthogonal, but for certain mixtures, separation by size may be correlated with RP retention. Similarly, ion-exchange (IEX) LC separates analytes mainly on the basis of charge, but might show hydrophobic interactions.

In contrast to the more obvious orthogonal combinations, intuitively one would expect hydrophilic interaction liquid chromatography (HILIC) to be inversely correlated to RPLC: that is, compounds displaying a large k value on HILIC are not well retained by RPLC, and vice versa. Investigations have, however, shown that this is not the case, leading to the conclusion that the interactions in HILIC cannot be explained solely by hydrophilic interactions (Gilar et al. 2005).

Gilar et al., 2005 investigated the orthogonality off-line by evaluating normalized 2D LC plots using RPLC, SCX, and HILIC in combination with RPLC at low pH in the second dimension for the separation of peptides. Their results are shown in Figure 8.2. Orthogonality was measured by dividing the normalized 2D plots evenly into a number of bins that equal the number of data points ($14 \times 14 = 196$). The area of bins covered by the data points represents the surface coverage or orthogonality. The 2D separations presented in Figure 8.2A and 8.2C have a similar theoretical peak capacity, but taking the surface coverage into account, the practical peak capacity is dramatically different. As discussed in Chapter 3, this example illustrates how easily optimistic conclusions can be drawn when the orthogonality is not taken into account.

The orthogonality of the RPLC–RPLC (Figure 8.2C) and the SCX–RPLC (Figure 8.2F) systems are nearly identical, but the latter has a lower peak capacity, due to the lower peak capacity of the SCX column over the RPLC column (51

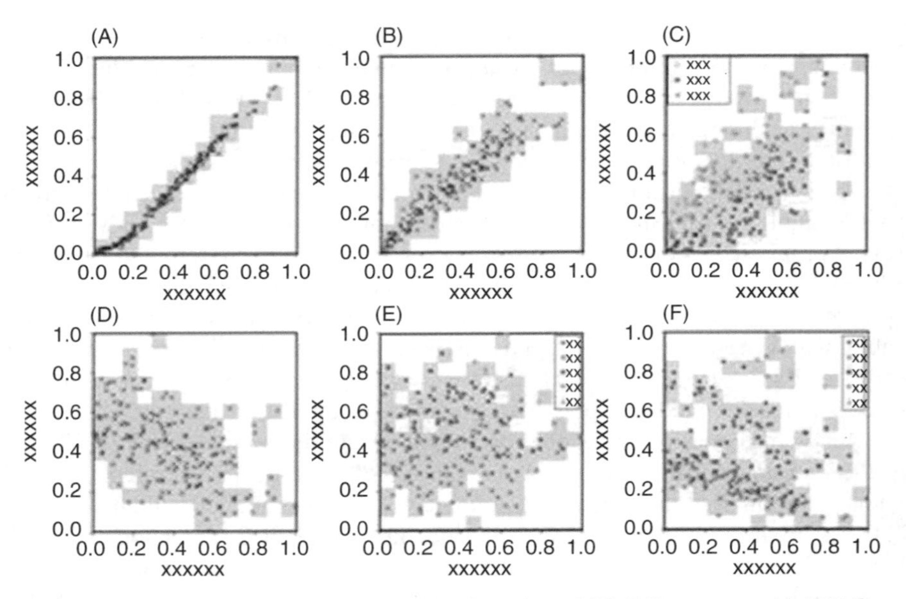

Figure 8.2 Normalized retention-time plots for selected 2D LC systems with RPLC at pH 2.6 in the second dimension. First-dimension separations were RPLC on phenyl silica at pH 2.6 (A), on perfluorophenyl silica at pH 2.6 (B), on hybrid silica C_{18} at pH 10 (C), SEC on diol silica at pH 4.5 (D), HILIC on silica at pH 4.5 (E), and SCX on poly(2-sulfoethyl aspartamide) silica at pH 3.25 (F). For detailed experimental conditions, see Gilar et al. (2005). The plots are the result of independent 1D LC–MS measurements of a test mixture of digested proteins and peptides. [From Gilar et al. (2005), with permission. Copyright © 2005 by The American Chemical Society.]

vs. 115). Based on these investigations, François et al. (2009b) have successfully combined two RPLC dimensions in an online configuration for the separation of tryptic digests of bovine serum albumin (BSA) (Figure 8.3) and blood serum. The high orthogonality was obtained by operating the two dimensions at different pH values; the first dimension was characterized by a high efficiency by coupling four fused-core columns in series at a temperature of 45°C and was operated at acidic pH (1.8), whereas the alkaline medium (pH 10) ensured a significantly different elution pattern in the second dimension. Despite the disappointing results for the SCX × RPLC combination, it is still the most commonly used multidimensional (MD) liquid separation approach in the field of proteomics, although other combinations are gaining in popularity (Sandra et al., 2009).

Another parameter that has a significant influencing on the success of comprehensive 2D LC is the sampling frequency. To maintain the most resolution obtained in the first dimension, a sufficient number of fractions should be transferred along a peak eluting from the first dimension. When the modulation cycle is significantly larger than the peak widths eluting from the primary column, the first-dimension resolving power may be reduced substantially, since resolved peaks can be captured in the same fraction prior to transfer to the second

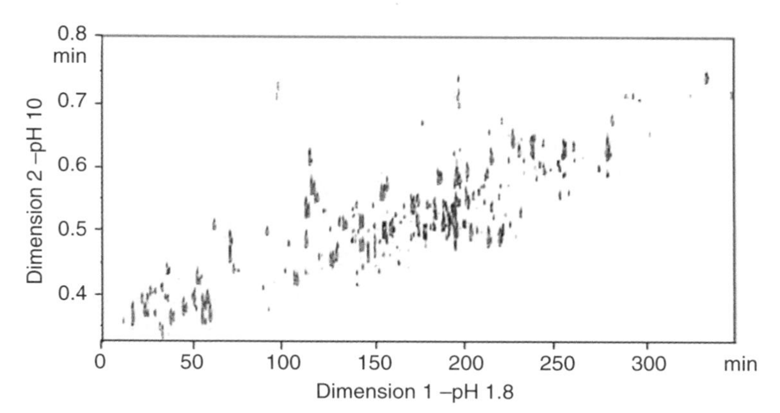

Figure 8.3 RPLC × 2RPLC separation of a BSA tryptic digest. The first dimension consisted of four serially coupled HALO C_{18} columns operated at a temperature of 45°C and pH 1.8, while the second dimension was operated on two columns in parallel under basic conditions. For detailed experimental conditions, see François et al. (2009b). [From François et al. (2009b), with permission of the copyright owner.]

dimension. Fast second-dimension analyses are, in other words, prerequisite in comprehensive techniques. Unfortunately, this implies that little time is available for separation in the second dimension, and hence the peak capacity of the secondary column is relatively low. The influence of sampling frequency on the practical peak capacity has been investigated by various research groups (Davis, 2005; Davis et al., 2008a,b; Horie et al., 2007; Li et al., 2009; Murphy et al., 1998a; Seeley, 2002) and two or three fractions transferred per first-dimension peak seems to be sufficient.

In proteomics experiments, the number of fractions (or samplings) is often less than the peak capacity of the first dimension. In this case, the overall peak capacity is calculated more adequately by multiplying the number of fractions by the peak capacity of the second dimension. This is illustrated by multidimensional protein identification technology (MudPIT), described by Washburn et al. (2001) and Wolters et al. (2001), in which 15 fractions are displaced via a stop-flow approach from an SCX onto an RPLC stationary phase, packed sequentially in one capillary. In that particular study, the second dimension had a peak capacity of 216; hence, the overall peak capacity was estimated at 3240. It is obvious that the full potential of the 2D setup is not used and that a higher sampling rate could lead to an increased peak capacity—of course, at the expense of analysis time. The number of samplings was limited since a salt step gradient was employed.

Nevertheless, this hypothesis might also produce misleading results. There is definitely a mismatch in the calculation of the first-dimension peak capacity (Stoll et al., 2007). Collecting 40 fractions from a first-dimension technique with a peak capacity of 10 would result in an overestimation of the overall peak capacity when multiplying the number of fractions by the second-dimension peak capacity, but would give a more precise calculation when the first- and second-dimension peak

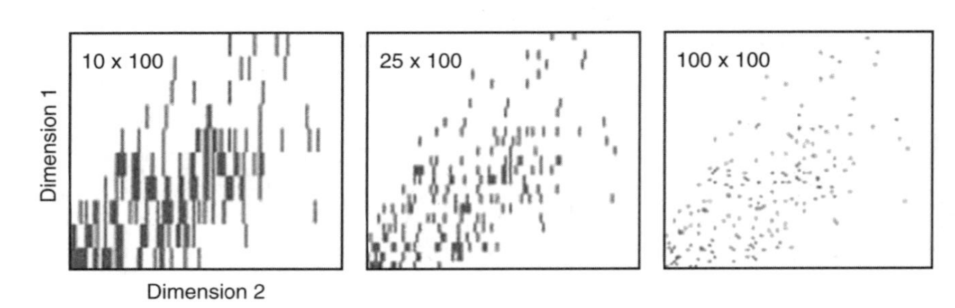

Figure 8.4 Effect of sampling rate on chromatographic resolution. From left to right, the first dimension is sampled 10, 25, and 100 times. Second-dimension peak capacity is 100. Highest gain in performance is achieved at the highest sampling rate. [From Gilar et al. (2005), with permission. Copyright © 2005 by The American Chemical Society.]

capacity values are multiplied. This shows, once again, how easily optimistic conclusions (ignorantly or deliberately) can be drawn in MD separations. Figure 8.4 illustrates the improvement in a 2D separation upon increasing the number of samplings.

8.1.2 Second Dimension

For the second dimension, the analysis speed is of fundamental importance. The time for analysis (and regeneration when a gradient is applied) is determined by the sampling rate and the flow rate used in the first dimension. As discussed in Chapter 3 and Section 8.1.1, the first-dimension resolution is maximally maintained when the first-dimension peaks are sufficiently fractionated. This implies that fast second-dimension analyses should be performed at high flow rates within sampling periods ranging from 15 s to several minutes, depending on the application. Evidently, resolution in the second dimension is compromised in this way. In the majority of applications, RPLC is selected in the second dimension. This is also especially beneficial when using MS detection.

The internal diameter (i.d.) of the column is usually 4.6 mm, to allow the introduction of relatively large fraction volumes (generally equal to the mobile-phase quantity per modulation time eluting from the first dimension). Another favorable feature of using secondary columns with conventional diameters is that high-flow-rate operation permits fast repetitive gradients without excessive pressure drop. As a rule, the second-dimension column i.d. is always larger than that of the column used for the primary separation (four to eight times larger; Schoenmakers et al., 2006). This is not the case for the proteomics configurations. As sample quantities in proteomics research are often limited, the entire comprehensive system is miniaturized in most investigations, and the columns used in both dimensions are of micro or nano scale. As a consequence, the configuration shown in Figure 8.1 is typically not used in the proteomics domain.

In high-performance liquid chromatography (HPLC), various strategies are available to increase the analysis speed. Monolithic columns are ideally suited for implementation in the second dimension, due to their high permeability and short regeneration characteristics (Eeltink et al., 2006a; van Nederkassel et al., 2003) and have therefore been exploited widely in comprehensive LC (Cacciola et al., 2007; Dugo et al., 2004, 2006a–c, 2008c; François et al., 2006; Hu et al., 2005; Ikegami et al., 2006; Leitner and Kampfl, 2008; Murahashi, 2003; Pól et al., 2006; Tanaka et al., 2004; Zhang et al., 2006). Furthermore, they allow operation at elevated flow rates without loss in efficiency and resolution. A drawback of monolithic columns is the upper limit of temperature, which is significantly lower than that of fused-core stationary phases.

Another way to speed up second-dimension analysis is to use conventional columns packed with small-particle stationary phases. Despite the longer regeneration times, these columns are considered beneficial in terms of analysis time for separations requiring less than 30,000 theoretical plates (Eeltink et al., 2006b; Schoenmakers et al., 2006), as is the case for the second dimension in comprehensive LC (Blahová et al., 2006; Cacciola et al., 2006; Chen et al., 2004; Eggink et al., 2008; Gilar et al., 2005; Haefliger, 2003; Jandera et al., 2008; Kivilompolo and Hyötyläinen, 2007; Kivilompolo et al., 2008; Murphy et al., 1998b; Pól et al., 2006; van der Klift et al., 2008; Venkatramani and Patel, 2006; Venkatramani and Zelechonok, 2003). Obviously, a decrease in the particle size results in enhanced resolution but places strict requirements on instrumentation as well as on stationary phase stability. The operation of sub-2-μm-particle packed columns at high flow rates in the second dimension of a comprehensive separation needs sophisticated instrumentation capable of delivering high pressures (600 to 1200 bar).

The use of elevated temperatures in the second dimension to raise the second-dimension speed has been promoted by Carr and co-workers (Dixon et al., 2006; Stoll and Carr, 2005; Stoll et al., 2006), who performed ultrafast gradients on narrow-bore wide-pore carbon-coated zirconia columns by increasing the temperature above 100°C. The reduced viscosity of the mobile phase leads to low-pressure drops through which the flow rate (and hence column reequilibration) can be strongly accelerated without a loss in efficiency, as a favorable consequence of the effect of temperature on the mass transfer properties. The stabilities of the stationary phase and analytes are, however, of major concern when temperature is used to speed up the second-dimension separation.

Recently, partially porous or fused-core materials have experienced a revival for hyperfast separations with high resolution. The stationary-phase particle consists of a thin layer of porous shell fused to a solid particle. The small path for diffusion of the solutes into and out of the stationary phase reduces the residence time of the sample components inside the particles compared to conventional porous-particle stationary phases. Due to this column technology, the efficiency and separation speed rival those of sub-2-μm-columns, with the advantage of lower pressure drops, through which conventional instrumentation can be employed (Cunliffe and Maloney, 2007; DeStefano et al., 2008; Gritti et al., 2007; Marchetti and Guiochon, 2007; Marchetti et al., 2007). This type of stationary phase has been used in the second dimension of off-line MD analyses of peptides (Marchetti et al., 2008) as well as in comprehensive separations for the analysis of polyphenolic antioxidants (Dugo et al., 2008d) and pharmaceuticals (Alexander and Ma, 2009). Fused-core materials have been used in the second dimension of comprehensive RPLC × RPLC analyses of a tryptic digest using a combination of totally porous and partially porous stationary phases (Mondello et al., 2010).

For second-dimension analyses, both isocratic and gradient programs are used. Isocratic elution facilitates, on the one hand, handling the fast runs and avoids column conditioning steps and the presence of negative effects due to the rapid changes in modifier concentration on the background signals of ultraviolet or MS detectors. On the other hand, gradient elution in the second dimension is more favorable, since it reduces the number of wraparounds. A wraparound occurs when the retention time of a component in the second-dimension separation exceeds the sampling time. In that case, the compound elutes together with components from the next modulation cycle, at a retention time different than its real retention time. This leads to coelution problems, and due to relatively long residence inside the column, the component elutes in a significantly broader peak, which adversely affects the detection limit. Gradient elution in the second dimension is also beneficial to better exploit the separation time available, even at the cost of the need for a reconditioning step.

8.1.3 First Dimension

The separation in the first dimension is generally characterized by optimum resolution for the given application since the analysis time is not limited. To reach a high efficiency, the first dimension sometimes consists of several serial coupled columns. The first dimension can be operated under isocratic or gradient conditions, depending on the application. When a gradient is applied in the first dimension, problems might arise from the fact that the composition of the fractions is changing throughout the separation. As the solvent strength of the fractions injected onto the second dimension is altering, peak distortion might occur. This is obviously of higher concern in the case of a 2D combination that employs incompatible mobile phases in the two dimensions. François et al. (2006) investigated the NPLC × RPLC separation of a citrus oil extract with a first-dimension gradient using *n*-hexane and ethyl acetate on a microbore diol column. Figure 8.5 represents separation under optimized and nonoptimized conditions. Severe peak broadening is observed for components eluting at a high ethyl acetate concentration when starting the second dimension at moderately polar conditions (40% acetonitrile) (Figure 8.5C), which had been defined as the optimal initial conditions of the secondary mobile phase during off-line method development prior to coupling. On the contrary, Figure 8.5B depicts the off-line separations of the fractions collected between 44 and 67 min under the same conditions as in Figure 8.5C, but evaporated to dryness and redissolved in a solvent containing a high *n*-hexane content. The presence of *n*-hexane (immiscible in both RP solvents) is clearly beneficial to create a focusing effect, while a high ethyl acetate

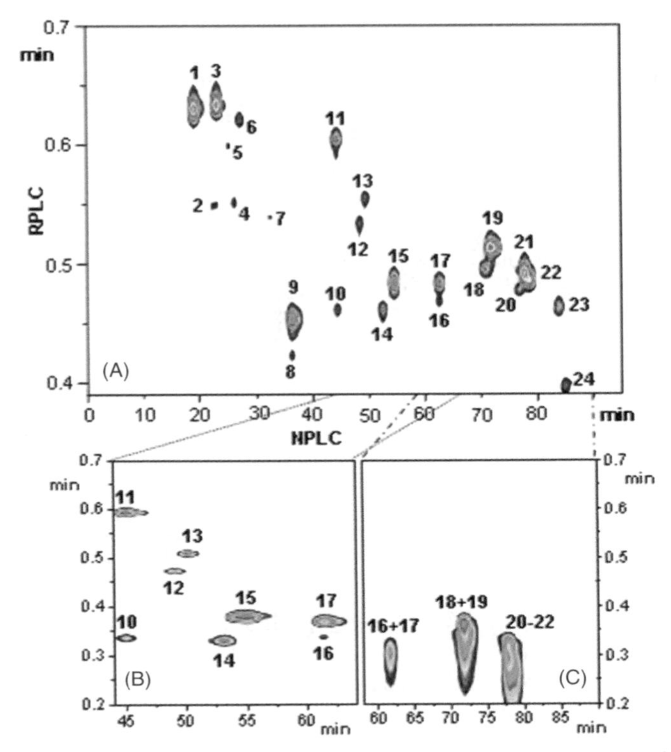

Figure 8.5 NPLC × RPLC separation of citrus oil under optimized (A) and nonoptimized conditions (C). (B) Off-line chromatograms of the fractions collected between 44 and 67 min, after evaporation to dryness and reconstitution in a sample solvent containing a higher n-hexane content. For detailed experimental conditions, see François et al. (2006). [From François et al. (2006), with permission of the copyright owner.]

content, miscible in acetonitrile, loses this positive effect. By starting the gradient with pure water, the focusing was reestablished (Figure 8.5A).

To significantly reduce band broadening and diminish signal interferences caused by solvent incompatibilities (Section 8.2.2), the first dimension is usually a micro- or narrow-bore column. The low flow rates typically employed on these columns provide fraction volumes that are compatible with the conventional bore columns of 4.6 mm i.d. in the second dimension. An alternative is to use a wider-bore column and to split the effluent prior to the interface, but in this way a large part of the sample components is wasted through which the sensitivity is jeopardized. When a wider-bore column is used in the first dimension, the flow rate is occasionally reduced to lower values than the optimum deduced from the Van Deemter curve in order to allow a sufficient number of samplings per first-dimension peak. The first-dimension resolution and hence overall peak capacity are sacrificed in this case. A better alternative is the implementation of another interface that allows a higher sampling frequency.

8.1.4 Interface

The heart of any comprehensive LC system is the interface, since this device automates the continuous transmission of the primary column effluent to the second dimension. Especially for those combinations incorporating separation modes that are rather difficult to combine, the interface design is particularly important. Throughout the years, many arrangements have been reported, each representing advantages and disadvantages. One of the future directions in LC × LC method development lies in improvement of the present interfaces, which will together with system ameliorations undoubtedly lead to even more spectacular separations.

Due to the importance of the interface in comprehensive LC, various interface designs are presented in this paragraph, with attention to their pros and cons. For an overview of the various applications according to the combined mechanisms or sample characteristics, the reader is referred to the reviews available on this subject (Berek, 2010; Dixon et al., 2006; Dugo et al., 2008a,b; Evans and Jorgenson, 2004; François et al., 2009a; Guiochon et al., 2008; Guttman et al., 2004; Herrero et al., 2009; Jandera, 2006; Liu and Lee, 2000; Pól and Hyötyläinen, 2008; Sandra et al., 2009; Shalliker and Gray, 2006; Shellie and Haddad, 2006; Stoll et al., 2007; Stroink et al., 2005; Tranchida et al., 2007) and to Chapter 10.

Loop Interface The loop interface is the most widely used interface and is based on a two-position 10-port (Cacciola et al., 2006, 2007; Dugo et al., 2004, 2006a,b, 2006c, 2008c; Eggink et al., 2008; Jandera et al., 2006, 2008; Jiang et al., 2005; Kivilompolo and Hyötyläinen, 2007; Kivilompolo et al., 2008; Mondello et al., 2005; Pól et al., 2006; Stoll and Carr, 2005; Van der Horst and Schoenmakers, 2003; van der Klift et al., 2008; Zhang et al., 2006) or eight-port (Bushey and Jorgenson, 1990; Chen et al., 2004, 2007; Erni and Frei, 1978; Hu et al., 2005; Murphy et al., 1998b; Opiteck et al., 1997a; Van der Horst and Schoenmakers, 2003) switching valve equipped with two storage loops with identical volumes. Other configurations have been developed using the same principle through a combination of several six-port valves (Gray et al., 2004, 2005; Ikegami et al., 2006; Leitner and Kampfl, 2008). A schematic representation of a comprehensive system employing a 10-port valve was given in Figure 8.1 (Van der Horst and Schoenmakers, 2003). The loop sizes are determined by the mobile-phase quantity per sampling period eluting from the first dimension. The two loops are alternately filled with primary column effluent and emptied toward the second dimension in a continuous way. The time available for analysis (and regeneration when a gradient is applied) in the second dimension is for this arrangement equal to the sampling period. As an example, if the flow rate on the primary column is set at 30 μL/min and the sampling period is tuned at 1 min, the valve should correspondingly be equipped with storage loops of 30 μL.

Van der Horst and Schoenmakers (2003) evaluated the influence of different flow paths through the sampling loops on a two-position 10-port switching valve during loading with primary column effluent and reinjection into the second dimension of the fractions collected. They discovered severe retention time differences in an "asymmetrical" configuration (Figure 8.1B), where one of the loops is emptied in the forward-flush mode and the other loop in the back-flush mode. In the symmetrical arrangement shown in Figure 8.1A, identical peak shapes and retention times were obtained. Furthermore, they pointed out that to obtain true comprehensiveness, the loop size should be significantly larger than the volume of the fraction captured caused by the parabolic flow profile inside the loops. This rule is, however, rarely followed. When an eight-port valve is used as a modulator, the flow directions in the loops are always dissimilar.

Packed Loop Interface The sampling loop interface can be modified by replacing the empty storage loops with loops packed with stationary phase, which focuses the solutes prior to their analysis in the second dimension. In practice, guard columns often function as packed loops. While one guard column collects the first-dimension effluent, the compounds trapped in the previous cycle are released from the second guard column loop to the secondary column. The development of such a configuration is complicated by the fact that trapping efficiency and rapid desorption are inversely related. The characteristics of the adsorbent in the loops are dependent on the properties of the sample components and the solvents used in both dimensions. The first-dimension mobile phase is preferably a weak solvent for the stationary phase in the loop to allow focusing, while fast desorption by a strong solvent should be performed by the secondary mobile phase. Furthermore, since the guard columns are flushed in the reverse direction during transfer of the fractions to the second dimension and because of chromatographic separation inside the guard columns, changes in the retention order of the sample components should be considered. In this case, sampling frequency plays a role not only in maintenance of the first-dimension resolution, but in addition also in the retention behavior of the solutes.

Holm et al. (2005) and Pepaj et al. (2006) used a two-position 10-port switching valve equipped with two guard columns containing the same material as the second-dimension column for the IEX–LC × RPLC separation of proteins. The first dimension separation was achieved through a salt gradient on an SCX column (Holm et al., 2005) or a pH gradient on a strong anion-exchange (SAX) column (Pepaj et al., 2006). Since the mobile phase consisted of a higher organic modifier content throughout the salt gradient, the primary column effluent was diluted with water containing 0.1% trifluoroacetic acid (TFA) prior to entrance in the packed loops in order to focus the analytes on the stationary phase, whereas in the latter approach, the addition of a weak second-dimension solvent was not necessary. The trapping efficiency was examined and was not sufficient for the dipeptides and neurotransmitters, not even after addition of heptafluorobutyric acid (HFBA) as a counter ion. In both applications, the columns in the first and second dimensions were miniaturized, which was beneficial for the detection of sample compounds present in lower quantities.

An identical setup was used by Cacciola and co-workers (Cacciola et al., 2006, 2007) for the RPLC × RPLC analyses of phenolic and flavone antioxidants. In this research, two short X-Terra C_{18} columns served as loops, but excessive modulation times were employed (ranging from 3 to 9 min). In the majority of the separations, an average of two fractions per first-dimension peak [on a poly(ethylene glycol) (PEG) column] was sent to the second dimension [monolithic C_{18} (Cacciola et al., 2007) or zirconia–carbon (Cacciola et al., 2006) column], thereby satisfying the sampling frequency criterion of comprehensive separations. However, the contour plots display first-dimension peak widths of up to 20 min, which poses severe questions as to the quality of the first-dimension separation.

Wilson et al. (2007) anticipated the disadvantage of long modulation times by designing an interface consisting of a six-port switching valve with a solid-phase extraction (SPE) column selector that incorporated 18 trapping columns collecting the same number of fractions throughout the first-dimension chromatogram employing HILIC and RPLC in the first and second dimensions, respectively. The major drawback of this approach was the extremely long analysis time. The second-dimension runs of 53 min each were performed in series after finishing the first-dimension analysis. In fact, this separation cannot be considered as truly comprehensive, but as an automated off-line approach.

In the setup proposed by Venkatramani and Patel (2006), a two-position 12-port switching valve equipped with three identical packed loops was employed for the RPLC × RPLC separation of a drug mixture. A schematic diagram of the system is shown in Figure 8.6. One of the three packed loops is not used to trap

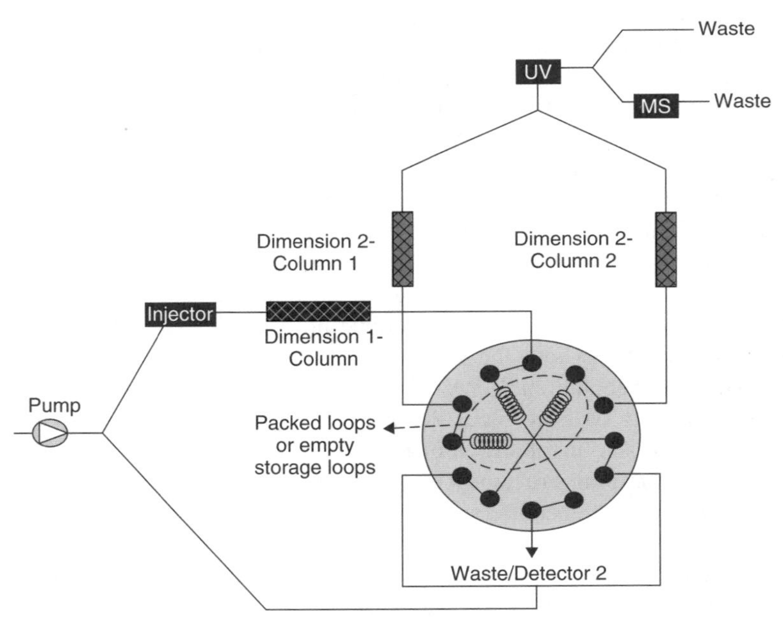

Figure 8.6 System with packed-loop interface. An identical setup was used with empty storage loops (Venkatramani and Zelechonok, 2003, 2004). The interface furthermore allows operation of two columns in parallel in the second dimension. [From Venkatramani and Patel (2006).]

the primary column effluent, but ensures constant flow through the two second-dimension columns, which are used in parallel. Both dimensions are operated with the same pump, which implies that the second dimension is performed in "progressive solvent strength increment" mode.

Packed loop interface configurations have been used extensively in the proteomics domain for the SCX × RPLC separations of peptides. These configurations, which can be categorized partially as stop-flow interfaces, allow the use of the favorable linear salt or pH gradients in the first dimension. Le Bihan et al. (2003) have shown that the alternative method using a stop-flow approach (described below) in combination with salt steps of increasing salt concentration was harming the separation. The authors demonstrated that peak broadening occurs when salt steps are used to displace peptides. This resulted in the distribution of peptides over multiple fractions and a potential dilution below their detection levels. Linear gradients, on the other hand, gave rise to dramatically improved chromatographic performance and less carryover. Some typical setups are presented in Figures 8.7 and 8.8. While peptides are displaced from the SCX column and trapped on a first trap column, an RP gradient, delivered by a second pump, elutes a previous batch of peptides off the second trap for separation

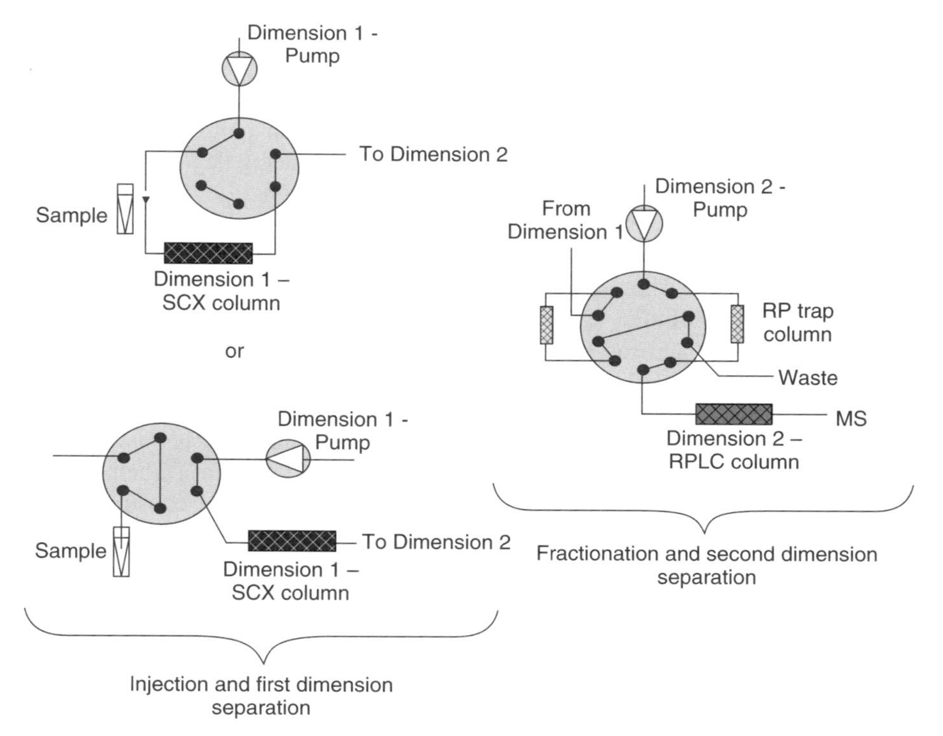

Figure 8.7 SCX × RPLC column switching setups using linear gradients in the first dimension. [From Nägele et al. (2004); Winnik (2005).]

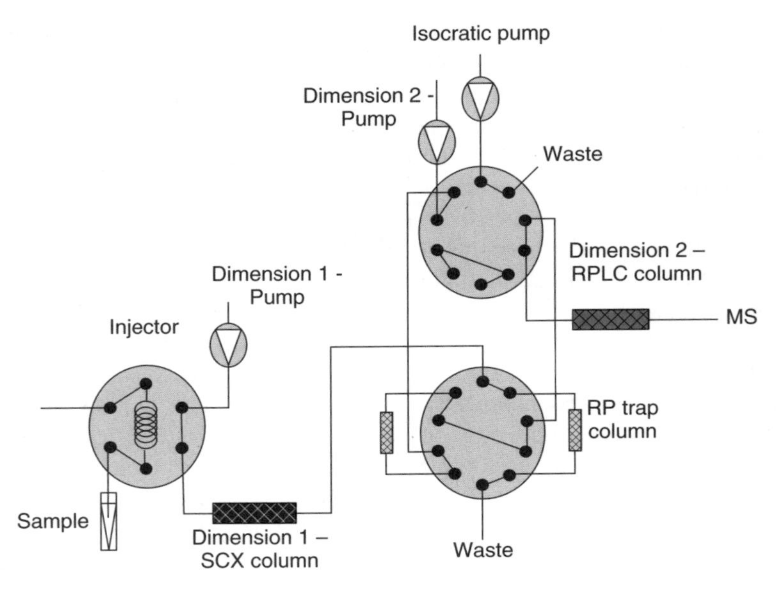

Figure 8.8 SCX × RPLC column switching setup using a linear gradient in the first dimension. [From Mitulovic et al. (2004a).]

onto the RPLC column. Following the separation, the valve is switched and the RP gradient displaces peptides from the first trap column onto the RPLC column, while the next batch of SCX-separated peptides is being deposited onto the second trap column.

Nägele et al. (2004) described such an online system (Figure 8.7, top) making use of a semicontinuous salt gradient and demonstrated that it compared favorably to a salt step gradient. The salt gradient is pumped in steps beginning at 0 to 2.5% of a 500 mM NaCl solution. The following steps start with the end concentration of the foregoing step and end with the starting concentration of the following step. The SCX gradient elution segment takes only a fraction of the RPLC gradient time, and the SCX column, which is attached to a six-port valve, is placed offline once the gradient has developed, and the trap column is washed (with pure water) by the SCX pump to remove all remaining NaCl residues. Their design resembles the stop-flow approach and thereby removes the time constraints for the second dimension and allows the use of narrower second-dimension columns, which is highly beneficial toward mass spectrometric detection.

The setup presented in Figure 8.8 was used by Mitulovic et al. (2004a) and makes use of a separate isocratic pump to wash the enrichment columns. The authors stated that the use of linear salt gradients instead of salt plugs nearly doubled the number of proteins identified and reported the absence of coeluting highly abundant peptides over multiple fractions.

Winnik (2005) combined a salt gradient with a pH gradient to elute peptides from an SCX column. The same layout as the one described by Nägele was used

(Figure 8.7, bottom), but in contrast to that report, the SCX column remained in-line with the pump and the flow was parked at a level close to zero (resembles the stop-flow approach) following every SCX gradient segment. Since volatile salts (ammonium acetate) were used, there was no immediate need for an extensive washing step.

Stop-Flow (or Stop-and-Go) Interface In the stop-flow approach, the columns are connected via an interface without sampling loops. The conventional loop interface is also usable in this approach, but then the sampling loops have no function and negatively influence the extra-column band broadening. Other configurations have been described using one or several two-position six-port switching valves, all following the same principle. In one position, the interface connects the two columns directly and directs the primary column effluent to the second dimension. When the transfer of the desired fraction is finished, the interface is switched and, simultaneously, the first-dimension flow is interrupted, allowing second-dimension separation of the fraction. In this configuration, the second-dimension analysis time, through which the secondary peak capacity can be significantly higher, is not limited. However, the analysis time of the overall 2D separation is severely compromised. The mobile phases in both dimensions play important roles since a focusing of the solutes at the head of the second-dimension column is necessary to avoid peak shape deterioration. This design is particularly popular in the proteomics field since it puts fewer constraints on the second dimension, which allows the use of columns with a small internal diameter, thereby allowing sensitive mass spectrometric detection.

Köhne and co-workers have designed stop-flow interfaces using two (Köhne and Welsch, 1999) or three (Köhne et al., 1998) six-port valves, both of which accomplished complete isolation of the primary column during analysis of the fraction collected previously in the second dimension. No peak broadening was observed caused by the flow interruption in the first dimension.

One six-port valve was used in the stop-flow setup by Blahová et al. (2006) for the RPLC × RPLC separation of beer and hop extracts, with the stopped first-dimension flow path open to the atmosphere (Figure 8.9). A combination of a PEG and an octadecyl silica (ODS) column offered the highest orthogonality. A similar setup using SEC and RPLC in the first and second dimensions, respectively, for peptide separation was suggested by Bedani et al. (2006). A monolithic column was used in the second dimension and peaks were sampled four times, resulting in a peak capacity of 300, achieved in over 500 min. Peak capacity values of 300 are now easily obtained in a much shorter period of time using 1D RPLC, explaining the absence of this combination in the proteomics literature.

Another example of the stop-flow interface was developed by Yates and co-workers. In their approach, SCX and RP materials were packed sequentially into a single microcapillary column which is connected directly to ESI tandem MS. In fact, their configuration can be referred to as "interface-less." They have introduced this technique specifically for the analyses of peptides as multidimensional protein identification technology (MudPIT) (Washburn et al., 2001),

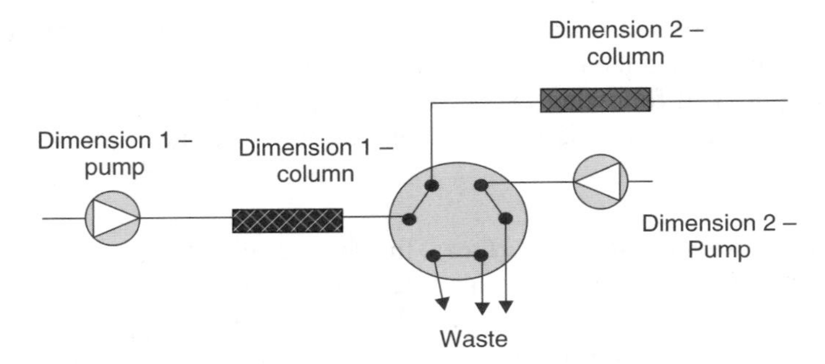

Figure 8.9 Comprehensive LC system with stop-flow interface. During second-dimension analysis, the flow rate on the first dimension is interrupted with the primary flow path open to the laboratory atmosphere. [From Blahová et al. (2006).]

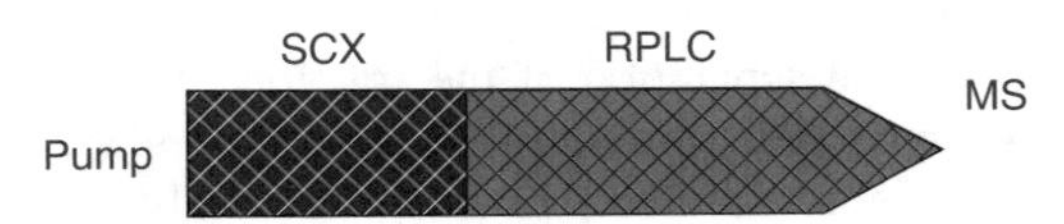

Figure 8.10 Biphasic MudPIT. [From Washburn et al. (2001); Wolters et al. (2001).]

shown in Figure 8.10. The outlet of the column is a pulled tip, which maintains the stationary phase in position and serves as the ESI needle. After loading a sample onto the biphasic capillary, a step gradient of buffer with increasing ionic strength is applied to elute the peptides from the SCX portion of the column onto the RP material. In between these salt steps, a linear acetonitrile gradient is applied for the elution of peptides into the tandem mass spectrometer. It has been reported that the repeated RPLC elution gradients do not affect the retention of peptides in the SCX phase. Salt step gradients followed by acetonitrile gradients appear to be required to detect peptides from the biphasic column (Link et al., 1999).

The MudPIT approach can be regarded as the ultimate online system, since it removes all sample-handling steps once the sample is loaded onto the system. The transfer from SCX to RPLC to MS(/MS) is fully automated and the outcome of the experiment is a list of identified, and potentially quantified, peptides and proteins. In a first study, MudPIT, at that time termed direct analysis of protein complexes (DALPC) due to the nature of the samples under investigation, was used for the analysis of the *Saccharomyces cerevisiae* 80S ribosome tryptic digest, and it allowed the mass spectrometric identification of 75 of the 78 predicted ribosomal proteins. In comparison, only 56 and 64 proteins were identified with 1D RPLC–MS–MS and 2D-PAGE, respectively. DALPC initially involved two 1-mm i.d. columns in series, but was soon refined to a biphasic 100-μm i.d. capillary to reduce sample load and increase sensitivity. Wolters et al., 2001 further refined the technique and subsequently applied MudPIT to the large-scale analysis of the *S. cerevisiae* proteome (Washburn et al., 2001). They were able to identify 5440 unique peptides originating from 1484 proteins in 83 h. To achieve this, proteins were loaded on a 100-μm i.d. microcapillary packed with 4 cm of SCX material and 10 cm of RPLC material. Peptides were displaced from the SCX stationary phase in 15 steps using increasing ammonium acetate concentrations. The volatile nature of ammonium acetate allowed uninterrupted electrospraying. Proteins were identified by means of tandem MS. Due to the unbiased nature of this MudPIT approach, proteins from all subcellular portions of the yeast cell with extremes in p*I* value, molecular weight, abundance, and hydrophobicity were identified, representing a significant advantage over 2D-PAGE.

A dynamic range of 10,000 could be demonstrated, allowing the identification of a protein at 100 and at 10^6 copies per cell in a single experiment (Wolters et al., 2001). Evidently, this can be achieved only if the initial sample load is sufficient so that the low-abundance protein is present within the detection limit of the MS instrument. Fortunately, an SCX column has an inherently high loading capacity. The chromatographic peak capacity of the biphasic system was calculated as 3240. The practical peak capacity, however, is lower, due to the existence of a certain degree of correlation between SCX and RPLC (see section 8.1.1). A subtle increase in RPLC retention per SCX cycle was noted (Wolters et al., 2001). In combination with tandem MS, this peak capacity increased further, to 23,000. This value was obtained by multiplying the chromatographic peak capacity by 7, which is the number of MS–MS spectra that the 3D ion trap could acquire over a chromatographic peak (25 s in the case presented). Upon excluding this MS–MS step, the peak capacity would increase substantially, depending on the resolution of the MS instrument used.

Later, a triphasic design was proposed (Figure 8.11). Peptide interaction with the SCX sorbent can be jeopardized by the presence of chaotropic agents such as urea and guanidinium hydrochloride, which are usually added for efficient protease digestion or relatively high concentrations (>10 mM) of buffer salts. Therefore, samples must be desalted before loading onto the standard biphasic MudPIT column. Off-line desalting is prone to sample loss, since it requires sample manipulation and vacuum drying. The triphasic capillary, with an additional section of RP material upstream of the SCX material, allows online desalting prior to 2D LC.

McDonald et al. (2002) compared this triphasic MudPIT with a biphasic MudPIT and with 1D RPLC–MS–MS and compared the performance in terms of protein identifications for a complex sample. The results were 147, 341, and 431 peptide and 26, 55, and 62 protein identifications for the 1D, two-phase, and

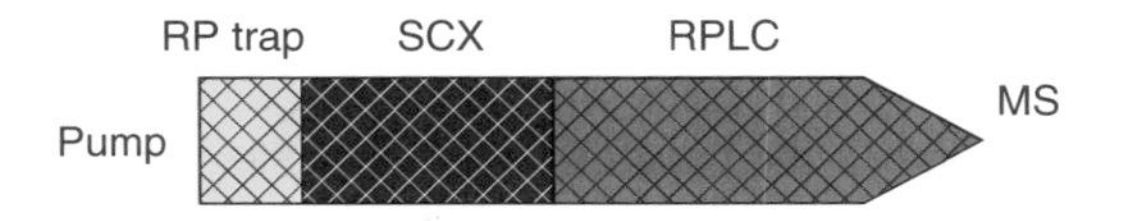

Figure 8.11 Triphasic MudPIT. [From McDonald et al. (2002).]

three-phase experiments, respectively. The gain in identifications upon switching from 1D to MD separations is to be expected given the increased resolution and the sixfold-greater time period. The better performance of the triphase MudPIT over its biphasic counterpart is explained by the sample loss during the off-line desalting step performed in the latter case.

Following this initial improvement, several advancements have taken the MudPIT technique to a substantial higher level. Guzetta and Chien (2005) described a vented serial tetraphasic capillary column approach. The heart of their device consisted of a triphasic MudPIT trap located upstream of a venting tee. The trap was followed by a longer RPLC capillary column. A biphasic trap upstream of a vent was reported at nearly the same time (Kang et al., 2005). Licklider et al. (2002) first described the vented column to fully automate nanoscale microcapillary LC–MS–MS. Such a design allows the flow to be directed to a waste line by controlling a two-position valve (which is blocked in one position and open or vented in the other position). As a consequence, fast sample loading and the injection of larger volumes become possible. In addition, such a vent makes it possible to perform fast washes, fast salt steps, and since the SCX column is decoupled from the back-end RPLC column and MS, it relieves them from the salt burden and allows the use of nonvolatile salts such as KCl. In this way, the entire MudPIT procedure speeds up and becomes highly automated. Guzetta and Chien (2005) compared the performance of a 10 cm length × 150 μm i.d. × 5 μm d_p RPLC back-end column with a 60 cm length × 150 μm i.d. × 5 μm d_p column in a seven-step MudPIT on a serum tryptic digest and concluded that for the same gradient time, the longer column outperforms the shorter column in terms of chromatographic efficiency (sharper and more intense peaks) and peptide identifications following MS (25% gain).

Shortly thereafter, Motoyama et al. (2006) described a vented automated ultrahigh-pressure (UHP) triphasic MudPIT procedure to further address the needs in the field (i.e., the tackling of even more complicated samples). The increased pressure (15 kpsi) was used to drive a 50-cm-long capillary packed with 3–μm RP particles and with a peak capacity of 400 when applying a 350-min gradient. They compared the performance of UHP MudPIT with that of traditional MudPIT (10-cm back-end column, 3 kpsi) in terms of protein identifications resulting from the injection of 20 μg of a yeast soluble fraction digested with both Lys-C and trypsin. To keep the total cycle time approximately constant (ca. 1 day), 15 salt steps, combined with a 70-min RPLC gradient in the standard MudPIT and four salt steps, each succeeded by a 350-min RPLC gradient, were performed in the ultrahigh-pressure method. The four-step UHP MudPIT resulted in 30% more protein identifications, and the authors concluded that the use of highly efficient second-dimension separations greatly reduces the number of salt steps. If the second-dimension separation has insufficient separation power for the sample analyzed, fewer salt steps will result in fewer identifications. It would have been interesting to see the capabilities of a 15-salt-step UHP MudPIT. One might expect a drastic improvement over the four-step UHP MudPIT, and to compensate for the spread of peptides over adjacent fractions, the initial sample load could have been increased to the optimal level of the back-end column. Very recently, a biphasic monolithic capillary column with a 10-cm segment of SCX monolith and 65 cm of RP monolith has been described as a low-pressure alternative to the UHP MudPIT (Wang et al., 2008).

The ion-exchange part has itself been the subject of further investigation. Motoyama et al. (2007) demonstrated that an anion and cation mixed-bed exchange resin results in improved peptide recovery and better orthogonality than those due to. SCX alone. A Donnan effect (Helfferich, 1995) taking place due to the introduction of opposite charges into one column is believed to be the mechanism responsible for this improved recovery. The Donnan effect is equal to that of an organic modifier and thus is ideally suited to be combined with MudPIT where the addition of organic modifier has to be minimized. The increase in orthogonality (Figure 8.12) can be explained by a combination of increased retention of acidic peptides and a moderately reduced retention of neutral to basic peptides by the added anion-exchange resin. When applied to a tryptic digest of a yeast whole-cell lysate, the increased peptide recovery and orthogonality of the mixed-bed ion-exchange resin gave rise to a 100% increase in the number of peptides identified over those with SCX alone. In addition, the use of this improved MudPIT in the analysis of a phosphopeptide-enriched HeLa nuclear extract resulted in an increase in phosphopeptide identifications by 94% over SCX alone.

The applications of MudPIT are already numerous and have recently been reviewed (Fournier et al., 2007). To date, MudPIT has been limited to

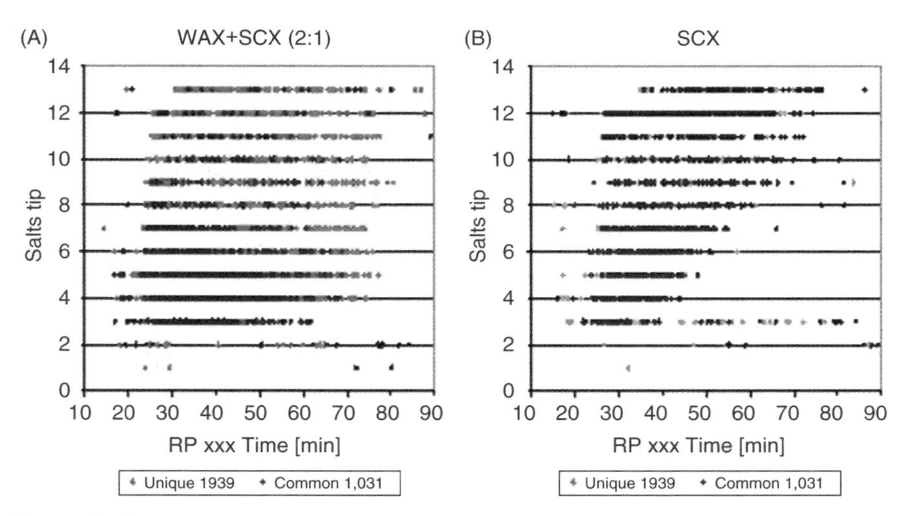

Figure 8.12 Orthogonality increase when using an anion (WAX) and cation (SCX) mixed bed (A) compared to SCX resin (B). The 2D separation space is used more efficiently with the mixed bed. The different colors represent the peptides that are commonly and uniquely identified in the two formats. [From Motoyama et al. (2007), with permission. Copyright © 2007 by The American Chemical Society.]

the combination of SCX with RPLC due to the compatibility between both approaches. MudPIT can be simulated in valve-switching approaches making use of separate columns (Davis et al., 2001; Fujii et al., 2004; Mitulovic et al., 2004b; Nägele et al., 2003; Xiang et al., 2004), preferably with a trapping column in between (Davis et al., 2001; Fujii et al., 2004; Mitulovic et al., 2004b; Nägele et al., 2003). A typical setup is presented in Figure 8.13. Fractions eluting from the SCX column are trapped on top of a small RP enrichment column and washed free of salts. Following a short period of time, the valve switches and the trapping column is placed in the flow path of the RPLC pump and peptides are flushed backward onto the analytical RPLC column.

One negative feature over traditional MudPIT is the possibility of sample loss onto the enrichment column. As with the vented column approach, there are fewer constraints on the use of nonvolatile salts, and fast sample loading becomes possible. In addition, by column switching, columns with a larger i.d. in the first dimension can be used, allowing the injection of high sample amounts.

As has been the case with the MudPIT strategy, efforts have also been conducted to increase the efficiency of the second dimension. Luo et al. (2007) recently reported an online 2D porous-layer open-tubular (PLOT)/LC–ESI–MS system using a 3.2 m length × 10 μm i.d. polystyrene divinylbenzene (PS-DVB) capillary for ultrasensitive proteomics. The impact of using a five-step SCX fractionation prior to RPLC analysis was clearly demonstrated.

Liu et al. (2006) encountered two problems when setting up a column-switching 2D approach according to Figure 8.13: (1) the inability to observe many hydrophobic peptides and (2) the degradation of peak quality due to peak broadening and tailing as compared to 1D RPLC. The former problem

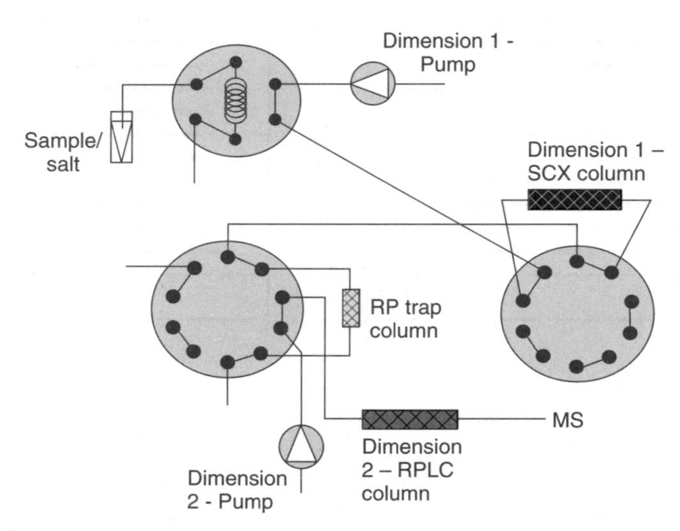

Figure 8.13 SCX × RPLC column switching setup with stop-flow approach using salt steps in the first dimension. [From Liu et al. (2006).]

could easily be addressed in an off-line approach by adding large amounts of organic modifier in the buffer system. The latter problem was due to the decoupling of the columns, an issue typically not encountered when using biphasic columns. A band refocusing method, in which an analytical RP column with more hydrophobicity than the RP trap column, was implemented and used successfully to generate 1D-like peaks.

To recover hydrophobic peptides still unreleased from the SCX phase, following a conventional salt step gradient, a soft RP step gradient at high salt concentrations was used. As demonstrated in a 2D analysis of an *Escherichia coli* digest, the latter step gave rise to 19% of all peptide identifications. The peak capacity for this 2D setup was calculated to be 1400 (7×200—nonideal orthogonality not taken into account) and the time required was 350 min. The peak capacity of the 2D run was 3.4 times higher than that of a 1D run performed in approximately the same period of time. This can be attributed to the increase in peak width upon switching from a 30-min to a 300-min gradient. In principle, this gives rise to increased peak concentrations and an augmented potential to fall within the detection limits of the MS system used. The authors, however, stated that each peptide was split among two or more SCX fractions.

Interface with Parallel Second Dimension In contrast to the conventional LC × LC system, where each dimension consists of only one column (in the first dimension, a set of serially coupled columns might be used to increase resolution), the parallel column system implies the operation of two (or more) columns in the second dimension to perform analyses of consecutive fractions in parallel. Obviously, such configurations suffer more from practical problems. Recently, the application of multiple columns in parallel has been investigated theoretically by Fairchild et al. (2009a), and the authors raise some practical aspects for consideration. Since the chromatograms achieved on the two (or more) secondary columns are combined to obtain an overall separation chromatogram or contour plot, the two (or more) second-dimension columns must deliver identical retention times and efficiencies. As suggested by the authors and confirmed in earlier studies (Kele and Guiochon, 1999a–c, 2000, 2001, 2002), tight specifications are met when commercially available columns are used in the second dimension. Using columns of the same batch is advised. Evidently, a thorough system optimization (adaptation of tubing lengths) is also of primordial importance when a parallel second dimension is configured. The extra-column volumes need to be minimized and identical through all pathways.

As a practical example, the standard procedure in our laboratory for the implementation of a parallel second-dimension configuration, as shown in Figure 8.14 (François et al., 2008a, 2009b), starts with purchasing second-dimension columns of the same batch. Prior to the implementation of these two columns in the comprehensive configuration, the columns are compared off-line in terms of efficiency by injecting a standard mixture under isocratic conditions. When similar results are obtained, the optimum second-dimension method conditions are determined off-line for the application considered, after which the online combination is

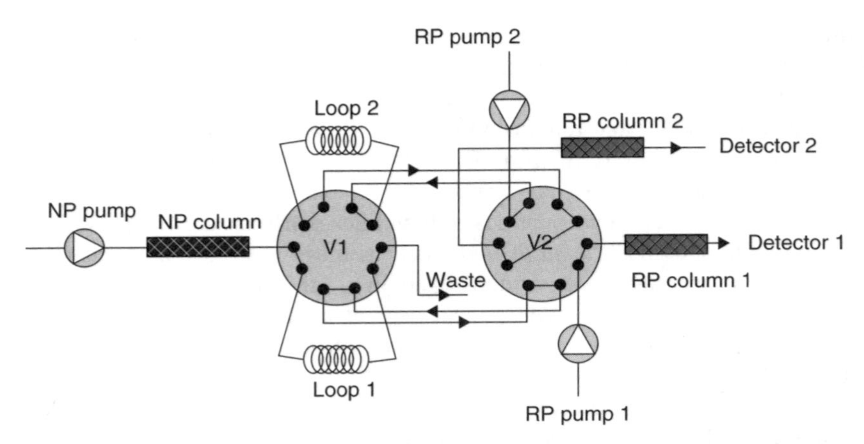

Figure 8.14 LC × 2LC system with parallel second dimension. [From François et al. (2008a, 2009b), with permission of the copyright owner.]

established. In the configuration shown in Figure 8.14, the two loops on the first switching valve are then used as injection loops, with the first-dimension effluent entrance foreseen with a needle port. The sample is injected and the separations via both pathways are compared by overlaying the chromatograms. In case the retention times differ, the tubing lengths are adapted accordingly, until identical profiles are obtained. Once the system is designed such that delay volumes are identical, this optimization procedure does not need to be repeated, except when two new second-dimension columns are implemented on which the analytes display different retention factors. Fairchild et al. (2009a) also suggested using software corrections if the characteristics of the columns used are close but slightly different.

A parallel second-dimension setup can be achieved by substituting the storage loops on a two-position 10-port switching valve by the secondary columns themselves (Cacciola et al., 2006; Haefliger, 2003; Kimura et al., 2004; Murahashi, 2003; Opiteck et al., 1997b, 1998; Unger et al., 2000; Wagner et al., 2000). While the first secondary column is loaded with a fraction of the effluent from the first dimension, the analytes eluted in the previous sampling are separated on, and eluted from, the second secondary column, and this cycle is repeated throughout the entire analysis. To minimize peak-broadening effects due to introduction of the solutes in plugs of strong second-dimension mobile phase, the first-dimension effluent is occasionally diluted with water before reaching the interface valve (Haefliger, 2003; Murahashi, 2003).

In most applications, the two second-dimension columns are identical, but Haefliger (2003) described the NPLC × RPLC separation of complex surfactant mixtures using two different RP stationary phases in the second dimension: C_2 (dimethyl) and C_4 (butyl). In principle, the realization of this NPLC × RPLC combination is difficult, since the two dimensions generally employ immiscible solvents as mobile phases. Problems related to solvent incompatibility will be described in Section 8.2.2 but were not an issue in this investigation, since the first dimension employed an aqueous mobile phase to perform an ion chromatographic type of separation on a microbore diol column. The C_2 column enabled separation of the nonionics and sulfonates, while the cationics, amphoterics, and sulfates were analyzed on the C_4 column. Nevertheless, during the 54-min analysis, only four fractions could be transferred, due to the relatively long gradients applied in the second dimension. This weakness is ascribed to the lack of short and efficient C_2 and C_4 columns used in the second dimension, through which shortening of the second-dimension analysis time would be impossible to achieve.

Excessive modulation times of several minutes were also used for the RPLC × RPLC separation of phenolic antioxidants by Cacciola et al. (2006). Among other configurations, two zirconia–carbon columns were installed on a two-position 10-port switching valve and were operated at 120°C, in combination with a PEG column in the first dimension. Modulation times of 5 and 6 min were used, corresponding to two samplings per first-dimension peak. However, since peak widths of more than 10 min are observed for the first-dimension peaks, the quality of the first-dimension separation is questionable.

Other applications use very short sampling times. Wagner et al. (2000) operated very short nonporous RPLC columns of 14 mm length for the IEX × RPLC separation of proteins at a flow rate of 2.5 mL/min with a sampling time of 1 min. Murahashi (2003) used two 5-cm C_{18} monolithic columns isocratically at 100% of organic modifier at a flow rate of 16 mL/min for the RPLC × RPLC separation of polycyclic aromatic hydrocarbons with a sampling frequency of 0.2 min. Unger et al. (2000) used short columns packed with small particles to speed up the second dimension, necessary for optimal operation of this configuration, while Kimura et al. (2004) used monolithic columns to achieve fast and efficient separations. Stoll and Carr (2005) exploited temperature to obtain efficient second-dimension separation. They operated their second-dimension RPLC column at 100°C in an analysis time of 21 s. The 2D system, which employed SCX in the first dimension, generated a peak capacity of 1350 in 20 min. The authors also stated that the generation of such a high peak capacity in 1D LC would require a 5-m-long column and an analysis time of 16 h.

Another configuration using a dual four-port switching valve system has been described by Opiteck et al. (1997b). In fact, these authors were the first to describe the use of parallel LC columns instead of loops for the SEC × RPLC separation of tryptic digests. Simultaneously, their report represented one of the first contributions of online MS detection combined with multidimensional chromatography. As the analytes eluted from the first dimension, they were alternately injected, with a 4-min interval, corresponding roughly to two samplings per peak onto either of a pair of RPLC columns. During the loading of one column, a peptide resolving gradient was developed over the other column. Since no organic modifier was used in the first dimension, the online coupling of both dimensions was relatively straightforward, and peptides were trapped at the head of the RPLC columns during the 4-min loading. Given the short period of time available for the second-dimension analysis, two short columns packed with small nonporous

C_{18} particles were used, thereby allowing an efficient separation (peak capacity of 33). The combined peak capacity was 495 in 160 min, meaning that the first dimension only possessed a peak capacity of 15 despite the connection of eight SEC columns in series (total length of 2.4 m × 7.8 mm i.d.). The setup was subsequently applied to resolve protein mixtures (Opiteck et al., 1998) with two short polystyrene divinylbenzene second-dimension columns.

Wagner et al. (2002) reported an impressive configuration applying four second-dimension columns in parallel for the IEX × RPLC separation of peptides and proteins. Their system is presented schematically in Figure 8.15. During the loading of the first-dimension effluent on one RP column, the analyses of the two previous fractions were carried out on two other RP columns, while the fourth was regenerated. Of course, a large number of second-dimension columns is beneficial for the peak capacity, but unfortunately leads to an increased complexity of the system, and a vast amount of instrumentation is required. The authors reported a peak capacity of 3000 in 96 min.

In another attempt to increase throughput, Gu et al. (2006) described an MD capillary array LC system with 18 parallel RPLC capillaries in the second dimension. Their ingenious design allowed the sequential trapping of 18 SCX fractions on 18 monolithic precolumns and the subsequent parallel second-dimension

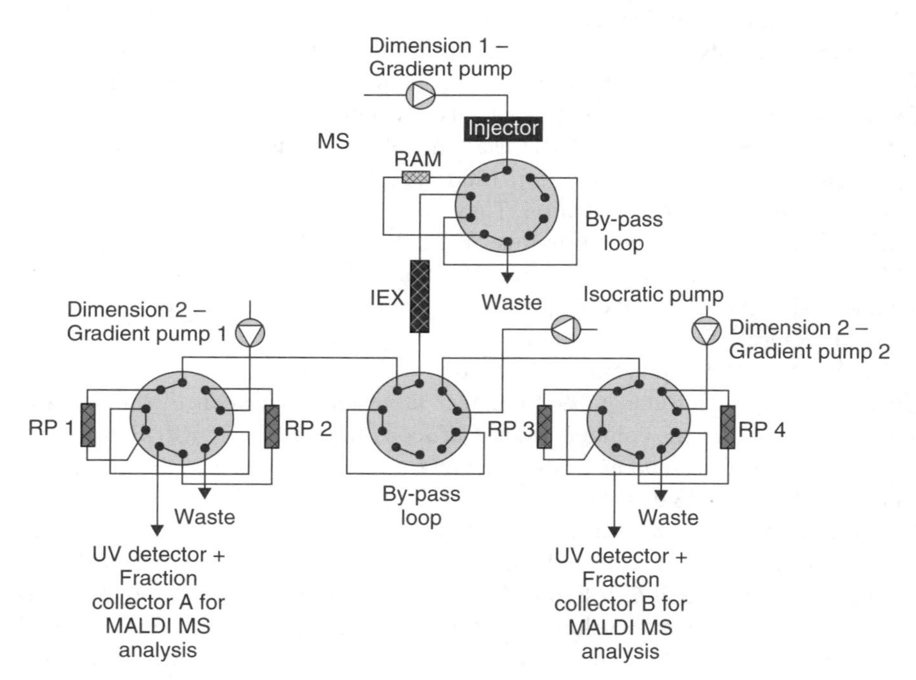

Figure 8.15 Flowchart of a comprehensive IEX × RPLC system for the separation of peptides applying four C_{18} columns in parallel in the second dimension. [From Wagner et al. (2002).]

RPLC separations by back-flushing the precolumns by identically split solvent-gradient flows. The RPLC effluent was deposited directly onto MALDI targets through an array of capillary tips. Both the precolumns and second-dimension columns were homemade. Second-dimension repeatability was ensured by packing 40 columns in one batch and placing 18 columns in the array that were most identical. Evidently, the system allowed an 18-fold increase in throughput compared to the serial designs. A more simplified version with online matrix mixing was subsequently described by the same group (Liu and Zhang, 2007).

Since a large number of samplings along the first-dimension separation is highly beneficial for the peak capacity, the second-dimension analysis time in a comprehensive LC configuration is limited, resulting unavoidably in reduced secondary separation space. Furthermore, signal interferences inherent in the combination of incompatible separation dimensions (Section 8.2.2) diminish the available secondary separation space. The development of interfaces with a second dimension employing two columns in parallel in combination with storage loops (Alexander and Ma, 2009; François et al., 2008a, 2009b; Tanaka et al., 2004; Venkatramani and Zelechonok, 2003, 2005) or guard-column loops (Venkatramani and Patel, 2006) offers a number of advantages from this point of view. In contrast to all other configurations where the time available for analysis and regeneration in the second dimension is equal to the sampling time (with the exception of the stop-flow interface), interfaces exploiting this principle ensure the collection of two first-dimension cuts during the analysis of a specific fraction and regeneration of the secondary column. Alternate injection into the two secondary columns allows the separation time to be twice as long as the fractionation interval. In this way, the peak capacity of the second dimension is doubled and the comprehensive LC system accomplishes a separation with drastically increased peak capacity.

Venkatramani, Zelechonok, and Patel were the first to explore the possibilities of such a parallel column system in the second dimension for the RPLC × RPLC separations of standard mixtures (Venkatramani and Zelechonok, 2003, 2005) and a drug sample (Venkatramani and Patel, 2006). The interface was a two-position 12-port switching valve equipped with either three storage loops (Venkatramani and Zelechonok, 2003, 2005) or three guard-column loops (Venkatramani and Patel, 2006) (see Figure 8.6). They used identical second-dimension columns, but additionally described a setup employing secondary columns packed with different stationary phases (cyano- and aminopropyl silica) (Venkatramani and Zelechonok, 2003). When each first-dimension peak is sampled several times (modulation time of 10 s), the latter separation provides a high amount of information within the same analysis time. The peak capacity in the applications described was, however, limited as a consequence of the application of only one pump for the two secondary columns or for both dimensions, forcing the second dimension to the isocratic or "progressive solvent strength increment" mode.

Tanaka et al. (2004) described the RPLC × RPLC separation of a test mixture by means of a comprehensive system with a parallel second dimension employing two six-port valves, each equipped with a storage loop (Figure 8.16).

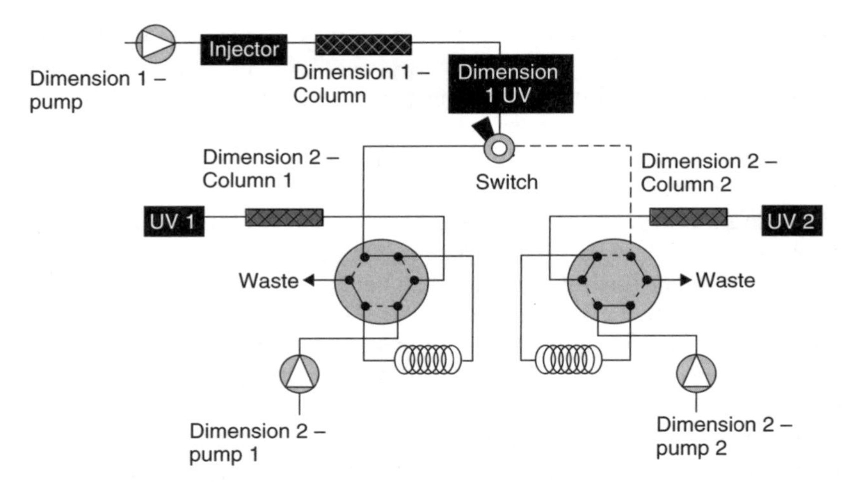

Figure 8.16 RPLC × RPLC system with parallel second dimension according to Tanaka et al. (2004).

An additional three-port valve was used for fractionation of the primary column effluent and was switched every 15 s. Two homemade ODS or ODS and dimethyl-2,3,4,5,6-pentabromobenzyloxypropylsilyl monolithic columns were used in the second dimension to complete analysis of the fractions within 30 s.

Recently, François et al. 2008a, 2009b presented a novel configuration with two two-position 10-port switching valves (Figure 8.14) for the NPLC × RPLC separations of a test mixture and citrus oil extracts and the RPLC × RPLC separations of steroids, sulfonamides, and tryptic peptides, applying two parallel columns in the second dimension. The authors suggested naming the comprehensive separations using two parallel columns in the second dimension as LC × 2LC. In their configuration, the first valve had a function comparable to that of the conventional loop interface, while the second valve guaranteed increased separation power in the second dimension. The peak capacities for the separation of a lemon oil extract by NPLC × 2RPLC were compared with NPLC × RPLC using identical first- and second-dimension columns. The conventional loop interface, consisting of one 10-port valve provided a peak capacity of 437, while the addition of the extra equipment and secondary column resulted in a value of 1095, clearly illustrating the large gain in separation power.

Alexander and Ma (2009) have recently optimized the configuration of Venkatramani (Venkatramani and Patel, 2006; Venkatramani and Zelechonok, 2003, 2005) in a way that allows two independent gradients (or isocratic flows) to be directed to the dual secondary columns. They used two fused-core second-dimension columns in parallel for the separation of an active pharmaceutical ingredient and its degradation products. Orthogonality in the RPLC × RPLC separation was obtained by using different pH values in the two dimensions.

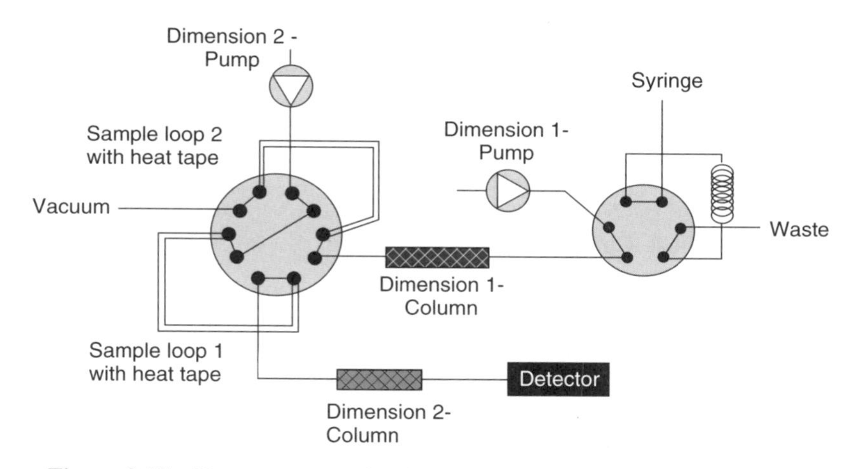

Figure 8.17 Vacuum evaporation interface according to Tian et al. (2006).

Vacuum Evaporation Interface A novel interface, abbreviated as VEI, was recently designed specifically for the coupling of NPLC and RPLC by Tian et al. 2006, 2008. The interface valve not only enabled collection of the fractions eluting from the first dimension, but additionally, accomplished evaporation of the mobile-phase solvents by employing high temperatures under vacuum conditions. This innovative interface presented in Figure 8.17 is highly beneficial for NPLC × RPLC separations since immiscibilities of the mobile phases, together with large viscosity differences between the two dimensions, are in other arrangements the origin of peak dispersion and signal interferences (Section 8.2.2). Furthermore, the relatively low boiling point of the mobile phase in the first dimension facilitated the evaporation. However, the interface was not able to provide full recovery of all sample compounds. Solutes with a low boiling point logically suffered most, but even the recovery of nonvolatile compounds never exceeded 60%.

8.2 TECHNICAL PROBLEMS IN COMPREHENSIVE LIQUID CHROMATOGRAPHY

8.2.1 Column Connections, Valves, and Delay Volumes

The design and implementation of column switching is a delicate and critical procedure in comprehensive LC. Transfer of the first-dimension effluent to the second dimension has to be pursued in a fast and reliable manner. Furthermore, the volumes of tubing, column connections, and the internal parts of the valve ports should be minimal and lead to only a small contribution to peak broadening. Configurations with a parallel second dimension suffer even more from this augmentation in delay volume, due to the increased complexity of the system. The description of such a setup was given in Section 8.1.4, and careful attention to the reproducibility of the second-dimension columns used is required. Separation of the sample components on both columns should be identical in terms of retention times and bandwidths in order to avoid problems in data handling, peak identification, peak capacity determination, and quantification. The reproducibility increases severely, however, when commercially available columns rather than homemade stationary phases (Kele and Guiochon, 1999a–c, 2000, 2001, 2002) are used.

8.2.2 Mobile-Phase Compatibility

An important issue in comprehensive LC is the compatibility of the mobile phases. Note that this holds true for online combinations only. It is also important to note that use of nonvolatile buffers (as in IEX) may be detrimental if hyphenation with MS is carried out. The mobile phase eluting from the primary column preferably consists of a weak solvent constituent of the second-dimension mobile phase in order to create a focusing effect (Hoffman et al., 1989). Furthermore, if the solvents or solvent mixtures that are used as mobile phases are not completely miscible, serious difficulties arise, resulting in the complication of the combination of various separation modes. As an example, this is the case when one of the separation dimensions is RPLC, HILIC, or IEX while the other is either NPLC or SEC. The solvent in the former step is usually an aqueous solution, while that used in the latter modes is generally an organic solvent that is not necessarily miscible in aqueous solvent mixtures.

These incompatibility problems were observed in 1D LC in the past when sample solvents were significantly different from the initial conditions of the mobile phase. This was thoroughly investigated for polymer samples by various research groups. Shalliker et al. (1991a,b) and Shalliker and Kavanagh (1997) observed double peaks in gradient LC separations of high-molecular-mass polystyrenes using mobile phases containing dichloromethane and methanol. However, with dichloromethane and acetonitrile as solvents, the sample eluted with only one symmetrical peak. Jandera and Guiochon (1991) investigated the effect of dichloromethane as a sample solvent for cholesterol injected on a Nucleosil 500 ODS column with acetonitrile mobile phase on a preparative scale. The appearance of "breakthrough" peaks during the separation of narrow molecular sizes of poly(methyl methacrylate) was examined extensively by Jiang et al. (2002). They showed that the breakthrough peak increased as the injection volume, column temperature, strength of the sample solvent injected, and strength of the initial mobile phase in gradient LC increased, or as the polymer concentration decreased. This phenomenon was also strongly influenced by the properties of the polymer itself. These observations in 1D LC give a clear indication of what can be expected in MD separations when coupling dimensions using immiscible solvents as mobile phases.

As an example, the coupling of NPLC and RPLC is highly beneficial in terms of orthogonality, but solvent immiscibilities originating from the apolar NP solvents (e.g., *n*-hexane with the aqueous second-dimension mobile phases) can cause the separation to deteriorate and create signal interferences (François et al., 2006). Figure 8.18 shows the influence of the introduction of incompatible

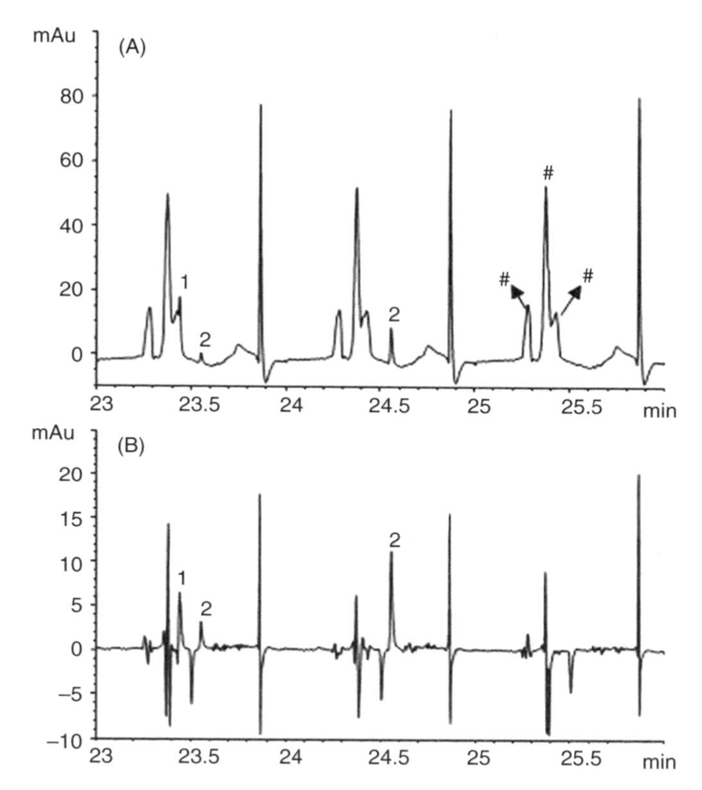

Figure 8.18 Influence of the introduction of immiscible solvents in the second dimension of a comprehensive LC system in the NPLC × RPLC separation of a mixture of pharmaceuticals; without (A) and with (B) background subtraction. #, system peaks; "bump" originating from poor mixing in the loops; 1, 2, pharmaceutical compounds. [From François et al. (2006), with permission of the copyright owner.]

solvents for the NPLC × RPLC separation of a mixture of pharmaceuticals. The ultraviolet (UV) chromatograms of three consecutive fractions suffer from system peaks and an additional "bump" originating from the poor mixing of solvents in the sampling loops at the high flow rate employed. This effect is evidently more pronounced for UV detection at lower wavelengths. Notwithstanding the problems encountered, these signal interferences are highly reproducible, and blank subtraction is sufficient to allow correct data interpretation, as shown in Figure 8.18B.

In contrast to these negative effects, the incompatibility possibly leads to focusing (François et al., 2006). Figure 8.5 showed the NPLC × RPLC separation of a citrus oil extract (this application was discussed in Section 8.1.3). By adapting the initial conditions of the second-dimension gradient, immiscibility of the solvent fraction introduced was obtained, and focusing of the analytes was achieved throughout the entire separation.

8.2.3 Viscous Fingering

Viscous fingering (VF) (Catchpoole et al., 2006; Keunchkarian et al., 2006; Mayfield et al., 2005; Shalliker et al., 2007) is a phenomenon that takes place at the interface of two solvents with different viscosities and is of particular interest in MD chromatography. In a column packed with porous particles, flow instability occurs when a fluid with low viscosity displaces and penetrates a high-viscosity solvent. This results in a complex pattern resembling fingers, from which the name of the event was derived. This observation becomes more pronounced with larger viscosity differences and has a detrimental effect on chromatographic performance. The peak shape is completely distorted, and elution in multiple bands is likely to occur in the case of excessive VF. Mayfield et al. (2005) have investigated this phenomenon in 17-mm i.d. glass columns chromatographically and visually, by taking photographs in which solvents with dissimilar refractive indices can be distinguished by different colors. Their results are reproduced in Figure 8.19. In this example, the sample plug and mobile-phase viscosities were 0.38 and 0.86 cP, respectively.

In analytical 1D LC, the occurrence of VF is almost nonexistent, since the injection volumes are rather small and the injection solvent has been chosen carefully to match the initial conditions of the mobile phase. In contrast, when first-dimension fractions are directed to the secondary column in LC × LC, the viscosity of the aliquots practically always differs from the mobile phase used in the second dimension, and VF appears at the end or beginning of the sample plug introduced.

In NPLC × RPLC, the viscosity discrepancy, next to immiscibility and solvent strength problems, is clear, but even for the coupling of compatible and

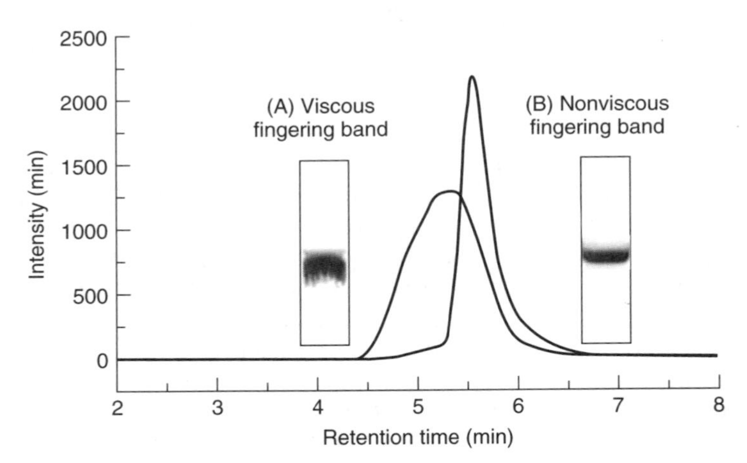

Figure 8.19 Viscous fingering observations during the injection of a less viscous sample solvent than the mobile phase, including photographs and post-column detector elution profiles at 575 nm. [From Mayfield et al. (2005), with permission. Copyright © 2005 by Elsevier.]

seemingly identical separation modes, such as RPLC × RPLC, VF is very likely to negatively influence the chromatographic performance. To achieve highly orthogonal separations, the choice of solvents is as important as the different surface chemistries of the stationary phases. When using acetonitrile and methanol as mobile phases in the two dimensions of an RPLC × RPLC separation, VF will appear when the less viscous acetonitrile penetrates into methanol.

Extreme VF results in important loss of resolution. When performing comprehensive separations, the analyst should be aware of and intervene in these interferences, since the metal tubes of analytical columns are not transparent and the occurrence of this flow instability is difficult to identify.

8.2.4 Dilution and Sensitivity

Dilution factors are important characteristics from the point of view of analyte detectability. Chromatography is a separation technique that is always accompanied by dilution, and in LC × LC, this dilution takes mostly place at the interface during injection of the fraction to the second dimension. It is the main cause of loss in sensitivity and decreasing detection limits in comprehensive LC.

Comprehensive configurations with interfaces that enable focusing or concentration as an intermediate step between collection and reinjection of the fractions alleviate the dilution effect. The choice of the appropriate focusing mechanism or medium is, however, not that straightforward. The nature of the fractions changes significantly when performing a gradient in the first dimension, and no loss of sample components is tolerated. Selectivity differences caused by the intermediate processing are not acceptable. Furthermore, the occurrence of wraparounds should always be avoided.

Different interface designs with trapping capabilities have been described in the literature and were described briefly earlier, so are merely repeated here. One of the possibilities for analyte focusing is to use reverse osmosis, but this has never been done in practice (Guiochon et al., 2008). The first-dimension effluent can then be forced through a semipermeable membrane by applying pressure. Differential pressures in the two dimensions are of major concern in terms of membrane stability, and the implementation of such a device on an interface valve is highly questionable.

Second, the concentration step can be achieved by partial evaporation. The solvents used as mobile phases usually have a higher volatility than the sample compounds and can be partially removed from the fractions. However, the exclusion of artifact formation and compound loss are again of fundamental importance. For more information on this type of interfacing device, refer to Section 8.1.4.

As a third option, the analytes can be trapped in loops packed with stationary phase rather than in empty storage loops. The interface is then responsible for the collection and temporary storage of the fractions, after which the analytes are released toward the secondary column. The characteristics of the adsorbent are dependent on the properties of the sample components and the solvents used in both dimensions. The first-dimension mobile phase is preferably a weak solvent for the stationary phase in the loop, to allow focusing, while fast desorption by a strong solvent should be performed by the secondary mobile phase. This type of interface was described thoroughly in Section 8.1.4 and illustrated with practical examples.

8.3 DETECTION

A comprehensive 2D LC separation is easily combined with all conventional LC detectors, including photodiode arrays (PDAs), evaporative light scattering detectors (ELSDs), and mass spectrometers, offering additional information on the sample components. The detector is usually placed behind the second-dimension, although an additional detector can be used to collect first-dimension data, notwithstanding that this is leading to extra-column broadening. Due to the high speed of the second-dimension analysis, the detector acquisition rate is required to be very fast, since no loss in resolution can be tolerated resulting from a low number of data points. Opiteck et al. (1997a) described the SEC × RPLC separation of peptides with ultraviolet (UV) and MS detection. They highlighted the effect of inadequate sampling rates offered by their MS detector (3 s per scan), which resulted in a severe loss in chromatographic resolution compared to the UV signal (two data points per second). Figure 8.20 shows the separation of a single fraction by UV (Figure 8.20A) and MS (Figure 8.20B), illustrating the tremendous separation loss with MS. Today, fast-scanning MS devices are available that do not compromise chromatographic separation. Assuming that in the fast second-dimension analyses, peak widths of a few seconds are common, a detector acquisition rate of 5 Hz is minimally required to allow a reliable peak description (six to 10 data points per peak). Obviously, a higher acquisition rate results in an improved peak description and resolution. Time-of-flight (TOF) MS is therefore an excellent choice for comprehensive LC detection, due to the high scanning rate.

When using MS for detection, the 2D effluent must allow ionization of the compounds. Therefore, the number of possible combinations is limited, but the fact that most applications employ RPLC in the second dimension alleviates this problem. MS detection essentially adds a third dimension to the 2D system, because the mass spectrometer can identify the presence of coeluting nonisobaric peaks when they are not resolved by chromatography. The flow is commonly split to values of 1 mL/min and lower prior to entrance in the ion source, since the extremely high flow rates used in the second dimension to accelerate the analysis speed are not tolerated in the ion source. The flow splitting negatively influences the sensitivity and induces peak broadening through the addition of the extra volume of the splitter. ESI and atmospheric-pressure chemical (APCI) ionization are the most frequently used ionization techniques for online analyses, whereas MALDI can be applied to off-line collected fractions. Pól and Hyötyläinen (2008) recently published a review on comprehensive LC with special attention to MS

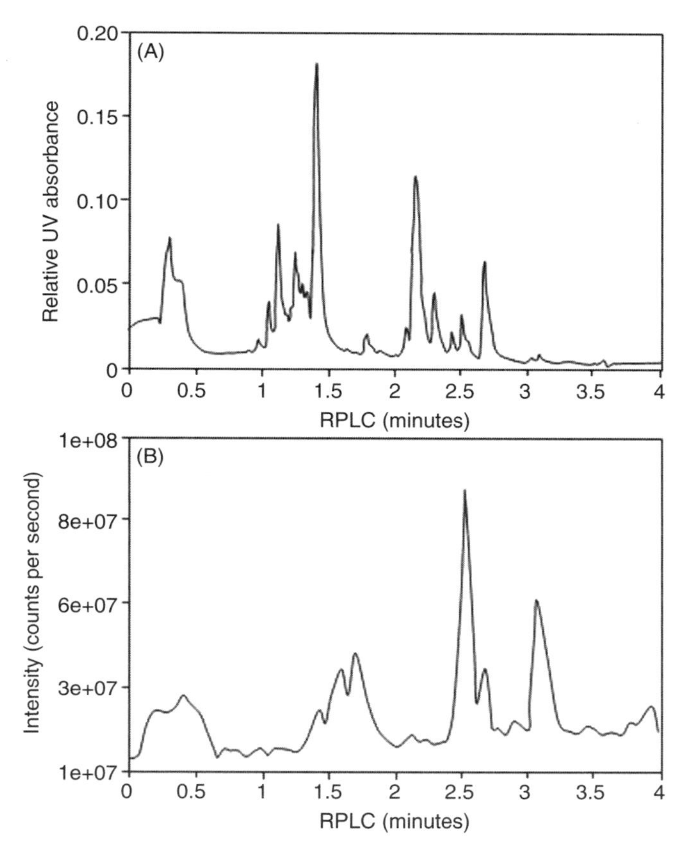

Figure 8.20 Influence of the detector acquisition rate on the chromatographic resolution in the second dimension of a SEC × RPLC separation of peptides: (A) UV at 215 nm; (B) MS. [From Opiteck et al. (1997a), with permission. Copyright © 1997 by The American Chemical Society.]

detection. Comprehensive LC in combination with mass spectrometry is discussed in greater detail in Chapter 9.

8.4 DATA REPRESENTATION

In comprehensive LC, a considerable amount of data is produced within a relatively short period. When fast-scanning devices are used for the data acquisition, the chromatograms are composed of a huge number of data points and generate huge files; this is especially true when a mass spectrometer is used for detection. The processing of the data should convey the largest possible amount of information to the analyst in a convenient way. Next to a visual display, the availability of tables including retention times (for identification) and sizes (for quantification) of all peaks detected is highly desirable.

The data are commonly represented visually in 2D plots or contour plots, where the retention times in the first and second dimensions are plotted along the x- and y-axes, respectively. The color of the spots is a measure of the intensity of the peaks. Unfortunately, LC × LC currently suffers from the limited availability of dedicated software, essential for both instrument control and data visualization. With increased interest and more users, growth is expected here. Due to the lack of appropriate programs, the instrumentation in most laboratories is controlled by software written in house that is capable of starting the pumps and signal acquisition and switching of the interface valve(s). Commonly, a series of individual second-dimension chromatograms is produced or the fractions generate one long second-dimension chromatogram that is broken down afterward into the chromatograms of the individual runs. The data are exported to construct a matrix containing absorbance values as a function of first- and second-dimension retention times, after which the contour plots are constructed using software that is able to represent three-dimensional (3D) data, such as Origin (OriginLab Corporation, Northampton, MA), Fortner Transform (Fortner, Inc., Savoy, IL) or homemade software programmed by means of Matlab (Natick, MA).

For the same reason, quantification in LC × LC remains problematic and has rarely been reported. Hyötyläinen and co-workers (Kivilompolo and Hyötyläinen, 2007; Pól et al., 2006) considered all individual extracted ion chromatograms of the second dimension and summed the peak areas belonging to the same compound in consecutive modulation periods. The quantification process was lately automated by Mondello et al. (2008) using in-laboratory developed software that enabled the automatic and correct integration of each 2D peak. A loss in sensitivity was observed in comparison with conventional 1D LC due to the sample dilution in LC × LC. Automatic quantification was also achieved by Kivilompolo et al. (2008) for the antioxidant phenolic compounds in wines and juices. They used special software developed earlier in their laboratory for the treatment of 2D GC data. Recently, this niche in the chromatography field was discovered by software writers and new programs, supported by experience gained in GC × GC, were developed, such as Chromsquare 1.2 (Chromaleont, Messina, Italy) and LC Image (GC Image, Lincoln, NE).

The data interpretation is often facilitated by chemometric handling, in which mathematical or statistical methods are applied to the chromatographic data. The general aims of all methods are the removal of artifacts, corrections for the baseline, and the ability to detect and quantify peaks. Critical for all these models is their robustness, when the data act differently than assumed in the model. Chemometric data handling is a research area on its own and is discussed in more detail in Chapter 12.

8.5 INSTRUMENTATION

Nowadays, multidimensional chromatography systems designed for proteomic applications are commercially available, most often combining SCX with RPLC according to the designs shown in Figures 8.7, 8.8, and 8.13. Other LC × LC applications described until present have been established on home-made configurations, since no commercial instrumentation was available. This is in contrast to comprehensive GC, for which several vendors are offering packages, including instrumentation and software programs. This is, of course, due to the fact that the comprehensive GC technique is much more mature and the modulation principles are limited primarily to thermal and flow modulation. In comprehensive LC, however, the configuration varies by application and capabilities desired. Today, the individual modules [i.e., pumps, injector, valve(s), thermostat(s), detector(s)] are combined by the operator him- or herself to create the appropriate configuration. In a routine laboratory, the time available to play around with different configurations is limited and therefore the availability of commercial instruments would be very convenient. As an example, up to now the majority of non-proteomics-related comprehensive LC separations have been performed by means of the loop interface. This configuration would hence be an ideal starting point for instrument manufacturers to build LC × LC systems. Prototypes are now assembled and evaluated by various manufacturers in collaboration with researchers who have built up outstanding experience in this area during the last 10 years (e.g., Shimadzu Scientific Instruments, Kyoto, Japan; Dionex, Bannockburn, IL; Agilent Technologies, Waldbronn, Germany). Progress toward instrument availability is thus expected in the near future.

8.6 MILESTONES IN COMPREHENSIVE LIQUID CHROMATOGRAPHY

Erni and Frei (1978) have been pioneers in the field of comprehensive LC and were followed by Bushey and Jorgenson (1990), who modified their configuration. The main difference in their approach was the fact that the eight-port valve enabled an increased fractionation of the primary column effluent. In addition, they removed from the system the detector acquiring the first-dimension chromatogram, which is highly advantageous in terms of reducing extra-column broadening. Their separation of a protein sample shown in Figure 8.21 was achieved through the coupling of a microbore IEX column in gradient elution mode to a semipreparative SEC column (250 mm × 9.4 mm i.d.). The flow rates were 5 and 2.1 mL/min in the first and second dimensions, respectively. The valve was equipped with two sampling loops of 30 μL, which were filled alternately with first-dimension effluent and their content was transferred and analyzed by the secondary column every 6 min. The separation showed high orthogonality and offered significantly more information on the sample components, but a large part of the available separation space was not used. The adaptation of the first-dimension flow rate and gradient, together with an increased sampling frequency established a complete exploitation of the 3D plane. Furthermore, Bushey and Jorgenson were the first to plot the data of a comprehensive separation in a 3D representaeased sampling frequency established a complete exploitation of the 3D plane. Furthermore, Bushtion, providing a more reliable and easier data interpretation.

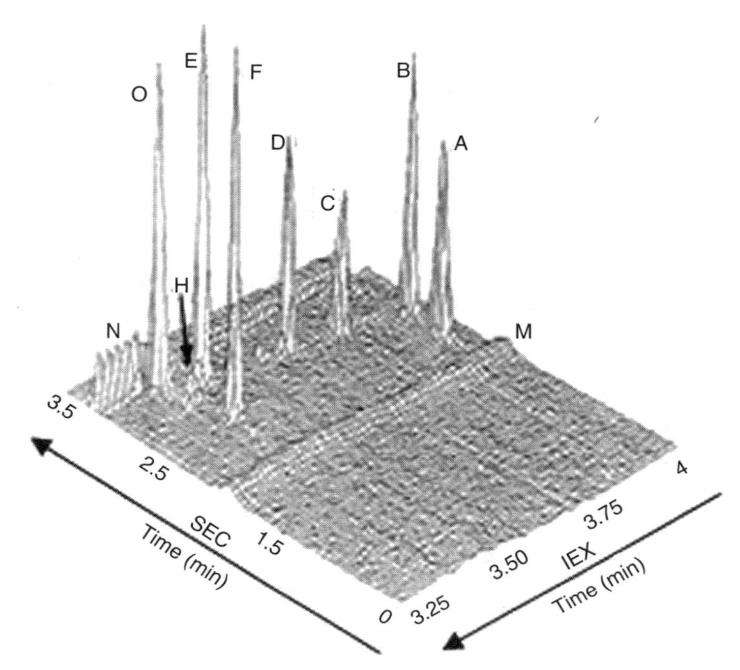

Figure 8.21 IEX × SEC separation of a protein sample. A, glucose oxidase; B, ovalbumin; C, β-lactoglobulin A; D, trypsinogen; E, α-chymotrypsinogen A; F, conalbumin; G, ribonuclease A; H, hemoglobin; M, exclusion volume "pressure" ridge; N, inclusion volume "salt" ridge. [From Bushey and Jorgenson (1990), with permission. Copyright © 1990 by The American Chemical Society.]

Another important improvement in the development of comprehensive LC was the hyphenation of the system to MS detection. Opiteck et al. were the first to use ESI–MS in addition to UV detection for the peak identification of proteins and peptides by IEX–LC × RPLC (Opiteck et al., 1997a) and SEC × RPLC (Opiteck et al., 1997b) in an online configuration. Prior to entering the ion source, the flow was split to avoid excessive gas generation.

The development of comprehensive LC systems capable of performing NPLC × RPLC is much more challenging than combinations of dimensions utilizing miscible mobile phases. Despite the inherent problems, the coupling of NPLC to RPLC is considered one of the most orthogonal arrangements. Murphy et al. (1998a) were the first to combine NPLC and RPLC for the separation of alcoholethoxylates, but avoided the incompatibility dilemma by applying an aqueous mobile phase in the NP dimension. Dugo et al. (2004) were the first to achieve a true comprehensive separation by NPLC × RPLC of the oxygen heterocyclic fraction of cold-pressed lemon oil. A microbore bare silica column was used in the first dimension at a flow rate of 20 μL/min in isocratic mode, while the secondary gradient was performed on a monolithic ODS column at 4 mL/min. The 10-port interface valve was switched every minute, permitting the collection and analysis of 20 μL fractions. In the first dimension, the separation was achieved based on polarity, while the sample components were analyzed according to hydrophobicity by the secondary column. A similar setup was used by François et al. (2006), but they also suggested a modification of the system by the addition of extra tools, including a second-dimension pump, column, detector, and most important, a supplementary 10-port switching valve (François et al., 2008a) to allow operation of the second dimension by a parallel column system. This system accomplished separation with a larger peak capacity and an improved total comprehensive analysis, due to the fact that the second-dimension analysis time was extended to twice the modulation time of 1 min. In this way, signal interferences resulting from solvent immiscibilities contributed less to the separation space available than described earlier (François et al., 2006).

Another approach to minimizing the negative properties of NPLC × RPLC separation and fully exploiting the separation power and orthogonality is the development of interfaces that remove the incompatible solvents. Tian et al. 2006, 2008 have built an interface that makes it possible to remove hazardous NP solvents by evaporation under vacuum conditions. An innovative alternative was developed by François et al. 2008b, 2010 and François and Sandra (2009c) by using supercritical fluid chromatography (SFC) as the first-dimension separation mechanism, which enabled the exploitation of the high orthogonality of the NPLC × RPLC combination with fewer disadvantages than when a liquid-based first-dimension mechanism is used. Using SFC in the first dimension of a comprehensive configuration is highly beneficial compared to NPLC, since the apolar constituent of the mobile phase is supercritical carbon dioxide, simply evaporating after expansion. These configurations are described in more detail in Chapter 11.

Carr and co-workers have stressed the beneficial effects on the analysis speed of the second dimension of supra-ambient temperatures in combination with high linear velocities (Stoll et al., 2006). The overall comprehensive 2D LC separation time was shortened significantly, to about 30 min, by the second-dimension gradient elution separations of 20 s. The interface was a two-position 10-port switching valve with two identical storage loops, and the second-dimension column was provided with a heating jacket at a temperature of 110°C.

Triacylglycerols (TGs) are the most abundant constituents of food lipids and their complex nature renders separation by 1D LC very difficult. Triglycerides consist of a glycerol backbone to which three long-chain fatty acids (FAs), differing in chain length and number of double bonds (DBs), are attached. The general approach used for the comprehensive separation of TGs is the combination of silver-ion chromatography (SIC) and nonaqueous RPLC coupled by means of a loop interface consisting of a 10-port switching valve (Dugo et al., 2006b,c; Mondello et al., 2005; van der Klift et al., 2008). In the silver-ion dimension, separation occurs according to the degree of saturation and the position or configuration of the DBs in every FA. A solute with a higher degree of unsaturation is characterized by a higher retention time. Sample components with the same number of DBs coelute in the first dimension and are captured in the same fraction after which they are analyzed by nonaqueous RPLC on the basis of their partition number (PN). PN is defined as the combined carbon number of the three FAs (CN) minus twice the number of DBs.

In the proteomics arena, MudPIT (Washburn et al., 2001; Wolters et al., 2001) and all its variations can certainly be regarded as extremely innovative. The fact that two types of stationary phases are packed sequentially into a single capillary, of which the outlet serves as an ESI needle is a breakthrough. However, in the perspective that recent investigations have shown that the combination of two RPLC dimensions, rather than the SCX × RPLC combination, is capable of delivering higher peak capacities, the peptide RPLC (pH1.8) × 2RPLC (pH10) separations by François et al. (2009b) are remarkable, as this was the first time that such an online configuration was described.

HILIC is a variation of NPLC, but employs an aqueous mobile phase with a high percentage of organic solvent. The retention behavior of the solutes is suspected to be determined by a partitioning process of the analyte, which takes place between a water-enriched layer of a stagnant solvent on a hydrophilic stationary phase and a relatively hydrophobic bulk mobile phase. The difference in selectivity together with the aqueous mobile phases make HILIC separations potential partners of RPLC, SEC, and IEX–LC in an MD configuration. However, very few reports in this field have been reported (Jandera et al., 2006; Wilson et al., 2007). Jandera (2008) recently reviewed HILIC stationary phases with additional attention to their implementation into MD concepts.

8.7 APPLICATIONS

Comprehensive LC systems can consist of various combinations of separation modes, preferably showing distinct retention mechanisms to the sample components. The selectivity differences are based primarily on size, shape, polarity, hydrophobicity, degree of saturation, acidity/basicity, or charge. According to these properties, SEC, NPLC, RPLC, HILIC, IEX, SIC, or ion-pairing chromatography has been employed in the two dimensions of an LC × LC system.

Various reviews have described in detail the various applications of comprehensive LC analyses (Berek, 2010; Dixon et al., 2006; Dugo et al., 2008a,b; Evans and Jorgenson, 2004; François et al., 2009a; Guiochon et al., 2008; Guttman et al., 2004; Herrero et al., 2009; Jandera, 2006; Liu and Lee, 2000; Pól and Hyötyläinen, 2008; Sandra et al., 2009; Shalliker and Gray, 2006; Shellie and Haddad, 2006; Stoll et al., 2007; Stroink et al., 2005; Tranchida et al., 2007). For a more detailed description of LC × LC applications, we refer to Chapter 10.

8.8 BEYOND TWO-DIMENSIONAL CHROMATOGRAPHY

Despite the fact that LC × LC provides higher peak capacities than 1D LC for highly complex samples such as mammalian proteomes, the resolution is still too limited. This raises the idea of the implementation of a third dimension to increase the peak capacity even more. The complexity of a chromatographic system employing this concept would be extremely high. First of all, a combination of three independent separation mechanisms needs to be found, and the sequence in which these modes are best configured needs to be defined. Furthermore, separation speed in the second and third dimensions will be even more critical than in 2D LC. Problems related to connectivity and compatibility will be more challenging. These types of separations will also create enormous data-handling problems.

Only a few studies have reported on online three-dimensional (3D) LC separations. As early as 1995, Moore and Jorgenson made such an attempt for the separation of the peptide digests of hen ovalbumin. They combined a very slow SEC separation in the first dimension, a faster secondary separation by RPLC, and a third, very fast separation by capillary zone electrophoresis (CZE). The authors reported peak capacities of 5 (SEC), 23 (RPLC), and 24 (CZE) for the three respective separations, resulting in a total peak capacity of 2800. During the development, some difficulties needed encountering. First, due to the significant sample dilution during the three separation steps, the injection of large sample amounts was mandatory to achieve reasonable detection limits. Furthermore, the temperatures in the three systems needed thorough control to avoid retention time drifts. Third, the RPLC–CZE coupling was problematic. Finally, the data representation of such a 3D separation is not straightforward to design in such a way that allows easy understanding and interpretation.

Wei et al. (2005) described a triphasic MudPIT type of approach (see Figure 8.11) in which the first RP segment was used effectively for separation and not purely for desalting. Twelve SCX salt steps were preceded by five RPLC gradients, resulting in 60 fractions separated on the last RP segment. The RPLC gradient used to displace peptides from the first RP segment onto the SCX sorbent was repeated onto the second RP segment following SCX displacement. This online 3D LC–MS–MS approach resulted in the identification of 3019 unique yeast proteins (representing 48% of the total proteins in the yeast database) with an average of 5.5 peptides per protein from the sequential analysis of 200 μg of the yeast-soluble, urea-solubilized peripheral membrane and SDS-solubilized membrane proteins. Upon comparison with the classic biphasic MudPIT, 4.5 times more proteins could be identified from the same sample with more peptide per protein identifications and with a reduced percentage of proteins identified by one peptide, all this evidently at the expense of time. To demonstrate that it was the 3D setup and not the separation time that resulted in the performance increase, the authors repeated the MudPIT analysis using a longer RPLC gradient. Despite improvements over the classic MudPIT procedure, proteome coverage was still much lower.

8.9 COMPARISON OF LC × LC AND OFF-LINE 2D LC

As evidenced by all the impressive achievements that have been established in this field, online comprehensive LC is a very powerful approach for the separation of complex samples. In recent years, the technique was propagated as a panacea for all types of difficult separations, but this very optimistic conclusion needs serious mitigation. Since the procedure is characterized by a number of practical limitations, the full potential behind the theoretical idea is reduced significantly. Orthogonality and sampling frequency are critical, but the major factor that is limiting the resolving power of LC × LC remains the low peak capacity of the second dimension. As discussed in Section 8.1.2, this is caused by the limited time available for the separation. Within one modulation cycle, only a fraction of this time (gradient time t_G) is used effectively for the separation. Other contributions are the dwell time, the equilibration or regeneration time, and the injection time (valve switching time), the summation of which is referred to as additional time, t_{add} (Horvath et al., 2009a). Horvath et al. (2009b) have recently investigated the generation and limitations of peak capacity in online 2D LC. The authors concluded that method development in online 2D LC is cumbersome. The optimization procedure implies the simultaneous consideration of many parameters, and thorough brainstorming is required as little room is available for empirical adjustments. In their theoretical investigation, the authors predicted that online comprehensive 2D LC probably will never be capable of delivering peak capacities higher than 10,000, a value that, as a matter of fact, has never been reported until present.

In another contribution by the same authors (Fairchild et al., 2009b), the stop-and-go method is considered to be beneficial when intermediate peak capacities (500 to a few thousands) are required, at the cost of a significant increase in analysis time. This setup allows the operation of longer and hence more efficient second-dimension columns or columns using slower gradients. However, this stop-flow principle is also limited for generating higher peak capacities due to the axial diffusion of the peaks during their parking time, which results in a loss of separation power.

Off-line 2D LC is the only technique that is truly capable of delivering extremely high peak capacities, at the expense of very long analysis times. One of the largest benefits is the independence of the two dimensions. In both dimensions, the mode, the composition and flow rate of the mobile phase, the gradient program, and the temperature of the separation can be carefully chosen by the analyst, with no time constraints. This offers a number of advantages: the second-dimension separation is carried out with higher efficiency, the sampling frequency is ideal for the application considered, and the complexity of the system is reduced significantly. In fact, one HPLC instrument is sufficient to perform all analyses. The major bottleneck is the longer analysis time, together with problems arising from the management and storage of a large number of collected fractions in such a way that contamination or losses are avoided. Horvath et al. (2009a) have recently proposed a step-by-step strategy for the optimization of comprehensive off-line 2D LC separations, taking into account the characteristics of the first- and second-dimension columns and the number of fractions of the first-dimension effluent that should be collected. Two different approaches toward optimization can be followed, whether the goal is to achieve a given resolution (a target peak capacity) in the shortest possible time or to reach the highest possible resolution in a given analysis time. The most striking observations in this research were the following: (1) an increase in the first-dimension peak capacity, combined with the collection of larger volume fractions, permits a significant reduction in the time needed to achieve the desired peak capacity, and (2) there is an optimum sampling frequency which yields the target peak capacity in the minimum time. The authors again stressed the importance of t_{add}, which should be maintained to a minimum in order to maximize the second-dimension peak capacity and reduce its effect on method optimization.

A few examples of off-line procedures have been discussed. The majority of these applications are logically focusing on separations in proteomics, lipodomics, and metabolomics research, since these samples consist of thousands of components in a very wide concentration range. In continuation of the work performed by Yates et al. (Washburn et al., 2001; Wolters et al., 2001), the group of Gygi (Peng et al., 2003) studied the yeast proteome using SCX combined off-line with RPLC instead of using the MudPIT approach. The authors confirmed the advantages of the off-line procedure: (1) a linear gradient can easily be applied, which results in superior peptide separation; (2) high amounts of organic modifier can be used, providing increased peptide recovery; (3) more peptide fractions can be collected (high sampling rates); and (4) the fractions can be separated on highly efficient second-dimension columns operated at optimal flow rate and gradient slope. The authors loaded 1 mg of whole yeast protein digest onto a 2.1-mm i.d. SCX column and collected 1-min fractions during an 80-min elution. The 80 fractions were subsequently reduced in volume and analyzed using nano-RPLC–MS–MS. The amount of material injected onto the second-dimension column and the gradient slope was adjusted according to first-dimension ultraviolet absorbance. This flexibility, which is not offered by online methods, both prevents sample overloading and the concomitant MS suppression effects and allows less abundant fractions to be injected entirely, separated by a shorter gradient, and to present fairly concentrated peptides to the detector. More than 162,000 MS–MS spectra were acquired, resulting in 26,815 identified peptides (of which 7537 were unique) and 1504 highly confident protein identifications. The data were compared with the data generated by Yates and colleagues (Washburn et al., 2001; Wolters et al., 2001) and 858 proteins (about 60%) were common in both data sets.

The authors demonstrated further that analyzing every other fraction would reduce the analysis time by 50% but would still allow the identification of 84% of the same proteins. Analyzing only every third fraction resulted in 74% of the same proteins being identified. This illustrates that at a certain level, major efforts are required to further increase the number of proteins identified (i.e., a twofold-longer analysis time is required to obtain an increase in protein identifications of 16%). Nevertheless, the data suggest that even greater separation capacity is required to meet the complexity challenge of a higher eukaryotic proteome. This is precisely what Smith and co-workers tried to address in their work describing ultrahigh efficiency 2D SCX–RPLC–MS–MS for the high dynamic range characterization of the human plasma proteome (Shen et al., 2004). By combining SCX fractionation (15 fractions) with highly efficient second-dimension nano-RPLC separations ($n_c \sim 1000$), chromatographic peak capacities in excess of 10^4 were generated, enabling the detection of proteins over a dynamic concentration range of 10^8 from the injection of 150 μg of undepleted blood plasma. The setup allowed the identification of cytokines (present at pg/mL concentrations in blood plasma) and albumin in a single experiment, and of course, this is the Holy Grail in blood-based protein biomarker discovery. Using the same mass spectrometer, this represents a 10^4 improvement in dynamic range over the classic MudPIT strategy (Washburn et al., 2001; Wolters et al., 2001) and over the use of their highly efficient ($n_c \sim 1000$) nano-RPLC–MS in the 1D mode.

The strength of the off-line mode was exploited fully by Guiochon and colleagues in a recent contribution (Marchetti et al., 2008). The authors described the off-line separation of BSA using SCX and RPLC columns with peak capacities of 54 and 142, respectively. Peaks were sampled 3.4 times, giving rise to 182 fractions that were subsequently separated using a 4-min gradient (9-min run time). A peak capacity in excess of 7000 was estimated in a total analysis time of 27 h. A comparable 1D peak capacity would cost a twofold increase in analysis time, since a column efficiency in excess of 20 million theoretical plates would be necessary. High peak capacity was achieved by using a short second-dimension column packed with shell particles. The use of a similar-sized column packed with fully porous particles (5 μm) generated a threefold lower peak capacity. The authors assumed a correlation factor between the two dimensions of 0.9, which is more optimistic than the orthogonality that arose from the geometric approach described by Gilar et al. (2005) and discussed earlier. It was concluded that the separation power of the method, by far exceeds the separation powers reported in the literature dealing with online 2D chromatography.

A fair comparison between online and off-line was made by Nägele et al. (2004). The authors analyzed the yeast proteome using online 2D LC–MS with injected salt steps and with a pumped semicontinuous salt gradient and off-line 2D LC–MS with a pumped continuous gradient. Sample, columns, and MS equipment were identical, and comparable salt concentration ranges were transferred to the second dimension. From the data it was evident that the off-line methodology was far superior to the online approach in terms of peptide and protein identifications. The same group also proposed guidelines to maximize peptide and protein identifications in off-line 2D LC–MS (Vollmer et al., 2004).

Non-bio-related applications have highlighted the power of off-line 2D LC. Kalili and de Villiers (2009) recently combined HILIC and RPLC off-line for the separation of procyanidins. In the HILIC mode, oligomeric procyanidins were separated according to molecular weight, while the RPLC second dimension was used to separate oligomers based on hydrophobicity. Experimental peak capacities in excess of 2300 were obtained.

François et al. (2010) recently compared off-line and comprehensive approaches for the separation of the triglycerides in fish oil. They used SI-SFC and nonaqueous RPLC in the first and second dimensions, respectively, allowing orthogonal separation according to saturation and hydrophobicity. The off-line procedure showed that despite the good online attempt, the sample complexity rendered the baseline separation of all compounds impossible. Even the high efficient separation in the off-line procedure by serially coupling three Zorbax SB C18 15 cm long × 4.6 mmi.d. × 1.8 μm d_p columns was not sufficient and MS was necessary as a third separation dimension. This application is discussed in more detail in Chapter 11.

8.10 CONCLUSIONS

Comprehensive LC has exhibited a huge momentum since the 1990s, and when attending international conferences it is clear that the popularity of the technique is growing each year. The main advantage of the procedure is the increased resolving power, which massively outperforms the peak capacity of 1D LC. Through the combination of orthogonal dimensions in which the solutes experience diverse affinities, the peak capacity is increased to a value that theoretically approaches the product of the peak capacities of the individual dimensions. In contrast to the powerful off-line 2D LC, this enormous amount of information is often obtained within the relatively short analysis time of the first-dimension separation.

Method development and instrument design are more difficult and time consuming than in 1D LC or off-line 2D LC. However, the great potential of comprehensive LC justifies all efforts and will certainly attract an increasing number of researchers. Progresses in the development of commercially available instruments with dedicated software for the direct conversion of raw data into 2D or 3D chromatograms, together with easy quantitative and qualitative analysis, will facilitate comprehensive operations in the future. Nevertheless, innovation should be stimulated toward new strategies, including the design of new interfaces that allow combinations of highly orthogonal dimensions operated with optimum performance.

It is important to note that for extremely complex mixtures, off-line 2D LC offers much higher peak capacities than those of LC × LC at the price of a long analysis time.

Upon reviewing the literature, a discrepancy was observed between the setups typically employed in the life sciences field (e.g., proteomics) and the setups designed by analytical chemists. The aim of the latter is to approach the theoretical peak capacity, and therefore short second-dimension columns with wide internal diameters operated at high flow rates have to be used to obtain sufficient samplings per first-dimension peak. The proteomics researcher does not necessarily aim at reaching the theoretical peak capacity but, rather, aims at delivering peptides to the mass spectrometer as purified as possible and preferably highly concentrated. The high flow rates at which some comprehensive setups are operated are not compatible with highly sensitive mass spectrometric detection, which is a prerequisite, and would require massive amounts of sample, which are often lacking. In addition, very narrow peaks are generated that require fast-scanning MS devices or high-frequency spotting devices in case LC–MALDI is used. By combining our expertise in analytical chemistry and proteomics, we hope to have succeeded in providing the reader with a view of the history, evolution, and optimization aspects of LC × LC.

REFERENCES

Alexander AJ, Ma L. *J. Chromatogr. A* 2009; 1216:1338–1345.

Bedani F, Kok WT, Janssen HG. *J. Chromatogr. A* 2006; 1133:126–134.

Berek D. *Anal. Bioanal. Chem*. 2010; 396:421–441.

Blahová E, Jandera P, Cacciola F, Mondello L. *J. Sep. Sci*. 2006; 29:555–566.

Bushey MM, Jorgenson JW. *Anal. Chem*. 1990; 62:161–167.

Cacciola F, Jandera P, Blahová E, Mondello L. *J. Sep. Sci*. 2006; 29:2500–2513.

Cacciola F, Jandera P, Hajdu Z, Česla P, Mondello L. *J. Chromatogr. A* 2007; 1149:73–87.

Catchpoole HJ, Shalliker RA, Dennis GR, Guiochon G. *J. Chromatogr. A* 2006; 1117:137–145.

Chen XG, Kong L, Su XY, Fu HJ, Ni JY, Zhao RH, Zou HF. *J. Chromatogr. A* 2004; 1040:169–178.

Chen XG, Jiang ZL, Zhu Y, Tan JY. *Chromatographia* 2007; 65:141–147.

Cunliffe JM, Maloney TD. *J. Sep. Sci*. 2007; 30:3104–3109.

Davis JM. *J. Sep. Sci*. 2005; 28:347–359.

Davis JM, Stoll DR, Carr PW. *Anal. Chem*. 2008a; 80:461–473.

Davis JM, Stoll DR, Carr PW. *Anal. Chem*. 2008b; 80:8122–8134.

Davis MT, Beierle J, Bures ET, McGinley MD, Mort J, Robinson JH, Spahr CS, Yu W, Luethy R, Patterson SD. *J. Chromatogr. B* 2001; 752:281–291.

DeStefano JJ, Langlois TJ, Kirkland JJ. *J. Chromatogr. Sci*. 2008; 46:254–260.

Dixon SP, Pitfield ID, Perrett D. *Biomed. Chromatogr*. 2006; 20:508–529.

Dugo P, Favoino O, Luppino R, Dugo G, Mondello L. *Anal. Chem*. 2004; 76:2525–2530.

Dugo P, Škeříková V, Kumm T, Trozzi A, Jandera P, Mondello L. *Anal. Chem*. 2006a; 78:7743–7750.

Dugo P, Kumm T, Crupi ML, Cotroneo A, Mondello L. *J. Chromatogr. A* 2006b; 1112: 269–275.

Dugo P, Kumm T, Chiofalo B, Cotroneo A, Mondello L. *J. Sep. Sci*. 2006c; 29: 1146–1154.

Dugo P, Cacciola F, Kumm T, Dugo G, Mondello L. *J. Chromatogr. A* 2008a; 1184: 353–368.

Dugo P, Kumm T, Cacciola F, Dugo G, Mondello L. *J. Liq. Chromatogr. Relat. Technol*. 2008b; 31:1758–1807.

Dugo P, Herrero M, Kumm T, Giuffrida D, Dugo G, Mondello L. *J. Chromatogr. A* 2008c; 1189:196–206.

Dugo P, Cacciola F, Herrero M, Donato P, Mondello L. *J. Sep. Sci*. 2008d; 31:3297–3308.

Eeltink S, Desmet G, Vivó-Truyols G, Rozing GP, Schoenmakers PJ, Kok WT. *J. Chromatogr. A* 2006a; 1104:256–262.

Eeltink S, Gzil P, Kok WT, Schoenmakers PJ, Desmet G. *J. Chromatogr. A* 2006b; 1130:108–114.

Eggink M, Romero W, Vreuls RJ, Lingeman H, Niessen WMA, Irth H. *J. Chromatogr. A* 2008; 1188:216–226.

Erni F, Frei RW. *J. Chromatogr*. 1978; 149:561–569.

Evans CR, Jorgenson JW. *Anal. Bioanal. Chem*. 2004; 378:1952–1961.

Fairchild JN, Horvath K, Guiochon G. *J. Chromatogr. A* 2009a; 1216:6210–6217.

Fairchild JN, Horvath K, Guiochon G. *J. Chromatogr. A* 2009b; 1216:1363–1371.

Fournier ML, Gilmore JM, Martin-Brown SA, Washburn MP. *Chem. Rev*. 2007; 107:3654–3686.

François I, Sandra P. *J. Chromatogr. A* 2009; 1216:4005–4012.

François I, de Villiers A, Sandra P. *J. Sep. Sci*. 2006; 29:492–498.

François I, de Villiers A, Tienpont B, David F, Sandra P. *J. Chromatogr. A* 2008a; 1178:33–42.

François I, dos Santos-Pereira A, Lynen F, Sandra P. *J. Sep. Sci*. 2008b; 31:3473–3478.

François I, Sandra K, Sandra P. *Anal. Chim. Acta* 2009a; 641:14–31.

François I, Cabooter D, Sandra K, Lynen F, Desmet G, Sandra P. *J. Sep. Sci*. 2009b; 32:1137–1144.

François I, dos Santos-Pereira A, Sandra P. *J. Sep. Sci*. 2010; 33:1504–1512.

Fujii K, Nakano T, Hike H, Usui F, Bando Y, Tojo H, Nishimura T. *J. Chromatogr. A* 2004; 1057:107–113.

Gilar M, Olivová P, Daly AE, Gebler JC. *Anal. Chem*. 2005; 77:6426–6434.

Gray MJ, Dennis GR, Slonecker PJ, Shalliker RA. *J. Chromatogr. A* 2004; 1041:101–110.

Gray MJ, Dennis GR, Slonecker PJ, Shalliker RA. *J. Chromatogr. A* 2005; 1073:3–9.

Gritti F, Cavazzini A, Marchetti N, Guiochon G. *J. Chromatogr. A* 2007; 1157:289–303.

Gu X, Deng C, Yan G, Zhang X. *J. Proteome Res*. 2006; 5:3186–3196.

Guiochon G, Marchetti N, Mriziq K, Shalliker RA. *J. Chromatogr. A* 2008; 1189:109–168.

Guttman A, Varoglu M, Khandurina J. *Drug Discov. Today* 2004; 9:136–144.

Guzetta AW, Chien AS. *J. Proteome Res*. 2005; 4:2412–2419.

Haefliger OP. *Anal. Chem*. 2003; 75:371–378.

Helfferich FG. *Ion Exchange*. New York: Dover, 1995.

Herrero M, Ibanez E, Cifuentes A, Bernal J. *J. Chromatogr. A* 2009; 1216:7110–7129.

Hoffman NE, Pan SL, Rustum AM. *J. Chromatogr*. 1989; 465:189–200.

Holm A, Storbraten E, Mihailova A, Karaszewski B, Lundanes E, Greibrokk T. *Anal. Bioanal. Chem*. 2005; 382:751–759.

Horie K, Kimura H, Ikegami T, Iwatsuka A, Saad N, Fiehn O, Tanaka N. *Anal. Chem*. 2007; 79:3764–3770.

Horvath K, Fairchild JN, Guiochon G. *J. Chromatogr. A* 2009a; 1216:2511–2518.

Horvath K, Fairchild JN, Guiochon G. *Anal. Chem*. 2009b; 81:3879–3888.

Hu LH, Chen XG, Kong L, Su XY, Ye ML, Zou HF. *J. Chromatogr. A* 2005; 1092:191–198.

Ikegami T, Hara T, Kimura H, Kobayashi H, Hosoya K, Cabrera K, Tanaka N. *J. Chromatogr. A* 2006; 1106:112–117.

Jandera P. *J. Sep. Sci*. 2006; 29:1763–1783.

Jandera P. *J. Sep. Sci*. 2008; 31:1421–1437.

Jandera P, Guiochon G. *J. Chromatogr*. 1991; 588:1–14.

Jandera P, Fischer J, Lahovska H, Novotna K, Česla P, Kolářová L. *J. Chromatogr. A* 2006; 1119:3–10.

Jandera P, Česla P, Hájek T, Vohralik G, Vyňuchalová K, Fischer J. *J. Chromatogr. A* 2008; 1189:207–220.

Jiang XL, Van der Horst A, Schoenmakers PJ. *J. Chromatogr. A* 2002; 982:55–68.

Jiang XL, van der Horst A, Lima V, Schoenmakers PJ. *J. Chromatogr. A* 2005; 1076:51–61.

Kalili KM, de Villiers A. *J. Chromatogr. A* 2009; 1216:6274–6284.

Kang D, Nam H, Kim YS, Moon MH. *J. Chromatogr. A* 2005; 1070:193–200.

Kele M, Guiochon G. *J. Chromatogr. A* 1999a; 830:41–54.

Kele M, Guiochon G. *J. Chromatogr. A* 1999b; 830:55–79.

Kele M, Guiochon G. *J. Chromatogr. A* 1999c; 855:423–453.

Kele M, Guiochon G. *J. Chromatogr. A* 2000; 869:181–209.

Kele M, Guiochon G. *J. Chromatogr. A* 2001; 913:89–112.

Kele M, Guiochon G. *J. Chromatogr. A* 2002; 960:19–49.

Keunchkarian S, Reta M, Romero L, Castells C. *J. Chromatogr. A* 2006; 1119:20–28.

Kimura H, Tanigawa T, Morisaka H, Ikegami T, Hosoya K, Ishizuka N, Minakuchi H, Nakanishi K, Ueda M, Cabrera K, Tanaka N. *J. Sep. Sci*. 2004; 27:897–904.

Kivilompolo M, Hyötyläinen T. *J. Chromatogr. A* 2007; 1145:155–164.

Kivilompolo M, Obůrka V, Hyötyläinen T. *Anal. Bioanal. Chem*. 2008; 391:373–380.

Köhne AP, Welsch T. *J. Chromatogr. A* 1999; 845:463–469.

Köhne AP, Dornberger U, Welsch T. *Chromatographia* 1998; 48:9–16.

Le Bihan T, Duewel HS, Figeys D. *J. Am. Soc. Mass Spectrom*. 2003; 14:719–727.

Leitner T, Klampfl CW. *J. Liq. Chromatogr. Relat. Technol*. 2008; 31:169–178.

Li X, Stoll DR, Carr PW. *Anal. Chem*. 2009; 81:845–850.

Licklider LJ, Thoreen CC, Peng J, Gygi SP. *Anal. Chem*. 2002; 74:3076–3083.

Link AJ, Eng J, Schieltz DM, Carmack E, Mize GJ, Morris DR, Garvik BM, Yates JR. III *Nat. Biotechnol*. 1999; 17:676–682.

Liu Z, Lee ML. *J. Microcolumn Sep*. 2000; 12:241–254.

Liu CL, Zhang XM. *J. Chromatogr. A* 2007; 1139:191–198.

Liu H, Finch JW, Luongo JA, Li GZ, Gebler JC. *J. Chromatogr. A* 2006; 1135:43–51.

Luo Q, Yue G, Valaskovic GA, Gu Y, Wu SL, Karger BL. *Anal. Chem*. 2007; 79:6174–6181.

Marchetti N, Guiochon G. *J. Chromatogr. A* 2007; 1176:206–216.

Marchetti N, Cavazzini A, Gritti F, Guiochon G. *J. Chromatogr. A* 2007; 1163:203–211.

Marchetti N, Fairchild JN, Guiochon G. *Anal. Chem*. 2008; 80:2756–2767.

Mayfield KJ, Shalliker RA, Catchpoole HJ, Sweeney AP, Wong V, Guiochon G. *J. Chromatogr. A* 2005; 1080:124–131.

McDonald WH, Ohi R, Miyamoto DT, Mitchison TJ, Yates JR. III *Int. J. Mass Spectrom*. 2002; 219:245–251.

Mitulovic G, Swart R, van Ling R, Jakob T, Chervet JP. *LC-GC Eur*. 2004a; Suppl.: 16–20.

Mitulovic G, Stingl C, Smoluch M, Swart R, Chervet JP, Steinmacher I, Gerner C, Mechtler K. *Proteomics* 2004b; 4:2545–2557.

Mondello L, Tranchida PQ, Stanek V, Jandera P, Dugo G, Dugo P. *J. Chromatogr. A* 2005; 1086:91–98.

Mondello L, Herrero M, Kumm T, Dugo P, Cortes H, Dugo G. *Anal. Chem*. 2008; 80:5418–5424.

Mondello L, Donato P, Cacciola F, Fanali C, Dugo P. *J. Sep. Sci*. 2010; 33:1454–1461.

Moore AW, Jorgenson JW. *Anal. Chem*. 1995; 67:3456–3463.

Motoyama A, Venable JD, Ruse DI, Yates JR III. *Anal. Chem*. 2006; 78:5109–5118.

Motoyama A, Xu T, Ruse CI, Wohlschlegel JA, Yates JR. *Anal. Chem*. 2007; 79:3623–3634.

Murahashi T. *Analyst* 2003; 128:611–615.

Murphy RE, Schure MR, Foley JP. *Anal. Chem*. 1998a; 70:1585–1594.

Murphy RE, Schure MR, Foley JP. *Anal. Chem*. 1998b; 70:4353–4360.

Nägele E, Vollmer M, Hörth P. *J. Chromatogr. A* 2003; 1009:197–205.

Nägele E, Vollmer M, Hörth P. *J. Biomol. Tech*. 2004; 15:134–143.

Opiteck GJ, Lewis KC, Jorgenson JW, Anderegg RJ. *Anal. Chem*. 1997a; 69:1518–1524.

Opiteck GJ, Jorgenson JW, Anderegg RJ. *Anal. Chem*. 1997b; 69:2283–2291.

Opiteck GJ, Ramirez SM, Jorgenson JW, Moseley MA. *Anal. Biochem*. 1998; 258:349–361.

Peng J, Elias JE, Thoreen CC, Licklider LJ, Gygi SP. *J. Proteome Res*. 2003; 2:43–50.

Pepaj M, Wilson SR, Novotna K, Lundanes E, Greibrokk T. *J. Chromatogr. A* 2006; 1120:132–141.

Pól J, Hohnová B, Jussila M, Hyötyläinen T. *J. Chromatogr. A* 2006; 1130:64–71.

Pól J, Hyötyläinen T. *Anal. Bioanal. Chem*. 2008; 391:21–31.

Sandra K, Moshir M, D'hondt F, Tuytten R, Verleysen K, Kas K, François I, Sandra P. *J. Chromatogr. B* 2009; 877:1019–1039.

Schoenmakers PJ, Vivó-Truyols G, Decrop WMC. *J. Chromatogr. A* 2006; 1120:282–290.

Seeley JV. *J. Chromatogr. A* 2002; 962:21–27.

Shalliker RA, Gray MJ. *Adv. Chromatogr*. 2006; 44:177–236.

Shalliker RA, Kavanagh PE. *Chromatographia* 1997; 44:421–426.

Shalliker RA, Kavanagh PE, Russell IM. *J. Chromatogr*. 1991a; 543:157–169.

Shalliker RA, Kavanagh PE, Russell IM. *J. Chromatogr*. 1991b; 558:440–445.

Shalliker RA, Catchpoole HJ, Dennis GR, Guiochon G. *J. Chromatogr. A* 2007; 1142:48–55.

Shellie RA, Haddad PR. *Anal. Bioanal. Chem*. 2006; 386:405–415.

Shen Y, Jacobs JM, Camp DG II, Fang R, Moore RJ, Smith RD, Xiao W, Davis RW, Tompkins RG. *Anal. Chem*. 2004; 76:1134–1144.

Stoll DR, Carr PW. *J. Am. Chem. Soc*. 2005; 127:5034–5035.

Stoll DR, Cohen JD, Carr PW. *J. Chromatogr. A* 2006; 1122:123–137.

Stoll DR, Li X, Wang X, Carr PW, Porter SEG, Rutan SC. *J. Chromatogr. A* 2007; 1168:3–43.

Stroink T, Ortiz MC, Bult A, Lingeman H, de Jong GJ, Underberg WJM. *J. Chromatogr. B* 2005; 817:49–66.

Tanaka N, Kimura H, Tokuda D, Hosoya K, Ikegami T, Ishizuka N, Minakuchi H, Nakanishi K, Shintani Y, Furuno M, Cabrera K. *Anal. Chem*. 2004; 76:1273–1281.

Tian H, Xu J, Xu Y, Guan Y. *J. Chromatogr. A* 2006; 1137:42–48.

Tian H, Xu J, Guan Y. *J. Sep. Sci*. 2008; 31:1677–1685.

Tranchida PQ, Donato P, Dugo P, Dugo G, Mondello L. *Trends Anal. Chem*. 2007; 26:191–205.

Unger KK, Racaityte K, Wagner K, Miliotis T, Edholm LE, Bischoff R, Marko-Varga G. *J. High Resolut. Chromatogr*. 2000; 23:259–265.

Van der Horst A, Schoenmakers PJ. *J. Chromatogr. A* 2003; 1000:693–709.

van der Klift EJC, Vivó-Truyols G, Claassen FW, van Holthoorn FL, van Beek TA. *J. Chromatogr. A* 2008; 1178:43–55.

van Nederkassel AM, Aerts A, Dierick A, Massart DL, Vander Heyden Y. *J. Pharm. Biomed. Anal*. 2003; 32:233–249.

Venkatramani CJ, Patel A. *J. Sep. Sci*. 2006; 29:510–518.

Venkatramani CJ, Zelechonok Y. *Anal. Chem*. 2003; 75:3484–3494.

Venkatramani CJ, Zelechonok Y. *J. Chromatogr. A* 2005; 1066:47–53.

Vollmer M, Hörth P, Nägele E. *Anal. Chem*. 2004; 76:5180–5185.

Wagner K, Racaityte K, Unger KK, Miliotis T, Edholm LE, Bischoff R, Marko-Varga G. *J. Chromatogr. A* 2000; 893:293–305.

Wagner K, Miliotis T, Marko-Varga G, Bischoff R, Unger KU. *Anal. Chem*. 2002; 74:809–820.

Wang F, Dong J, Ye M, Jiang X, Wu R, Zou H. *J. Proteome Res*. 2008; 7:306–310.

Washburn MP, Wolters DA, Yates JR. *Nat. Biotechnol*. 2001; 19:242–247.

Wei J, Sun J, Yu W, Jones A, Oeller P, Keller M, Woodnutt G, Short JM. *J. Proteome Res*. 2005; 4:801–808.

Wilson SR, Jankowski M, Pepaj M, Mihailova A, Boix F, Vivó-Truyols G, Lundanes E, Greibrokk T. *Chromatographia* 2007; 66:469–474.

Winnik WM. *Anal. Chem*. 2005; 77:4991–4998.

Wolters DA, Washburn MP, Yates JR. *Anal. Chem*. 2001; 73:5683–5690.

Xiang R, Shi Y, Dillon DA, Negin B, Horvath C, Wilkins JA. *J. Proteome Res*. 2004; 3:1278–1283.

Zhang J, Tao DY, Duan JC, Liang Z, Zhang WB, Zhang LH, Huo YS, Zhang YK. *Anal. Bioanal. Chem*. 2006; 386:586–593.

9

COMPREHENSIVE TWO-DIMENSIONAL LIQUID CHROMATOGRAPHY COMBINED WITH MASS SPECTROMETRY

PAOLA DUGO AND LUIGI MONDELLO

University of Messina, Messina, Italy

FRANCESCO CACCIOLA

Chromaleont s.r.l., A spin-off of the University of Messina, Messina, Italy and University of Messina, Messina, Italy

PAOLA DONATO

University Campus Bio-Medico, Rome, Italy and University of Messina, Messina, Italy

One-dimensional liquid chromatography (1D LC) is used widely in the analysis of real-world samples in several fields. Recent progress has been made in optimizing the potential of 1D LC. The most promising approaches to achieving enhanced performance consist of the use of smaller particle-size materials (sub-2-μm particles) (McNair et al., 1999; de Villiers et al., 2006), with separation carried out under elevated temperatures. Unfortunately, the cost of improved efficiencies is higher column backpressure, which limits the use of optimum flow rates or long columns, to get high resolution. Hence, to fully utilize the benefits of sub-2-μm particles, high-performance liquid chromatographic (HPLC) instrumentation with improved pressure resistance is necessary. An alternative today can be represented by shell-packed HPLC columns, which offer efficiencies equal to those of sub-2-μm columns at half the backpressure (de Stefano et al., 2008; Gritti et al., 2007; Marchetti et al., 2007; Way and Campbell, 2007). However, even using such columns coupled in series and increasing the temperature to reduce the backpressure, the peak capacity could not be high enough for the separation of complex samples. A different approach to increasing the separation

Comprehensive Chromatography in Combination with Mass Spectrometry, First Edition.
Edited by Luigi Mondello.

power is represented by the use of a multidimensional (2D LC) system, where the two (or more) dimensions are based on different separation mechanisms. Multidimensional chromatographic techniques are characterized by greatly increased resolving power, generally measured in terms of peak capacity (Giddings, 1987), with respect to one-dimensional methods. Of course, when the separation mechanisms selected provide completely different selectivities (high orthogonality), the advantages of 2D approaches are maximized. Two-dimensional LC con be performed either off- or online and can be operated using either a "heart-cutting" or a comprehensive approach (LC × LC). The difference between them is that the first enables reinjection of a limited number of multicomponent effluent fractions from a primary to a secondary column, while the second extends the multidimensional advantage to the entire sample. Requirements of comprehensive 2D LC, as well as instrument descriptions and method developments have been treated extensively in many recent review papers (Dugo et al., 2008d,e; François et al., 2009; Jandera, 2006) as well as in Chapter 8 of this book, and so are not included in this chapter.

It is important to emphasize that apart from the enhanced resolving power, an additional advantage is represented by an increased identification power due to the formation of 2D chemical class patterns. This advantage can be exploited even further if a detector with identification capabilities is combined in the comprehensive LC system. In particular, the coupling of 2D LC to mass spectrometry combines the advantages of the multidimensional chromatographic step (high selectivity and separation efficiency) and of mass spectrometry (structural information and further increase in selectivity). In respect to conventional HPLC–mass spectrometry (MS), the combination of two different chromatographic separations before introduction of analytes into the mass spectrometer contributes to enhanced resolution by physical separation of components and thus also reduces the undesirable matrix effects due to coeluting components. Pól and Hyötyläinen (2008) recently published a review on comprehensive LC coupled to MS.

After a brief introduction to coupling between HPLC and mass spectrometry, in the first part of this chapter we cover the general aspects of LC × LC coupled to mass spectrometry, instrumental configurations, the main differences with respect to conventional HPLC–MS, and in the second part discuss the applications developed in different areas of research using both off- and online comprehensive 2D LC coupled to MS. For each application, the type of interface, column sets, modulation time, and MS parameters are specified.

9.1 HPLC–MS

As stated some years ago by Sandra et al. (2001), mass spectrometry is slowly but surely becoming the detection system of choice for LC analysis in many areas. The combination of HPLC with mass spectrometry is technically more complex than that of GC–MS for different reasons: one is the nature of the analytes (low volatility), while one of the prerequisites for MS analysis is the formation of volatile ions. The second is the necessary elimination of the liquid mobile phase. Third, salts and other additives of the mobile phase are often nonvolatile. Many coupling methods have been developed in the past decades to solve the problems mentioned above. The development of atmospheric-pressure ionization techniques has made LC–MS as reliable, versatile, and easy to use as capillary GC–MS. Two are the most common atmospheric-pressure ionization techniques: electrospray ionization (ESI) and atmospheric-pressure chemical ionization (APCI). In both cases, the eluent passes through a small capillary, forms a fine spray, and the solvent is evaporated. Ions formed in this process are then analyzed by the mass spectrometer. ESI is the technique of choice for highly polar compounds, which can be ionized easily, ranging in molecular weight from 100 to 150×10^3 Da. In fact, large molecules acquire more than one charge, allowing their analysis by MS instruments with a mass range of 3000 *m/z*. APCI is preferred for medium-polarity components, with a molecular weight of 100 to 2000 Da. Both techniques can be used in positive and negative ionization modes, on the basis of the chemical characteristics of the analytes. A third, recently introduced atmospheric-pressure photoionization (APPI) technique is applicable to many of the same solutes as APCI, but can give better response for highly apolar solutes.

Many limitations are related to the ionization process: non-volatile buffers (sodium phosphate) and salts (used to adjust the ionic strength) should be avoided, because they can easily block the capillary and the entrance to the mass spectrometer. Attention has to be taken in the correct choice of the flow rate: in the case of a conventional ESI ion source, the flow rate should be in the range 20 μL to 1 mL/min. If a high water content is present, the maximum flow rate may be even smaller. If higher flow rates are used, splitting is necessary, while in the case of very small flow rates, microspray or nanospray ion sources can be used. With APCI, higher flow rates are necessary for the ionization of the analytes (up to 2 mL/min). It is also important to note that relative sensitivity of ultraviolet (UV) and MS detection for various compounds may differ by many orders of magnitude in either direction. The best way to overcome this problem is to obtain both UV and MS spectra simultaneously by connecting the two detectors sequentially or by splitting the solvent at the exit of the HPLC column between the two detectors. Solvent quality is also of highest importance in HPLC–MS, because the chemical background (due to solvent impurities) is usually far higher in MS than in UV detection.

Of course, many advantages of HPLC–MS can be enlisted: first, the possibility of obtaining structural information in addition to information on molecular mass, as well as the possibility of detecting even compounds without good chromophores, due to the universality of MS detection. On the other hand, the specificity of a mass spectrometer (used in the SIM mode) can be used to reduce the chemical background or to increase the resolution between peak pairs with different molecular mass or with the same molecular mass but different mass spectra.

Both APCI and ESI are considered soft-ionization techniques—they produce spectra with low fragmentation. The degree of fragmentation can be varied by varying the voltage applied to the cone. At low cone voltage, only the molecular ion is usually observed, while at higher cone voltage, the internal energy of molecules increases, so the probability of fragmentation also increases. Of course, accurate mass measurements and tandem MS give further structural information. Unlike GC–MS, no ESI or APCI spectral libraries are commercially available, due to the fact that spectra are closely related to the experimental conditions used [e.g., mobile-phase composition, temperature and voltage of the capillary (ESI) or the corona needle (APCI), cone voltage].

Various mass analyzers can be used in HPLC–MS: quadrupole, ion trap, time-of-flight (TOF), sector, and Fourier transform (FT) instruments. However, the last two are rarely coupled to chromatography, due to their high cost and operational difficulties. At present, quadrupole-type instruments are the most widespread. Their advantages are simplicity of use, relatively low cost, and ruggedness. Using this instrument, one ion is detected at a time, and the mass spectrum is obtained by scanning over the mass range of interest. Sensitivity can be increased using single-ion monitoring when the abundance of only one or a few ions is monitored. Quadrupole instruments are low-resolution systems (typically, 1 Da) and the mass range is limited approximately to *m/z* 4000. Specifications of today's commercially available quadrupole systems present values more than sufficient for most applications. The last generation of quadrupole instrumentations allow for high-speed scanning (up to 10,000 or even 15,000 amu/s) and ultrafast polarity switching for ultrafast analyses. It is also easy to couple multiple quadrupoles to one another or to other MS analyzers. Together with the quadrupole instruments, ion traps are widely used mass analyzers. Their small size and the possibility of performing tandem MS make quadrupole mass analyzers ideal for benchtop applications and MS^n applications.

TOF can detect all ions in a mass range simultaneously, so there is no need for scanning. The full sensitivity is obtained irrespectively if one ion or a full mass spectrum is studied. TOF instruments present the advantage of high scan speed (up to 20,000 scans/s), high resolution (using reflectron), and virtually no limit on mass range. They are also ideal as a second stage in tandem MS experiments: for example, combined with quadrupole (qTOF).

More information on the LC–MS technique, and detailed descriptions of interfaces for LC–MS, mass analyzers, and their working principles, may be found in numerous books and scientific literature (Lemière, 2001a,b; Niessen, 1999; Sandra et al., 2001; Vékey, 2001).

9.2 LC x LC–MS INSTRUMENTATION AND METHOD DEVELOPMENT

A comprehensive 2D LC instrument can easily be combined with all traditional HPLC detectors, such as ultraviolet (UV), photodiode array (PDA), evaporative light scattering (ELS), and MS. Usually, only one detector is installed after the second-dimension column. An extra detector that monitors first-dimension separation results is useful during method optimization but is not necessary and contributes to extra-column broadening when the instrument is used for LC × LC analysis. Operating the second dimension in fast mode, fast detectors will be required to ensure a high data acquisition rate for proper reconstruction of the multidimensional chromatogram.

Nowadays, a PDA detector can be operated at acquisition rates up to 80 Hz and is most commonly used. However, a detector acquisition rate of 5 Hz is minimally required to achieve a reliable peak description (six to 10 data points per peak). Although a PDA detector can give useful information for component identification, the combination of a mass spectrometer as the detector is surely much more powerful in the identification process of analytes. It can add a third dimension to the 2D LC separation because it can "separate" nonisobaric peaks not resolved by the chromatographic process. LC × LC–MS utilizes the same MS systems described for 1D LC–MS. The same limitations and advantages described in the preceding section can also be applied to LC × LC–MS.

Several points have to be considered when developing a LC × LC method:

1. The 2D effluent must allow ionization of the components and must be compatible with the MS. This may limit the number of possible LC × LC combinations, but in practice the majority of LC × LC applications use reversed phase (RP) LC in the second dimension, making the coupling to the mass spectrometer not so problematic from the point of view of the choice of the mobile phase.
2. The detector acquisition rate should be fast enough to adequately sample D2 effluent. Quadrupole MS systems available today are capable of supplying sufficient spectra per peak, thanks to an acquisition speed as high as 10,000 or 15,000 amu/s. At a speed of 10,000 amu/s, with a *m/z* range of 1000 units, 10 scans/s will be obtained. Obviously, TOF MS possesses very high scanning speed (up to 20,000 scans/s) and represents a very good choice for comprehensive LC applications.
3. One requirement of LC × LC method development is represented by the need to perform at least three or four modulations for each first-dimension peak, to avoid a serious loss of information due to undersampling of first-dimension peaks. As a consequence, analysis time in the second dimension should be as short as possible, often using very high flow rates. In these cases, the flow rate must be split before the entrance of the MS. A possible solution should be the use of a second detector used simultaneously with MS, and split the flow between the two detectors.

ESI and APCI have been reported widely as ion sources in LC × LC online applications, while MALDI can be applied to fractions collected offline. As stated earlier, LC × LC–MS uses the same systems as described earlier for 1D LC–MS. The coupling between the second dimension of the LC and the mass spectrometer does not require the development of particular interfaces. Only a splitting device could be necessary to reduce the flow rate that enters the MS to values compatible with the ion source employed. For a detailed description of LC × LC configuration, with particular attention to the column used in the two dimensions, as well as valve configuration, the reader may refer to specific reviews and to Chapter 8 of this book.

Method development in LC × LC–MS follows the same rules as those described in general for the LC × LC technique, regarding the choice of the column set, column dimensions, type of packing, separation mode in the two dimensions (paying attention to obtaining a good degree of orthogonality), mobile phases and flow rates, isocratic or gradient programs, possible use of temperature higher than ambient, modulation time, and amount of sample transferred from the first to the second dimensions. In addition to this, it is necessary to follows the rules relative to solvent and flow rate compatibility with the MS, and the correct choice of sampling rate.

9.3 LC x LC–MS APPLICATIONS

The number of LC × LC applications in many different fields has increased very rapidly since the first example reported by Erni and Frei in 1978. With an approximate estimation we have counted about 150 papers on LC × LC published between 2000 and 2008, while "only" 40 were published before (in the time frame 1978–1999). Among them, about one-third used MS as a detector.

Tables 9.1 to 9.4 summarize comprehensive 2D LC applications in different fields, where at least one of the detectors used was MS. As can be seen, the majority of these applications deal with peptides and protein mapping (see Table 9.1). Others are related to the analysis of natural, pharmaceutical, and environmental components (see Tables 9.2 to 9.4).

9.3.1 LC x LC–MS Methods for Separating and Identifying Peptides and Proteins

The last decade has witnessed an increasing number of LC × LC–MS methods devoted to proteome analysis, which has constituted a major driving force toward the development of multidimensional separations with enhanced separation ability and subsequent identification. A high demand is, in fact, placed both on the power of separation techniques, given the extremely high complexity of the proteome samples, and on the sensitivity of detection methods, to enable probing of low-abundant proteins or peptides.

The proteome can be defined as the protein complement of a genome expressed in a tissue or cell population of a multicellular organism (Wasinger et al., 1995). The term *proteomics* encompasses expression proteomics and functional proteomics. Expression proteomics deals with the characterization (i.e., the identification and quantification) of proteins in cells, tissues, or biological fluids; that could mean such a complex mixture as one comprising more than 10,000 proteins in a single-cell population. The highly variable concentration range of proteins brings in added difficulty to their analysis; just to make an example, the concentration ratio between albumin and the least abundant species in serum can be as high as 1:10,000 (Wilkins et al., 1997).

The word *characterization*, in turn, encompasses determination of the function(s) of all expressed proteins (functional proteomics) and the assessment of their cellular localization and post-translational modifications. Since proteins rarely act alone at the biochemical level, functional proteomics also involves the assay of protein interactions either in performing a given cellular task or as key players in a number of diseases. To complicate the situation further, different environmental, biological (even individual), pharmacological, and disease factors will ultimately affect protein expression and determine statistically significant variations.

The enormous complexity, variability, and dynamic range (a serum proteome may contain up to 20,000 proteins with a concentration range of 10^{10}) make the proteome much more complicated a task for analysis than the genome, which shows little (if any) variations between cells and tissues (Anderson and Anderson, 2002). Over the years, different strategies have been developed to address the needs of modern proteomics and to allow researchers to investigate the complicated biological networks in which proteins are involved at different levels and ultimately lead to a better understanding of living organisms in both health and disease. Furthermore, complementary knowledge generated by integrated approaches (i.e., genomic and proteomic) is expected to provide a more comprehensive view of the molecular basis and disrupted pathways involved in several pathologies (among which, cancer is one of the more devastating), as well as contributing to the discovery of new biomarkers for early diagnosis of diseases, the prediction of drug response patterns, and hopefully, the development of safer and more effective alternatives to existing therapies (Hamdan and Righetti, 2005; Hanash, 2003).

Within the wide host of hyphenated techniques, liquid chromatography–mass spectrometry has emerged as playing a central role in the field of proteomic research. The recent progress made in this technology has, in fact, forced MS to keep pace with the needs of modern proteomics and integrate other analytical methods successfully to tackle various challenging tasks. Despite the unsurpassed versatility of its application in both basic and applied research, for more than half a century the limitations on the ionization side have excluded the use of MS to investigate large molecules, including proteins. Since its concealment and first use (Thomson, 1913) in the early twentieth century, MS ionization methods have been limited to nonpolar molecules, which are thermally stable and volatile.

Later, new, sensitive, and more versatile ionization techniques were introduced, making it possible to generate ions of a wide range of molecular masses. The first breakthrough consisted of the introduction of the ^{252}Cf plasma desorption ionization method (Macfarlane and Togerson, 1976), suitable for the ionization of nonvolatile, thermally unstable biomolecules such as proteins and peptides. In this approach, the radioactive element gives two fission fragments in the mass range 80 to 160 Da; one of these hits the sample, producing both volatilization and ionization, and both positive and negative ions emerging from the opposite side are accelerated onto a time-of-flight analyzer and detected. The second fragment travels in the opposite direction, to be detected and used to establish a precise zero-time marker. This technique permitted *m/z* measurements in the kilodalton range, with an order of accuracy of 10^{-4} Da. Despite these positive features, its use remained very limited for a number of reasons, including the short availability of ^{252}Cf, the need of a well-shielded ion source to ensure the safety of operation, and the early limitations in TOF technology.

Almost contemporarily, the more versatile fast-atom ion bombardment (FAB) technique was introduced (Barber et al., 1981), coupled to a variety of MS configurations. In this "static" approach, the analyte was dissolved into an excess of liquid matrices (usually, glycerol), deposited onto a probe tip, and then bombarded to produce secondary ions by a kiloelectron-volt inert-gas atomic beam or an ion (Cs^+) beam. In the "dynamic" approach developed later by Ito, the generation of a thin film of the sample–glycerol solution over a large surface area greatly reduced the chemical background noise and the formation of aggregates, which negatively affected the sensitivity of the detection (Ito et al., 1985). A more recent version of the dynamic FAB is continuous-flow (CF-FAB), in which the flow of solvent supplied by a microbore capillary provides cleansing of the bombardment surface area (Caprioli et al., 1986).

Since its introduction, this ionization technique has long been the key player in MS-based proteome analysis and has made a substantial contribution to the development of modern proteomics. FAB–MS and FAB–MS–MS techniques were commonly run on sector machines as a complementary approach to Edman chemistry in the analysis of both native and recombinant proteins, given the capability to provide sequence information and detect amino acid substitution as a result of abnormalities in translation, insertion, deletion, or mutation. The major fragmentation pathways observed for peptide ions generated by FAB involve the cleavage of three types of bonds in the peptide's backbone: —CHR— CO—, —CO—NH—, and —NH—CHR— (Dass and Desiderio, 1987; Seki et al., 1985; Williams et al., 1981). The fragmentation rules deducted and associated nomenclature are still used for MS–MS data interpretation (Biemann, 1988; Roepstorff and Folman, 1984).

For over a decade, MS and tandem MS based on FAB ionization have dominated the field of protein and peptide analysis; remarkably, the studies and developments matured within this technique have paved the way to the introduction of electrospray ionization (ESI) (Yamashita and Fenn, 1984) and matrix-assisted laser desorption ionization (MALDI) (Tanaka et al., 1988), which are the workhorses of current macromolecule analysis.

Gas-phase ions are formed into an ESI source from charged droplets, and the mechanism of transferring a macromolecular ion to the gas phase has been the subject of many speculations (Dole et al., 1968; Thomson and Iribarne, 1979). The capability of electrospray for proteins was demonstrated early (Meng et al., 1988);

since then, ESI has been the method of choice for protein and peptide analysis by LC–MS and LC–MS–MS. To address the needs of modern proteomics, considerable effort has been expended to enhance the sensitivity of the technique. The coupling of nanoscale LC to a new version of the ESI source realized with gold-coated capillaries allowed detection in the femtomole range and thus analysis of low-abundant peptides in complex mixtures such as cell lysates. At the time of its introduction (Wilm and Mann, 1996), nano-ESI was operated in the direct infusion mode; samples were delivered to the ion source via an injection valve, without prior separation, and therefore needed to be free of salts and other additives. In current approaches, complex peptide mixtures are typically delivered to the interface from a reversed-phase column and are therefore separated according to the different degree of hydrophobicity prior to ionization, analysis, and detection.

Meantime, new analyzers are replacing the older-sector machines, used in conjunction with the new simpler, more sensitive, and more versatile ionization methods: ion trapping instruments, quadrupoles, and TOF. Later, a variety of hybrid instruments became available commercially, providing the capability for high resolution, high sensitivity, and high mass accuracy over a wide dynamic range; among these, ion mobility TOF, quadrupole time of flight, ion trap–TOF, and linear ion trap–Fourier transform ion cyclotron resonance. Without going into a theoretical discussion, we can just notice how the unprecedented accuracy, speed, and resolution of the most advanced devices currently marketed (containing, e.g., a reflector–TOF, an Orbitrap, or a FT-ICR analyzer) have combined to bring mass spectrometry to a central role in present-day proteome research, overcoming most of the limitations associated with two-dimensional gel electrophoresis (2D GE) and Edman degradation techniques, which represent classical approaches to amino acid sequencing.

At the same time, adequate front-end separation techniques are needed in any proteomic analysis prior to detection, to avoid peak overlap and signal suppression in the case of low-abundant proteins. Traditionally, 2D GE followed by MS identification has been the core equipment in this field, given its excellent resolving power for intact proteins ($\approx$2000). In a classical gel-based experiment, high orthogonality can in fact be achieved for complex mixtures of proteins, which are separated according to their isoelectric point (pI) value and molecular weight. Excised proteins are then subjected to enzymatic digestion (usually with trypsin) into peptides, and subsequently analyzed by MS, usually via an ESI or MALDI interface (Aebersold and Mann, 2003; Delahunty and Yates, 2005; Eng et al., 1994; Peng et al., 2003; Perkins et al., 1999; Simpson, 2003).

Major drawbacks of these methods consist of difficulty of automation, low accessibility of membrane-bound proteins, problematic detection of proteins characterized by high molecular weight, high pI, strong hydrophobicity, or low abundance (Beranova-Giorgianni, 2003; Gorg et al., 2004; Rabilloud, 2002; Reinders et al., 2006). On the other hand, different separation techniques, such as capillary electrophoresis or liquid chromatography, could not long rival the resolution power afforded by gel electrophoresis. Over the last decade, a lot of effort has been put into the development of ultrahigh-efficiency LC methodologies, and many advances have been made in both HPLC column phase and technology, with the commercialization of new stationary phases with perfusive packing, shell-packed particles, inorganic–organic hybrids, monoliths, high-temperature columns, sub-2-μm particles, nanocolumns, and fast HPLC columns (Cavazzini et al., 2007; Gritti and Suiochon, 2007; Gritti et al., 2007; Marchetti et al., 2007; Meiring et al., 2002; Premstaller et al., 2005; Way, 2007; Way and Campbell, 2007). The needs for reproducibility and for robustness of chromatographic separations were encountered at the same time, with improvements in high-purity silica and novel bonding chemistries, resulting in more efficient and reliable HPLC columns, with less peak tailing and improved batch-to-batch reproducibility (Guillarme et al., 2004).

Higher throughput, relative speed, capability of quantitation, easiness of full automation, and the likelihood of straightforward hyphenation to mass spectrometry can easily explain why liquid-based separation techniques have gained ever wider acceptance as a convenient alternative to the more laborious, less comprehensive gel-based methods (Premstaller et al., 2001).

The more widespread method currently used for identifying proteins using MS data is the *bottom-up approach*, which involves proteolitic digestion of the protein mixture, followed by chromatographic separation and MS or MS–MS analysis. In this strategy, separation is therefore performed at the peptide level rather than at the protein level as it is in 2D GE; separate peptides are then used for database searching, through computer algorithms (such as MASCOT, Sonar, SEQUEST), and the matches are used to predict the identity and sequence of the protein. Peptide mass fingerprinting (PMF) uses data obtained from MALDI–MS analysis of intact peptides, while tandem MS data are obtained by further fragmentation of the peptides by collision-induced dissociation (CID). The workflow of an LC–MS–MS-based shotgun proteomics strategy is shown in Figure 9.1.

The *shotgun strategy* relies on both LC separation techniques and MS detectors; in this regard it has constituted a major driving force toward the development of multidimensional systems capable of affording enhanced resolution for complex proteome samples, and also enabling probing of low-abundant species. Far from being comprehensive of all the LC × LC–MS methods devoted to proteome analysis since the pioneering work of Opiteck's group (Opiteck et al., 1997a), Table 9.1 lists some significant applications in this field, which permits an overall discussion of the related features.

First is the need for efficient and highly resolving separations. Although intact proteins are difficult to handle, and their separation by HPLC and subsequent MS–MS detection is challenging, their digestion with a proteolitic enzyme (usually, trypsin) increases sample complexity dramatically. Assuming that trypsin yields, as an average, 30 peptides per protein and taking plasma as an example, which contains approximately 30,000 different proteins, the proteolitic digestion will, in fact, result in as many as 900,000 peptides, not taking into account any processing or modification. Highly efficient separations in proteomics have recently been reviewed by Sandra et al. (2008, 2009). Yet unlike proteins, which can be very diverse in their chemical nature, peptides show

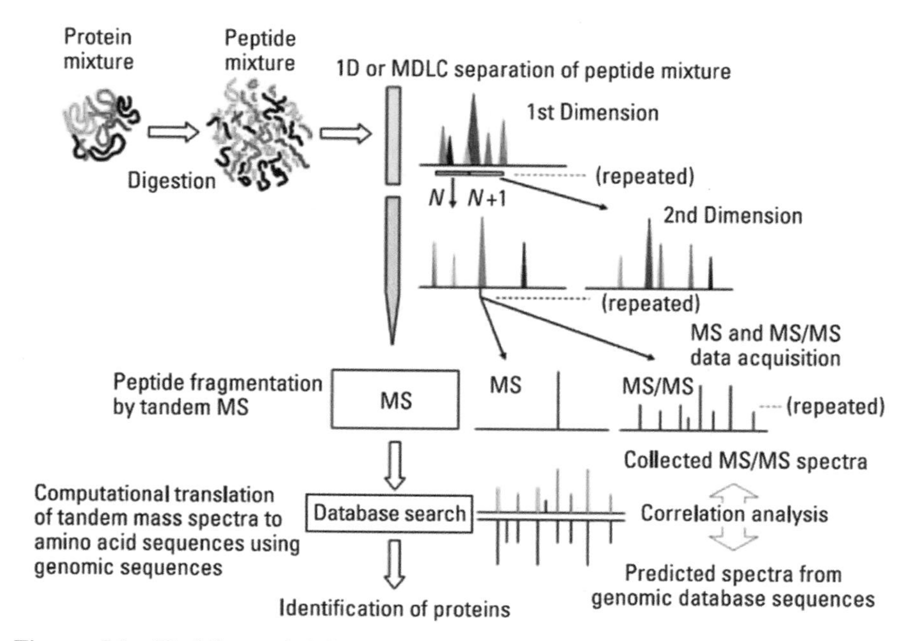

Figure 9.1 Workflow of LC–MS–MS-based shotgun proteomics strategy. [From Motoyama and Yates JR (2008), with permission. Copyright © 2008 by The American Chemical Society.]

uniform behavior, and thus setting suitable experimental conditions for LC–MS analysis is simpler in "peptidomics" than in "proteomics." The drawback of such an approach is the limited peak capacity affordable by any single-mode chromatographic separation; the need for multidimensional (comprehensive) techniques comes with different reasons. Better fractionation of the peptides according to their charge and hydrophobicity will improve ionization efficiency and reduce the complexity of the sample entering the mass spectrometer. Furthermore, resolving low-abundant peptides from more abundant species will ultimately result in an overall increase in dynamic range, addressing typical MS detection issues such as ion suppression and undersampling. This will definitely be beneficial for data-dependent data acquisition in a finite cycle time of the tandem MS process, leading to better representation of the entire protein.

In 2D GE, high orthogonality is provided by the various separation mechanisms: that is, separation according to isoelectric point (charge) and separation by size in the first and second dimensions, respectively. Comparable resolving power can be gained by comprehensive chromatographic approaches in which the two stationary phases involve different selectivities. As can be noted in Table 9.1, different separation modes have been exploited to deliver a wide range of selectivities in the first dimension (D1), employing ion-exchange chromatography (IEX), namely strong-cation exchange (SCX)

TABLE 9.1 Comprehensive 2D (LC x LC-MS) Separations of Proteins and Peptides

Ref.	Type of Interface	Column Packing and Dimensions (Length x i.d., mm, d_p, μm); Flow rate; Elution Mode[a]		Detection	MS and MS-MS Parameters	Application
		First Dimension, D1	Second Dimension, D2			
[1]	Two-position 8-port valve	IEC 125 × 0.75, 5 μm; 10 μL/min; GP	RP (R2/H) 100 × 0.5; 50 μL/min; GP	UV, ESI-MS	1000–2000 *m/z*, 0.1 amu steps, 0.2 ms dwell; 0.5 Hz	Mixture of standard proteins and *Escherichia coli* cell lysate
[2]	Two 2-position 4-port valves; connected to alternative D2 columns	Six SEC 300 × 7.8; 1 mL/min, 0–40 min run; 100 μL/min, 40–140 min run; IP	Two parallel C_{18} 33 × 4.6, non-porous; 1 mL/min; GP	UV, ESI-MS	600–2600 *m/z*, 0.125-amu steps; 0.190 ms dwell, 0.33 Hz	Ovoalbumin and serum albumin tryptic digests
[3]	Two-position 4-port valve; connected to alternative D2 columns	SEC 300 × 7.8; (a) 8 columns, 5 μm; 250 μL/min; GP (b) 12 columns, 8 μm; 150 μL/min; GP	Two RP (R2/H) (a) 33 × 2.1 (b) 100 × 2.1; 1.5 mL/min; GP	UV, ESI-MS, MALDI-TOF-MS	600–2600 *m/z*, 0.125-amu steps; 0.190 ms dwell, 0.33 Hz	Native and non-native proteins in *Escherichia coli* cell lysate
[4]	No interface; directly coupled-columns	Biphasic 140 × 0.1, 5 μm; SCX: 40 mm capillary inlet, C_{18}: 100 mm distal segment; 0.15–0.25 μL/min; GP		ESI-MS-MS	400–1400 *m/z*, CID 35% (30 ms activation time), repeat count and duration: 2 and 0.5 min	Peptides and proteins in *Saccharomyces cerevisiae* lysate

[5]	Two-position 10-port valve; connected to alternative D2 columns	IEC 35 × 4.6, 5 μm non-porous; 1 mL/min; GP	Four C_{18} 14 × 4.6, 5 μm non-porous; 2.5 mL/min; GP	UV, MALDI-TOF-MS	Reflectror and linear mode: 20 kV voltage, 150 ns delay time	Human emofiltrate and human fetal fibroblast cell lysate
[6]	Two-position 10-port valve; connected to alternative D2 columns	IEC 35 × 4.6; 400 μL/min; GP	(a) Two C_{18} 33 × 4.6, 1.5 μm; (b) $C_4$50 × 2.1, 3.5 μm; 0.5 mL/min; GP	UV, ESI-TOF-MS	3000–46,000 *m/z*; 65 μs flight time, MCP at 2700V	Mixture of standard proteins and yeast ribosomial fraction
[7]	Two-position 10-port valve	IEC 50 × 2.1, 5 μm; 50 μL/min; GP	C_{18} monolithic (a) 25 × 4.6; 5 mL/min (to UV) (b) 100 × 0.1; 3 μL/min (to MS); GP	UV, ESI-TOF-MS	400–2000 *m/z*; 2 Hz	Bovine serum albumin tryptic digest
[8]	Three 2-position 10-port valves; connected to alternative D2 columns	IEC 35 × 4.6, 5 μm non-porous; 0.5 mL/min; GP	Four parallel C_{18} 14 × 4.6, 1.5 μm non-porous; 2 mL/min; GP	UV, MALDI-TOF-MS	1300–4100 *m/z*; ion source voltage: 25 kV	Human emofiltrate tryptic digest
[9]	Two 2-position 10-port valves	IEC 150 × 0.3; 6 μL/min; GP	C_{18} 1500 × 0.075, 5 μm; 300 nL/min; GP	ESI-MS-MS	300–2000 *m/z*	Human tissue proteins and peptides

(*continued overleaf*)

TABLE 9.1 (*Continued*)

Ref.	Type of Interface	Column Packing and Dimensions (Length x i.d., mm, d_p, μm); Flow rate; Elution Mode[a]		Detection	MS and MS-MS Parameters	Application
		First Dimension, D1	Second Dimension, D2			
[10]	Two 2-position 10-port valves	IEC 150 × 0.3; 15 μL/min; IP	C_{18} 1500 × 0.075; 200 nL/min; GP	ESI-TOF-MS-MS	90–100000 *m/z*; CID energy: 51.2 V	Human hepatocellular carcinoma proteins and peptides
[11]	Two 10-port column selectors system with SPE columns	ZIC-HILIC 150 × 0.3, 5 μm; 70 μL/min; IP	PLRP-S C_{18} 150 × 0.3, 3 μm; 5 μL/min; GP	UV, ESI-TOF-MS	200–1300 *m/z*	Arg-bradykinin and bradykinin in rat muscle tissue
[12]	Two-position 6-port valve	IEC 50 × 0.5, 5 μm; 12 μL/min; GP	SB- C_{18} 150 × 0.5, 3.5 μm; 20 μL/min; GP	ESI-TOF-MS	350–1250 *m/z*; scan time: 0.88 s, interscan delay: 0.1 s, MCP at 2700V	Recombinant proteins tryptic digests

[1] Opiteck et al., 1997a; [2] Opiteck et al., 1997b; [3] Opiteck et al., 1998; [4] Wolters et al., 2001; [5] Wagner et al., 2002; [6] Liu et al., 2002; [7] Kimura et al., 2004; [8] Machtejevas et al., 2004; [9] Mitulovic et al., 2004; [10] Wang et al., 2005; [11] Wilson et al., 2007; [12] Kajdan et al., 2008.

[a] IP: isocratic program; GP: gradient program.

or strong-anion exchange (SAX), size-exclusion chromatography (SEC), or hydrophilic interaction liquid chromatography (HILIC). On the other hand, the second dimension (D2) usually consists of reversed-phase LC, due to its high compatibility to linkage to MS detection. Most common RP stationary phases are octadecylsilica (ODS), which are commercially available for this type of application with a choice of length, internal diameter, particle size and pore size, pH stability, and hydrophobicity. Hydrophobic interactions are responsible for peptide separation under gradient conditions, typically employing acetonitrile (ACN) as the organic modifier, and trifluoroacetic acid (TFA) or formic acid (FA) as ion-pair reagents (Hancock et al., 1978). The use of low percentages of organic solvent allows effective on-line desalting and concentration of peptides at the same time, providing excellent compatibility with ESI and MS detection.

Various strategies have been used by researchers in designing multidimensional LC systems for proteomic studies, in which separations need to be comprehensive to maximize the separation of all components. A fraction can be transferred from the first to the second dimension in the off- or online mode; both approaches are in widespread use, and both have advantages and disadvantages. In offline methods, the fraction is collected from the first dimension via a fraction collector and then reinjected into the second-dimension column. These approaches are more flexible than online separation and offer greater flexibility of separation modes in terms of mobile phases and buffer compatibility, time, and duration. Given the possibility to optimize each separation dimension independently, they can also afford an overall increase in the separation power. Disadvantages include the low likelihood for automation and those of potential loss of sample and artifact formation. Discontinuous offline methods using fraction collection have employed multidimensional approaches advantageously when nonvolatile buffers such as NaCl or KCl or high concentrations of organic solvent were needed to elute peptides in the first (SCX) dimension prior to a reversed-phase second dimension. They have also been used in the COFRADIC (combined fractionaction diagonal chromatography) procedure, which uses two RPLC dimensions, or whenever there is a need for isolation of specific classes of peptides (Gevaert et al., 2003; Gygi et al., 1999; Wehr, 2003). In the online approach, analyte transfer between the two separation dimensions is fully automated and requires no flow interruption. Online approaches are more convenient when dealing with limited sample amounts, since losses can be minimized, and have the advantage of easy automation. However, they are more technically challenging and pose more stringent requirements. Solvents used in the two dimensions must be compatible to avoid salt precipitation; in addition, the eluting strength of the mobile phase used for the first-dimension chromatography of the peptides must be weaker than that used in the second dimension.

A further distinction can be made between directly coupled column methods and column-switching methods. In the first approach, analysts use two columns with orthogonal separation selectivities, typically based on SCX and RP materials, which are packed in tandem into a single capillary. Fractions are eluted from the first column by applying a series of pulsed steps of buffer with increasing ionic strength. At each pulse, peptides are eluted from the RP second-dimension column using a linear acetonitrile gradient into the mass spectrometer. A pulled tip placed at the outlet end serves as the electrospray ionization needle, with minimum delay volume. An example of the first method is the MudPIT (multidimensional protein identification technology) developed by Yates and co-workers (Link et al., 1999; Washburn et al., 2001; Wolters et al., 2001b), who identified 75 or 78 predicted proteins in a *Saccharomyces cerevisiae* 80S ribosome tryptic digest, compared to 56 and 64 proteins identified by 1D RPLC–tandem MS and 2D-PAGE, respectively. A biphasic 140 × 0.1 mm fused-silica capillary was used, packed with 5-μm particles of a SCX phase 40-mm length at the capillary inlet, and C_{18} particles of the same size for the distal 100-mm segment. A complex gradient profile was employed for peptide elution from the SCX material, consisting of 15 steps of increasing ammonium acetate concentration, each followed by a reversed-phase gradient. The use of a volatile salt allowed continuous electrospraying and subsequent MS–MS analysis. The authors calculated a theoretical peak capacity of 23,000, given by the product of the chromatographic peak capacity of the biphasic system (3240) and the number of MS–MS spectra (seven) acquired by the 3D ion trap over a chromatographic peak duration of 25 s (Link et al., 1999; Wolters et al., 2001b).

Later, this technique was used in the analysis of the *S. cerevisiae* proteome. A database search performed by the SEQUEST algorithm of the tandem MS spectra generated from the proteome lysate enabled the identification of 5440 peptides, originating from 1484 proteins (Washburn et al., 2001). Significant advantages were demonstrated over 2D GE, detecting proteins with extremes in isoelectric point (p*I* <4.3 and p*I* >10), molecular weight (MW<10 kDa and MW>180 kDa), and hydrophobicity. A dynamic range of 10,000 was demonstrated, including both peripheral and membrane proteins, as well as low-abundant species, thanks to the higher loading capacity of the SCX column. The method implemented was therefore less biased against low-copy-number proteins then was 2D GE.

Column-switching techniques probably represent the most commonly used approaches to comprehensive 2D separations in the proteomic field. Even at the cost of higher complexity, comprehensive 2D LC systems based on column-switching techniques undoubtedly allow for greater flexibility. Different configurations have been developed, based on two-position ten-, eight-, or six-port switching valves equipped with two storage loops of identical volumes. The two loops are filled alternately with the effluent from the first column, and their size is fixed by the mobile-phase quantity per sampling period. This period must, in turn, be equal to the time available for analysis in the second dimension, since the content of the loops is reinjected continuously into the secondary column.

Both chromatographic dimensions can be operated under isocratic or gradient conditions. The analysis time is of minor concern in the first dimension, which is implemented to achieve the maximum resolution for a given application; with this aim, more columns can be coupled in series, to add extra length to the stationary phase. The use of narrow- or microbore primary columns operated at a low rate is helpful to alleviate problems of mobile-phase incompatibility between the two dimensions, to reduce band broadening, and to allow effective focusing. Wider-bore columns have higher loading capacities and require that the flow be split prior to the interface, to be compatible with the short columns generally used as the second dimension; alternatively, they can be operated at less-than-optimum flow rates. Drawbacks of such choices include waste of sample components in the first case and loss of efficiency in the second case. In contrast, miniaturization of the system is highly beneficial for subsequent MS detection; this holds particularly true for second-dimension columns, which can have internal diameters as small as 75 μm or even less. They are usually operated under gradient conditions, to minimize wrap-around phenomena and peak broadening. This requires column conditioning steps at the end of each modulation cycle and results in higher background detection noise, as the mobile phase composition is varied rapidly.

Jorgenson and co-workers were pioneers in this field, also reporting the first comprehensive 2D LC system coupled to MS detection, which yielded online molecular weight information regarding the protein mixtures analyzed (Opiteck et al., 1997a). In this respect, the mass spectrometer represents a third, added dimension to the 2D system, being capable of detecting coeluted peaks, which are not resolved chromatographically; in this respect, it alleviates the requirements as to LC resolution. They used a 0.75 mm i.d. × 125 mm length SCX column in the first dimension, coupled to a secondary 0.5 mm i.d. × 100 mm RP column by means of a two-position 10-port valve. Both dimensions were operated under gradient conditions at a flow rate of 10 μL/min in the first dimension and 50 μL/min in the second dimension. A 150 μm i.d. × 360 μm o.d. fused-silica capillary was used for on-capillary UV detection, while a 29 μm i.d. × 150 μm o.d. capillary supplied the electrospray interface. Coupling ion-exchange and reversed-phase chromatography provides complementary selectivities that are similar to those of 2D gel electrophoresis; in fact, IEX separation is roughly equivalent to isoelectric focusing, whereas in RP chromatography, retention tends to increase with increasing molecular weight, affording a quasi-molecular size separation. The system therefore comes close to Giddings' criteria for ideal 2D separation, since the resolving power of the coupled modes will approximate the product of the peak capacities of the individual modes (Giddings, 1987). Another advantage is that of solvent compatibility between the two dimensions, since analytes are eluted from IEX with increasing concentrations of a neutral salt in aqueous buffers, which are weak solvents for RP chromatography. The use of RP as a secondary chromatographic dimension further allows the concentrated salt and urea IEX mobile phase to be coupled directly to the mass spectrometer. Finally, both chromatographic modes were implemented on a microcolumn scale, gaining better efficiency and less sample dilution, the latter being highly beneficial for ESI–MS, which is a concentration-sensitive detection. The 2D LC–MS system implemented with a calculated peak capacity of 2500 allowed the complete separation of a mixture of 10 standard proteins without prior knowledge of their molecular weight, hydrophobicity, or isoelectric point. Molecular weights of the proteins were obtained online by Hypermass reconstruction of the charge envelope of each chromatographic peak. The mass spectrometer was presented with as little as 3.2 pmol of analyte but still having a lower acquisition rate than that for UV (i.e., 0.5 vs. 4 Hz).

Later, the same research group developed a novel LC–LC interface using two RPLC columns in parallel, rather than storage loops, to join the two chromatographic dimensions. The first dimension of the comprehensive 2D LC–MS system developed consisted of six serially coupled SEC columns (total length of 2.4 m) of 7.8 mm i.d., to obtain enhanced resolution for first-dimension separation. From the primary set of columns the effluent was alternately transferred to two short polystyrene divinylbenzene C_{18} columns (nonporous), by means of a dual four-port switching valve, to obtain fast D2 analysis (Opiteck et al., 1997b). The use of conventional-size columns made the system rugged and easy to implement, giving the good column-to-column reproducibility that is required to avoid any shifts in retention times in the 2D plots (less than 3 s in this case). Since the columns in the second dimension perform analyses of consecutive fractions in parallel, a stringent requirement of this type of system is that the two secondary columns deliver identical efficiency and retention behavior in order for the two chromatograms to be matched to produce a contour plot. The interfacing valve was switched every 4 min; during that time the peptides eluted from the SEC phase were trapped onto the RP columns, since the mobile phase lacked an organic modifier. The system was used for the separation, detection, and identification of the peptide fragments resulting from tryptic digestion of ovoalbumin and serum albumin; the effluent could be introduced directly into ESI–MS after splitting the flow. Nevertheless, the inadequate scanning rate of the MS detection (3 s for a complete scan) caused a fair loss in resolution compared to the higher sampling rate of the UV detector (2 Hz). While gradient elution cannot be employed to improve separation in SEC chromatography, column coupling proved to be an effective tool to increase the chromatography efficiency in the first dimension. Furthermore, it allows an estimation of protein molecular weights and is compatible with the high salt and buffer concentrations involved in the fermentation process, which would be troublesome in IEX operation. In another application, identical separation modes were employed in the analysis of a mixture of native and non-native from an *Escherichia coli* cell lysate (Opiteck et al., 1998). Following chromatographic separation (eight or 12 SEC serially coupled columns in the first dimension), fractions were deposited into six 96-well polypropylene plates (576 in total) via a fraction collector, and subjected to MALDI–TOF MS or ESI–MS, depending on sample concentration. The 2D LC–MS system was fully automated and provided quick isolation and separation and accurate (2 to 0.02%) molecular weight information on overexpressed proteins from the bacterial cell lysate. The sensitivity was comparable to that of the Edman sequencing technique, which followed to provide unambiguous sequence information.

In a later report, four second-dimension D2 columns were employed in parallel for online fractionation and separation in the IEX × RPLC separation of human emofiltrate and human fetal fibroblast cell lysate (Wagner et al., 2002). The system was totally automated, rugged, and showed good reproducibility in the resolution of about 1000 peaks within a 96-min analysis time, while avoiding sample losses during offline collection or handling. A sample preparation step was integrated in the setup using silica-based restricted-access materials (RAMs) for the separation of target analytes from the matrix. Following this size-selective fractionation step, low-molecular-weight proteins and peptides (>20 kDa) were subjected to cation- or anion-exchange chromatography. After valve switching (every 4 min), the first-dimension effluent was loaded on one RP column, the analyses of the two previous fractions were carried out on two other RP columns, and the fourth column was regenerated. Peaks were collected and analyzed by MALDI–TOF MS, revealing a molecular mass range of the peptides in the *m/z* range 80 to 5000.

Parallel second-dimension columns were later employed by Unger and coworkers, who developed a comprehensive two-dimensional HPLC system with integrated sample cleanup for the high-resolution separation and identification of peptides below 20 kDa from human hemofiltrate, which is one of the most complex natural peptide samples (Machtejevats et al., 2004).

After a size-selective sample fractionation step using silica-based materials with ion-exchange functionalities (SCX–RAM), peptides were separated using an SCX column with a conventional i.d. as the first dimension, interfaced by the column-switching technique (three two-position 10-port valves) to four parallel short RP columns. After a total analysis time of 96 min, the molecular weights of the peptides were obtained by MALDI–TOF MS, and their amino acid sequence was obtained using the Edman degradation technique. High sensitivity was obtained, making it possible to identify single components at concentrations below 1 pmol/L. In this respect, RAM columns are ideally suited to prefractionation of small biopolymers, since they make it possible to boost the sensitivity for low-abundant components by loading high amounts of sample. On the other hand, the use of capillary columns in the second dimension leads to better MS sensitivity and less solvent consumption than those of conventional-size columns.

Tanaka's research group (Kimura et al., 2004) performed D2-HPLC on a short monolithic C_{18} capillary (100-μm) column with split-flow injection to afford high sensitivity for ESI–MS detection of tryptic digested bovine serum albumin sample, in combination with a first-dimension ion-exchange column based on polymer beads (50 mm length × 2.1 mm i.d.). The split ratio was set at 3/2000 with a solvent delivery rate at the pump of 2 mL/min and a flow rate in the capillary column of 3.0 μL/min. Since the eluent was split after the injector, a very high flow rate was employed at the pump compared to the flow in the column, resulting in very little delay in gradient. A second RP column was used for D2-HPLC followed by UV detection, 25 mm length × 4.6 mm i.d., operated at 5.0 mL/min, and the system was operated as follows: the loop at the D2-HPLC was loaded with the effluent of the first-dimension column at 50 μL/min for 1 min 58 s; then the injection valve was turned to inject the 100-μL fraction for 2 s into the D2-HPLC at 5 mL/min, and turned back for loading for the next 1 min 58 s, resulting in fractionation in the first dimension every 2 min. With the capillary D2 column, the system was operated as follows: the loop at the HPLC was loaded with the effluent of the first-dimension column at 50 μL/min for 3 min 53 s; then the injection valve was switched to inject the 200-μL fraction (with a split ratio of 3/2000) for 7 s into the second-dimension column at 3 μL/min, and turned back for loading for the next 3 min 53 s. This resulted in fractionation in the first dimension every 4 min. In ESI–TOF MS detection, a *m/z* range of 400 to 2000 was scanned in 0.5 s for data collection, applying a flow rate of nebulizer gas (nitrogen) of 4.0 mL/min, to facilitate ionization (the ESI probe temperature was 140°C). Peptides were identified by searching the MASCOT database.

Due to the very high scanning rate allowed, TOF detection is the optimum choice for comprehensive LC, since a higher acquisition rate results in improved peak sampling and resolution. Electrospray is used more frequently for online analysis of separated samples, whereas MALDI is employed to analyze off-line collected fractions. The SEC × RPLC comprehensive 2D setup developed by Liu and co-workers allowed both online analyses of intact protein masses, using ESI–TOF MS detection, and off-line fraction collections for MALDI peptide mass fingerprinting for confirming the identifications (Liu et al., 2002). Separation in the first dimension was performed on a short SCX column (35 mm length × 4.6 mm i.d.); background contaminants from the urea used in buffers were removed through a C_{18} column. Reversed-phase chromatography was performed in the second dimension using either a C_{18} (33 mm length × 4.6 mm i.d.), or a C_4 (50 mm length × 2.1 mm i.d.) column, both operated under gradient conditions. A schematic representation of the experimental system is depicted in Figure 9.2. Two major benefits over a single second-dimension column configuration were a significant reduction in the overall analysis time (since efficient off-line loading and reconditioning steps are included) and the capability to distinguish between elution problems occurring in the two dimensions (i.e., fraction splitting in the first dimension and carryover in the second dimension, respectively). The system was used in the analysis of both standard proteins and of an enriched yeast ribosomal fraction containing around 100 proteins. In 2D LC–MS systems, a careful choice of the proper component connections is critical to minimize postcolumn band spreading, resulting in the loss of chromatographic resolution. Comparable peak widths were observed by the authors in the UV chromatogram (280 nm) and the ESI–TOF MS TIC. Furthermore, optimizing diversion valve timing allows the diversion of non-volatile components from the first IEX dimension prior to introducing RP effluent into the MS source, minimizing adduct formation and reduction in sensitivity. In this work, maximum relative signal was roughly higher for the raw mass spectra and more than higher for the deconvoluted spectra.

Deconvoluted masses of sample components were determined using a maximum entropy (MaxEnt1) deconvolution algorithm from neutral mass spectra acquired in the *m/z* range 3000 to 46,000 Da; ribosomal protein assignment was accomplished by comparing the masses observed to those predicted from the yeast genome sequence, assuming either loss or retention of the initiating

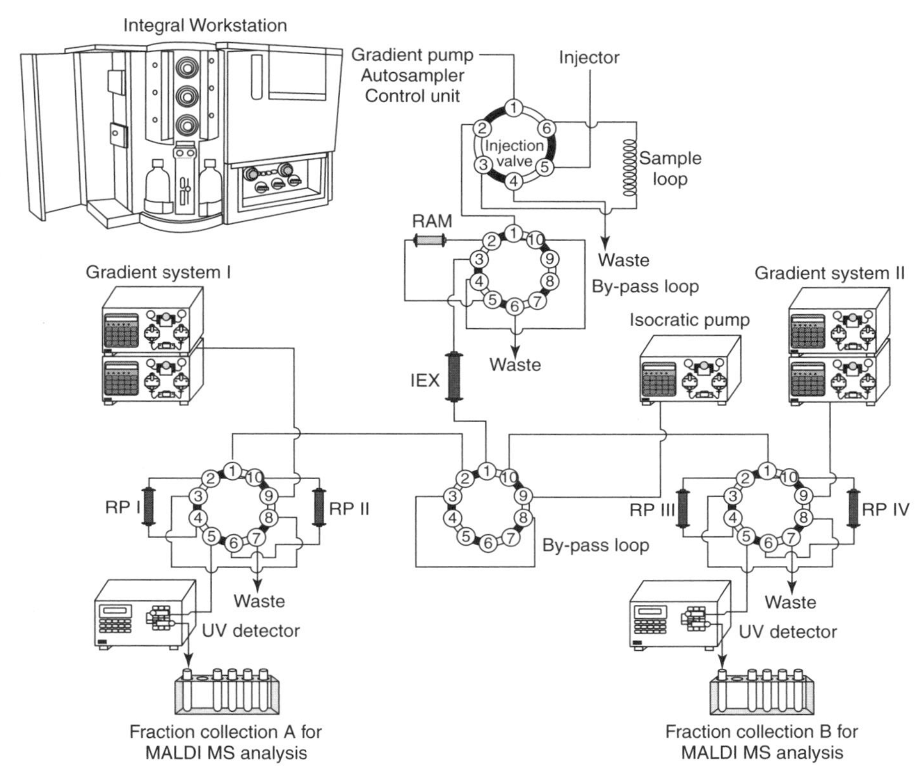

Figure 9.2 Online comprehensive two-dimensional HPLC–MS system, including an integrated sample preparation step. [From Wagner et al. (2002), with permission. Copyright © 2002 by The American Chemical Society.]

methionine and the presence or absence of N-acetylation. Additionally, identification of processed and post-translationally modified ribosomal subunits was based mainly on data obtained from the Yeast Protein Database. Assignment of nearly 80% of ribosomal subunits was accomplished at a load of 0.4 to 0.9 pmol per component, with a split of 50 to 100 fmol to the MS; an average mass measurement accuracy of 50 ppm was achieved using external calibration. Molecular mass identification was confirmed afterward by performing MALDI–TOF MS analysis of the fractions collected, demonstrating the interplay between intact and peptide digest information. Peptide mass fingerprinting yielded confirmation of all the subunits assigned, among them both isoforms of the ribosomal proteins rpL31 and rpL33, and also afforded high sequence coverage for the rpL20A/B protein, for which intact mass data have not been obtained.

A comprehensive 2D LC–ESI–TOF MS system was employed successfully by Cortes and co-workers to resolve structural alterations and/or post-translational modification of proteolytic peptides from recombinant proteins (Kajdan et al.,

2008). The chromatographic separation coupled SCX as the first dimension (50 mm length × 0.5 mm i.d. column) to RP (150 mm length × 0.5 mm i.d. column) as the second dimension using a two-position six-port valve as the interface. When the switching valve was in position A, the effluent from the first dimension was directed onto a C_{18} trapping column (0.5 mm length × 2 mm i.d.), then out to waste, while the flow from the second dimension pump was directed through the C_{18} separation column and into the second-dimension detector. When the valve was in position B, the first-dimension effluent bypassed the trap column and was diverted directly to waste, and the second-dimension flow was directed through the trap column and onto the C_{18} column for separation of the peptides. Three sets of recombinant proteins were analyzed, demonstrating the high level of efficiency of the orthogonal separation modes, in conjunction with high mass specificity and sensitivity. Bovine serum albumin (BSA) and its oxidized variant, ribonuclease A and its N-linked glycosylated analog, ribonuclease B, α-casein, and dephosphorylated α-casein were characterized by means of tryptic mapping and 2D LC–MS, achieving sequence coverage up to 99%. The increased performance of the 2D system is evident from Figure 9.3, showing the TIC obtained for optimized 1D RP separation of BSA tryptic digest compared to the optimized 2D LC separation of the same sample. The corresponding mass spectra of peptide fragments T47–48 and T57–58, which coelute in the 1D analysis and are separated in the 2D analysis, are shown as well. Remarkably, a unique set of conditions was suitable for the characterization of all the proteins, employing a spike gradient approach for the first dimension, for better recovery of the peptides (Figure 9.4).

Replicate analysis of BSA digests yielded good reproducibility of the entire system, with a relative standard deviation (RSD) for retention times of 0.007%. Figure 9.5 shows the 2D separation of RNase A, consisting of 124 amino acids (13,682 Da mass), and RNase B, which differs for an N-linked sugar having the heterogeneous structure GlcNAc2Man5–9 attached to asparagine 34 located in the N-linked Asn–Xaa–(Ser/Thr) motif (where Xaa can be any amino acid except proline). Peptides from both proteins are identical, with the exception of two glycopeptides eluting at 10 and 30 mM in RNase B which have an identical glycosilation pattern, with prominent ions separated by *m/z* 162, characteristic of the presence of repeating hexose units. The peptide eluting at 10 mM ammonium formate has no missed cleavage (peptide sequence: NLTK), whereas the one eluting at 30 mM ammonium formate has one missed cleavage (peptide sequence: SRNLTK). Multiple glycosylation patterns on the same peptide sequence result in corresponding chromatographic peaks broader than those of regular peptides. The full-length glycopattern was identified as (Man)9(GlcNAc)2, while T10 fragment observed only in the RNase A 2D contour plot was identified as deamidated. Further analysis by MS–MS revealed that Asn67 was deamidated, probably as an effect of the high temperature, pH, and/or ionic strength. Replacing the empty storage loops of the sampling interface with loops packed with stationary

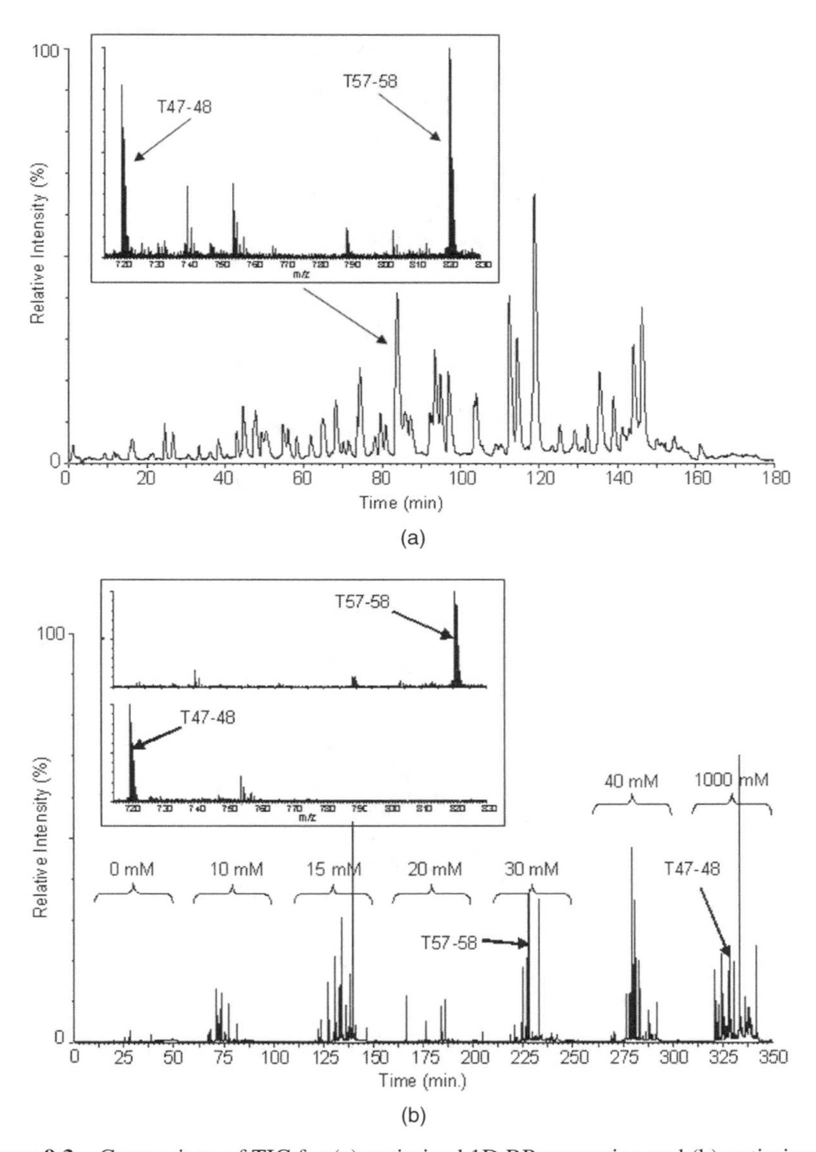

Figure 9.3 Comparison of TIC for (a) optimized 1D RP separation and (b) optimized 2D LC separation of bovine serum albumin tryptic digests. The insets show the corresponding mass spectra of peptide fragments T47–48 and T57–58 that (a) coelute at a retention time of 83.9 min, and (b) elute at retention times of 328.7 and 227.9 min, respectively. [From Kajdan et al. (2008), with permission. Copyright © 2008 by Elsevier].

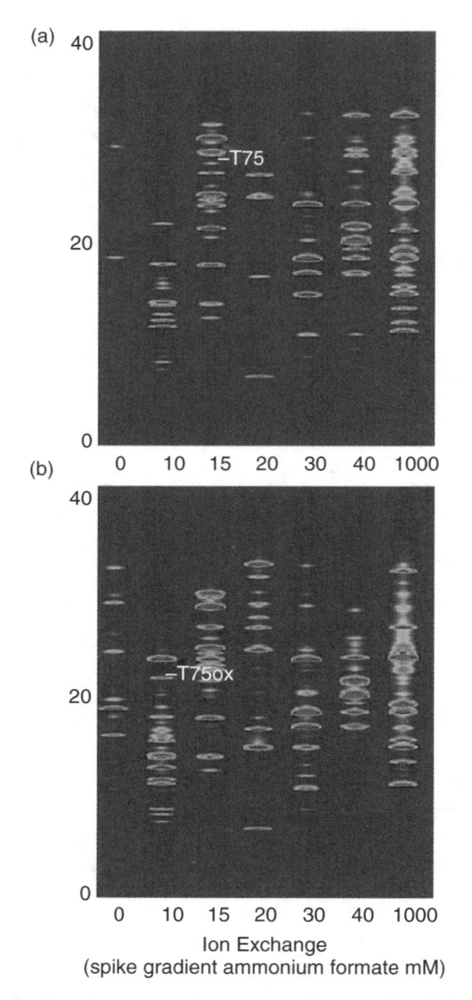

Figure 9.4 Two-dimensional contour plots comparing tryptic digests of (a) bovine serum albumin and (b) oxidized bovine serum albumin by 2D LC–MS at different ammonium formate concentrations. Methionine containing tryptic peptide fragment T75 either unoxidized or oxidized. [From Kajdan et al. (2008), with permission. Copyright © 2008 by Elsevier.]

phase has the beneficial effect of focusing the solutes prior to their second-dimension analysis; the drawback is that trapping efficiency and rapid desorption are inversely related.

In the system designed by Greibrokk's research group, the two chromatographic dimensions were connected through use of a column selector incorporating 18 C_{18}-based SPE columns of 1.0 mm i.d. The trapping columns collected the same number of fractions throughout the first-dimension chromatogram,

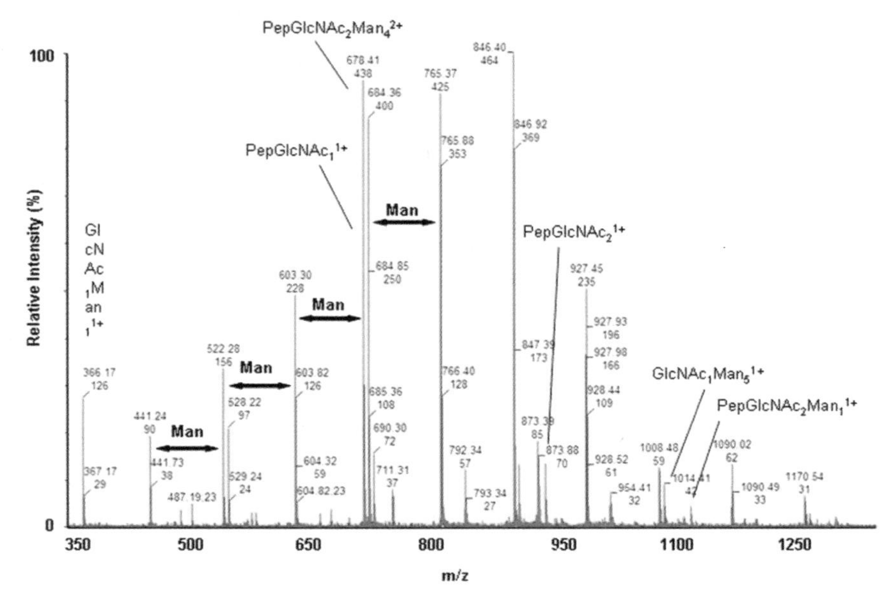

Figure 9.5 2DLC separation of RNase A and RNase B tryptic digests. Mass spectrum of the peak eluting at 62.59 min for RNase B of the glycopeptides with a pentammosidic *N*-glycan (*m/z* 846). Heterogeneity in mannose contents is indicated by double-headed arrows. [From Kajdan et al. (2008), with permission. Copyright © 2008 by Elsevier.]

allowing the direct injection of large aqueous sample volumes (e.g., 450 μL) onto a capillary (150 mm length × 0.3 mm i.d.) HILIC column. A reversed-phase column of the same size was used as the second dimension, and the overall system performance was evaluated for the separation of a tryptic digest of three model proteins, yielding good trapping accuracy (>95%) and retention-time RSDs (≈1%). A dialysis sample of rat muscle tissue was separated afterwards, and two target analytes (arg-bradykinin and bradykinin) were monitored by ESI–TOF MS and quantified using a post-column-injected internal standard (three-point calibration curve at 0.5, 5, and 10 ng). Unlike the other comprehensive systems described here, in this approach the second-dimension analyses only started after the end of the first-dimension separation (Wilson et al., 2007).

As far as dilution and detection sensitivity are concerned, miniaturization of the entire system is undoubtedly beneficial. The advantages of coupling nanoscale LC with nanospray MS–MS have been illustrated by Wang and co-workers, who reported a nanoflow 2D LC–ESI–MS–MS system for the proteome analysis of human hepatocellular carcinoma (HCC) proteins and peptides (Wang et al., 2005). They developed a fully automated high-sensitivity and high-throughput proteome analysis platform coupling an SCX column (150 mm × 0.3 mm i.d.) in first-dimension with a C_{18} (150 cm × 75 μm i.d.) column in the second dimension, the latter operated at 200 nL/min. Data obtained through ESI–TOF MS–MS analysis were searched against the SWISSPROT database by the MASCOT search engine, and enabled the identification of 229 proteins in HCC tissues.

Despite all the setups described above, which make use of salt steps to elute peptides from the first SCX column, Mitulovic and co-workers employed linear salt gradients in the first dimension for the comprehensive 2D nano-LC–MS analysis of proteins and peptides from human tissues. Avoiding the use of salt plugs, which results in peak broadening and carryover phenomena, the authors claimed to have nearly doubled the number of proteins identified (i.e., 98 vs. 53 for the given sample), and also reported the absence of coeluting peptides of high abundance over multiple fractions. But, this type of approach requires more sophisticated instrumental setup, with switching valves and a second pump capable of delivering an independent gradient. A continuous capillary gradient was used in this application, at 6 μL/min, over the SCX column, while two RP trap columns were used for trapping the peptides eluted prior to second-dimension RP separation on the nanocolumn, operated at 300 nL/min. A separate isocratic pump was needed to wash the enrichment columns (Mitulovic et al., 2004).

9.3.2 LC x LC–MS Methods for Separating and Identifying Natural Components

Natural products are very complex mixtures containing different classes of components present in a wide range of concentration. Their analysis may be directed to the assessment of their quality and authenticity or the detection of molecules with a possible beneficial or toxic effect on human health. In this field of analysis, the use of HPLC coupled to MS is a powerful tool for the unambiguous identification of nonvolatile and ionic substances and for the quantification of compounds in complex matrices. In the analysis of natural products, commercial standards are rarely available, and LC–MS can represent a great help in compound identification, greatly simplifying sample pretreatment procedures for the isolation of target analytes before structure elucidation studies by MS or nuclear magnetic resonance. Because of the complexity of samples to be analyzed, one-dimensional HPLC could not be sufficient for the separation of all the molecules of interest. In these cases, multidimensional LC can be used to increase the separation power before MS detection.

A relatively high number of applications of LC × LC coupled to MS to the analysis of natural components can be found in the literature. They are related primarily to the determination of triglycerides (TAGs) in fats and oils, carotenoids in Citrus products, and polyphenols in various plant extracts as well as in food products. Table 9.2 summarizes applications of LC × LC–MS in this specific field. The choice of MS as detector was fundamental for the reliable identification of target analytes. TAGs, the main constituents in fats and oils, are esters formed mainly by three long-chain fatty acids (FAs) bonded to a glycerol molecule. The possible number of FA combinations along the glycerol backbone is high, generating an enormous number of individual TAGs. Thus, the determination of TAG composition in real samples is not an easy task.

TABLE 9.2 Comprehensive 2D (LC x LC-MS) Separations of Natural Compounds

Ref.	Type of Interface	Column Packing and Dimensions (Length x i.d., mm, d_p, μm); Flow rate; Elution Mode[a]		Detection	MS and MS-MS Parameters	Application
		First Dimension, D1	Second Dimension, D2			
[1]	Two-position 10-port valve	Nucleosil 100–5 SA silvered in lab 150 × 1.0, 5 μm; 13 μL/min; IP	C_{18} monolithic 100 × 4.6; 4 mL/min; GP	UV, APCI-MS	500–1000 *m/z*; 0.1 sec, 12 Hz	TAGs in rice oil
[2]	Two-position 10-port valve	Nucleosil 100–5 SA; silvered in lab; 150 × 1.0, 5 μm; 11 μL/min; GP	C_{18} monolithic 100 × 4.6; 4 mL/min; GP	ELSD (D1), APCI-MS	400–900 *m/z*; 0.2 sec, 8 Hz	TAGs in soybean and linseed oils
[3]	Two-position 10-port valve	Nucleosil 100–5 SA silvered in lab 150 × 1.0, 5 μm; 11 μL/min; GP	C_{18} monolithic 100 × 4.6; 4 mL/min; GP	ELSD (D1), APCI-MS	300–900 *m/z*; 0.2 sec, 6 Hz	TAGs in donkey milk fat
[4]	Two-position 10-port valve	C_{18} 150 × 2.1, 3 μm; 100 μL/min; GP	CN 75 × 4.6, 3 μm; 1.9 mL/min; IP	UV, ESI-TOF-MS	100–800 *m/z*; 0.5 sec, 2 Hz	Antioxidant phenolic acids in *Lamiaceae* herbs
[5]	Two-position 10-port valve	RP C18 150 × 2.1, 3 μm; 100 μL/min; GP	$NH_2$50 × 2.0, 3 μm; 1.8 mL/min; IP	ESI-TOF	150–1400 *m/z*; 0.5 sec	Glycosides in *Stevia rebaudiana* extract

(*continued overleaf*)

TABLE 9.2 (*Continued*)

Ref.	Type of Interface	Column Packing and Dimensions (Length x i.d., mm, d_p, μm); Flow rate; Elution Mode[a]		Detection	MS and MS-MS Parameters	Application
		First Dimension, D1	Second Dimension, D2			
[6]	Two-position 10-port valve	Nucleosil 100 5-SA silvered in lab 250 × 2.1, 5 μm; 20 μL/min; GP	XDB C_{18} 30 × 4.6, 1.8 μm; 3 mL/min; IP	PDA, ELSD, APCI-MS	500–1000 *m/z*; 0.67 sec	TAGs in corn oil
[7]	Two-position 10-port valve	LC-SI 300 × 1, 5 μm; 10 μL/min; GP	C_{18} monolithic 100 × 4.6; 4.7 mL/min; GP	PDA, APCI-MS	350–950 or 250–1300 *m/z*; 0.5 sec, 2 Hz	Carotenoids in mandarin essential oil
[8]	Two-position 10-port valve	Cyano 250 × 1, 5μm; 10 μL/min; GP	C_{18} monolithic 100 × 4.6; 5 mL/min; GP	PDA, APCI-MS	250–1300 *m/z*; 0.5 sec, 2 Hz	Carotenoids in red orange essential oil
[9]	Two-position 10-port valve	PEG 150 × 2.1, 5 μm; 25, 50 or 100 μL/min; GP	C_8 or C_{18}; 1, 2 or 4 mL/min; GP	PDA, coul-array, offline APCI-IT-MS	50–500 *m/z*	Phenolic and flavone natural antioxidant test mixture
[10]	Two-position 10-port valve	Phenyl 250 × 1.0, 5 μm; 10 μL/min; GP	(a) C_{18} monolithic 25 (+5) x 4.6 (b) C_{18}30 × 4.6, 2.7 μm; 4 mL/min; GP	DAD, ESI-IT-TOF-MS	130–800 *m/z*; ion accumulation: 100 ms	Polyphenolic antioxidant in wine

[11] Two-position 10-port valve	C_{18}150 × 2.1; 100 μL/min; GP	C_{18}50 × 3.0; 1.35 mL/min; IP	DAD, ESI-TOF-MS	100–700 *m/z*; 0.5 Hz	Antioxidant phenolic acids in wines and juices
[12] Two-position 10-port valve	Cyano 250 × 1.0, 5 μm; 10 μL/min; GP	C_{18} monolithic 100 × 4.6; 4.7 mL/min; GP	DAD, APCI-IT-TOF-MS	*m/z*200–1200; ion accumulation: 10 ms	Epoxycarotenoid esters in orange juices
[13] Offline, fraction collectors equipped with 1.5-mL vials with 250-μL insert	HILIC 250 × 1.0, 5 μm; 50 μL/min; GP	Two SB- C_{18} 50 × 4.6, 1.8 μm; 0.8 mL/min; GP	DAD, Fluorescence, ESI-MS	*m/z* 285–2025	Procyanidins in apple and cocoa extracts

[1] Mondello et al., 2005; [2] Dugo et al., 2006a; [3] Dugo et al., 2006b; [4] Kivilompolo and Hyotylainen, 2007; [5] Pól et al., 2007; [6] Van der Klift et al., 2008; [7] Dugo et al., 2008a; [8] Dugo et al., 2008b; [9] Hájek et al., 2008; [10] Dugo et al., 2008c; [11] Kivilompolo et al., 2008; [12] Dugo et al., 2009; [13] Kalili and de Villiers, 2009.

[a]IP: isocratic program; GP: gradient program.

Due to the non-volatile nature of TAGs, HPLC is the technique most commonly used for their analysis. Both non-aqueous (NA) RP HPLC and silver-ion (Ag) HPLC have been employed widely. The first approach separates TAGs on the basis of increasing partition number ($PN = CN - 2\ DB$, where CN is the total carbon number of the three FAs and DB is the number of double bonds). TAGs with the same PN are considered critical pairs, and their separation under NARPLC is a very difficult task, especially for more complex samples. Moreover, another limitation of NARPLC is that the separation of positional isomers is not possible.

Ag-HPLC separates TAGs on the basis of an increasing degree of unsaturation and on the position and configuration of the double bonds within each FA. In this case, the separation of TAGs with the same number of DBs is critical. As far as detection is concerned, MS with APCI interface can give useful information, not only on which FAs are present in a TAG molecule, but also on the specific distribution of FAs in the TAG. APCI in positive mode produces a $[M + H]^+$ ion, and fragments related to the loss of an FA ($[M-(R-COO)]^+$), which is a diglyceride ion ($[DG]^+$). The intensity of the pseudomolecular ion is higher when the degree of unsaturation is higher. Diglyceride ion intensity is related to the position occupied by the FA that has been removed: loss of the FA from the *sn*-2 position (energetically less favorable) produces a less intense $[DG]^+$ ion than those produced by the loss of FAs from the external positions (*sn*-1 and *sn*-3).

LC × LC–MS has been used for the analysis of TAGs in rice oil (Mondello et al., 2005), soybean and linseed oils (Dugo et al., 2006a), donkey milk (Dugo et al., 2006b), and corn oil (van der Klift et al., 2008). For a comprehensive LC analysis of TAGs the approach was the use of Ag-HPLC in the first dimension and NARPLC in the second dimension. This approach is entirely orthogonal because the two separation modes are based on different retention mechanisms. The first-dimension column was a microbore column (1 mm i.d.) silvered in the laboratory, starting from a Nucleosil 5-SA (strong cation-exchange column). Interface was a two-position 10-port valve, equipped with two identical loops. Flow rates in the two dimensions, volume of the loops, modulation time, and other specific experimental conditions used by each research group have been described in Table 9.2. Three of the four papers mentioned above described a configuration where a mobile phase based on *n*-hexane modified with a small amount of acetonitrile was used as the solvent, with a gradient program using isopropanol and acetonitrile used in the second dimension. Problems with incompatibility of the first-dimension nonpolar eluent with the second-dimension separation and focusing of analytes were solved in the following way:

1. A microbore column with a very low flow rate was used in the first dimension.
2. A monolithic column operated at a relatively high flow rate was used in the second dimension.

3. The percentage of the weaker solvent (acetonitrile) of the second dimension was maintained high during the transfer, thus reducing eluent strength and allowing focusing of the analytes.

The flow rate of the second dimension was reduced by a splitting device placed prior to the MS interface. The 2D chromatograms obtained showed the formation of TAG group-type patterns located in characteristic positions of the 2D space in relation to their PN and DB values. MS detection greatly helped in peak identification, combining MS spectra information with those of PN and DB values obtained from the chromatogram. MS results were particularly clear, due to the absence of peak coelutions with respect to those obtained by one-dimensional HPLC coupled to MS. Figure 9.6 shows an example of how the attainment of pure MS spectra relative to totally separated analytes will provide more reliable information. Figure 9.6a shows an MS spectrum of some coeluting TAGs eluted from monodimensional HPLC, all of them having PN = 44 (LaPP, LaPO, and LnPP). Spectra of these three TAGs obtained after LC × LC separation are reported in Figure 9.6b–d. In this case, components were well resolved in the 2D space, because of the different DB number. Spectra were clear enough to allow reliable peak identification, supported by chromatographic data, even without using pure standard compounds.

An improved AgHPLC × NARPLC–MS system was developed by van der Klift et al. (2008). In this case, solvent used in the first dimension was MeOH-based, while a MeOH–MTBE (70 : 30) mixture was used in the second dimension. Solvent incompatibility was avoided, and as the second-dimension solvent was stronger than the first-dimension solvent, peak focusing was achieved at the head of the secondary column. Analysis of corn oil revealed the presence of 44 components: 34 TAGs, eight oxygenated TAGs, and two TAGs containing a *trans* double bond. The authors performed quantitative analysis of TAGs, obtaining area percent values both manually, from the integrated untransformed chromatogram, summing areas of peaks occurring in multiple slices, and by automatic calculation by a developed software. Values obtained after correction with a literature-based response factor (Holčapek et al., 2005) were compared with those reported in the literature for corn oil TAGs. The main trend coincided, but for individual components, significant deviations were detected. This could not be explained entirely by natural variation, because a comparison of LC × LC–APCI–MS quantitative data with those obtained for the same sample by GC–FID analysis of fatty acid methyl esters (FAMEs) revealed considerable deviations for many FAs. Such differences could probably be due to the use of correction factors obtained under different experimental conditions (both MS and chromatographic parameters). Due to the lack of pure TAGs and oils with an accurately known TAG content, quantitative determination remains very difficult, and for more accurate data, calibration with at least the major constituents of the oil should be performed under the same experimental conditions. However, in terms of peak overlap, LC × LC–APCI–MS results were superior to those of one-dimensional methods, and with further improvement, this technique will be

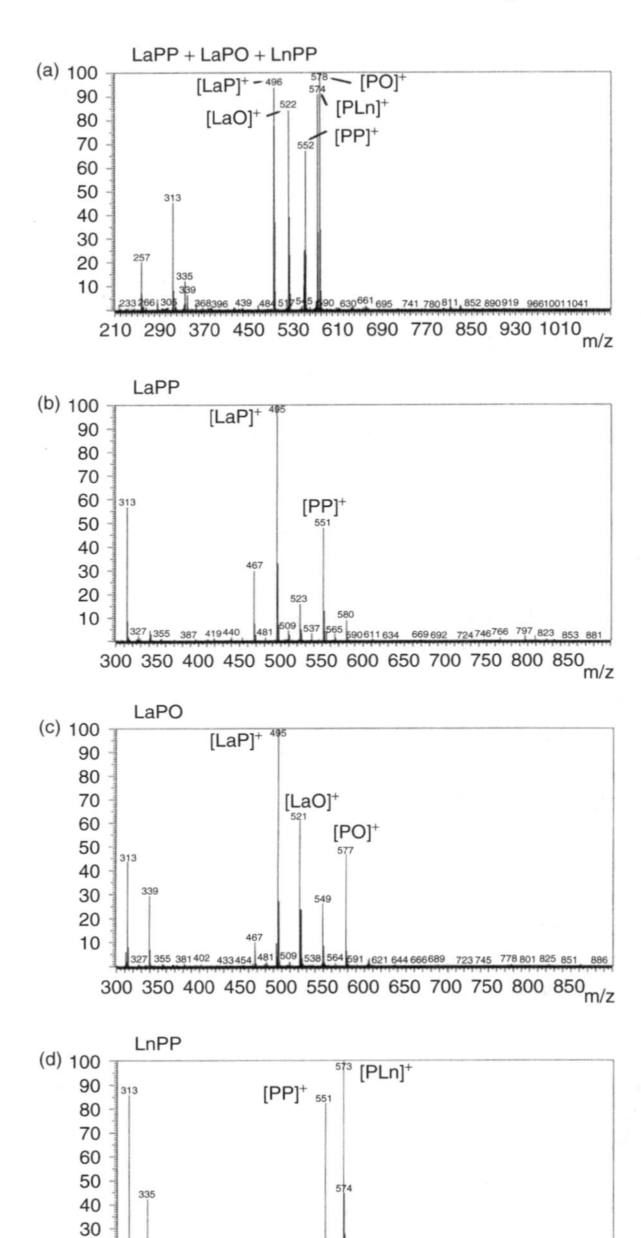

Figure 9.6 (a) MS spectrum relative to a one-dimensional triple coelution (LaPP, LaPO, and LnPP); (b–d) MS spectra derived from the comprehensive LC separation of the same TAGs. [From Dugo et al. (2006a), with permission. Copyright © 2006 by Elsevier.]

of great help in the qualitative and quantitative determination of TAG fraction in complex real samples.

Carotenoids are usually based on a C_{40}-tetraterpenoid structure with a symmetrical skeleton. From the basic structure, different modifications can occur, resulting in a great variety of structures. Carotenoids are usually divided into two groups: hydrocarbon carotenoids (carotenes) and oxygenated carotenoids (xanthophylls). The latter can be found in their free form or in a more stable fatty acid esterified form. The study of esterified carotenoids in natural samples is rather limited, due to the high degree of complexity that can be found, the lack of standard components, and the instability of the carotenoid structure. To reduce sample complexity, an approach commonly employed has been the use of a saponification step prior to the carotenoid analysis, so that all the components could be transformed in their free form. Citrus species are well known to possess a rich carotenoid pattern and are often mentioned as the most complex natural source of this type of compounds.

A group of papers reported the use of normal-phase (NP) LC × RPLC for the analysis of the free and esterified carotenoid composition in Citrus products (Dugo et al., 2008a, 2008b, 2009). Both PDA and MS were used as detectors. In NPLC, free carotenoids are separated into groups of different polarity starting from the nonpolar carotenes up to more polar polyols. In RPLC the carotenoids are eluted according to their increasing hydrophobicity and decreasing polarity. Both separation modes may present some difficulties, mainly in the analysis of complex real samples, and coelution of structurally similar components may occur.

The entire carotenoid composition of mandarin essential oil has been studied by developing two different LC × LC–PDA–APCI–MS methods, one for the determination of free carotenoids after saponification and the other for carotenoids in their native form, using the same instrumental setup and two different column sets. A silica microbore column operated in the NP mode was coupled to a C_{18} monolithic column operated in the RP mode to study the saponified fraction of a mandarin essential oil. The LC × LC chromatogram showed the presence of different classes of carotenoids in different zones of the 2D plot, according to their chemical structure. The use of two detection systems installed in parallel (PDA and MS) allowed the identification of 21 components, some of them for the first time. It is important to point out that although very informative, ultraviolet–visible (UV-vis) spectra of many different carotenoids are very similar, or even identical, and do not allow for reliable peak identification. The information provided simultaneously by MS spectra was of great help because it allowed differentiation of components of diverse molecular mass. However, the opposite can also be true, since carotenoids very similar in their chemical composition presented the same molecular weight and UV–vis spectrum was necessary to confirm their identification (see Figure 9.7) (Dugo et al., 2008a).

With regard to the analysis of carotenoid esters, the 2D plot showed that carotenoids eluted from the first dimension (cyano microcolumn operated under the NP mode) into groups formed by the esters of a certain carotenoid, between

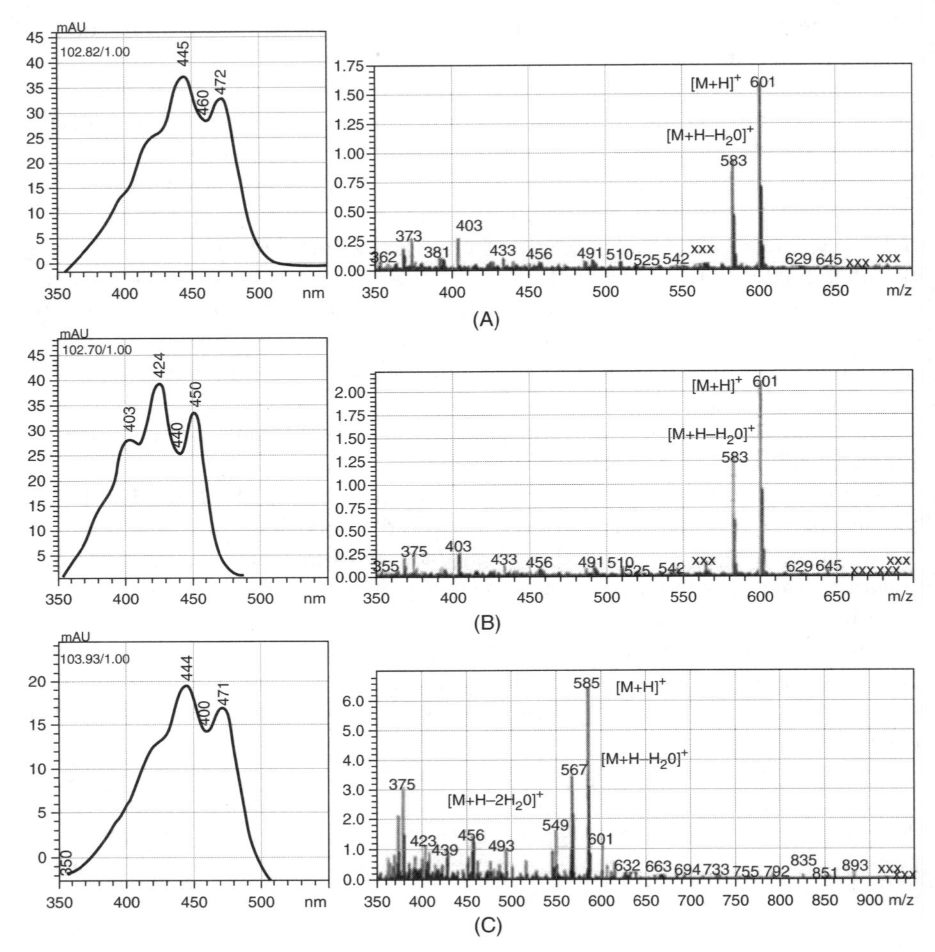

Figure 9.7 Identification of the different carotenoids contained in the sample based on the combined information provided by the two detection methods (PDA and APCI–MS) employed. (A) Violaxanthin; (B) luteoxanthin; (C) antheraxanthin. [From Dugo et al. (2008a), with permission. Copyright © 2006 by Elsevier.

hydrocarbons (eluting at the beginning) and free xanthophylls (at the end). Under RP conditions of the second dimension, each group was separated into its single carotenoid esters. Fundamental was the role of MS detection in peak identification, because different esters of the same carotenoid present identical UV–vis spectra. Also important is the chromatographic information, because different esters eluted from the second-dimension column on the basis of the FA esterified to the structure, from the shortest-chain FA to the longest.

A similar approach was taken to determine the native carotenoid composition of orange essential oil, an extremely complex matrix (Dugo et al., 2008b).

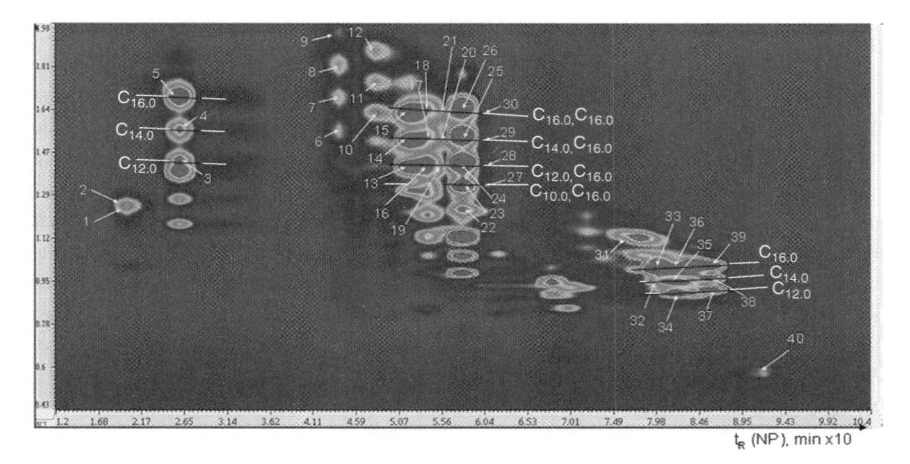

Figure 9.8 Two-dimensional plot of the intact red orange essential oil comprehensive 2D LC carotenoid analysis. Wavelength, 450 nm. Peak identification: 1, ζ-carotene; 2, phytofluene; 3–5, monoesters of β-criptoxanthin; 6–9, monoesters of lutein; 10–12, diesters of antheraxanthin; 13–15, diesters of luteoxanthin (a); 16–18, diesters of violaxanthin) 19–21, diesters of auroxanthin; 22–26, diesters of luteoxanthin (b); 27–30, diesters of an auroxanthin isomer; 31, antheraxanthin palmitate; 32, 33, monoesters of luteoxanthin (b); 34–36, monoesters of luteoxanthin (a); 37–39, monoesters of violaxanthin; 40, lutein. [From Dugo et al. (2008b), with permission. Copyright © 2008 by The American Chemical Society.]

Figure 9.8 shows the 2D plot of the orange sample. Identification was carried out combining information from PDA and MS, chromatographic data, literature data, and available standards, obtained from the analysis of an intact orange oil sample. In this application, the use of MS was compulsory, because carotenoid esters cannot be distinguished by their UV–vis spectra and from their corresponding free forms. MS spectra obtained by APCI in the positive mode were very informative, thanks to the presence of pseudomolecular ions and several losses and complementary ions that help in the identification. For example, for a carotenoid diester, typical ions formed by $[M + H]^+$ and losses of one or two FAs, as well as several losses of water, were produced, enabling the identification of the FAs bound to the carotenoid structure. In NP separation, hydrocarbons were eluted first, followed by different groups of mono- and diesters eluting in agreement with a combination between their esterification degree and the polarity of the carotenoid. Forty structures were identified: three free carotenoids, 16 carotenoid monoesters, and 21 carotenoid diesters, most of them for the first time.

Studies were also extended to the intact carotenoid fraction of orange juice, probably the most widely consumed fruit juice worldwide. LC × LC configuration was similar to that described in previous works carried out by the same research group. In this case, an IT–TOF mass spectrometer was used, and APCI was performed in both positive and negative ion mode. Special attention was paid to the epoxycarotenoid fraction, which composition could be a useful parameter to employ to evaluate the age and freshness of the juice. In fact, rearrangements from 5,6- (violaxanthin, antheraxanthin) to 5,8-epoxides (luteoxanthin, mutatoxanthin) can occur with time, due partially to the natural acidity of the juice. One-dimensional analysis using a C_{30} HPLC column and UV and APCI–MS detection of a valencia orange juice sample allowed detection of violaxanthin mono- and diesters, although several coelutions were observed and the resulting spectra were not very clear. Moreover, luteoxanthin esters were not detected at all. LC × LC–PDA–APCI–IT–TOF MS of the same sample of Valencia orange juice allowed the separation and identification of several *cis*-violaxanthin and all-*trans*-luteoxanthin diesters as well as some all-*trans*-violaxanthin monoesters. MS and UV spectra were clearer, as a result of higher separation in LC × LC than in conventional LC.

LC × LC–TOF MS equipment was used in the characterization of stevioside glycosides in candy leaf, *Stevia rebaudiana* Bertoni (Pól et al., 2007). Stevioside is a diterpene glycoside; both *Stevia* and stevioside have been used as substitutes for sucrose as sweeteners. Different combinations of strong cation-exchange, amino (NH_2), and C_{18} stationary phases were tested. A combination of C_{18} as the first-dimension column and NH_2 as the second-dimension column provided separation of all nine known glycosides contained in the matrix. ESI in the negative mode was used, and components were identified based on their exact masses obtained by TOF MS. The structure of the target analytes is very similar, thus not allowing efficient monodimensional LC separation. Moreover, two pairs of stevioside glycosides (stevioside and rebaudioside B, rebaudioside A, and rebaudioside E) present the same molecular weight, and their chromatographic separation is fundamental. These pairs could be separated using NH_2 phase, but coelution with matrix compounds may cause ion suppression during ionization in MS. Quantitative analysis was also carried out by using a calibration curve constructed for a stevioside standard using $[M - H]^-$ MS peak at *m/z* 803.4.

Polyphenols are natural compounds, found widely in fruits and vegetables, divided mainly into phenolic acids and flavonoids. Some of them exhibit strong antioxidant properties. Their presence in food products is important for various reasons: they inhibit oxidation of lipids; contribute to the sensorial properties, such as color, astringency, bitterness, and flavor; and may influence the nutritional value of a product. The profile of polyphenols in real samples may vary significantly and may be too complex for a conventional HPLC technique. The extra separation efficiency offered by LC × LC techniques has been considered for the study of real samples, such as herb extracts (Kivilompolo and Hyötyläinen, 2007), beer (Hájek et al., 2008), wine (Dugo et al., 2008c; Kivilompolo et al., 2008), juices (Kivilompolo et al., 2008), and apple and cocoa extracts (Kalili and de Villiers, 2009).

Phenolic acids extracted from basil, oregano, rosemary, sage, spearmint, and thyme of the *Lamiaceae* family using dynamic sonication-assisted ethanol extraction were analyzed by LC × LC interfaced to a micro-TOF mass spectrometer via an electrospray source operated under negative ionization (Kivilompolo and

Hyötyläinen, 2007). The RPLC × RPLC setup was optimized (as to selection of the column systems, choice of solvent systems, and modulation time). To obtain a fast modulation cycle, a new fraction was introduced to the column before the previous fraction had reached the detector. In this way, each peak was divided into three fractions, according to theory (Murphy et al., 1998). Quantitative analysis was carried out from the LC × LC chromatogram using ion chromatograms of the pseudomolecular ions $[M - H]^-$. Peaks of the same component modulated into successive cuts were summed before calculation. Quantitative analysis was performed using LC–TOF MS and compared with results obtained by LC × LC–TOF MS. In LC–MS, the degree of ion suppression or enhancement was studied and demonstrated by comparing the external calibration with the more laborious standard addition method. The results of standard addition calibration in LC–MS and external calibration by LC × LC–MS correlate well, indicating that due to the efficient separation by the 2D system, the matrix effects are insignificant (see Figure 9.9).

Hájek et al. (2008) optimized an LC × LC method for the analysis of polyphenols using UV, electrochemical coulometric, and MS detection. They used a polyethyleneglycol (PEG) column in the first dimension and tested different C_{18} and C_8 stationary phases (i.e., porous shell particles, monoliths, or totally porous particles) in the second dimension. The combination of PEG and C_{18} or C_8 provided low-selectivity correlation; thus, a high degree of orthogonality and non-correlated retention of 27 natural antioxidants tested were obtained. Gradients with matching profiles running in parallel in the two dimensions over the entire 2D separation time range were used. Superficially porous C_{18} particles

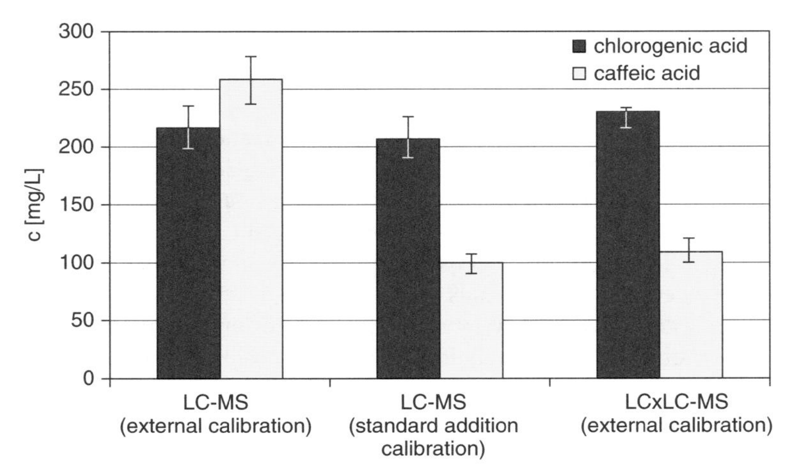

Figure 9.9 Comparison of quantitative results of chlorogenic and caffeic acids from sage extract obtained by LC–MS (external calibration), LC–MS (standard addition calibration), and LC × LC–MS (external calibration) methods. [From Journal of Chromatography A, 1145, Kivilompolo and Hyötyläinen (2007), with permission. Copyright © 2007 by Elsevier.]

and monolithic columns improved the resolution and speed of second-dimension analysis over that of totally porous C_{18} particles. However, better permeability of monolithic columns made it possible to use longer columns and higher flow rates than could be used with particle-based columns.

The use of different detectors in combinations allowed positive identification of components in real samples of beer. A coul-array detector provided significant improvement in the sensitivity and selectivity of detection compared with both UV and MS. MS was used in the off-line mode; 0.2-mL fractions were collected continuously (12 s per fraction); 225 of the total 1330 fractions were subjected to MS analysis under negative-ion APCI–MS conditions.

Polyphenols of red wine were analyzed using a RPLC × RPLC setup, connected online to PDA and ESI–IT–TOF–MS detectors (Dugo et al., 2008c). In this case, a microphenyl column was used in the first dimension, while in the second dimension a partially porous C_{18} short column and a monolithic C_{18} column of identical dimensions (30 × 4.6 mm) were compared. Although a high degree of correlation was determined, a comprehensive method was developed to compare the performance of the two C_{18} columns. Both columns were operated successfully under a fast repetitive gradient at high flow rates with very brief reconditioning times. However, the partially porous column provided increased peak capacity over that of the monolithic column, due to the minor peak width. Thanks to the great resolution and accuracy of the IT–TOF MS used, equipped with an ESI source operated under positive and negative ionization, it was possible to identify different polyphenol antioxidants in a real sample of red wine.

A novel RPLC × RPLC system was developed by Kivilompolo et al. (2008) for the quantification of polyphenolic antioxidants in wine and juice. Figure 9.10 shows the system configuration, where components well separated in the first column were directed to the detector, while the remaining part of the sample (more complex) proceeds to the second dimension via the 10-port valve. Two C_{18} columns, the latter with an ion-pair reagent (tetrapentylammonium bromide), were used in the two dimensions. Compound identification was carried out using LC–ESI–TOF MS after the first column, since the ion-pair reagent used in the second column is not suitable for MS analysis. Quantitative analysis was performed by using peak volumes in two-dimensional contour plots obtained by DAD, because LC × LC made possible good separation of the compounds of interest from the matrix compounds.

Kalili and de Villiers (2009) developed an offline comprehensive two-dimensional liquid chromatographic HILIC × RP method for the analysis of procyanidins in cocoa and apple extracts. Procyanidins comprise oligomeric or polymeric phenols composed of flavan-3-ol monomeric units joined through interflavanoid linkages. Their chromatographic separation remains a challenge, owing to the complexity of their structures, particularly their degree of polymerization. Oligomeric procyanidins were separated according to molecular weight, using HILIC in the first dimension. One-minute fractions of the HILIC separation were collected and stored in vials kept under N_2 until analyzed by RPLC (within 2 days of collection). ESI–MS analysis was performed after

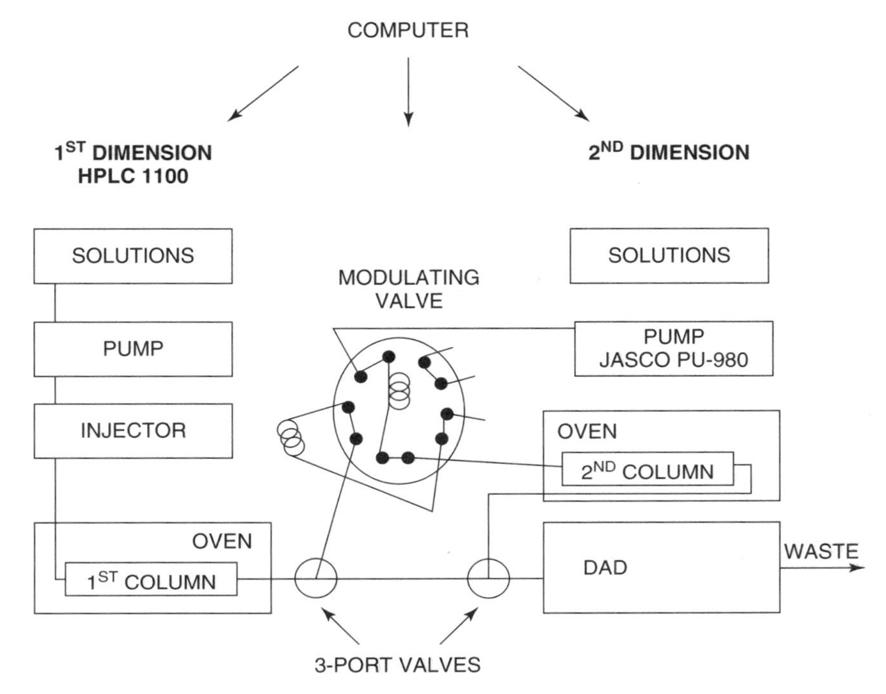

Figure 9.10 Setup of comprehensive two-dimensional liquid chromatograph. [From Kivilompolo et al. (2008), with permission. Copyright © 2008 by Springer-Verlag.]

the first HILIC separation and after RP separation of the fractions transferred. MS data indicated that cocoa and apple extracts contained only catechin and epicatechin monomeric units. The combination of HILIC and RP LC separations provided very high orthogonality and made it possible to obtain significantly improved resolutions of oligomeric procyanidin isomers, due to the pre-separation according to the degree of polymerization in the HILIC mode, followed by separation in the RP mode, that separated oligomers according to their hydrophobicity. Figure 9.11 shows how a combination of molecular weight information (y-axis, first dimension) as well as the isomeric distribution (x-axis, second dimension) of procyanidins is obtained in a single contour plot.

9.3.3 LC x LC–MS Methods for Separating and Identifying Pharmaceutical Components

The pharmaceutical industry has a growing interest in isolating and assessing the active pharmaceutical compounds from all potential impurities and/or degradation products (Table 9.3) (ICH 3QA, 1996; ICH 3QB, 1997). For regulatory purposes, pharmaceutical impurities can be classified as organic, inorganic, and residual solvents (ICH 3QA, 1996). In particular, organic impurities can originate

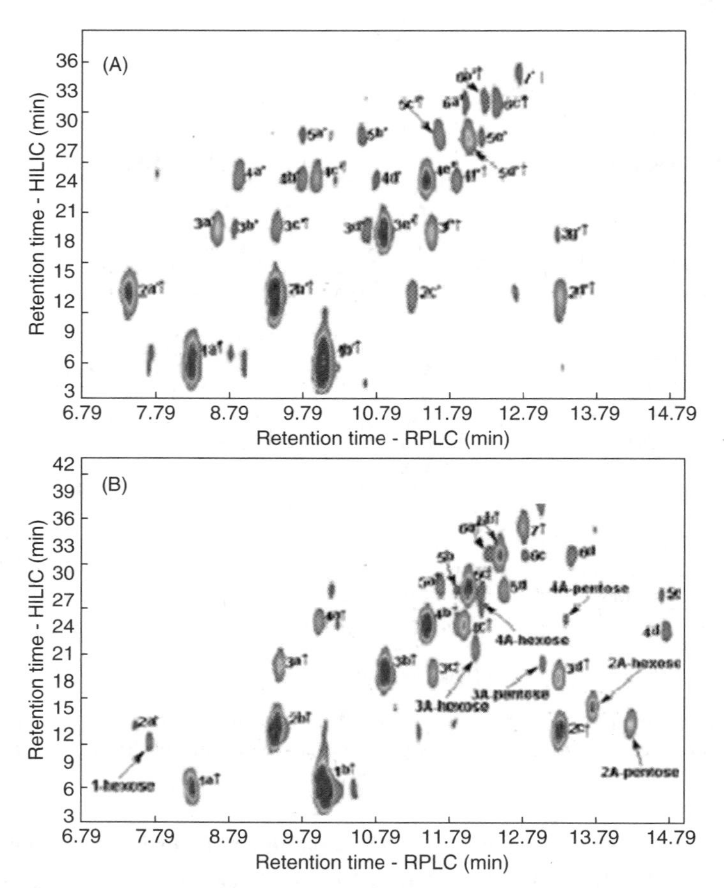

Figure 9.11 Comparison of fluorescence contour plots obtained for the HILIC × RPLC analysis of apple (A) and cocoa (B) procyanidins. Numbers correspond to the degree of polymerization of procyanidin isomers as identified by ESI–MS. Compounds common to both samples are marked with †. [From Kalili and de Villiers (2009), with permission. Copyright © 2009 by Elsevier.]

from alterations of raw material or synthetic intermediates under either reaction conditions (temperature, pH) or storage conditions (hydrolysis, oxidation, ring opening, etc.), leading to very complex mixtures. For this reason, a conventional LC–MS system often fails in the detection of impurities, thus requiring feasible alternatives such as comprehensive LC to overcome this problem. A nice example is reported by Huidobro et al. (2008), who developed a comprehensive offline method for the impurity profiling of pharmaceutical samples in stability and stress studies. The outcome of this study was the combination of SB–CN × BEH – C_{18} columns used in the first and second dimensions, respectively. The two-dimensional chromatogram resulting from the combination of these chromatographic methods is represented in Figure 9.12. The use of an

TABLE 9.3 Comprehensive 2D (LC × LC-MS) Separations of Pharmaceutical, Antioxidants, Biological, Organic, Natural and Environmental Compounds

		Column Packing and Dimensions (Length x i.d., mm, d_p, μm); Flow rate; Elution Mode[a]				
Ref.	Type of Interface	First Dimension, D1	Second Dimension, D2	Detection	MS and MS-MS Parameters	Application
[1]	Two-position 8-port valve	CN 200 × 2.0, 5 μm; 40 μL/min	C_{18} 50 × 3.0; 0.7 mL/min	UV, APCI-MS	50–1000 *m/z*; 0.5 scan/sec	Characterization of *Rhizoma chuanxiong*
[2]	Two-position 8-port valve	(a) CN 150 × 4.6, 5 μm; 0.133 mL/min; GP (b) IEC 150 × 4.6, 5 μm; 0.133 mL/min; GP	C_{18} Chromolith 50 × 4.6; 3.0 mL/min; GP	PDA, APCI-MS	50–1000 *m/z*; 0.5 scan/sec	Chinese medicines
[3]	Two-position 12-port valve with guard trap columns	C_{18} 150 × 4.6, 5 μm; 0.8 mL/min	SB-Phenyl 50 × 3.9, 5 μm; 1.3 mL/min	UV, ESI-MS	50–800 *m/z*	Drug and barbiturate test mixture
[4]	Two-position 8-port valve	IEC 200 × 2.0, 5 μm; 40 μL/min	C_{18} 100 × 4.6, 5 μm; 2 mL/min	PDA, APCI-MS	50–1000 *m/z*; 0.5 scan/sec	Characterization of *Flos Lonicera*
[5]	Double two-position 10-port valves with micro-electric actuators	Betasil Diol 250 × 1.0, 5 μm; 30 μL/min; IP	SB C_{18} 50 × 4.6, 5 μm; 4 mL/min; GP	UV, APCI-MS		Lemon oil extract, sulphonamide drugs, and steroid mixture
[6]	Two position 10-port valve	C_{18} 50 × 1.0, 3 μm; 50 μL/min; GP	Primesep A 30 × 4.0, 5 μm; 1.5 mL/min; GP	PDA, APCI-TOF-MS		Complex mixtures (alkaloids, amino acids)
[7]	Offline	SB-CN 150 × 3.0, 5 μm; 250 μL/min; GP	BEH-C18 50 × 2.1, 1.7 μm; 1 mL/min; GP	UV, ESI-MS	50–800 *m/z*	Pharmaceutical samples
[8]	Two position 8-port valve	ILC 150 × 4.6; 50 μL/min; IP	ODS; 2 mL/min; GP	UV, APCI-MS	50–1000 *m/z*; 0.5 scan/sec	Extracts of Longdan Xiegan Decoction

[1] Chen XG et al., 2004; [2] Hu et al., 2005; [3] Venkatramani and Patel, 2006; [4] Chen et al., 2007; [5] François et al., 2008; [6] Eggink et al., 2008; [7] Huidobro et al., 2008; [8] Wang et al., 2009.

[a]IP: isocratic program; GP: gradient program.

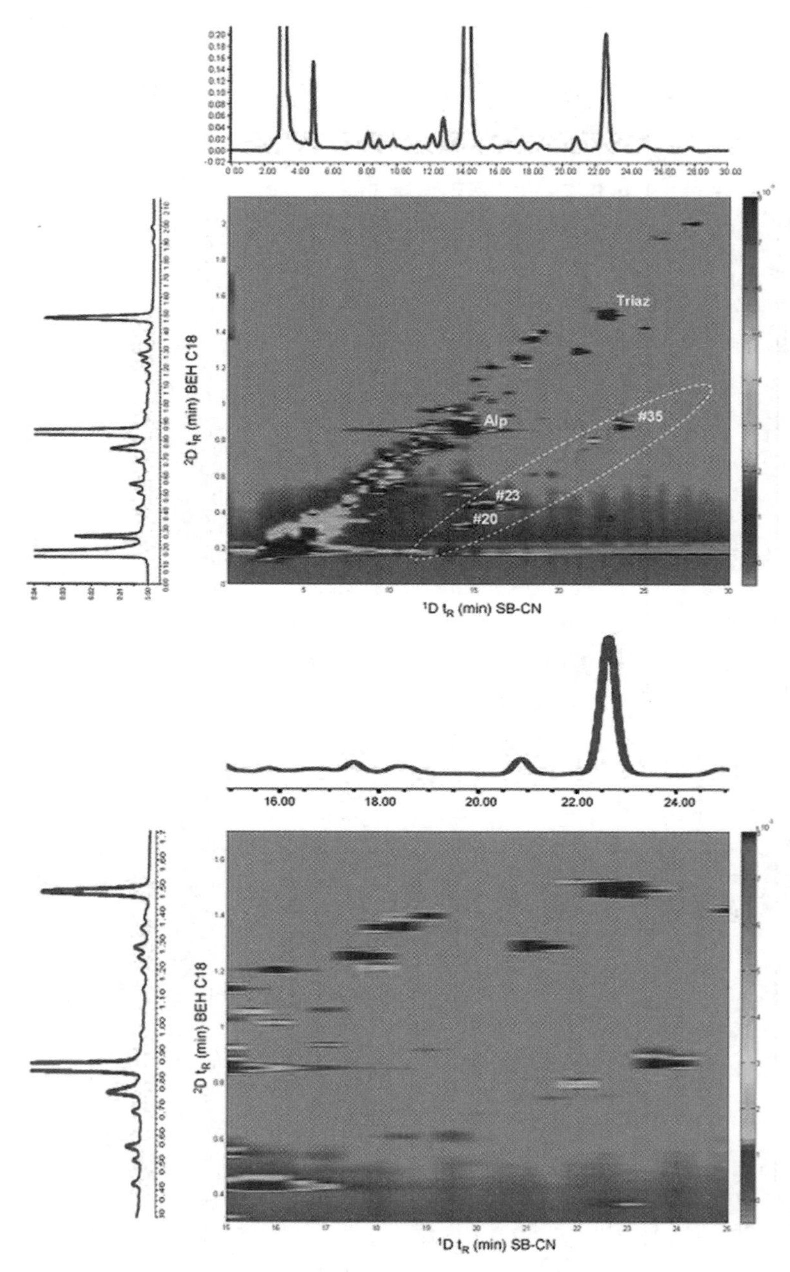

Figure 9.12 Contour plot for comprehensive LC × UPLC separation of degraded alprazolam tablet sample and its corresponding one-dimensional profiles (top) and an enlargement (bottom). [From Huidobro et al. (2008), with permission. Copyright © 2008 by Elsevier.]

ultrahigh-pressure liquid chromatography system (UPLC) as the second chromatographic separation increased the sensitivity and speed, completing the entire chromatographic separation in a reasonable time frame. The chromatographic methods developed were transferred to LC–MS to confirm the different selectivities between the one-dimensional methods.

Another application in the pharmaceutical analysis field involved the separation of a 32-component mixture differing widely in physicochemical properties (Eggink et al., 2008). To tackle such task, a 2D LC × LC system consisting of an RPLC column in the first dimension and a mixed-mode cation–IEC/RPLC column in the second dimension were tested. From this point of view, the alternative approach of RPLC × IEC, although attractive for its comprehensive two-dimensionality, would be less favorable, because IEC generally requires relatively high salt concentrations and is therefore less compatible with MS detection. The multistep isocratic elution in the second dimension yielded a higher peak capacity and improved peak resolution of the system without jeopardizing the stability and background of the UV and APCI–MS detectors employed. The combination of the two detectors was found to be more favorable, since depending on the sample and/or conditions employed, some compounds could not ionize, whereas others could not show UV activity. The 2D system was validated by the analysis of a spiked human urine sample showing the potential in the screening of biological samples of the technology developed.

An SEC–RP–HPLC method for a pharmaceutical mixture analysis has been reported by Winther and Rebauset (2005). An additional flow was employed in this work to avoid severe band broadening and solute loss of hydrophilic compounds connected to the sample transfer from the SEC to RP columns. In fact, to achieve peak focusing on the top of the second-dimension column, a post-column flow was made up to a total of 1 mL/min, thus decreasing the concentration of the organic modifier in the eluate. The eluate was then fed in large volume onto one of the two second-dimension columns before the valve switched to form an aqueous–organic gradient to eluate the focused compounds. During the gradient the first-dimension eluate was loaded onto the other column. Detection was performed by either UV or MS. However, the effect of post-column dilution was not taken into account, and loss of non-focused analytes precluded true comprehensive analyses.

Drug mixtures have been chosen by Venkatramani and Patel (2006) and François et al. (2008) for the assessment of novel interfaces for comprehensive two-dimensional liquid chromatography (LC × LC). In the first study (Venkatramani and Patel, 2006), a two-dimensional system using only one pump operated in both dimensions has been developed by coupling RP columns of different selectivity and using an electronically controlled two-position 12-port valve with guard columns that enabled continuous, alternating sampling of the primary column eluate onto dual secondary columns. While one column was loaded with eluate, the other two columns were in line with the two second-dimension columns. An X-Terra C_{18} column was employed in the first dimension, whereas two Zorbax SB-phenyl columns in parallel were used for the second-dimension

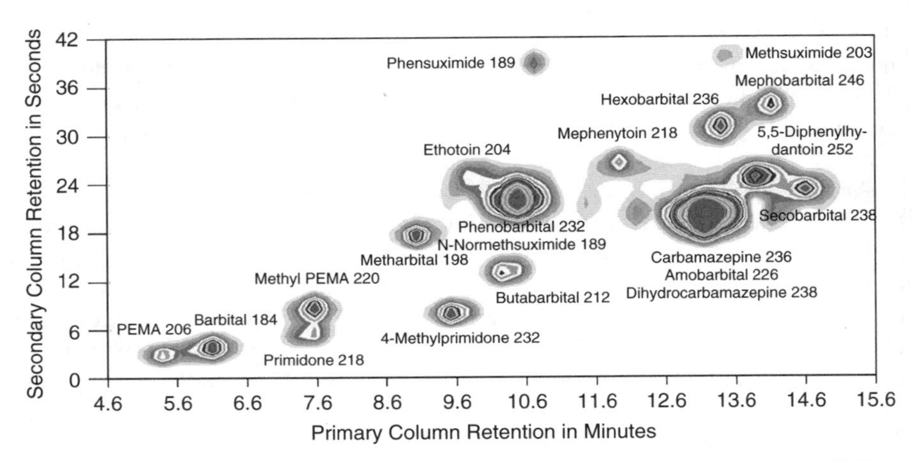

Figure 9.13 Orthogonal 2D LC–MS separation of a drug mixture on a Waters X-Terra MS C_{18} column in primary and dual SB-phenyl columns in parallel in the secondary dimension. [From Venkatramani and Patel (2006), with permission. Copyright © 2006 by Wiley-VCH Verlag Gmbh & Co. KGaA.]

separation (Figure 9.13). In this plot, the primary column eluate was sampled onto the secondary column every 30 s. The orthogonality of the 2D system was obtained thanks to the different retention mechanisms in both dimensions (predominantly hydrophobic in the first-dimension, and polar, such as $\pi-\pi$ or dipole–dipole interactions, in the second dimension). This work demonstrated how commercially available LC–MS systems can readily be converted into a comprehensive 2D LC system by incorporating electronically controlled valves and guard columns. On the other hand, in the second work (François et al., 2008), a conventional LC × LC system with a loop-type interface consisting of a two-position 10-port switching valve equipped with two loops was enriched with an additional two-position 10-port switching valve, a detector, a pump, and a second column placed in parallel with the second-dimension column. The main features of the interface are related to the enlargement of the separation space in the second-dimension separation, allowing the transfer of a much higher number of fractions from the first to the second dimension without losing resolution in the first-dimension separation. RPLC × 2RPLC analyses were carried out using a CN column in the first dimension and two octadecylsilica (ODS) columns in the second dimension. Despite the low orthogonality, the setup was found to be useful for the separation of steroids and sulfonamide samples, thanks to the faster modulation time (30 s). With regard to the sulfonamide drugs, whose analysis is of great importance for the pharmaceutical industry, baseline separation was achieved for all sulfonamides and sulfadimethoxine impurities.

As an important resource for drug development, traditional Chinese medicines (TCMs) are gaining more and more attention in modern pharmaceutical institutes. TCMs have a long history, dating back several thousand years. With the development of theory and clinical practice, China has accumulated a rich body of empirical knowledge about the use of medicinal plants for the treatment of various diseases. Such broad Chinese natural medicinal resources provide valuable materials for the discovery and development of new drugs of natural origin. As a consequence, separation and analysis of the components in TCMs is an important subject to ensure the reliability and repeatability of biological, pharmacological, and clinical research. However, these medicinal plants usually contain complex constituents and a low content of possible bioactive ingredients. Therefore, in past decades, efficient methods such as thin-layer chromatography (Zschocke et al., 1998), gas chromatography (Guo et al., 2004), high-performance liquid chromatography (Gu et al., 2004), biochromatography (Kong et al., 2001; Wang et al., 2000), and micellar electrokinetic chromatography (MEKC) (X. F. Chen, et al., 2004) have been developed for this purpose. In the recent years a large number of papers have dealt with the use of comprehensive two-dimensional liquid chromatography for the analysis of these extracts, and a concise summary of recent developments and contributions coming from Chinese scientists in the field of TCMs has been provided by Gao et al. (2007).

In most approaches to a comprehensive two-dimensional system, in addition to a diode array detector, an atmospheric-pressure chemical ionization (APCI) mass spectrometer is used for the detection of compounds and yields online molecular weight information. This adds a third dimension to this two-dimensional separation system because the overlapped peaks can be identified by mass spectrometry even though they are not resolved by chromatography.

A great deal of work effort in the fingerprint analysis of TCMs has been done by Zou and co-workers, who employed various comprehensive two-dimensional liquid chromatography modes successfully in the separation and identification of components in *Rhizoma chuanxiong* (X. G. Chen, et al., 2004), *Ligusticum chuanxiong* and *Angelica sinensis* (Hu et al., 2005), *Rheum palmatum* (Hu et al., 2006), and *Longdan xiegan decoction* (LXD) (Wang et al., 2009). X. G. Chen et al. 2004 used an LC × LC system based on the combination of a CN column and an ODS column for the separation and identification of the components in *R. chuanxiong* to develop one of the most commonly used drugs in the prescriptions of TCMs. The second-dimension effluent was detected by both the diode array detector and APCI mass spectrometer. Figure 9.14 shows a typical two-dimensional chromatogram for the methanol extract of *R. chuanxiong*. The heights of peaks in the two-dimensional plot are determined by the relative UV absorbance in Figure 9.14a and the counts per second in Figure 9.14b and 9.14c. A total of more than 52 components were separated in less than 215 min, and 11 of them were identified simultaneously in the methanol extract without tedious pretreatments by a combination of UV and mass spectra. A similar setup, except for the employment of a silica monolithic ODS column for the second-dimension separation, was used for the separation of two methanolic extracts of Umbelliferae herbs, *L. chuanxiong* and *A. sinensis* (Hu et al., 2005). More then 100 components were separated and detected by UV, APCI, and MALDI–TOF MS. Through normalization of the peak heights, the low-abundant components showed up in the three-dimensional chromatograms for both TCM extracts. Normalization of

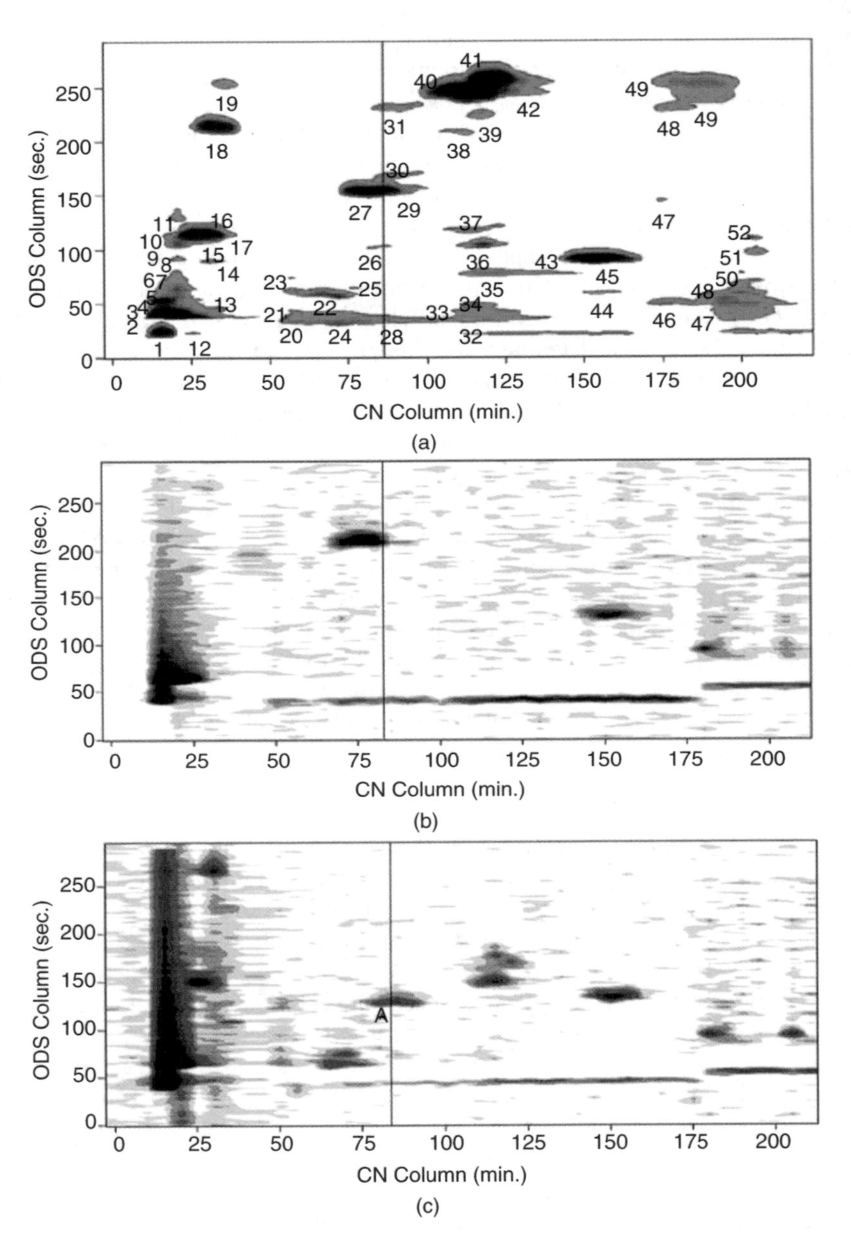

Figure 9.14 Two-dimensional chromatogram of *Rhizoma chuanxiong*. The heights of the two-dimensional plot are determined by (a) the relative UV absorbance; (b) the counts per second using APCI positive ion mode; (c) the counts per second using APCI negative ion mode. [From X. G. Chen et al. 2004, with permission. Copyright © 2004 by Elsevier.]

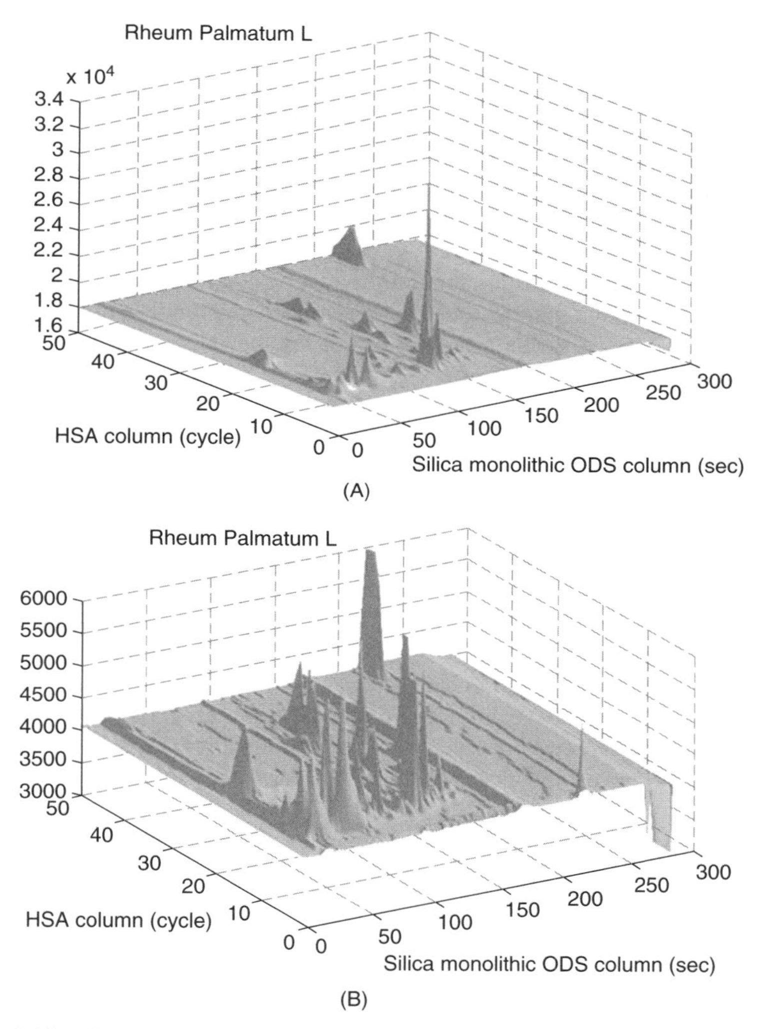

Figure 9.15 Three-dimensional chromatogram for the extract of *Rheum palmatum* L. (a) without normalization and (b) with normalization by setting a value of one-eighth of the highest peak height. [From Hu et al. (2006), with permission. Copyright © 2006 by Wiley-VCH Verlag Gmbh & Co. KGaA.]

the peak heights was also carried out for the visualization of 2D LC separation of *R. palmatum* extracts as in Figure 9.15 (Hu et al., 2006). In this work, the affinity chromatography with HSA-immobilized stationary phase was used as the first dimension to probe the interaction of components in the TMC extract, and a silica monolithic ODS component was used to further analyze the fractions eluted from the HSA column. In the most recent work, Zou and co-workers explored a comprehensive two-dimensional biochromatography system with an immobilized liposome chromatography column (ILC) and reversed-phase column in the two dimensions for separation of complex traditional Chinese medicine LXD in tandem aided by UV and APCI–MS (Wang et al., 2009). Usually, the biochromatography column is used as the first-dimension column, due to the ability of separation on the basis of interactions between the protein and the compounds, whereas the second-dimension column employs a reversed-phase column because of its powerful resolution and facility for MS detection. In this application more than 50 compounds in LXD were separated, while eight flavonoids and two iridoids were identified among the resolved components. Thanks to the non-specific, non-polar interactions of the target compounds with immobilized liposomes, ILC proved to be a powerful tool for the fingerprint analysis of the complex TCM prescription and can be employed for the study of drug–membrane interactions in LC, offering complementary selectivity when hyphenated with an ODS column.

The same approach has been tested by Liu et al. (2008) for the separation and identification of components in extracts of *Swertia franchetiana*. In this application, a longer column was used in the second dimension to enhance the separation capability and an *on-column focusing method* to reduce sample loss and contamination on concentrating fractions transferred from the first dimension. A total of more than 118 and 575 components were detected in the chloroform and *n*-butanol extracts respectively by 2D-HPLC–MS with respect to the 20 and 95 components detected by 1D-HPLC–MS, thus making the former a powerful tool for more comprehensive and more correct interpretation of complex samples in relation to the presence of eventually occurring low-abundant components, which may also have high pharmaceutical and/or toxic activity. An exhaustive example is provided in Figure 9.16, which shows clearly how a 2D-HPLC–MS approach allows us to resolve several peaks buried under main-component peaks in the traditional one-dimensional method, thus decreasing impurities and noise signals.

Another TCM application has been recently tested for the analysis of compounds in *Flos lonicera* (Chen et al., 2007). In this study, a combination of ion-exchange and RP chromatography has been employed successfully to separate more than 58 components and to identify six of them. For the separation of complex hydrophylic samples, a novel HILIC × HILIC qTOF MS system has been reported (Wang et al., 2008). For this task a TSKgel Amide-80 column and a short PolyHydroxyethyl A column have been used for the first and second dimensions, respectively. The separation capability of the 2D system developed was tested by separating saponins in a *Quillaria saponaria* extract. The employment of a high-speed qTOF MS detector as a third complementary dimension improved the peak capacity, significantly allowing the identification of more than 46 saponins by means of $[M - H]^-$ ions, characteristic product ions. Several pairs of isomers, which were often coeluted on conventional LC–MS methods and had similar fragmentation characteristics in MS–MS spectra, were well separated on the two-dimensional system based on their different hydrophilicity, demonstrating satisfactory orthogonal selectivity of the comprehensive 2D-HILIC setup.

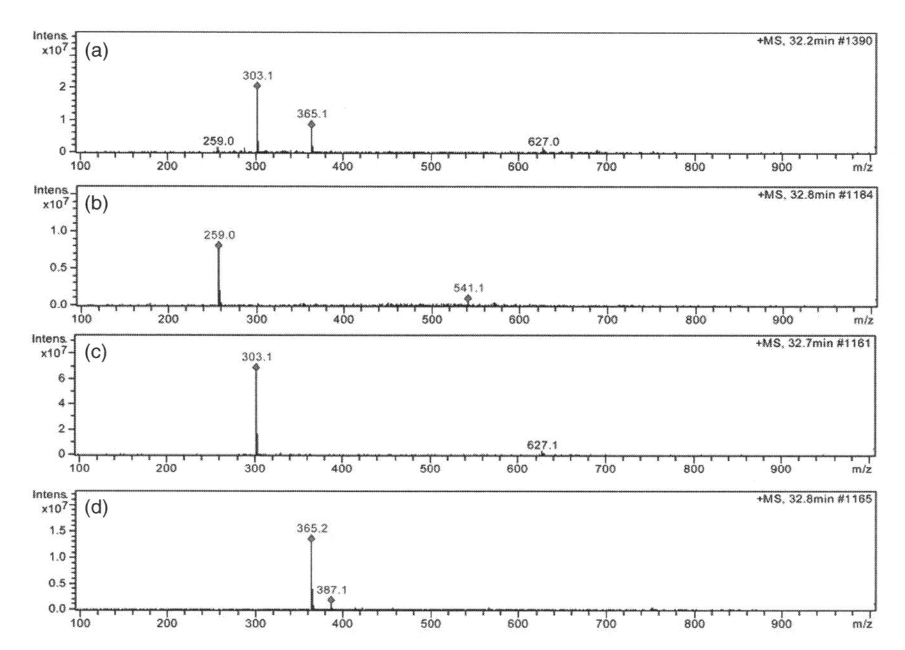

Figure 9.16 Mass spectra of peak 2 (a) obtained by one-dimensional analysis and related three compounds [(b) compound 61, (c) compound 22, (d) compound 15] obtained by two-dimensional analysis. [From Liu et al. (2008), with permission. Copyright © 2008 by Wiley-VCH Verlag Gmbh & Co. KGaA.]

9.3.4 LC × LC–MS Methods for Separating and Identifying Environmental Components

Environmental components include a variety of substances, such as polymers or pesticides, that can be found consistently indoors and outdoors, some of which may have short- or/and long-term adverse health effects. Comprehensive LC in combination with TOF MS proved to be an ideal tool for screening of such compounds because of the combined information of the elution pattern and the sensitive and accurate mass spectral data (Table 9.4).

Liquid chromatography represents a powerful tool in the molecular characterization of polymers, since they often have multivariate distribution in molecular characteristics such as molecular weight, chain architecture, chemical composition, and functionality (Berek, 2000; Chang, 2003, 2005; Glockner, 1991; Macko and Hunkeler, 2003; Pasch and Trathnigg, 1997; Trathnigg, 1995; Yau et al., 1979). In particular, size-exclusion chromatography (SEC) is the most widely employed technique in the molecular characterization of polymers, due to its high speed, facility to use, and wide applicability (Yau et al., 1979). The common configuration of 2D LC in the analysis of copolymers or functional polymers has been interaction chromatography or LC at critical conditions (LC–CC) for the first-dimension separation in order to separate in terms of the molecular

TABLE 9.4 Comprehensive 2D (LC x LC-MS) Separations of Synthetic and Natural Polymers and Oligomers

		Column Packing and Dimensions (Length x i.d., mm, d_p, μm); Flow rate; Elution Mode[a]				
Ref.	Type of Interface	First Dimension, D1	Second Dimension, D2	Detection	MS and MS-MS Parameters	Application
[1]	Two 2-position 6-port valves equipped with two alternative C_{18} columns	SEC (Macrosphere GPC) 250 × 4.6, 7 μm; GP	C_{18} 7.5 × 4.6, 5 μm; 1.0 mL/min; GP	UV, ESI-IT-MS		Proteins and small molecules
[2]	Two position 10-port valve	NP-GPEC Hypersil silica 250 × 3.0, 5 μm; 10 μL/min; IP	SEC HSPgel-RTMB-L/M 150 × 6.0, 3 μm; 0.8 mL/min; IP	UV, MALDI-TOF-MS		Poly Bisphenol A carbonate
[3]	Two position 10-port valve	LC-CC Inertsil platinum silica 150 × 3.2, 5 μm; IP	SEC PL 300 × 7.6; 1 mL/min; IP	UV, MALDI-TOF-MS		Poly Bisphenol A carbonate
[4]	Two position 10-port valve	RP-TGIC Kromasil C_{18}150 × 4.6; 20 μL/min; IP	SEC Plgel Mixed-C 300 × 7.5; 4 mL/min; IP	UV, MALDI-TOF-MS		Branched polystyrene
[5]	Two position 10-port valve	IEC 150 × 1.0, 10 μm; 40 μL/min; GP	C_{18} 50 × 4.6, 2.5 μm or 50 × 3.0, 2.5 μm; GP	UV, ESI-TOF-MS	80–800 *m/z*	Acidic compounds

[1] Winther and Reubsaet, 2005; [2] Coulier et al., 2005; [3]Coulier et al., 2006; [4] Im et al., 2006; [5] Pól et al., 2006.

[a] IP: isocratic program; GP: gradient program.

characteristics other than the molecular weight, primarily the composition or functionality, and SEC for the second-dimension separation to separate according to the molecular weight (Adrian et al., 2000; Pasch and Trathnigg, 1997; Trathnigg et al., 2002; Van der Horst and Schoenmakers, 2003). This combination of columns has been used by Coulier et al. (2005) to monitor the degradation of poly(bisphenol A) carbonate (PC). In particular, LC–CC was revealed to be a useful tool for observing differences in functionality as a result of hydrolytic degradation. The identification of peaks was accomplished by both MALDI–TOF MS and UV, coupled semi-online with LC–CC and comprehensive LC systems, respectively. In this regard, in the LC–CC–UV chromatogram of PC aged for 6 and 12 weeks, an additional peak at higher retention times showed up next to the peak at $t \sim t_0$ increasing in height with increasing degradation time. Through coupling of LC–CC to MALDI–TOF MS, it was possible to assign it to PC, with one OH end group and one t-butyl end group clearly demonstrating successful hyphenation of the two techniques (Figure 9.17). On the other hand, to fully

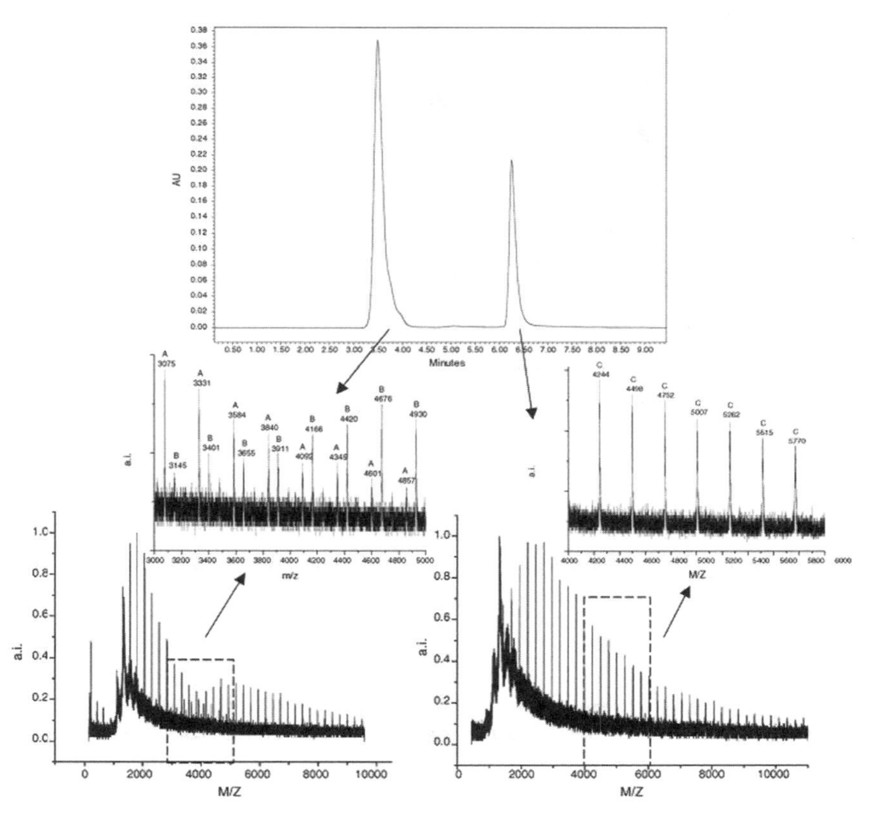

Figure 9.17 Identification of the peaks observed with LC–CC using semi-online coupling of MALDI–TOF MS. [From Coulier et al. (2005), with permission. Copyright © 2005, by Elsevier.]

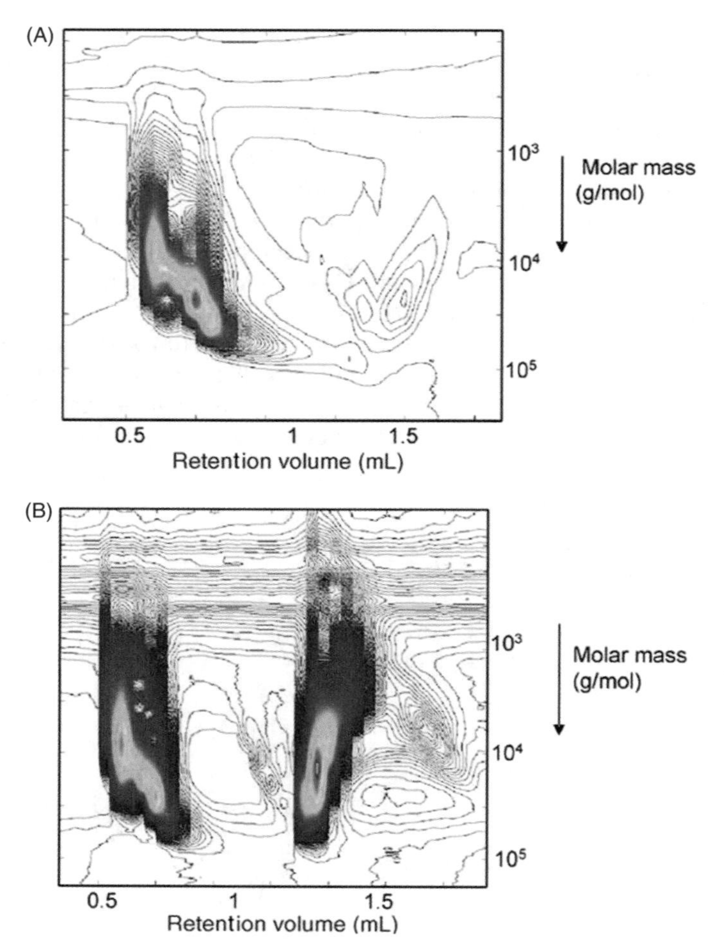

Figure 9.18 LC–CC × SEC–UV ($\lambda = 265$ nm) chromatograms of (A) virgin PC and (B) PC aged for 12 weeks. [From Coulier et al. (2005), with permission. Copyright © 2005 by Elsevier.

profit from the advantages of 2D LC, LC – CC × SEC was applied to the PC samples. Figure 9.18 shows the LC–CC × SEC chromatograms obtained with UV detection of virgin PC and PC aged for 12 weeks. The separation carried out in the second dimension by means of SEC revealed that there was a small molecular mass effect in the additional peak; the molecular mass increased with increasing retention times, thus achieving nearly critical separation for PC. Furthermore, as can be appreciated from Figure 9.18, PC with one OH end group and one *t*-butyl end group is present over the entire molecular mass distribution rather than, for example, only in the low molecular range. The latter information could not be obtained from the separate SEC or LC–CC chromatograms of the complete sample, thus illustrating the complementary information that can be obtained by using LC – CC × SEC. As an alternative, for the first-dimension separation, an NP-GPEC (gradient polymer elution chromatography) instrument has been compared. Such a technique, which performs separation on both molecular mass and polarity, has been employed successfully in the separation of copolymers or polymer blends based on their chemical composition conditions (Cools, 1999; Glockner, 1991; Philipsen, 1998; Staal et al., 1993). Also in this case, the information on molecular mass cannot be obtained from separate SEC and NP-GPEC measurements and hence show the additional information that can be obtained with 2D LC. In a more recent work, Coulier et al. (2006) developed the same 2D LC systems for the analysis of PC aged for different amounts of time. Using direct deposition of the eluent and subsequent analysis by MALDI–TOF MS, it was possible to obtain well-resolved mass spectra up to about 20,000 Da that included more specific information on the microstructure of the polymer as detected by LC–CC. From the masses and calibration of the MALDI–TOF MS obtained with PS standards, various mass distributions were identified as PC with either two *tert*-butyl end groups or cyclic oligomers. For PC aged for 12 weeks, the same pattern was found at other mass ranges, revealing the disappearence of cyclic PC oligomers as a result of degradation conditions. For polymer analysis, another interesting configuration was studied by Im et al. (2006), who employed reversed-phase temperature gradient interaction chromatography (RP-TGIC) in the first dimension to separate a branched polymer primarily according to molecular weight and liquid chromatography at critical conditions (LC–CC) in the second dimension, carrying out separation in terms of the number of branches (Figure 9.19). The 2D LC resolution of RP-TGIC × LC-CC combination worked better than common LC-CC × SEC, due to higher resolution of molecular weight in RP-TGIC than in SEC. An additional advantage exhibited by the system was the use of common eluent for the two LC separations, making them free of possible breakthrough or solvent plug effects. Figure 9.20 displays MALDI–TOF mass spectra of the two polystyrenes shown in Figure 9.19 (BS49 and BS65). The intensity of the peaks decreases as a consequence of the increase in molecular weight. This behavior is due to the fact that the mass spectrum reflects both the number distribution of the polymers and the mass discrimination phenomenon in MALDI–TOF MS analysis (Montaudo, 2002).

With regard to contaminant analyses, the only application comes from Pol et al., who developed a comprehensive two-dimensional liquid chromatography system interfaced to electrospray ionization TOF mass spectrometry for the determination of organic acids in atmospheric aerosols (Pol et al., 2006). Aerosol particulate material in the atmosphere affects not only the climate but has also been implicated in human disease and mortality. Chemical composition of aerosols can vary appreciably, often leading to very complex mixtures for which one-dimensional separations do not provide sufficient separation efficiency. For this application a micro strong cation-exclusion column and a C_{18} packed column were employed in the first and second dimension, respectively. Figure 9.21

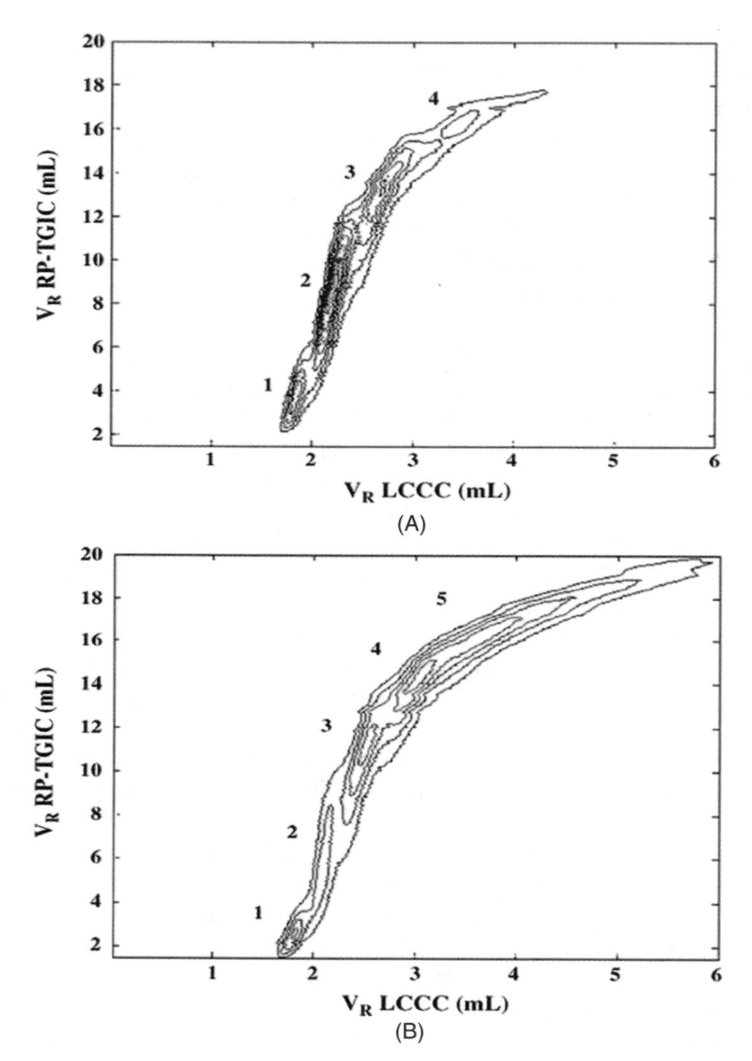

Figure 9.19 Contour plots of RP-TGIC × LCCC 2D LC chromatograms of BS49 (A) and BS65 (B). The numbers indicate the number of branches. [From Im et al. (2006), with permission. Copyright © 2006 by Elsevier.]

shows a contour plot of the 2D separation of acidic compounds from an extract of an urban background aerosol sample. The enhanced power of two-dimensional separation was demonstrated in the analysis of compounds in both rural and urban samples. The quantification was performed from a two-dimensional chromatogram using a selected ion chromatogram of the deprotonized ions $[M - H]^-$. The LC × LC–TOF MS system proved to be reproducible, linear, and sensitive, and thus challenging for complex environmental samples.

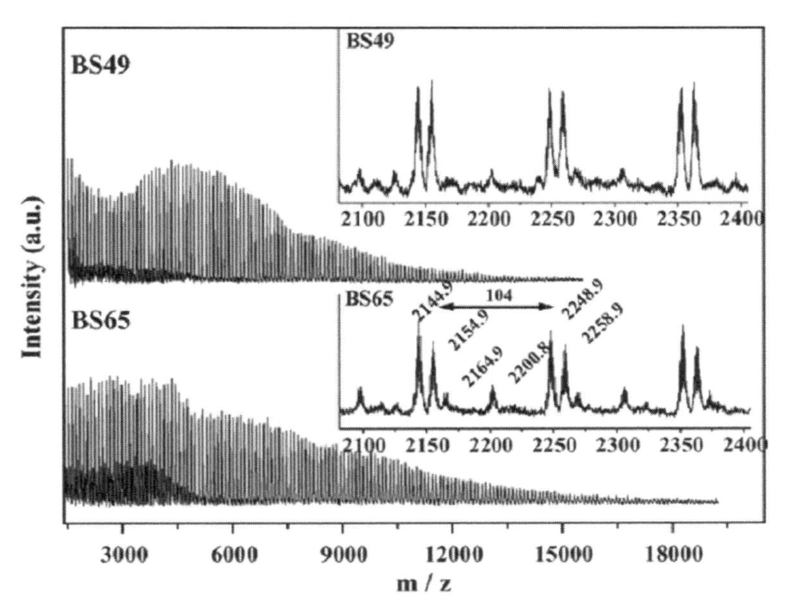

Figure 9.20 MALDI–TOF MS spectra of BS49 and BS65. In the insets, magnified spectra of BS49 and BS65 are shown (matrix, dithranol; salt, silver trifluoroacetate). [From Im et al. (2006), with permission. Copyright © 2006 by Elsevier.]

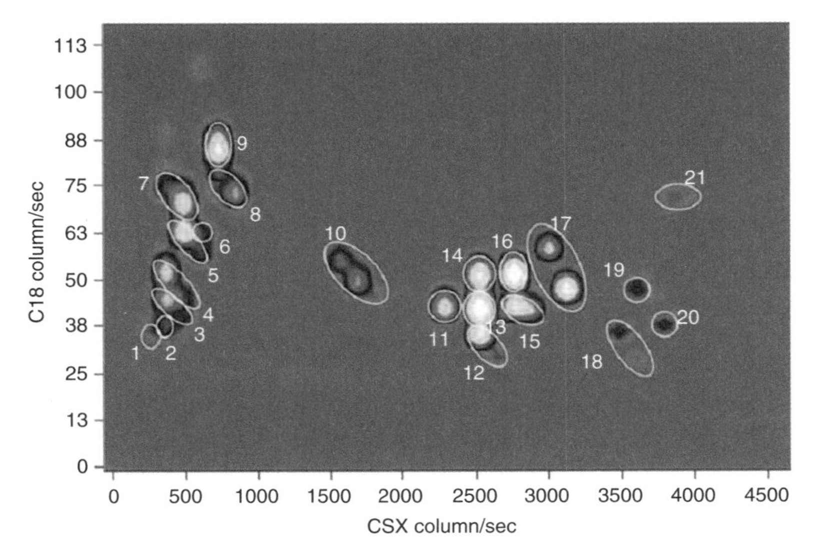

Figure 9.21 Contour plot of two-dimensional LC separation of rural forest sample from Hyytiala, Finland. 1, Not identified, m/z = 146.9648752; 2, not identified, m/z = 294.0645838; 3,4, pinonic acid; 5, 311.1660828; 6, 415.0678374; 7, n-octanoic acid; 8, n-nonanoic acid; 9, capric acid; 10, lauric acid; 11, n-pentadecanoic acid; 12, 339,2349723; 13, palmitic acid; 14, stearic acid. The zoomed areas show separation of azelaic acid (EIC 187.1) and isomers of pinic acid (185.1). [From Pól et al. (2006), with permission. Copyright © 2006 by Elsevier.]

REFERENCES

Adrian J, Esser E, Hellmann G, Pasch H. *Polymer* 2000; 41:2439–2449.

Aebersold R, Mann M. *Nature* 2003; 422:198–207.

Anderson NL, Anderson NG. *Mol. Cell. Proteom.* 2002; 1:845–867.

Barber M, Bordoli RS, Sedwwick RD, Tetler LW. *Org. Mass Spectrom.* 1981; 16:256–260.

Beranova-Giorgianni S. *Trends Anal. Chem.* 2003; 22:273–281.

Berek D. *Prog. Polym. Sci.* 2000; 25:873–908.

Biemann K. Biomed. *Environ. Mass Spectrom.* 1988; 16:99–111.

Caprioli RM, Fan T, Cottrell JS. *Anal. Chem.* 1986; 58:2949–2954.

Cavazzini A, Gritti F, Kaczmarski K, Marchetti N, Guiochon G. *Anal. Chem.* 2007; 79:5972–5979.

Chang TJ. *Adv. Polym. Sci.* 2003; 163:1–60.

Chang TJ. *Polym. Sci. Polym. Phys. Ed.* 2005; 43:1591–1607.

Chen XF, Zhang JY, Xue CX, Chen XG, Hu ZD. *Biomed. Chromatogr.* 2004; 18:673–680.

Chen XG, Kong L, Su XY, Fu HJ, Ni JY, Zhao RH, Zou HF. *J. Chromatogr. A* 2004; 1040(2):169–178.

Chen XG, Jiang ZL, Zhu Y, Tan JY. *Chromatographia* 2007; 65:141–147.

Cools PJCH. Ph.D. dissertation, Eindhoven Univeristy of Technology, Eindhoven, The Netherlands, 1999.

Coulier L, Kaal ER, Hankemeier T. *J. Chromatogr. A* 2005; 1070(1–2):79–87.

Coulier L, Kaal ER, Hankemeier T. *Polym. Degrad. Stab.* 2006; 91:271–279.

Dass C, Desiderio DM. *Anal. Biochem.* 1987; 163:52–66.

Delahunty C, Yates Jr. *Methods* 2005; 35:248–255.

de Stefano JJ, Langlois TJ, Kirkland JJ. *J. Chromatogr. Sci.* 2008; 46:254–260.

de Villiers A, Lestremau F, Szucs R, Gelebart S, David F, Sandra P. *J. Chromatogr. A* 2006; 1127:60–69.

Dole M, Mack LL, Hines RL, Mobley RC, Ferguson LD, Alice MB. *J. Chem. Phys.* 1968; 49:2240–2249.

Dugo P, Kumm T, Crupi ML, Cotroneo A, Mondello L. *J. Chromatogr. A* 2006a; 1112:269–275.

Dugo P, Kumm T, Chiofalo B, Cotroneo A, Mondello L. *J. Sep. Sci.* 2006b; 29:1146–1154.

Dugo P, Herrero M, Kumm T, Giuffrida D, Dugo G, Mondello L. *J. Chromatogr. A* 2008a; 1189:196–206.

Dugo P, Herrero M, Giuffrida D, Kumm T, Dugo G, Mondello L. *J. Agric. Food Chem.* 2008b; 56:3478–3485.

Dugo P, Cacciola F, Herrero M, Donato P, Mondello L. *J. Sep. Sci.* 2008c; 31:3297–3308.

Dugo P, Cacciola F, Kumm T, Dugo G, Mondello L. *J. Chromatogr. A* 2008d; 1184:353–368.

Dugo P, Cacciola F, Kumm T, Dugo G, Mondello L. *J. Liq. Chromatogr. Relat. Technol.* 2008e; 31:1758–1807.

Dugo P, Giuffrida D, Herrero M, Donato P, Mondello L. *J. Sep. Sci.* 2009; 32:973–980.

Eggink M, Romero W, Vreuls RJ, Lingeman H, Niessen WMA, Irth HJ. *J. Chromatogr. A* 2008; 1188:216–226.

Eng JK, McCormack AL, Yates JR III. *J. Am. Soc. Mass Spectrom.* 1994; 5:976–989.

François I, de Villiers A, Tienpont B, David F, Sandra P. *J. Chromatogr. A* 2008; 1178:33–42.

François I, Sandra K, Sandra P. *Anal. Chim. Acta* 2009; 641:14–31.

Gao M, Deng C, Lin S, Hu F, Tang J, Yao N, Zhang X. *J. Sep. Sci.* 2007; 30:785–791.

Gevaert K, Goethals M, Martens L, Van Damme J, Staes A, Thomas GR, Vandeckrckhove J. *Nat. Biotechnol.* 2003; 21:566–569.

Giddings JC. *J. High Resolut. Chromatogr.* 1987; 10:319–323.

Glockner G. *Gradient HPLC of Copolymers and Chromatographic Cross-Fractionation*. New York: Springer-Verlag, 1991.

Gorg A, Weiss W, Dunn M. *J. Proteomi.* 2004; 4:3665–3685.

Gritti F, Guiochon G. *J. Chromatogr. A* 2007; 1166:30–46.

Gritti F, Cavazzini A, Marchetti N, Guichon G. *J. Chromatogr. A* 2007; 1157:289–303.

Gu M, Zhang GF, Su ZG, Ouyang F. *J. Chromatogr. A* 2004; 1041:239–243.

Guillarme D, Heinisch S, Rocca JL. *J. Chromatogr. A* 2004; 1052:39–51.

Guo FQ, Liang YZ, Xu CJ, Huang LF, Li XN. *J. Chromatogr. A* 2004; 1054:73–79.

Gygi SP, Rist B, Gerber SA, Turecek F, Gelb MH, Aebersold R. *Nat. Biotechnol.* 1999; 17:994–999.

Hájek T, Škeříková V, Česla P, Vyňuchalová K, Jandera P. *J. Sep. Sci.* 2008; 31:3309–3328.

Hamdan M, Righetti PG, Eds. *Proteomics Today: Protein Assessment and Biomarkers Using Mass Spectrometry, 2D Electrophoresis, and Microarray Technology*. Hoboken, NJ: Wiley, 2005.

Hanash S. *Nature* 2003; 422:226–232.

Hancock WS, Bishop CA, Prestige RL, Harding DRK, Hearn MTW. *Science* 1978; 200:1168–1170.

Holčapek M, Lísa M, Jandera P, Kabátová N. *J. Sep. Sci.* 2005; 28:1315–1333.

Hu L, Chen X, Kong L, Su X, Ye M, Zou H. *J. Chromatogr. A* 2005; 1092:191–198.

Hu L, Li X, Feng S, Kong L, Su X, Chen X, Qin F, Ye M, Zou H. *J. Sep. Sci.* 2006; 29(16):881–888.

Huidobro AL, Pruim P, Schoenmakers P, Barbas C. *J. Chromatogr. A* 2008; 1190:182–190.

ICH 3QA. *Guideline for Industry: Impurities in New Drug Substances*. Washington, DC: U.S. Food and Drug Administration, Jan. 1996.

ICH 3QB. *Guideline for Industry: Impurities in New Drug Substances*. Washington, DC: U.S. Food and Drug Administration, Nov. 1997.

Im K, Kim Y, Chang T, Lee K, Choi N. *J. Chromatogr. A* 2006; 1103(2):235–242.

Ito Y, Takeuchi T, Ishi D, Goto M. *J. Chromatogr.* 1985; 346:161–166.

Jandera P. *J. Sep. Sci.* 2006; 29:1763–1783.

Kajdan T, Cortes H, Kuppannan K, Young SA. *J. Chromatogr. A* 2008; 1189:183–195.

Kalili KM, de Villiers A. *J. Chromatogr. A* 2009; 1216:6274–6284.

Kimura H, Tanigawa T, Morisaka H, Ikegami T, Hosoya K, Ishizuka N, Minakuchi H, Nakanishi K, Ueda M, Cabrera K, Tanaka N. *J. Sep. Sci.* 2004; 27:897–904.

Kivilompolo M, Hyötyläinen T. *J. Chromatogr. A* 2007; 1145:155–164.

Kivilompolo M, Obůrka V, Hyötyläinen T. *Anal. Bioanal. Chem.* 2008; 391:373–380.

Kong L, Li X, Zou H, Wang HL, Mao X, Zhang Q, Ni J. *J. Chromatogr. A* 2001; 936:111–118.

Lemière F. Interfaces for LC-MS. In: *Guide to LC-MS*. Chester: Advanstar, 2001a.

Lemière F. Mass analyzers for LC–MS. In: *Guide to LC-MS*. Chester: Advanstar, 2001b.

Link AJ, Eng J, Schieltz DM, Carmack GJ, Mize GJ, Morris DR, Garvick BM, Yates JR III. *Nat. Biotechnol.* 1999; 17:676–682.

Liu H, Berger SJ, Chakraborty AB, Plumb RS, Cohen SA. *J. Chromatogr. B* 2002; 782:267–289.

Liu Y, Xu Q, Xue X, Zhang F, Liang X. *J. Sep. Sci.* 2008; 31(6):935–944.

Macfarlane RD, Togerson DF. *Int. J. Mass Spectrom. Ion Phys.* 1976; 21:81–92.

Machtejevas E, John H, Wagner K, Ständker L, Marko-Varga G, Forssmann W-G, Bischoff R, Unger KK. *J. Chromatogr. B* 2004; 803:121–130.

Macko T, Hunkeler D. *Adv. Polym. Sci.* 2003; 163:61–136.

McNair JE, Patel KD, Jorgenson JW. *Anal. Chem.* 1999; 71:700–708.

Marchetti N, Cavazzini A, Gritti F, Guichon GJ. *J. Chromatogr. A* 2007; 1163:203–211.

Meiring HD, van Der Heeft E, ten Hove GJ, de Jong A. *J. Sep. Sci.* 2002; 25:557–568.

Meng CK, Mann M, Fenn JB. *Z. Phys. D* 1988; 10:361–368.

Mitulovic G, Swart R, van Ling R, Jackob T, Chervet JP. *LC-GC Eur.* 2004; 16:61–64.

Mondello L, Tranchida PQ, Stanek V, Jandera P, Dugo G, Dugo P. *J. Chromatogr. A* 2005; 1086:91–98.

Montaudo MS. *Mass Spectrom. Rev.* 2002; 21:108–144.

Motoyama A, Yates JR III. *Anal. Chem.* 2008; 80:7187–7193.

Murphy RE, Schure MR, Foley JP. *Anal. Chem.* 1998; 70:1585–1594.

Niessen WM. *Chromatographic Science Series*. New York: Marcel Dekker, 1999, p. 79.

Opiteck GJ, Lewis KC, Jorgenson JW, Anderegg RJ. *Anal. Chem.* 1997a; 69:1518–1524.

Opiteck GJ, Jorgenson JW, Anderegg RJ. *Anal. Chem.* 1997b; 69:2283–2290.

Opiteck GJ, Ramirez SM, Jorgenson JW, Moseley MA. III *Anal. Biochem.* 1998; 258:349–361.

Pasch H, Trathnigg B. *HPLC of Polymers*. Berlin: Springer-Verlag, 1997.

Peng J, Elias JE, Thoreen CC, Licklider LJ, Gygi SP. *J. Proteome Res.* 2003; 2:43–50.

Perkins DM, Pappin DJ, Creasy DM, Cottrell JS. *Electrophoresis* 1999; 20:3551–3567.

Philipsen HJA. Ph.D. dissertation, Eindhoven University of Technology, Eindhoven, The Netherlands, 1998.

Pól J, Hohnová B, Jussila M, Hyötyläinen T. *J. Chromatogr. A* 2006; 1130(1):64–71.

Pól J, Hohnová B, Hyötyläinen T. *J. Chromatogr. A* 2007; 1150:85–92.

Pól J, Hyötyläinen T. *Anal. Biochem.* 2008; 391:21–31.

Premstaller A, Oberacher H, Walcher W, Timperio AM, Zolla L, Chervet JP, Cavusoglu N, van Dorsselaer A, Huber CG. *Anal. Chem.* 2001; 73:2390–2396.

Premstaller A, Oberacher H, Walcher W, Timperio AM, Zolla L, Chervet JP, Cavusoglu N, van Dorsselaer A, Huber CG. *Anal. Chem.* 2005; 77:6426–6434.

Rabilloud T. *Proteomics* 2002; 2:3–10.

Reinders J, Zahedi RP, Pfanner N, Meisinger C, Sickmann A. *J. Proteome Res.* 2006; 5:1543–1554.

Roepstorff P, Fohlman J. *Biomed. Mass Spectrom.* 1984; 11: 601.

Sandra P, Vanhoenacker G, Lynen F, Li L, Schelfaut M. Considerations on column selection and operating conditions for LC-MS. In: *Guide to LC-MS*. Chester: Advanstar, 2001.

Sandra K, Moshir M, D'hondt F, Verleysen K, Kas K, Sandra P. *J. Chromatogr. B* 2008; 866:48–63.

Sandra K, Moshir M, D'hondt F, Tuytten R, Verleysen K, Kas K, Françis I, Sandra P. *J. Chromatogr. B* 2009; 877:1019–1039.

Seki S, Kambara H, Naoki H. *Org. Mass Spectrom.* 1985; 20:18–24.

Simpson RJ, Ed. *Proteins and Proteomics: A Laboratory Manual*. Cold Spring Harbor, NY: Cold Spring Harbor Laboratory Press, 2003.

Staal WJ, Cools PJCH, van Herk AM, German AL. *Chromatographia* 1993; 37:218–220.

Tanaka K, Ido Y, Akita S, Yoshida T. *Rapid Commun. Mass Spectrom.* 1988; 2:151–156.

Thomson JJ. *Rays of Positive Electricity and Their Application to Chemical Analyses*. London: Longmans, Green, 1913.

Thomson BA, Iribarne JV. *J. Chem. Phys.* 1979; 71:4451–4463.

Trathnigg B. *Prog. Polym. Sci.* 1995; 20:615–650.

Trathnigg B, Rappel C, Raml R, Gorbunov A. *J. Chromatogr. A* 2002; 953:89–99.

van der Horst A, Schoenmakers PJ. *J. Chromatogr. A* 2003; 1000:693–709.

van der Klift EJC, Vivó-Troyols G, Claassen FW, Van Holthoon FL, Van Beek TA. *J. Chromatogr. A* 2008; 1178:43–55.

Vékey K. *J. Chromatogr. A* 2001; 921:227–236.

Venkatramani CJ, Patel A. *J. Sep. Sci.* 2006; 29:510–518.

Wagner K, Miliotis T, Marko-Varga G, Bischoff R, Unger KK. *Anal. Chem.* 2002; 74:809–820.

Wang HL, Zou HF, Ni JY, Kong L, Gao S, Guo B. *J. Chromatogr. A* 2000; 870:501–510.

Wang Y, Zhang J, Liu C-L, Gu X, Zhang X. *Anal. Chim. Acta* 2005; 530:227–235.

Wang Y, Lu X, Xu G. *J. Chromatogr. A* 2008; 1181:51–59.

Wang Y, Kong L, Lei X, Hu L, Zou H, Welbeck E, Bligh SWA, Wamg Z. *J. Chromatogr. A* 2009; 1216:2185–2191.

Washburn MP, Wolters DA, Yates JR III. *Anal. Chem.* 2001; 73:5683–5690.

Wasinger VC, Cordwell SJ, Cerpa-Poljak A, Yan JX, Gooley AA, Wilkins MR, Duncan MW, Harris R, Williams KL, Humphery-Smith I. *Electrophoresis* 1995; 16:1090–1094.

Way WK. *The Reporter Europe.* Supelco, Oct. 28, 2007, pp. 3–4.

Way WK, Campbell W. *The Application Notebook.* Supelco, Feb. 2007, p. 55.

Wehr T. *LC-GC Eur.* Mar. 2003, pp. 2–8.

Wilkins MR, Williams KR, Hochstrasser DF, Eds. *Proteome Research: New Frontiers in Functional Genomics*. New York: Springer-Verlag, 1997.

Williams DH, Bradely C, Bojesen G, Santikaran S, Taylor LCE. *J. Am. Chem. Soc.* 1981; 103:5700–5704.

Wilm M, Mann M. *Anal. Chem.* 1996; 68:1–8.

Wilson RS, Jankowski M, Pepaj M, Mihailova A, Fernando B, Truyols GV, Lundanes E, Greibrokk T. *Chromatographia* 2007; 66:469–474.

Winther B, Reubsaet JLE. *J. Sep. Sci.* 2005; 28:477–482.

Wolters DA, Washburn MP, Yates JR III. *Nat. Biotechnol.* 2001a; 19:242–247.

Wolters DA, Washburn MP, Yates JR III. *Anal. Chem.* 2001b; 73:5683–5690.

Yamashita M, Fenn JB. *J. Phys. Chem.* 1984; 88:4451–4459.

Yau WW, Kirkland JJ, Bly DD. *Modern Size-Exclusion Liquid Chromatography: Practice of Gel Permeation and Gel Filtration Chromatograph*. New York: Wiley, 1979.

Zschocke S, Liu JH, Stuppner H, Bauer R. *Phytochem. Anal.* 1998; 9:283–290.

10

COMPREHENSIVE TWO-DIMENSIONAL LIQUID CHROMATOGRAPHY APPLICATIONS

PAOLA DUGO AND LUIGI MONDELLO
University of Messina, Messina, Italy

FRANCESCO CACCIOLA
Chromaleont s.r.l., A spin-off of the University of Messina, Messina, Italy and University of Messina, Messina, Italy

PAOLA DONATO
University Campus Bio-Medico, Rome, Italy and University of Messina, Messina, Italy

Multidimensional liquid chromatography has been used widely in many different combinations of LC techniques, generating increased peak capacity, selectivity, and resolution, especially in the comprehensive LC mode. Since 1978, when the first comprehensive two-dimensional (2D) LC system was presented by Erni and Frei, the number of comprehensive 2D LC applications has grown very fast, mainly in the past decade. Comprehensive 2D LC × LC methods have been developed and applied mainly to the separation of peptides and proteins, polymers, natural molecules such as polyphenols or lipids, and pharmaceutical samples. Based on the difference in size, polarity, shape, and acidity of the target analytes, many different approaches have been developed.

Most of the time, the reversed-phase (RP) separation mode represents one of the dimensions of the 2D LC system. Many different detection systems can be used in LC × LC applications. Although MS detection is the most informative, many other detectors can be used successfully in such a system: DAD (diode array detection) is also very informative if target analytes contain chromophores, but ultraviolet (UV), fluorescence, and coulometric detection can also be used in many applications.

Comprehensive Chromatography in Combination with Mass Spectrometry, First Edition.
Edited by Luigi Mondello.

The number of applications has increased very rapidly in the last decade, mainly because of the developments in the field of high-performance liquid chromatographic (HPLC) instrumentation, new stationary phases and column technologies, the development of dedicated software for data processing, and transformation into a two-dimensional chromatogram, all giving the necessary information to perform qualitative and quantitative analysis. LC × LC applications can be classified according to the LC techniques used in the two dimensions or to the type of sample analyzed.

For our discussion, applications have been gathered into four sections according to the various areas of research. Tables 10.1 to 10.4 summarize the principal information relative to the instrumental setup (type of interface, column sets, detector) and the type of analyte or sample, respectively, for synthetic and natural polymers and oligomers (Table 10.1), natural components (Table 10.2), pharmaceutical components (Table 10.3), and biological samples (Table 10.4). In the corresponding four sections of this chapter we describe the applications in more detail.

10.1 COMPREHENSIVE 2D LC SEPARATION OF SYNTHETIC AND NATURAL POLYMERS

Most comprehensive LC separations of synthetic and natural polymers have been carried out using LC × size-exclusion chromatography (SEC), where the LC can be represented in both the normal and the reversed phase (NP and RP) (Table 10.1). This combination permits separation based on the various functionalities (end groups with different polarities) in the first dimension and molar mass distribution in the second dimension. Under NP conditions, a silica column is generally used, whereas under RP conditions, a C_{18} column is employed, in both cases under isocratic or gradient mode.

As reported by Murphy et al. (1998a), one of the most challenging applications of two-dimensional LC separation systems are the separation of polymers, copolymers, and large molecules. They investigated the effect of sampling time on two-dimensional resolution by using 2D LC (RPLC–GPC, gel permeation chromatography) connected via an eight-port two-position valve equipped with loops for the separation of poly(ethylene glycol) (PEG) and surfactants with different alkyl and ethylene oxide chain lengths. They changed the sampling time into the second dimension by changing the split ratio of the first-dimension column effluent keeping the sample loop volume and column flow rate constant. The authors demonstrated that to obtain high two-dimensional resolution, each peak in the first dimension should be sampled at least three times into the second dimension when the sampling is in phase, and at least four times if the time is maximally out of phase. Keeping these results in mind, method development for 2D LC should be carried out as follows: the second-dimension method should be developed first, to provide the fastest analysis time with adequate resolution. Then the first dimension should be operated, to give an 8σ peak with four times the second-dimension elution window. This will make it possible

TABLE 10.1 Comprehensive 2D (LC x LC) Separations of Synthetic and Natural Polymers and Oligomers

Ref.	Type of Interface	Column Packing and Dimensions (Length x i.d., mm, d_p, μm); Flow rate; Elution Mode[a]		Detection	Application
		First Dimension, D1	Second Dimension, D2		
[1]	Two-position 8-port valve	C_{18} 150 × 3.0, 5 μm; 0.1 mL/min; GP	SEC 50 × 8.0, 3 μm; 1 mL/min; IP	ELSD	Polyethylene glycols and surfactants
[2]	Two-position 8-port valve	Amino 150 × 4.6, 3 μm; 0.05 mL/min; GP	C_{18} 33 × 4.6, 3 μm; 1.5 mL/min; IP	ELSD	Alcohol ethoxylates
[3]	Two-position 8-port valve	(a) Silica 250 × 1.0, 3 μm; 3 or 4 μL/min; IP (b) C_{18} 150 × 1.0, 5 μm; 4 μL/min; GP	SEC (a) 50 × 7.5, 3 μm; 0.4 mL/min; IP (b) 75 × 4.6, 5 μm; 0.4 mL/min; IP	UV (254 nm)	Functional polymers and co(polymers)
[4]	Two-position 6-port valve fitted with microelectric 2-position valve actuators	C_{18} 250 × 4.6, 5 μm; 0.1 mL/min; GP	Zirconia-carbon 30 × 4.6, 3 μm; 2 mL/min; GP	UV (262 nm)	Isomers of oligostyrenes
[5]	Four two-position 6-port valves fitted with microelectric 2-position valve actuators	C_{18} 250 × 4.6, 5 μm; 0.1, 0.7, or 1.0 mL/min; IP	Zirconia-carbon 30 × 4.6, 3 μm; 2.0 mL/min; IP	UV (262 nm)	Separation of isomers of oligostyrenes

(*continued overleaf*)

TABLE 10.1 (*Continued*)

Ref.	Type of Interface	Column Packing and Dimensions (Length x i.d., mm, d_p, μm); Flow rate; Elution Mode[a]		Detection	Application
		First Dimension, D1	Second Dimension, D2		
[6]	Two-position 10-port valve	Two Silica 150 × 1.0, 3 μm; 8 μL/min; IP	SEC One or two 50 × 4.6, 5 μm; 0.9 mL/min; IP	ELSD, PDA (220–300 nm)	Functional acrylate polymers
[7]	Two-position 10-port valve	Silica 150 × 3.9, 4 μm; 40 μL/min; GP	SEC (a) 50 × 20, 3 μm; 6.0 mL/min; IP (b) 150 × 6.0, 3 μm; 0.8 mL/min; IP (c) 50 × 4.6, 5 μm; 0.3 mL/min; IP	FTIR, PDA	Styrene-methylacrylate copolymers
[8]	Two-position 6-port valve	SEC (a) 250 × 4.6; 10 μL/min; IP (b) 250 × 1.0; 4 μL/min; IP	SEC (a) 50 × 4.6; 0.6 ml/min; IP (b) 150 × 4.6; 1.5 mL/min; IP	UV (260 nm)	Polystyrenes standards
[9]	Two-position 8-port valve	C_{18} 250 × 4.6, 5 μm; 0.018–0.064 mL/min; IP	SEC 50 × 2.0, 5 μm; 2.5–8 mL/min; IP	ELSD	PEG-g-PVAc copolymers
[10]	Two-position 10-port valve	Silica Two 150 × 1.0, 3 μm	SEC One or two 50 × 4.6, 5 μm; 0.9 mL/min; IP	UV (220 nm)	Poly(methyl-methacrylate) polymers

[11]	Two-position 10-port valve	C_{18} 150 × 1.0, 5 μm; 10 μL/min; GP	Amino-propyl silica (a) 150 × 1.0, 5 μm (b) 50 × 1.0, 3 μm; 0.5 mL/min; IP	ELSD	Co(oligomers)
[12]	Four two-position 6-port valves fitted with microelectric 2-position valve actuators	C_{18} 150 × 4.6, 5 μm; 0.1 mL/min; IP	Zirconia-carbon 50 × 4.6; 2.0 mL/min; IP	UV (262 nm)	Isomers of oligostyrenes
[13]	Two-position 10-port valve	C_{18} 250 × 1.0, 5 μm; 50 μL/min; GP	SEC 300 × 7.5, 5 μm; 1 mL/min; IP	UV (285 nm)	Sulphonated lignins

[1] Murphy et al., 1998a; [2] Murphy et al., 1998b; [3] Van der Horst and Schoenmakers, 2003; [4] Gray et al., 2003; [5] Gray et al., 2004; [6] Jiang et al., 2005; [7] Kok et al., 2005; [8] Popovici et al., 2005; [9] Knecht et al., 2006; [10] Vivó -Troyols and Schoenmakers, 2006; [11] Jandera et al., 2006; [12] Toups et al., 2006; [13] Brudin et al., 2008.

[a] IP: isocratic program; GP: gradient program.

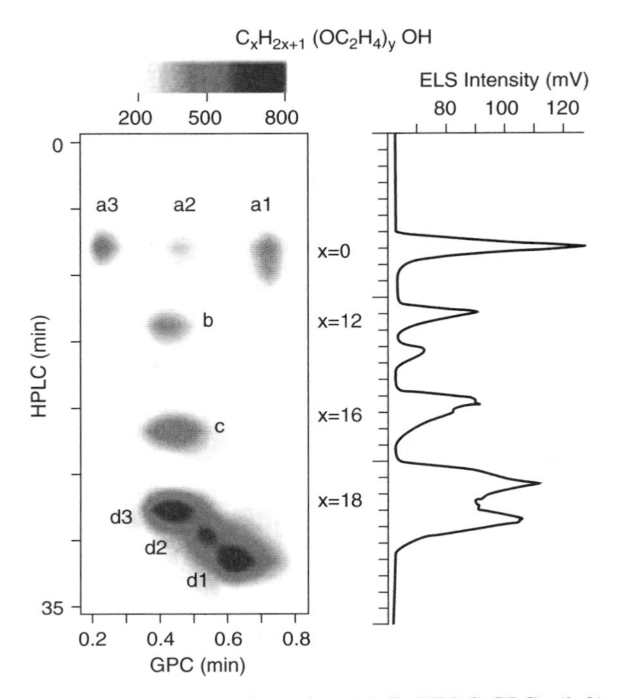

Figure 10.1 One- (right) and two-dimensional LC (HPLC-GPC) (left) contour chromatograms of the PEG and Brij mixture. (a) PEG 200 (a1, $x = 0$, $y = 4.5$), PEG 1000 (a2, $x = 0$, $y = 23$), and PEG 8000 (a3, $x = 0$, $y = 182$); (b) Brij 35 ($x = 12$, $y = 23$)); (c) Brij 58 ($x = 16$, $y = 20$)); (d) Brij 72 (d1, $x = 18$, $y = 2$), Brij 76 (d2, $x = 18$, $y = 10$), and Brij 78 (d3, $x = 18$, $y = 20$). The HPLC flow rate is 0.1 mL/min and the solvent composition is 98 : 2 (methanol/water). The GPC flow rate is 1 mL/min tetrahydrofuran. The sampling time is 1.5 min and only 0.67 min of the GPC axis is shown. [From Murphy et al. (1998a), with permission. Copyright © 1998 by The American Chemical Society.]

to obtain the highest resolution in the shortest amount of time. Figure 10.1 shows the one- and two-dimensional LC contour chromatograms for the analysis of a mixture of PEG and Brij. A decrease in R_s is observed when an increase in the sampling time occurs. Regarding the separation, ethylene oxide chains of different length cannot be separated by RPLC, whereas it is accomplished by GPC.

Synthetic polymers are very complex mixtures made up of many different compounds built up from one or a few different repeat units. Properties of polymers depend on the type of repeat unit(s) or monomer(s) used, the average molecular size and structure, and the variation around the average size and structure. NPLC × SEC has been used in different applications presented by Schoenmakers' group on the analysis of synthetic polymers.

The usefulness of the LC × SEC system for the separation and characterization of synthetic polymers has been demonstrated by a study of retention behavior in the vicinity of the critical conditions (where retention is independent of the polymer molecular mass), and by studying the separation of functional polystyrenes (PSs) and of copolymers of polystyrene and poly(methyl methacrylate) (PMMA) (van der Horst and Schoenmakers, 2003). A similar setup was used to determine the mutually dependent molar mass distributions and functionality-type distribution of functional PMMA polymers (Jiang et al., 2005) using both qualitative and quantitative information obtained using chromatographic data and UV and evaporative light scattering detection (ELSD). Figure 10.2 shows the LC × SEC chromatogram of a real sample (VL37A) (Jiang et al., 2005). It is easy to see that if only one-dimensional LC or SEC was used, peaks 3 and 4 were difficult to separate completely. Results obtained showed that accurate molar mass information [M_n, M_w, M_p, and the polydispersivity index (PDI)] could be obtained for individual peaks or a combination of peaks. Original ELSD chromatograms could be used to obtain the approximate molar mass values (M_n and M_p), but PDI and M_w were somewhat underestimated.

Many different detectors can be used in combination with LC or SEC of polymers. However, UV detection can be used only when chromophores are present in the structure (UV active polymers); refractive index (RI) detection exhibits a low sensitivity, while ELSD yields a nonlinear response. In contrast, infrared (IR) spectroscopy is better suited for selective detection in functional groups. An LC × SEC system was coupled to an IR flow cell and used for the functional group analysis of a series of styrene–methacrylate (SMA) copolymers with varying styrene content (Kok et al., 2005). UV detection allowed only the styrene part of the copolymer to be studied. Use of an IR flow cell made it possible to obtain contour plots for the carbonyl-stretching vibration characteristic of MA (1708 to 1748 cm^{-1}), and the C=C-stretching vibration characteristic of styrene (1510 to 1482 cm^{-1}). LC × SEC coupled to IR spectroscopy makes it possible to detect differences in functional groups between polymer samples, useful, for example, in the determination of changes in chemical composition due to polymer aging.

The analysis of data obtained from comprehensive two-dimensional chromatography using a chemometric approach has been the object of a further study carried out by Vivó-Truyols and Schoenmakers (2006). The method was designed for the analysis of families of compounds. It is based on the application of retention models that relate the retention of the sample to the properties that vary among a particular family of compounds. The method describes the data using a new pair of axes (replacing the retention time axes); one axis collects all the variance of a particular family of compounds (the chemical variance) while the other axis is constant. The method was applied to the analysis of functional acrylate polymers by LC × SEC, and it performed correctly. Although less frequent, RP × SEC separation has been applied to the analysis of poly(ethylene glycol)–poly(vinyl acetate) graft copolymers (PEG-g-PVAC) (Knetcht et al., 2006) and sulfonated lignins (Brudin et al., 2008).

The SEC separation method can give the complete molarmass distribution of a polymer. For this, it has been used as one of the two dimensions in a multidimensional LC system to obtain information on different distributions corresponding to

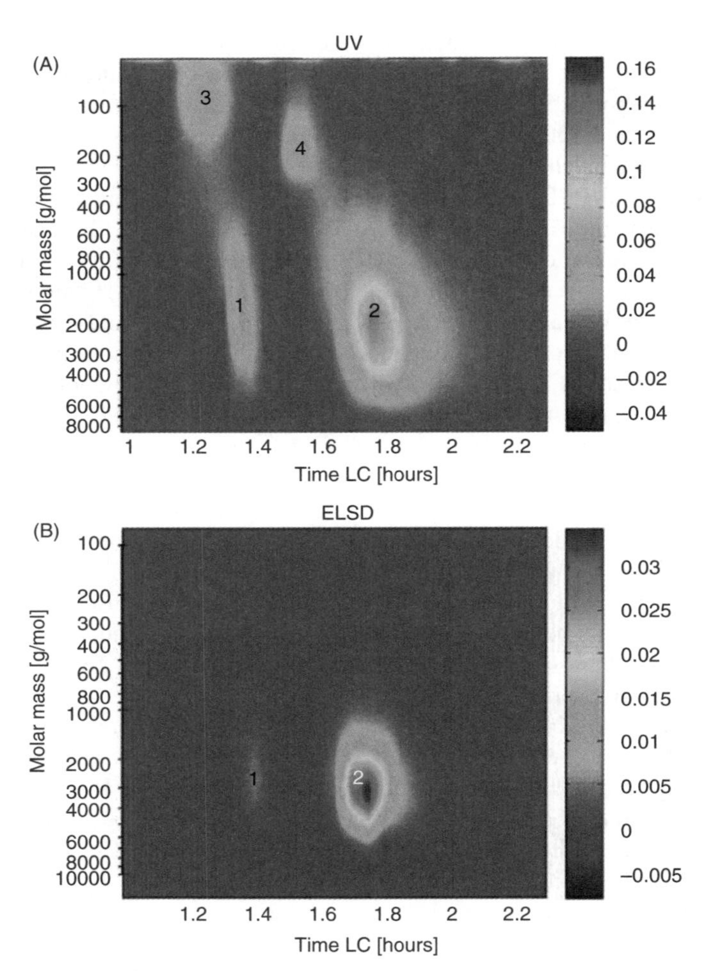

Figure 10.2 LC × SEC chromatogram of sample VL37A: (A) UV detection at 300 nm; (B) ELSD. Peak 1 represents nonfunctional PMMA and peak 2 represents mono-OH PMMA. Peak 3 in (a) represents molecules with a very low molar mass and without OH end groups; peak 4 may represent residual (unreacted) RAFT agent used in the polymerization process. LC columns: two 150 mm × 1.0 mm i.d., 3 μm, 100 Å bare silica; 50% ACN in DCM and the flow rate was 4 μL/min. [From Jiang et al. (2005), with permission. Copyright © 2005 by Elsevier.]

the relevant properties of the molecules. The coupling of SEC in both dimensions of a comprehensive 2D LC system is not a useful technique for characterizing complex polymers. However, a SEC × SEC system has been realized and used to study band-broadening phenomena in SEC (Popovici et al., 2005). The first dimension is used for fractionation purposes. If narrow fractions are collected and transferred into the second SEC column, the band broadening is due only to chromatographic dispersion.

When developing a two-dimensional chromatographic separation, attention must be paid so that ideally each separation dimension must be selective to a particular sample attribute. In this step, attention should also be focused on the appropriate mobile phases, since this can give vast differences in the selectivity of any given stationary phase. An RP × RP approach has been developed by Gray et al. (2003, 2004) and Toups et al. (2006) using the combination of a C_{18} column (operated with methanol) in the first dimension and a carbon-clad zirconia column (operated with acetonitrile) in the second dimension, for the separation of isomers in a mixture of oligostyrene with various numbers of styrene units and various alkyl end groups. The separation system was selected on the basis of data on the chromatographic orthogonality evaluated by using information theory (IT) and factor analysis.

Multidimensional samples containing two or more types of repeat structural units with different size and polarities can be separated by a combination of RP and NP LC systems. Jandera et al. (2006) developed an RP × NP system for the separation of ethylene oxide–propylene oxide (EO-PO) (co)-oligomers. Because of the different polarities of EO (polar) and PO (nonpolar) units, RPLC with a C_{18} column is suitable for the separation of (co)oligomers according to the PO unit distribution, whereas the species with different EO units coelute. Under an NPLC mode, retention varies according to the number of EO units and is little affected by the number of PO units. Hydrophilic interaction liquid chromatography (HILIC) with an aminopropyl silica column (APS) was used as the second dimension because it is more resistant than classical nonaqueous NP systems against adsorbent deactivation with aqueous solvents transferred from the first RP column.

The opposite combination, NP × RP LC, was developed by Murphy et al. (1998b) for the separation of alcohol ethoxylates with a dual distribution of non-polar alkyl and polar ethylene oxide units by using either a silica or an amino column in the first dimension and a conventional C_{18} column in the second dimension. Solvents used in the two dimensions were miscible, so no problems of mobile-phase incompatibility typical using NP–RP were encountered. Figure 10.3 shows the 2D LC analysis of a real sample (Neodol 25-12), where four distributions with different alkyl end groups are clearly seen (Murphy et al., 1998b).

10.2 COMPREHENSIVE 2D LC SEPARATION OF NATURAL PRODUCTS AND ANTIOXIDANTS

Many different classes of components may be present as main constituents and minor components in the natural products. These consist of fruit and herbs (raw agricultural products) but other agricultural and food products as well, such as citrus essential oils, wine, and beer. Minor components may contribute in an important way to the organoleptic characteristics and also demonstrate biological activities that affect human health.

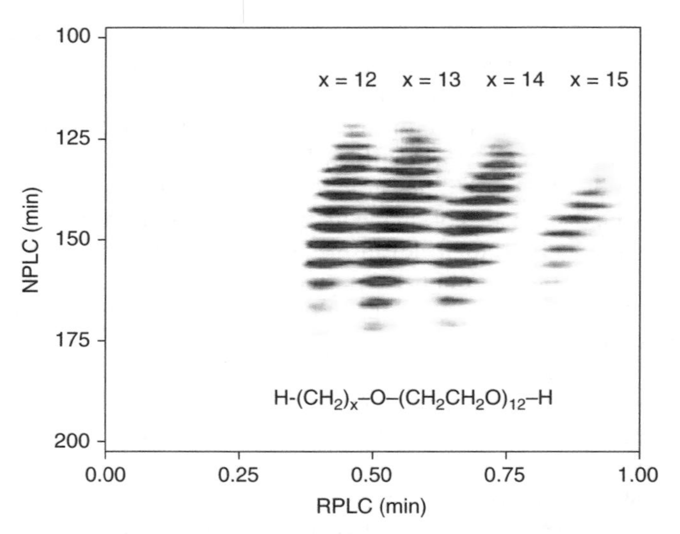

Figure 10.3 Two-dimensional HPLC (NPLC/RPLC) chromatogram of Neodol 25-12 with the corresponding chemical structure and average EO as supplied by the manufacturer. [From Murphy et al. (1998b), with permission. Copyright © 1998 by The American Chemical Society.]

As is clear from Table 10.2, applications of LC × LC to the analysis of constituents of natural products have been developed using both NP × RP and RP × RP approaches. An exception is the first application reported by Erni and Frei (1978), where a combination of SEC and RP was used to analyze a senna glycoside extract. NP × RP is considered the most orthogonal combination of HPLC separation modes, but its use is limited due to the "incompatibilities" presented by the coupling of the two LC modes. However, this combination can be very useful in the separation of complex mixtures that contain uncharged molecules of comparable dimensions, differing in polarity and hydrophobicity.

All the comprehensive NPLC × RPLC applications reported in Table 10.2 have been developed using a similar setup. A microbore column has been used in the first dimension, operated at a low flowrate, while a C_{18} column has been used in the second dimension, operated at a very high flow rate for fast analysis. The interface was a 10-port two-position valve equipped with sampling loops, because under these working conditions, direct entrapment of analytes at the head of the secondary column is not possible.

The first "true" NPLC × RPLC system was reported by Dugo et al. (2004) for the analysis of oxygen heterocyclic components of a cold-pressed lemon essential oil. The first dimension was operated with *n*-hexane/ethyl acetate, 75 : 25 in the isocratic mode at 20 μL/min. Fractions were transferred into the second-dimension C_{18} monolithic column operated under gradient conditions using water and acetonitrile (ACN) as mobile phase at 4 mL/min. The problem of mobile-phase immiscibility was solved using a microbore column in the first dimension

TABLE 10.2 Comprehensive 2D (LC x LC) Separations of Antioxidants and Natural Products

Ref.	Type of Interface	Column Packing and Dimensions (Length x i.d., mm, d_p, μm); Flow rate; Elution Mode[a]		Detection	Application
		First Dimension, D1	Second Dimension, D2		
[1]	Two-position 8-port valve	SEC 2000 × 4.0;1.2 mL/h; GP	C_{18} 250 × 4.0; 2.0 mL/min; GP	UV (254 nm)	Senna glycoside extract
[2]	Two-position 10-port valve	Silica 300 × 1.0, 5 μm;20 μL/min; IP	C_{18} monolithic 25 × 4.6;4 mL/min; GP	PDA (240–360 nm)	Oxygen heterocyclic components of a cold-pressed lemon oil
[3]	Two-position 10-port valve	Diol 250 × 1.0, 5 μm; 40 μL/min; GP	C_{18} 50 × 4.6, 3.5 μm; 5 mL/min; GP	PDA (190–400 nm)	Oxygen heterocyclic components of lemon oil and *Citrus* oil extracts
[4]	Two-position 10-port valve	Silica 300 × 1.0, 5 μm; 15.4 μl/min; IP	C_{18} monolithic 25 × 4.6; 4 mL/min; GP	PDA (190–360 nm)	Allergens
[5]	Two-position 10-port valve	Silica 300 × 1.0, 5 μm; 10 μL/min; GP	C_{18} monolithic 100 × 4.6; 4.7 mL/min; GP	UV (450 nm) PDA (250–550 nm)	Orange essential oil and juice carotenoids
[6]	Two-position 6-port valve (stop-flow mode)	PEG 50 × 2.1, 5 μm; 0.4 mL/min; IP	C_{18} 125 × 2.0, 5 μm; 0.4 mL/min; GP	UV (280 nm)	Phenolic antioxidants

(*continued overleaf*)

TABLE 10.2 (*Continued*)

Ref.	Type of Interface	Column Packing and Dimensions (Length x i.d., mm, d_p, μm); Flow rate; Elution Mode[a]		Detection	Application
		First Dimension, D1	Second Dimension, D2		
[7]	Two-position 10-port valve equipped with: (a) two 100 μL sampling loops (b) two C_{18} trapping columns (c) two alternative D2 columns	PEG (a) 150 × 4.6, 5 μm; 0.1 mL/min; GP (b) 150 × 4.6, 5 μm; 0.3 mL/min; IP (c) 150 × 4.6, 5 μm; 0.2 mL/min; GP (d) 50 × 2.1, 3 μm + C_{18} 150 × 4.6, 5 μm; 0.2 mL/min; GP	(a) C18 10 × 2.1, 3 μm;1.0 mL/min; IP (b) C18 monolithic 50 × 4.6;2.0 mL/min; GP (c) Zirconia-carbon 50 × 2.1, 5 μm; 1.0 mL/min; IP (d) Zirconia-carbon 50 × 2.1, 5 μm;1.0 mL/min; IP	PDA (254–320 nm)	Phenolic antioxidants in beer
[8]	Two-position 10-port valve equipped with two alternating Zirconia columns	C_{18} 125 x 4.6, 5 μm; 5 μL/min; GP	ZR-carbon 50 × 2.1, 5 μm;1.0 mL/min; IP at 120°C	PDA (254–320 nm)	Phenolic antioxidants in beer and wine
[9]	Two-position 10-port valve equipped with: (1) two 100 μl sampling loops	PEG 150 × 4.6, 5 μm; 0.1 mL/min; GP	C_{18} monolithic 50 × 4.6; 2.0 mL/min; GP		

	(2) two C18 trapping columns (30 × 4.6 mm, 2.5 μm)	PEG 150 × 4.6, 5 μm; 0.3 mL/min; GP	C_{18} monolithic 100 × 4.6; 2.0 mL/min; GP		
	(3) two C18 trapping columns (30 × 4.6 mm, 2.5 μm)	PEG 50 × 2.1, 3 μm; C_{18} 250 × 3.0, 5 μm; 0.3 mL/min; GP	C_{18} monolithic 100 × 4.6; 2.0 mL/min; GP		
	(4) two C18 trapping columns (30 × 4.6 mm, 2.5 μm)	PEG 150×4.6, 5μm; + SB Aq 50×4.6, 5μm; 0.3 mL/min; GP	C_{18} monolithic 100 × 4.6; 2.0 mL/min; GP	PDA (254–320 nm)	Phenolic and flavone antioxidants in beer and wine
	(5) two C18 trapping columns (30 × 4.6 mm, 2.5 μm)	Phenyl 50 × 3.9, 5 μm; 0.3 mL/min; GP	C_{18} monolithic 100 × 4.6; 2.0 mL/min; GP		
[10]	Two-position 10-port valve equipped with two 100 μL sampling loops	PEG 150 x 2.1, 5 μm 50 μL/min; GP	C_{18} monolithic 50 × 4.6; 3.5 mL/min; GP	PDA (254–320 nm)	Phenolic and flavone antioxidants

(*continued overleaf*)

TABLE 10.2 *(Continued)*

Ref.	Type of Interface	Column Packing and Dimensions (Length x i.d., mm, d_p, μm); Flow rate; Elution Mode[a]		Detection	Application
		First Dimension, D1	Second Dimension, D2		
[11]	Two-position 10-port valve	LC-SI 300 x 1.0, 5 μm; 18 μL/min; IP	C_{18} 25 × 4.6; 4 mL/min; GP	UV (315 nm)	Orange and grapefruit essential oil
[12]	Two-position 10-port valve	C_{18} 150 × 2.1, 3 μm; 100 μL/min; GP	C_{18} monolithic 50 × 4.6; 3 mL/min; IP C_{18} 50 × 2.1, 2.5 μm; 1.2 mL/min; IP C_{18} 50 × 2.1, 1.7 μm; 0.8 mL/min; IP	PDA (220–330 nm)	Phenolic acids in wines and juices
[13]	Two-position 10-port valve	Phenyl 250 x 1.0, 5 μm;10 μL/min; GP	C_{18} 30 × 4.6, 2.7 μm; 4 mL/min; GP	PDA (190–400 nm)	Phenolic antioxidants in wine

[1] Erni and Frei, 1978; [2] Dugo et al., 2004; [3] François et al., 2006; [4] Dugo et al., 2006a; [5] Dugo et al., 2006b; [6] Blahová et al., 2006; [7] Cacciola et al., 2006; [8] Cacciola et al., 2007a; [9] Cacciola et al., 2007b; [10] Jandera et al., 2008; [11] Mondello et al., 2008; [12] Kivilompolo and Hyötyläinen, 2008 [13] Dugo et al., 2009.

[a]IP: isocratic program; GP: gradient program.

at a flow rate compatible with the sample volume injected into the second conventional bore column, operated at a high flow rate. Under these conditions, the dilution of the first-dimension solvent occurs more rapidly through the second-dimension column, and therefore, band spreading is minimized.

To overcome a further complication due to the fact that the mobile phase injected into the second-dimension column is stronger than the mobile phase at the head of the second-dimension column, effective focusing was achieved maintaining the second-dimension initial eluent strength very low. After the injection, a repetitive gradient was necessary to elute all the components in the short analysis time and recondition the column for the successive injection. Peak identification was achieved observing the relative location in the 2D plane, and the UV spectra, which presented characteristic absorption maxima in relation to the substituted positions. When dealing with natural components it is important to remember that difficulties in peak identification are also due to the lack of standards. UV spectra, unless being less informative than MS spectra, could be very helpful and present the advantages of lower cost and very wide diffusion in analytical laboratories.

A similar setup was used some years later by François et al. (2006) and Mondello et al. (2008), again for the analysis of oxygen heterocyclic components in citrus oils and extracts. François et al. (2006) analyzed a lemon oil and a lemon–orange oil extract. The latter contained more polar components (polymethoxylated flavones), and a gradient with *n*-hexane and ethyl acetate was run in the first dimension. The authors noted severe peak distortion and broadening of the compounds eluting at high ethyl acetate levels. This problem was solved by increasing the amount of water in the initial second-dimension mobile phase, thus reducing the eluent strength. This effect also reduced the available second-dimension separation space. They demonstrated that the higher the *n*-hexane concentration, immiscible with the RP solvent, the better was the focusing of the solutes, pointing out the importance of solvent immiscibility in the NPLC × RPLC technique.

Mondello et al. (2008) developed an NPLC × RPLC method applied to the analysis of a standard mixture and grapefruit oil, with the aim of quantifying target analytes using dedicated laboratory-constructed LC × LC software. The software enabled both the conversion of the instrumental data into two- and three-dimensional chromatograms and direct peak integration. The software, using a novel algorithm, was able to simultaneously identify and integrate analyte bands relative to the same solute. The effectiveness of the software was evaluated in the LC × LC quantitative analysis of aurapten in grapefruit oil, using both internal and external standard calibration techniques. Calibration curves were determined at six concentration levels using coumarin as the internal standard, both using the LC × LC setup and a conventional LC setup. The limits of detection (LOD) and quantitation (LOQ) were determined by using the untransformed raw LC × LC chromatogram, with results about two times higher than those obtained by conventional LC. This because of sample dilution during the secondary separation, and also because of the modulation of a component in more than one slice. Peak area precision was evaluated [the relative standard deviation (RSD) ranged from 0.1 to 3.0%] as well as method accuracy with and without a matrix effect. In the latter case, solutions spiked with a known amount of aurapten were prepared, while in the former case an orange oil was spiked with aurapten. Relative error (%) values were always below 3%. Then the aurapten concentration was determined by LC × LC and LC in a sample of grapefruit oil, obtaining results in very good agreement.

The performance of a NPLC × RPLC system was evaluated by Dugo et al. (2006a) in a study carried out on a standard mixture made up mainly of allergens. Systematic changes in flow rate and first-step gradient composition in the second dimension operated using a monolithic column were explored. The column was operated at flow rates as high as 8 mL/min, with low back-pressure and without loss of resolution. The use of a low percentage of strong solvent in the first step of the gradient of the second dimension improved the peak focusing. The conditions used for the LC × LC analysis were also optimized under the assumption that the number of modulations per first-dimension peak should be at least three to avoid loss of information from the first dimension.

The same group has recently developed a NPLC × RPLC method for the analysis of carotenoids in orange essential oil and juice (Dugo et al., 2006b). Food carotenoids are an important class of pigments whose composition in nature is widely variable. The largest number of carotenoids found in any fruit are those of citrus fruits. Their determination is complicated by their great structural diversity, extreme instability, and lack of commercial standards. Free carotenoids obtained after a saponification step from orange oil and juice were subjected to NPLC × RPLC analysis in combination with a DAD (see Figure 10.4). Carotenoids were nicely located in the 2D plot according to their structural characteristics; hydrocarbons (carotenes) were fully separated from oxygenated components (xanthophylls). These components were separated into mono-, di-, and triols and their corresponding epoxides according to their polarity and hydrophobicity. Retention data and UV–visible spectra were used to tentatively identify a large number of carotenoids, due to the characteristic absorption maxima relative to different classes of carotenoids.

A wide number of RPLC × RPLC applications have been developed for the analysis of polyphenolic components. Blahová et al. (2006) studied the selectivity correlations between RP chromatographic systems with pentafluorophenylpropyl (F_5), PEG, and C_{18} stationary phases in the separation of polyphenols. Results demonstrated a high degree of correlation between F_5 and C_{18}, while a low correlation was found between PEG and C_{18}. These two stationary phases were employed in a LC × LC system developed for the analysis of phenolic antioxidants not fully separated on a single column. The interface was a six-port valve connected directly to the second-dimension column operated under stop-flow conditions. This combination has the advantage of on-column focusing of the fractions transferred, because due to the high polarity of the bonded phase, a PEG–silica column uses a weaker mobile phase for component elution.

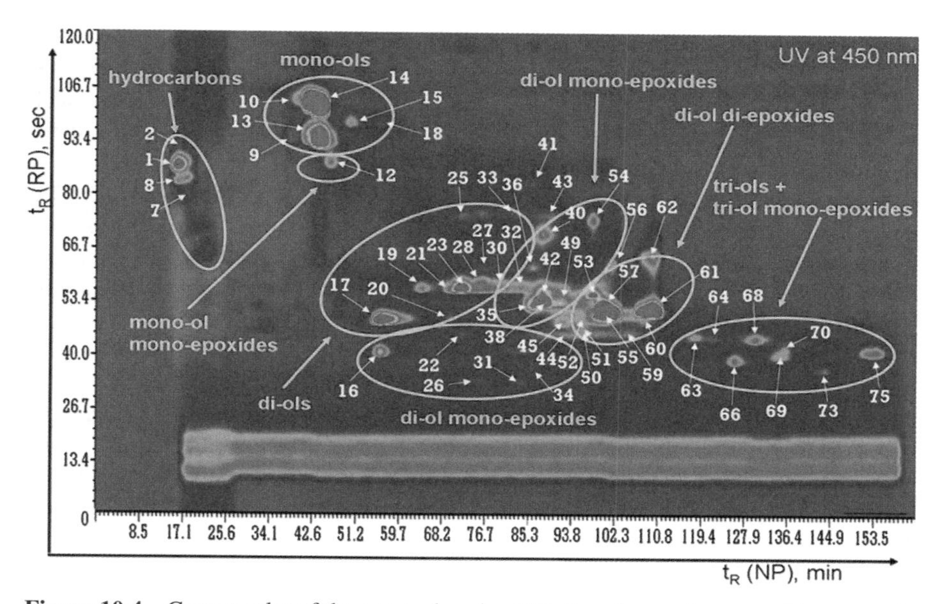

Figure 10.4 Contour plot of the comprehensive HPLC analyses of carotenoids present in sweet orange essential oil with peaks and compound classes indicated. 1, ζ-Carotene, phytoene, phytofluene; 2 to 6, 7, α-carotene; 8, β-carotene; 9, α-cryptoxanthin isomer; 10, β-cryptoxanthin–isomer; 11, unknown; 12, cryptoxanthin-5,6-epoxide; 13 α-cryptoxanthin; 14, β-cryptoxanthin; 15, unknown; 16, flavoxanthin; 17, lutein isomer; 18 to 20, unknown; 21, lutein isomer; 22, unknown; 23, lutein; 24, unknown; 25, zeaxanthin; 26, 27, unknown; 28, lutein isomer; 29, unknown; 30, lutein isomer; 31, unknown; 32, lutein isomer; 33, 34, unknown; 35, mutatoxanthin isomer; 36, 37, unknown; 38, mutatoxanthin; 39, unknown; 40, mutatoxanthin isomer; 41, unknown; 42, mutatoxanthin isomer; 43, mutatoxanthin isomer; 44, to 48, 49, mutatoxanthin isomer; 50, unknown; 51, 52, auraxanthin isomer; 53, violaxanthin; 54, *cis*-antheraxanthin; 55, auraxanthin; 56, antheraxanthin; 57, luteoxanthin; 58, 59 unknown; 60, auraxanthin–isomer; 61, 62) luteoxanthin isomer; 63 to 67, unknown; 68, neoxanthin; 69, trollichrome-like; 70, neoxanthin-like; 71 to 74, unknown; 75, trollichrome. [From Dugo et al. (2006b), with permission. Copyright © 2006 by The American Chemical Society.]

Cacciola et al. (2006) developed three different RPLC × RPLC systems based on the use of a PEG–silica column in the first dimension and a packed or monolithic C_{18} column or zirconia–Carbon column in the second dimension. A fourth setup was equipped with a combination of PEG + C_{18} columns and a zirconia–Carbon column, respectively, in the two dimensions. Different interfaces based on the use of a two-position 10-port valve were designed. The valve was equipped with sampling loops or with trapping columns or directly with two second-dimension columns operated in parallel at elevated temperature. Selectivity correlations study demonstrated that the PEG column presented a low correlation and high degree of orthogonality with the C_{18} and the zirconia–Carbon stationary phases. The same was found for the PEG + C_{18} combination with respect to the zirconia–Carbon column. The mobile phase in the first dimension had a lower elution strength than in the second dimension, allowing band compression of solute at the head of the secondary column. This effect was enhanced in the setup using trapping columns as the interface. The systems was used for the analysis of polyphenols in beer samples.

Based on the results obtained in their earlier work, Cacciola et al. (2007a) developed different LC × LC systems with different column combinations and interface configurations for the analysis of phenolic and flavone natural antioxidants. Because of the low degree of correlation between the column combinations employed in the two dimensions (described in Table 10.2), the authors were able to increase the peak capacity and resolution of antioxidants over that of single-dimension separation. The main features of the system developed in this work are the use of parallel matching 2D gradients (shown in Figure 10.5) (Cacciola et al., 2007a) and the substitution of sampling loops with short trapping columns. A parallel matching 2D gradient enhanced the regularity of the coverage of the 2D retention plane for samples with partially correlated elution time. The use of trapping columns allowed significant suppression of the band broadening, so that a narrow sample zone adsorbed on the top of the trapping column can be back-flushed to the second dimension column in a low volume of mobile phase. Figure 10.6 reports an example of the analysis of a beer and a Merlot red wine using a phenyl-monolith C_{18} setup and trapping columns in the interface.

The high thermal stability of a zirconia stationary phase was used to develop a comprehensive LC system using an isocratic program at elevated temperature (120°C) in the second dimension for the separation of polyphenols in beer and wine (Cacciola et al, 2007b). Two parallel zirconia–carbon columns were operated in alternating cycles in the second dimension, in combination with a C_{18} column in the first dimension, obtaining an almost orthogonal 2D system. Isocratic separation with temperature programming was compared to a solvent gradient at low temperature, resulting in a shorter analysis time without loss of resolution. However, the short switching cycle frequency in the second dimension did not allow temperature programming (due to the long time required for post-run column temperature equilibration). Isocratic separation at elevated temperature provided good resolution and peak capacity for the separation of target analytes.

Rational selection of columns and mobile phases is indispensable in the development of a comprehensive LC system with the highest orthogonality, able to maximize peak capacity with respect to a monodimensional analysis. This was the object of a study carried out by Jandera et al. (2008), in which the authors developed a systematic approach to the selection of columns and mobile phases suitable for orthogonal LC × LC separation of natural phenolic antioxidants based on the correlations between the retention and molecular structure descriptors of representative phenolic acid and flavone standards. Chemometric methods (multiple linear regression, cluster analysis, and window-diagram optimization strategies) were combined to classify different types of columns, selected on the basis of previous experience with the separation of natural phenolic antioxidants. A column

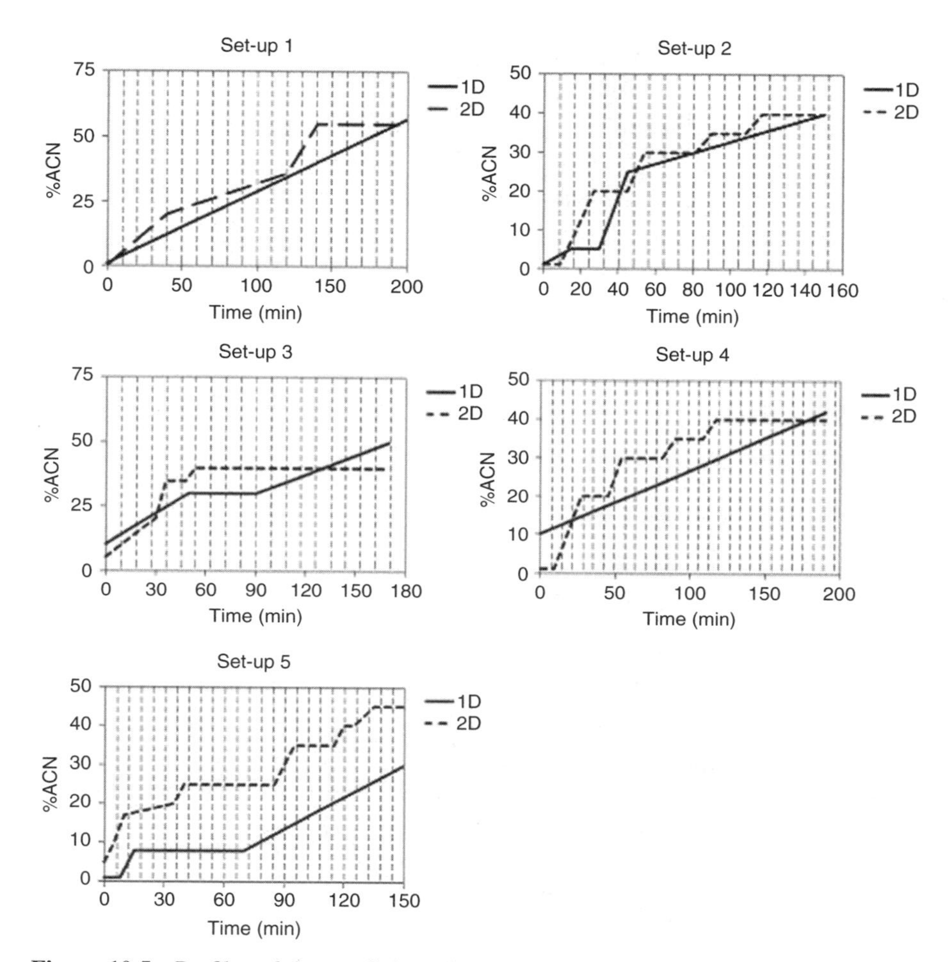

Figure 10.5 Profiles of the parallel gradients of acetonitrile in the first (solid lines) and second (dashed lines) dimensions in the five setups tested. The intervals between the vertical lines correspond to the individual fractions in setups 2 to 5 and to 10 consecutive fractions in setup 1. [From Cacciola et al. (2007b), with permission. Copyright © 2007 by Elsevier.]

set contained monolithic and fully porous particle C_{18} column, amide, PEG and a special column where alkyl chains were bonded to the silica gel support via a hydrophilic functional group for better solvation of the stationary phase in mobile phase with low concentration of organic modifier. The optimized setup described in Table 10.2 was used for the separation of phenolic acids and flavones with parallel gradients of acetonitrile (pH 3) in the two dimensions (see Figure 10.7).

Kivilompolo and Hyötyläinen (2008) compared different separation systems for the analysis of phenolic acids in wine and juices: one-dimensional HPLC and UPLC, as well as comprehensive LC systems using RPLC in the first dimension

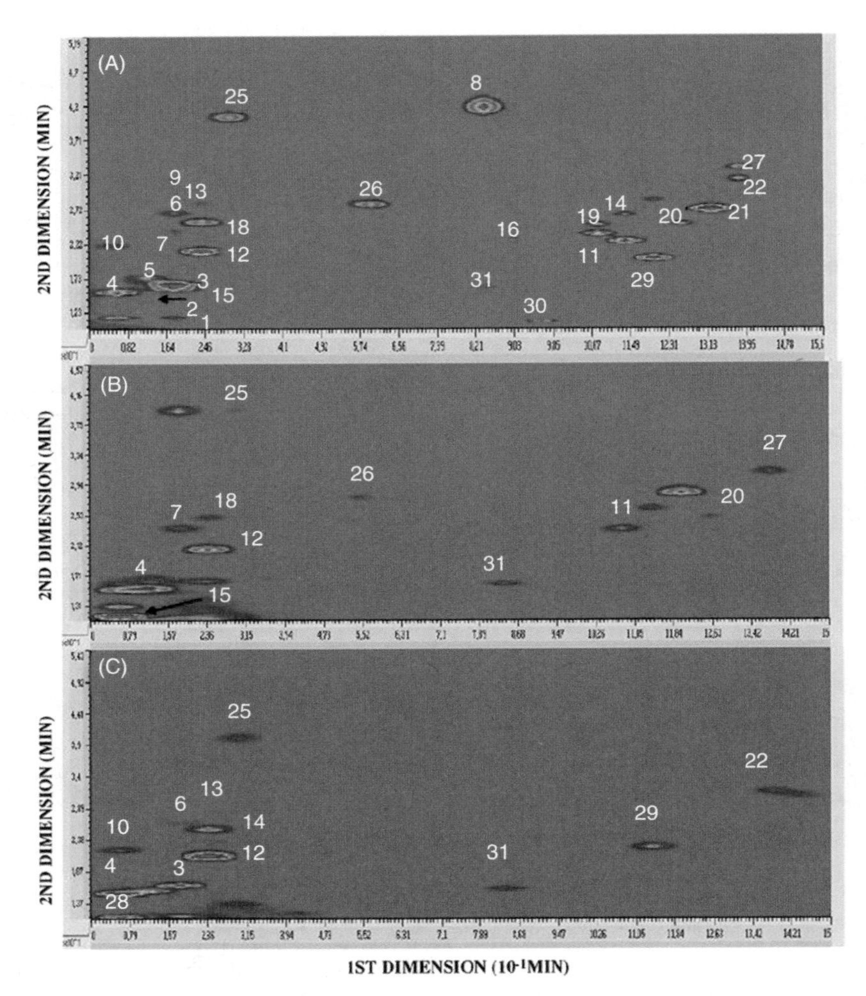

Figure 10.6 Comprehensive LC × LC separation of a mixture of phenolic and flavone standards (A), of a pilot beer sample 2 (B), and of an Andrè red wine sample (C) using setup 2. *X*-axis, first-dimension retention time since the sample injection; *y*-axis, second-dimension retention time since the fraction transfer; transfer cycle frequency, 8 min. For conditions see Table 10.3. [From Cacciola et al. (2007b), with permission. Copyright © 2007 by Elsevier.]

(C_{18} narrow-bore HPLC column) and ion-pair chromatography in the second dimension, with three column combinations (two short columns packed with either 2.5- or 1.7-μm particles and a monolithic column). The 1D UPLC separation resulted clearly better than that with the HPLC system, although baseline separation was not obtained for all the components in both cases. Moreover, coelution of matrix components could not be avoided in the analysis of real

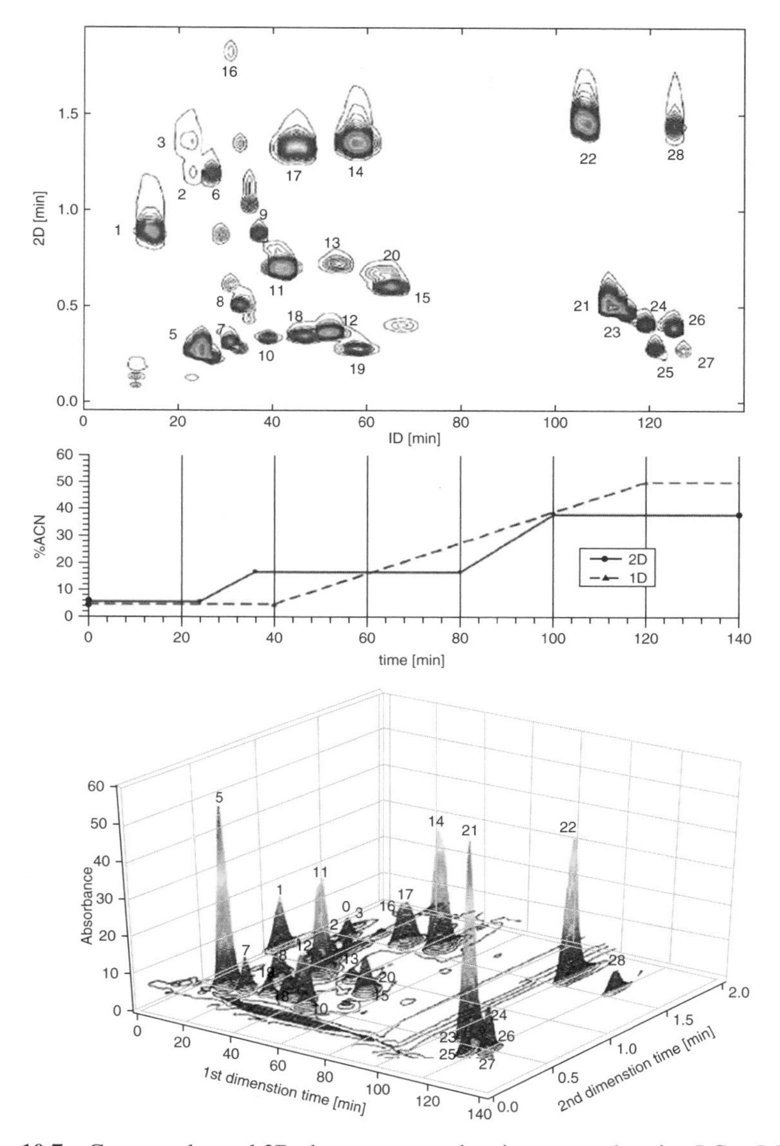

Figure 10.7 Contour plot and 3D chromatogram showing comprehensive LC × LC separations of phenolic acids and flavones on a PEG column in the first dimension and on a Chromolith RP-18e column in the second dimensions with parallel gradients of acetonitrile in the two dimensions. 1, Phenol; 2, benzonitrile; 3, nitrobenzene; 4, 2-bromonitrobenzene; 5, 3-chlorotoluene; 6, *p*-dichlorobenzene; 7, gallic acid; 8, 4-hydroxyphenylacetic acid; 9, protocatechuic acid; 10, syringic acid; 11, 4-hydroxybenzoic acid; 12, salicylic acid; 13, vanillic acid; 14, caffeic acid; 15, ferulic acid; 16, *p*-coumaric acid; 17, sinapic acid; 18, chlorogenic acid; 19, (−)-catechin, 20, (−)-epicatechin, 21, rutin; 22, naringin; 23, myricetin, 24, quercetin; 25, apigenin; 26, biochanin A; 27, luteolin; 28, naringenin; 29, 7-hydroxyflavone; 30, hesperidin; 31, morin; 32, hesperetin; 33, flavone. [From Jandera et al. (2008), with permission. Copyright © 2008 by Elsevier.]

sample. These results pointed out the need for more resolution than can be obtained using a two-dimensional comprehensive LC system. Preliminary results on the three columns to be used in the second dimension demonstrated the superiority of the column that worked under UPLC conditions (with 1.7-μm particles). However, in the 2D mode, the best results were obtained when a monolithic column was employed. This result is linked to the specific experimental conditions optimized for each second-dimension column, assuming that first-dimension conditions were always maintained unchanged.

Dugo et al. (2009) compared two-dimensional liquid chromatography with conventional liquid chromatography for the analysis of polyphenols in a real sample of wine, in terms of linearity, sensitivity, detection and quantification limits, and separation performance. Good agreement between quantitative results in conventional and comprehensive LC was attained. Moreover, due to the higher degree of separation obtained in comprehensive LC, components coeluted in conventional LC were easily quantified. Sensitivity results were lower in comprehensive LC, due to the dilution caused by the high second-dimension flow rate, but higher resolution power made it possible to obtain clearer spectra for a positive peak identification.

10.3 COMPREHENSIVE 2D LC SEPARATION OF PHARMACEUTICAL AND ENVIRONMENTAL COMPOUNDS

One of the most common approaches to LC × LC analysis is use of the RP mode in both dimensions. It has been demonstrated that orthogonality can be achieved by using either two different sets of mobile phases and one type of RP column, or a single mobile phase and two HPLC columns packed with different RP stationary phases. Many RP × RP separations have been developed and used in the determination of antioxidants, as shown in section 10.2. Moreover, the RP × RP approach has been applied to the analysis of other classes of analytes, such as traditional Chinese medicines (TCMs), explosives and by-products in water, polycyclic aromatic hydrocarbons (PAHs), metabolites, and others, as summarized in Table 10.3.

Köhne et al. (1998) developed a comprehensive RPLC × RPLC system for the determination of mixtures of explosives and their by-products. The 2D system shown in Figure 10.8 was equipped with an alkyl-modified silica (C_{18}) and a safrole-modified silica column in the two dimensions, presenting different retention characteristics. The higher peak capacity was realized employing a flexible switching technique and using different elution strengths of the mobile phases. Peak compression at the head of the secondary column was also optimized, increasing the differences in mobile-phase strength between the two dimensions.

To reduce the total analysis time in comprehensive 2D HPLC, a solution is to accelerate the separation in the second dimension. Köhne and Welsch (1999) realized this goal with the use of a short column packed with 1.5-μm nonporous particles in the second dimension. The instrumental setup was designed so that a

TABLE 10.3 Comprehensive 2D (LC x LC) Separations of Pharmaceutical and Environmental Compounds

Ref.	Type of Interface	Column Packing and Dimensions (Length x i.d., mm, d_p, μm); Flow rate; Elution Mode[a]		Detection	Application
		First Dimension, D1	Second Dimension, D2		
[1]	Three valves (stop flow)	C_{18} 150 × 4.0, 5 μm;1 mL/min; IP	Safrole Silica[1] 125 × 4, 5 μm;1 mL/min; IP	UV (254 nm)	Explosives mixture
[2]	Two 2-position 6-port valves	TCP Silica[2] 200 × 0.32, 5 μm; 5 μL/min; IP	C_{18} 30 × 4.6, 1.5 μm;1 mL/min; IP	UV	Phenols mixture
[3]	Two-position 10-port valve equipped with two second dimension parallel columns	PBB[3] 150 × 4.6, 5 μm; 1 mL/min; IP	C_{18} monolithic 50 × 4.6; 16 mL/min; IP	UV (254 nm)	Polycyclic aromatic hydrocarbons
[4]	Two 2-position 6-port valves	H5 F5[4] 50 × 2.1, 5 μm; 0.10 mL/min; GP	ZR Carbon 50 × 2.1, 3 μm;3.0 mL/min; GP	PDA (200–350 nm)	Indolic metabolites
[5]	Offline Fractions collected and stored for successive injection in the second column	Click OEG 150 × 4.6, 5 μm; 1 mL/min; GP	C18 150 × 2.1, 5 μm; 0.2 mL/min; GP	UV (280 nm)	*Lignum Dalbaergiae Odoriferae* extract

TABLE 10.3 (*Continued*)

Ref.	Type of Interface	Column Packing and Dimensions (Length x i.d., mm, d_p, μm); Flow rate; Elution Mode[a]		Detection	Application
		First Dimension, D1	Second Dimension, D2		
[6]	Offline fractions collected and stored for successive injection in the second column	C_{18} 250 × 4.6, 3.5 μm; 1 mL/min; GP	Click β-CD 150 × 4.6, 5 μm;GP	PDA (200–400 nm)	*Carthamus tinctorius* extract
[7]	Two-position 10-port valve	Cyano 160 × 0.53, 5 μm; 8 μL/min; GP	C_{18} monolithic50 × 4.6, 5 μm;4 mL/min; GP	UV (265 nm)	*Radix salviae miltiorrhiza* bage extract

[1] Köhne et al., 1998; [2] Köhne and Welsch, 1999; [3] Murahashi, 2003; [4] Stoll et al., 2006; [5] Liu et al., 2008a; [6] Liu et al., 2008b; [7] Tian et al., 2008.
[a] IP: isocratic program; GP: gradient program.
[1] (3′,4′-methylendioxyphenyl)propyl)silica
[2] tetrachlorophthalimidopropyl phase
[3] (pentabromobenziloxy)propylsilyl-bonded phase
[4] pentafluorophenylpropyl phase

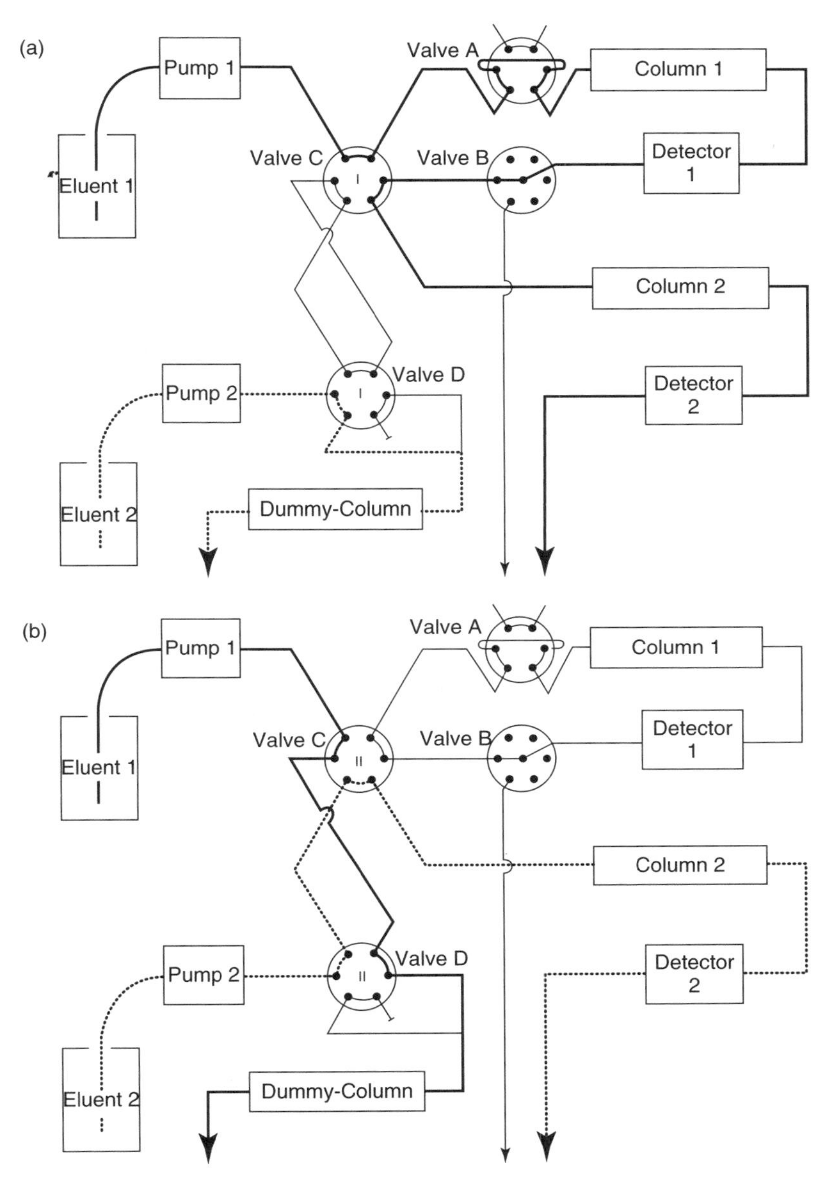

Figure 10.8 (a) Scheme of the 2D system and the valve positions for the separation of a sample on column 1 and transfer of fractions to column 2; (b) scheme of the 2D system and the valve positions for the separation of the transferred fraction on column 2 while the partly developed chromatogram on column 1 is "parked." [From Köhne et al. (1998), with permission. Copyright © by Springer.]

microcolumn was used in the first dimension, to reduce the injection volume into the second column. A mixture of amino-, chloro-, and nitrophenols was analyzed using a tetrachlorophthalimidopropyl (TCP) silica microcolumn (charge transfer phase retention mechanism) in the first dimension and an NPS C_{18} column in the second dimension, in a total analysis time of 70 min, transferring 27 fractions on the second dimension and analyzing each of them between 1 and 4 min under stop-flow conditions.

An RPLC × RPLC system was also developed for the analysis of PAHs using a pentabromobenzyl column in the first dimension and two short monolithic C_{18} columns in the second dimension, connected directly to the 10-port valve used as an interface (Murahashi, 2003). The choice of the stationary phases used in the two dimensions was carried out by measuring the retention behavior of PAHs, alkanes, alkylbenzenes, and nitro-PAHs on 12 different stationary phases, including alkyl, alkylamide, phenyl, nitrated, and alogenated stationary phases.

A more recent approach to high-speed comprehensive online dual-gradient 2D LC has been proposed by Stoll et al. (2006). They used a column combination with a very low correlation, where both columns had the same internal diameter. Ultrafast (< 30 s) high-temperature ($>100^\circ$C) gradient elution RP-HPLC was used in the second dimension. The RP gradient used in each second dimension was developed under high-pressure mixing conditions, with a very low dwell volume (50 μL, most of which is contributed by the volume of sample loops (34 μL each). Under these conditions, only 1 s was required to flush the strong solvent out of the tubings at the end of the gradient elution run before reequilibration of the column between two consecutive gradient runs. Peaks eluting from the first dimension were sampled at least twice across their width, avoiding any loss of resolution gained in the first-dimension separation.

The system was used in the analysis of low-molecular-weight components of wild-type and mutant maize seedlings, obtaining a separation of about 100 peaks on a time scale of 25 min.

One limitation in the development of 2D RPLC–RPLC systems has been the paucity of sufficiently orthogonal pairs of RPLC phases. Liu and co-workers (2008a,b) have developed new stationary phases orthogonal to C_{18} to be used in two-dimensional chromatographic systems. A new click oligo(ethylene glycol) (click OEG) was prepared by immobilization of oligo(ethylene glycol) on silica support via click chemistry (Liu et al., 2008a). The orthogonality measured between click OEG and C_{18} resulted in about a 69% higher value than that in more conventional combinations. An off-line 2D LC system using click OEG and C_{18} was used in the analysis of a traditional Chinese medicine (*lignum dalbergiae odoriferae*), with good results. The 2D system was able to separate components not separated using one-dimensional HPLC, and also detected minor-component neighbors of a very big peak. An additional advantage is represented by the good compatibility of mobile phases used in the two dimensions.

In a successive study (Liu et al., 2008b), an off-line 2D RPLC–HILIC system based on C_{18} and click β-cyclodextrin (β-CD) was developed and used in the analysis of polar and medium-polarity components in TCM (*Carthamus tinctorius* Linn.). The system showed excellent orthogonality, allowing the separation of components not resolved in one-dimensional chromatography. Orthogonality was calculated using a novel geometric approach. Other main advantages were the high compatibility of the mobile phases used in the two dimensions (the same mobile phases in different proportions) and the compatibility with MS detection. Use of the HILIC mode in the second dimension makes possible faster separation at lower pressure, due to the high organic modifier content.

Tian et al. (2008) developed a comprehensive two-dimensional liquid chromatographic system incorporating a vacuum-evaporation interface for online coupling of NPLC and RPLC. The 10-port valve equipped with two identical sampling loops used as an interface realized fast solvent evaporation coming from the first dimension at 25°C under vacuum conditions. In this way, the injection volume of the second-dimension was reduced and band compression at the head of the second dimension column was achieved. The system was used for the analysis of a traditional Chinese medicine, *Radix salvieae miltiorhiza bage* extract. The authors demonstrated that the system offered high peak capacity, true orthogonality, and high speed. Mobile-phase incompatibility was avoided and peak shape improved. Recovery of nonvolatile compounds were above 50%, so analyte evaporation occurred under vacuum conditions. Although this phenomenon is inevitable with the present interface, quantitative analyses can be achieved by adding internal standard.

10.4 COMPREHENSIVE 2D LC SEPARATION OF PROTEINS AND PEPTIDES

As pointed out in Chapter 9, the emerging field of proteomics (and, especially, peptidomics in the case of tryptic digestion of proteins) has always represented a major challenge for separation science, due to the great sample complexity. Strong attempts have been made to achieve an ever-higher degree of sample fractionation through the use of hyphenated electrophoretic and chromatographic techniques, or sometimes a combination of both. The complexity of proteome samples, in fact, often overwhelms the separation capabilities of any single 1D separation technique, and this has caused a strong push toward the development of a variety of 2D separation methods in which different orthogonal techniques are combined, both off- and online.

Two-dimensional polyacrylamide gel electrophoresis (2D-PAGE), followed by enzymatic digestion of the separated protein spots, has long been the workhorse in analytical proteomics, due to its ability to separate thousands of proteins on a time scale of a few hours. Despite their unique advantages, these methods suffer from some limitations, such as difficulty of automation and low accessibility of membrane-bound proteins. Problematic detection of proteins characterized by large molecular weight, high isoelectric point (p*I*), strong hydrophobicity, or low abundance also represents a major drawback.

These limitations have strongly motivated the evolution of gel-free separation techniques as a workable alternative to the more troublesome 2D-PAGE. Among these, multidimensional and comprehensive LC (LC × LC) has been growing very fast, mainly in the past decade, and considerable progress has been made both in hardware and in column technology. The development and availability of dedicated software for data processing have finally concurred to make this technique emerge to play a central role in the proteomic field. Multidimensional liquid chromatography can be used in many different combinations, generating increased peak capacity, selectivity, and resolution, especially in the comprehensive LC mode. In addition, some major limitations of gel-based methods are redressed, since LC techniques offer ease of full automation and the likelihood of straightforward hyphenation to mass spectrometry; furthermore, sample losses associated with poor recovery from gels are eliminated since the entire separation process takes place in solution.

A variety of comprehensive LC schemes have been developed for protein and peptide analysis, most of which are based on column-switching techniques; some representative examples are reported in Table 10.4 and discussed below. Bushey and Jorgenson (1990) were the first to show that it is possible to obtain two-dimensional separations of protein samples chromatographically. Their system employed two orthogonal types of separations, capable of providing complementary information, using a cation-exchange column under gradient conditions as the first column, coupled to a size-exclusion column through an eight-port computer-controlled switching valve. In addition, the entire first column effluent was analyzed on the second column without the use of stopped-flow methods, and also eliminated the need for prior knowlege of peak elution times that is required by heart-cutting techniques and diverted flow methods. Hence, they termed this technique comprehensive 2D LC to distinguish it from other 2D LC approaches.

Two main advantages were introduced over the earlier pioneering work done by Erni and Frei (1978). First, they sampled the first column only seven times over a 10-h run, therefore precluding 3D data presentation. Second, the large injection volume (each fraction volume was approximately 1.5 mL) necessitated a preconcentration step (peak compression), obtained through the use of gradient methods, at the head of the second column. The first dimension column (IEX, 250 × 1.0 mm) was operated at a 5-μL/min flow rate, while a flow rate of 2.1 mL/min was used through the secondary column (SEC, 250×9.4 mm), resulting in a 6-min SEC chromatogram. Using 30-μL loops, one injection on the second column was made every 6 min, the full run time of the SEC column. For the separation of a protein standard mixture, the method allowed for the analysis of the total first-column effluent on the second column without significant loss of first column resolution. At the same time, frequent sampling of the first column effluent provided peak profiles in both column dimensions; reliable 3D data representation in fact provided for easy peak identification from run to run. Remarkably, separation was attained in a fully automated system, using commercially available equipment, and analysis time was no longer than that of the first-dimension separation.

TABLE 10.4 Comprehensive 2D (LC x LC) Separations of Proteins and Peptides

Ref.	Type of Interface	Column Packing and Dimensions (Length x i.d., mm, d_p, μm); Flow rate; Elution Mode[a]		Detection	Application
		First Dimension, D1	Second Dimension, D2		
[1]	Two-position 8-port valve	IEX 250 × 1.0; 5 μL/min; GP	SEC 250 × 9.4; 2.1 mL/min	UV (215 nm)	Standard protein mixture
[2]	Two-position 8-port valve	IEX 900 × 0.1, 5 μm; 33 nL/min; GP	C_{18} 30 × 0.1, 5 μm; 6 μL/min; GP	Laser induced fluorescence (543.5 nm)	Porcine thyroglobulin tryptic digest
[3]	Two electronically controlled valves	IEX 1000 × 0.1, 5 μm; 33 nL/min; GP	C_{18} 35 × 0.15, 10 μm; 6 μL/min; GP	Laser induced fluorescence (543.5 nm)	Porcine thyroglobulin tryptic digest
[4]	Two-position 10-port valve	IEX 50 × 2.1, 5 μm; 0.1 mL/min; GP	C_{18} 50 × 2.1, 3.5 μm; 3.0 mL/min; GP	UV (214 nm)	Bovine serum albumin tryptic digest
[5]	6-port valve allowing stop-flow operation	SEC 300 × 7.8, 6 μm; 0.37 or 0.50 mL/min; IP	C_{18} monolithic 25 × 4.6; 2 mL/min; GP	UV (214 nm)	Peptide mixture
[6]	Offline fractions collected and stored for successive injection in the second column	IEX 110 × 4.6, 5 μm; 0.5 mL/min; GP	C_{18} 50 × 4.6, 2.7 μm; 3 mL/min; GP	UV (210 nm)	Myoglobin and bovine serum albumin tryptic digests

(*continued overleaf*)

TABLE 10.4 (***Continued***)

Ref.	Type of Interface	Column Packing and Dimensions (Length x i.d., mm, d_p, μm); Flow rate; Elution Mode[a]		Detection	Application
		First Dimension, D1	Second Dimension, D2		
[7]	Two-position 10-port valve	Four serially coupled C_{18} 150 × 2.1, 2.7 μm; 100 μL/min; GP	Two parallel C_{18} 50 × 4.6; 3.5 μm; 4 mL/min; GP	UV (215 nm)	Human and bovine serum albumin tryptic digests
[8]	Two-position 10-port valve	Four serially coupled C_{18} 150 × 2.1, 3 μm; 100 μL/min; GP	C_{18} 30 × 4.6, 2.7 μm; 4 mL/min; GP	UV (215 nm)	Human serum albumin tryptic digest

[1] Bushey and Jorgenson, 1990; [2] Holland and Jorgenson, 1995; [3] Holland and Jorgenson, 2000; [4] Stoll and Carr, 2005; [5] Bedani et al., 2006; [6] Marchetti et al., 2008; [7] Francois et al., 2009; [8] Mondello et al., 2010.
[a] IP: isocratic program; GP: gradient program.

Subsequently, Holland and Jorgenson (1995) coupled an anion-exchange microcolumn 90 cm long and 5 μm i.d. to a reversed-phase microcolumn 3 cm long and 100 μm i.d., both stationary phases having a pore size of 300 Å, to allow diffusion of the peptides in the pores. The two dimensions were interfaced using two electronically controlled valves, and operation of the system was entirely automated, so that effluent from the first column was collected in a sample loop and then concentrated onto the head of the secondary column.

The system described combined IEX chromatography, a separation mechanism based on charge, with RP, a separation mechanism based on hydrophobicity. The mobile phase used in the first dimension contained 50% acetonitrile, to minimize the reversed-phase characteristics introduced by the carbon content of the anion-exchange stationary phase and allowed the first dimension to be based mainly on charge separation rather than on a mixed-mode charge–hydrophobicity separation. Furthermore, sampling frequency was carefully chosen so that components separated by the first dimension remained separated in the second dimension. The comprehensive 2D LC system hence satisfied both of Giddings' requirements of a multidimensional system; its resolving power was demonstrated by the separation of the peptides obtained from a tryptic digest of porcine thyroglobulin. Compared to that of a conventional-sized column of similar length, the low mobile-phase volume of the microcolumn allowed small sample volumes to be loaded without diluting the contents of the injected sample and affect detection. Furthermore, the system was operated without interruption of flow, which is advantageous over stopped-flow methods because interruption in the flow of mobile phase in a chromatographic column will increase longitudinal band broadening and, as a consequence, decrease the resolution. Given a peak capacity of 200 and 7 for the first and second dimensions, respectively, the total peak capacity of the comprehensive system was estimated to be 1400.

Later, the same research group further improved the performance of the previous system (Holland and Jorgenson, 2000), through replacement of the second-dimension RP microcolumn packed with a silica support coated with an octadecylsilyl stationary phase with an equivalent perfusion-based stationary phase. This modification allowed the use of a high mobile-phase flow rate, with the advantage of less band broadening of the second-dimensional separation due to the rate of mass transfer. The total peak capacity of the modified two-dimensional system was nearly double the peak capacity of the two-dimensional system described previously; given a peak capacity of 169 for the first dimension and a peak capacity of 12 for the second dimension, the total peak capacity of the two-dimensional system was in fact calculated as 2028 peaks.

As a result of modification of the salt gradient of the anion-exchange separation, the peak capacity of this dimension was increased (from 150 for the earlier system to 169 for the modified system) by expanding the retention window in which the majority of the peaks eluted. The peak capacity of the second, RP separation was increased from 7 peaks for the earlier system to 12 peaks, through an increase in the retention window from 95 s to 120 s; the expansion of the retention window was again attained by modification of the solvent gradient used for the separation. After the peaks have eluted, the second-dimension column was subjected to high organic content to prevent sample carryover, followed by column equilibration with low organic content in the mobile phase to prevent sample breakthrough during the subsequent injection. The decrease in average peak width of the RP separation (i.e., from 13.5 s to 10 s) also contributed to the increase in peak capacity. The capability of the system was demonstrated on a tryptic digest from porcine thyroglobulin; the detection limit for glycine was 0.78 attomole.

Trying to overcome the limitation of long analysis time often requested for 2D LC analysis of proteomic samples, Stoll and Carr (2005) focused on improving the speed of gradient separations, typically used as the second dimension for peptide separations. By taking into account both the gradient time and the reequilibration time, which both control gradient cycle time, they found that excellent repeatability of retention time (<0.002 min) could be realized with narrow-bore (2.1 mm) wide-pore (300 Å) columns at higher temperature (100°C) with only about a one-column volume of flushing with the initial eluent. Furthermore, the decrease in viscosity at high temperatures (a factor of 3.5 for a 20 : 80 acetonitrile/water mixture at 120°C compared to 25°C) allowed a much higher eluent linear velocity through the column and hence faster gradient development with a reduction of the loss in efficiency that occurs at high linear velocity. A tryptic digest of bovine serum albumin was separated using a strong cation-exchange column (SCX, PO_4–ZrO_2, 5 μm, 250 Å pore size, 50 mm length × 2.1 mm i.d.) in the first dimension and a reversed-phase column in the second dimension (SB-C18, 3.5 μm, 300 Å pore size, 50 mm length × 2.1 mm i.d.). The total peak capacity was calculated using this method as 1350 for a 20-min analysis, which means approximately 1 peak/s (4000 per hour).

In contrast to what developed so far, the comprehensive SEC × LC system designed by Bedani et al. (2006) was run in stop-flow mode and applied to the analysis of complex mixtures of peptides derived from whey proteins. The authors demonstrated that this operation mode greatly facilitated instrument design and did not result in additional band broadening. They also showed that to simultaneously meet the two basic requirements of a certain desired total peak capacity and of four 2D runs over a 1D peak, only one set of conditions can be adopted for the single dimension in a situation where the 2D separation is already fixed. A 7.8 mm i.d. × 300 mm TSKGEL G2500PW_{XL} column packed with 6-μm particles was employed in the first dimension, while the second dimension consisted of a monolithic C_{18} modified silica 4.6 mm i.d. × 25 mm Chromolith Flash column. The SEC column outlet was connected to a six-port valve; in the transfer mode the effluent from the 1D column was directed to the 2D column, while the 2D gradient pump was connected to a flow restrictor to keep the pressure constant. In the 2D analysis mode, the 1D pump was connected to a stop-flow valve: before the first peak eluted, the stop-flow valve was left open, letting the dead volume go to waste. When the transfer of the first fraction started, the stop-flow valve was switched to the stop-flow position so that when switching the 1D pump off, the pressure in the 1D column stayed constant.

High orthogonality was achieved by Marchetti et al. (2008) for the separation of peptide digests from myoglobin and bovine serum albumin, performed with an off-line combination of two conventional commercial HPLC columns (Figure 10.9). The first separation step was attained on a 110 × 4.6 mm i.d. SCX column (packed with 5-μm particles), while one of two 50 × 4.6 mm i.d. columns packed with different C_{18}-bonded silica particles were used for the

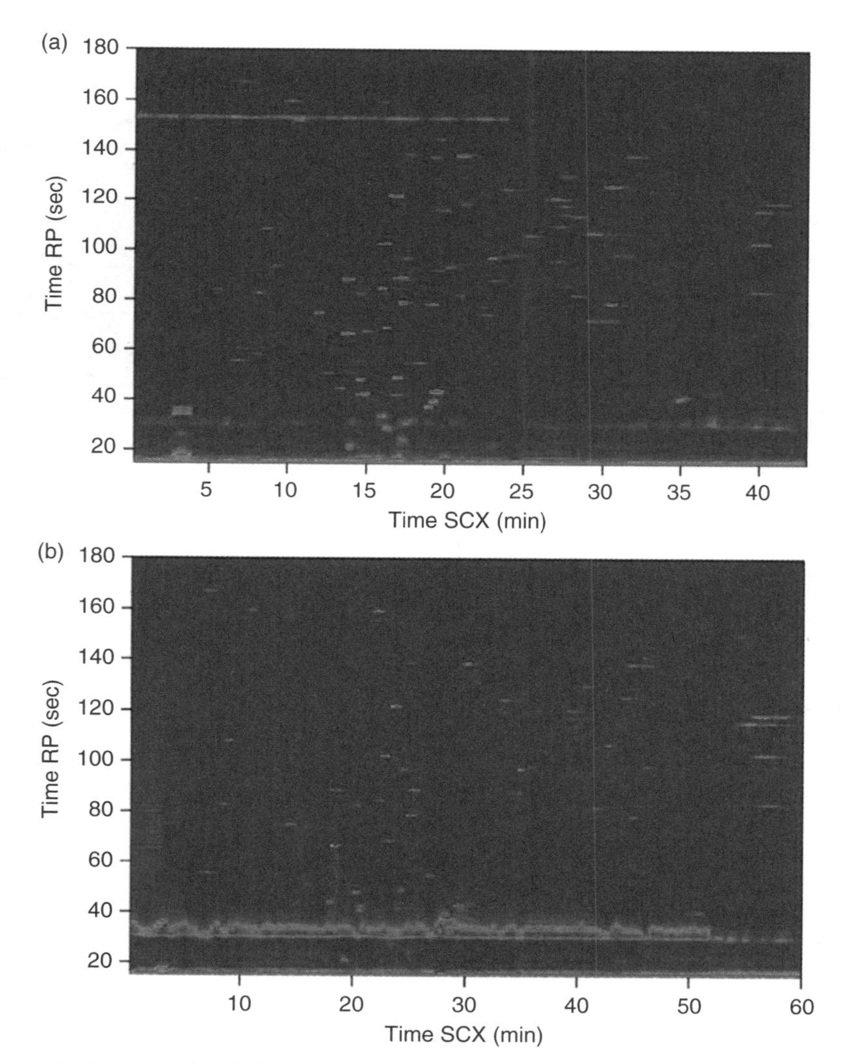

Figure 10.9 Reproduced [From Marchetti et al. (2008), with permission. Copyright © 2008 by The American Chemical Society.]

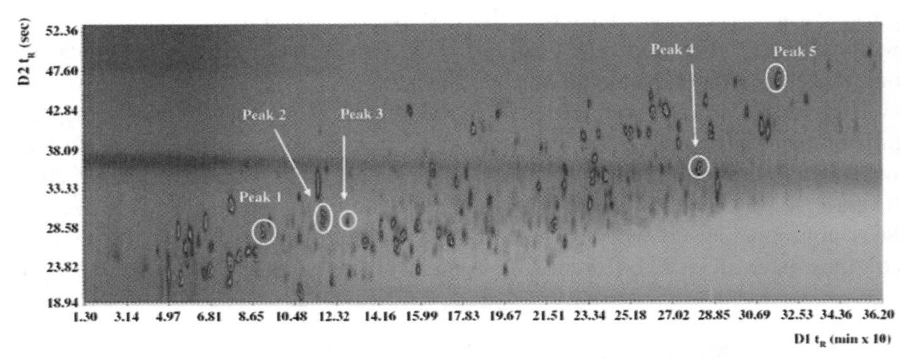

(a)

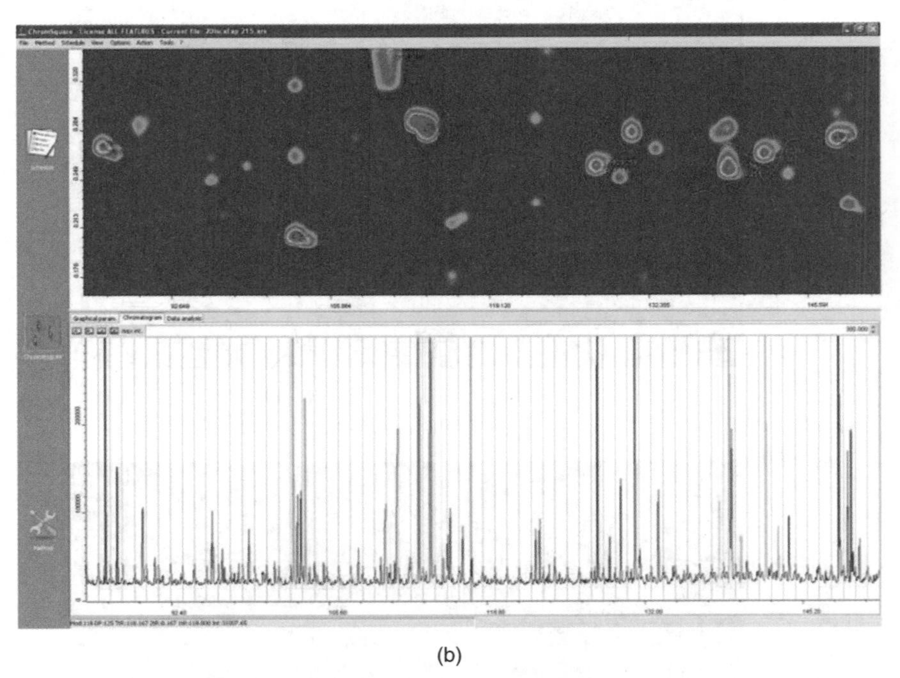

(b)

Figure 10.10 (Top) Comprehensive RPLC × RPLC analysis of a tryptic digest (0.01 M). First dimension: four serially linked Discovery BWP C_{18}, 15 cm × 2.1 mm, 3.0-μm particle-size columns (total length: 60 cm), operated at pH 9. Second dimension: Ascentis Express C_{18}, 3 cm × 4.6 mm, 2.7-μm particle size (fused-core) column, operated at pH 2. (Bottom) Enlargement of the contour plot. Upper software window: some peaks integrated for qualitative/quantitative data analysis. Lower software window: visualization of the entire modulation. [From Mondello et al. (2010), with permission. Copyright © 2010 by Wiley-VCH Verlag GmbH & Co. KGÅ.]

second separation dimension. The first column was packed with totally porous 5-μm particles; the second was packed with a new type of shell particle (average diameter 2.7 μm, shell thickness 0.5 μm, average mesopore size 9 nm). Their comprehensive system provided a theoretical peak capacity exceeding 7000, even at the cost of 28 h of analysis time.

The first column was packed with a strong ion exchanger and eluted with a continuous KCl gradient that allowed the collection of a large number of fractions of the first column eluent. The second column, eluted with an acetonitrile gradient, was packed with particles of C18-bonded silica based on a recent technology (i.e., fused-core). These 2.7 – μm particles, consisting of a 1.7-μm solid core and a 0.5-μm porous shell, have a small diffusion path compared to the approximately 1.5-μm diffusion path of a 3-μm totally porous particle, which reduces axial dispersion of solutes and minimizes peak broadening.

The same fused-core particles have been used by François et al. (2009), who described the first online comprehensive RPLC × RPLC separation of bovine serum albumin and human blood serum tryptic digest. High orthogonality and peak capacity were achieved through the application of a significantly different pH in the two dimensions.

The coupling of fused-core columns in series ensures high efficiency in the first dimension, while a specially designed interface with parallel second-dimension columns further enhanced the separation capability of the comprehensive system. Later, a similar configuration was developed by Mondello et al. (2010), who achieved enhanced resolution in the first dimension by serially coupling Bio wide-pore C_{18} columns at basic pH at moderate pressure. Selectivity was tuned using one short (3-cm) C_{18} column packed with partially porous particles for fast second-dimension analysis, operated at acidic pH due to its likelihood of hyphenation to mass spectrometry. A theoretical peak capacity of 10,686 was attained by serial coupling of up to four columns, even at the expense of longer analysis time. The system was employed for the separation of a human serum albumin tryptic digest (Figure 10.10), and newly designed software was used for data elaboration and visualization, which also makes it possible to perform peak integration in the 2D plane for quantitative or validation purposes.

REFERENCES

Bedani F, Kok WT, Janssen H-G. *J. Chromatogr. A* 2006; 1133:126–134.

Blahová E, Jandera P, Cacciola F, Mondello L. *J. Sep. Sci*. 2006; 29:555–566.

Brudin S, Berwick J, Duffin M, Schoenmakers P. J. *J. Chromatogr. A*. 2008; 1201:196–201.

Bushey M, Jorgenson JW. *Anal. Chem*. 1990; 62:161–167.

Cacciola F, Jandera P, Blahová E, Mondello L. *J. Sep. Sci*. 2006; 29:2500–2513.

Cacciola F, Jandera P, Mondello L. *J. Sep. Sci*. 2007a; 30:462–474.

Cacciola F, Jandera P, Hajdú Z, Česla P, Mondello L. *J. Chromatogr. A* 2007b; 1149:73–87.

Dugo P, Favoino O, Luppino R, Dugo G, Mondello L. *Anal. Chem*. 2004; 76:2525–2530.

Dugo P, del Mar Ramírez Fernández M, Cotroneo A, Dugo G, Mondello L. *J. Chromatogr. Sci*. 2006a; 44:561–565.

Dugo P, Škeříková V, Kumm T, Trozzi A, Jandera P, Mondello L. *Anal. Chem*. 2006b; 78:7743–7755.

Dugo P, Cacciola F, Donato P, Airado-Rodriguez D, Herrero M, Mondello L. *J. Chromatogr. A* 2009; 1216:7483–7487.

Erni F, Frei RW. *J. Chromatogr*. A 1978; 149:561–569.

Fraga C, Carley CA. *J. Chromatogr. A* 2005; 1096:40–50.

François I, de Villiers A, Sandra P. *J. Sep. Sci*. 2006; 29:492–498.

François I, Cabooter D, Sandra K, Lynen F, Desmet G, Sandra P. *J. Sep. Sci*. 2009; 32:1137–1144.

Gray MJ, Sweeney AP, Dennis GR, Slonecker PJ, Shalliker RA. *Analyst* 2003; 128:598–604.

Gray MJ, Dennis GR, Slonecker PJ, Shalliker RA. *J. Chromatogr. A* 2004; 1041:101–110.

Holland LA, Jorgenson JW. *Anal. Chem*. 1995; 67:3275–3283.

Holland LA, Jorgenson JW. *J. Microcol. Sep*. 2000; 12:371–377.

Jandera P, Fischer J, Lahovská H, Novotná K, Česla P, Kolařová L. *J. Chromatogr. A* 2006; 1119:3–10.

Jandera P, Česla P, Hájek T, Vohralík G, Vyňuchalová K, Fischer J. *J. Chromatogr. A* 2008; 1189:207–220.

Jiang X, Van der Horst A, Lima V, Schoenmakers PJ. *J. Chromatogr. A* 2005; 1076:51–61.

Kivilompolo M, Hyötyläinen T. *J. Sep. Sci*. 2008; 31:3466–3472.

Knecht D, Rittig F, Lange RFM. *J. Chromatogr. A* 2006; 1130:43–53.

Köhne AP, Welsch T. *J. Chromatogr. A* 1999; 845:463–469.

Köhne AP, Dornberger U, Welsch T. *Chromatographia* 1998; 48:9–16.

Kok SJ, Hankemeier T, Schoenmakers PJ. *J. Chromatogr. A* 2005; 1098:104–110.

Liu Y, Xue X, Guo X, Xu Q, Zhang F, Liong X. *J. Chromatogr. A* 2008a; 1208:133–140.

Liu Y, Guo Z, Jin Y, Xue X, Xu Q, Zhang F, Liong X. *J. Chromatogr. A* 2008b; 1206:153–159.

Marchetti N, Fairchild JN, Guiochon G. *Anal. Chem*. 2008; 80(8):2756–2767.

Mitulovic G, Swart R, van Ling R, Jakob T, Chervet JP. *LC Packings* 2004; 61–64

Mondello L, Herrero M, Kumm T, Dugo P, Cortes H, Dugo G. *Anal. Chem*. 2008; 80:5418–5424.

Mondello L, Donato P, Cacciola F, Fanali C, Dugo P. *J. Sep. Sci*. 2010; 33:1454–1461.

Murahashi T. *Analyst* 2003; 128:611–615.

Murphy RE, Schure MR, Foley JP. *Anal. Chem*. 1998a; 70:1585–1594.

Murphy RE, Schure MR, Foley JP. *Anal. Chem*. 1998b; 70:4353–4360.

Popovici ST, Van der Horst A, Schoenmakers PJ. *J. Sep. Sci*. 2005; 28:1457–1466.

Stoll DR, Carr PW. *J. Am. Chem. Soc*. 2005; 127:5034–5034.

Stoll DR, Cohen JD, Carr PW. *J. Chromatogr. A* 2006; 1122:123–137.

Tian H, Xu J, Guan Y. *J. Sep. Sci.* 2008; 31:1677–1685.

Toups EP, Gray MJ, Dennis GR, Reddy N, Wilson MA, Shalliker RA. *J. Sep. Sci*. 2006; 29:481–491.

van der Horst A, Schoenmakers PJ. *J. Chromatogr. A* 2003; 1000:693–709.

Vivó-Troyols G, Schoenmakers PJ. *J. Chromatogr. A* 2006; 1120:273–281.

11

OTHER COMPREHENSIVE CHROMATOGRAPHY METHODS

Isabelle François and Pat Sandra
University of Gent, Gent, Belgium
Danilo Sciarrone and Luigi Mondello
University of Messina, Messina, Italy

Combining two different forms of chromatography into a comprehensive system has been far less studied than GC × GC and LC × LC. The primary reasons are, on the one hand, that the mobile phases exist in a different physical state, seemingly rendering interfacing more complex, and on the other hand, that considering the present state of the art of GC × GC and LC × LC, good reasons must be advanced to run specific applications on a multimode comprehensive system.

Combinations such as LC [or in its solid-phase extraction (SPE) mode]–GC, supercritical fluid chromatography (SFC) [or in its (SFE) mode]–GC, or SFC (SFE)–LC have proven to be very efficient in sample preparation and/or sample fractionation. As an example, off-line LC separation before GC analysis enables the separation of chemical classes or targets out of a complex matrix that can then be analyzed by one-dimensional high-resolution GC. Off-line LC–GC is a two-dimensional technique that combines the primary-column selectivity (often low efficiency) and the secondary-column efficiency (often low selectivity). In this chapter, only techniques in which all parts of the first-dimension separation are online subjected to a second-dimension separation are discussed.

Comprehensive Chromatography in Combination with Mass Spectrometry, First Edition.
Edited by Luigi Mondello.

11.1 ONLINE TWO-DIMENSIONAL LIQUID CHROMATOGRAPHY–GAS CHROMATOGRAPHY

Liquid chromatography coupled with high-resolution GC in an online mode to heart-cut selected compounds or fractions has been described extensively in the literature. LC–GC is a very powerful analytical technique because of the combination of the selectivity features of LC with the high efficiency of GC. Furthermore, the drawbacks related to sample manipulation in off-line approaches are eliminated. Online LC–GC methods are particularly well suited to the separation of compounds with similar physicochemical features, in samples characterized by several chemical classes. The coupling of the second dimension to a mass spectrometer generates a very powerful three-dimensional analytical method (LC–GC–MS) in which spectra are usually highly pure, enabling a much easier interpretation compared to those generated from GC–MS analysis.

In LC–GC, a relatively large amount of liquid mobile phase must be eliminated before the GC analysis and several interfaces have been developed in this respect. On the other hand, a prerequisite in LC–GC analysis is that the target analytes must be volatiles or semivolatiles. Practically all LC–GC methods are derived from a previous GC method (e.g., essential oils, motor fuels). For efficient solvent removal and chromatography band transfer, on-column, loop-type, and vaporizing interfaces have been developed. For a detailed description of LC–GC interfaces and instrumental features, some excellent contributions are available in the literature (Cortes, 1991; Grob, 1991; Hyötyläinen and Riekkola, 2003; Mondello et al., 1999).

All interfaces have advantages and disadvantages, and selection is based primarily on the analytical problem at hand and on the LC conditions selected. The amount of solvent that has to be eliminated depends on the column dimensions and can vary between 0.01 and 1 mL. On the other hand, the nature of the solvent is of utmost importance, and very polar solvents (e.g., methanol, water) are difficult to remove effectively. This is the reason that, most often, relatively high volatile apolar to medium polar solvents are used. The programmable-temperature vaporizing (PTV) interface is, from a practical point of view, the easiest to use. In the PTV interface the solvents are removed via the split vent while involatiles remain in the liner, which can easily be cleaned. This is also the reason that the PTV is the most commonly used solvent-removal device in comprehensive two-dimensional liquid–gas chromatography (LC × GC) field.

LC × GC instrumentation is much more complex than that of either GC × GC or LC × LC. The technological step from LC–GC to LC × GC is a rather large one, in both hardware and software terms. However, the most important requisites that must be considered are essentially the same as in LC–GC. An additional aspect requiring consideration is related to the separation times in both dimensions. In GC × GC, the 1D/2D analysis-time ratio is very high, typically in the range 600 to 800 (e.g., 1 h/0.01 h). Hence, accumulation of the first-dimension effluent and the second-dimension GC separation process can be carried out simultaneously. In LC × GC techniques, the 1D/2D analysis-time ratio is very low (usually less than 10), and thus accumulation of the LC effluent and the second-dimension separation step are not carried out at the same time. In terms of MS detection requirements, there are also differences. In GC × GC experiments, thermal modulation and high second-dimension gas linear velocities generate very short chromatography bands, in both space and time. Generally, GC × GC peak durations vary in the range 100 to 500 ms and are related to the modulator type, the second-dimension phase chemistry, and film thickness, together with the temperature program rate and linear gas velocity. As seen in Chapter 4, use of time-of-flight (TOF) MS is most suited to meet the demands for correct GC × GC peak reconstruction. In LC × GC experiments, chromatography in the second dimension is normally within the range 5 to 15 min. Peak widths are much wider than their GC × GC counterparts, so high data acquisition frequencies are not necessary. However, wide mass ranges are required because analytes are commonly characterized by relatively high molecular weights (see the discussion of applications below). Peak capacities generated in LC × GC experiments are much lower than in GC × GC, and thus the availability of good deconvolution capabilities in MS is preferable. TOF MS has some unique features in this respect for LC × GC experiments (de Koning et al., 2004a).

At present, only a handful of true LC × GC papers have been published, and in our opinion there is much room for development and applications in this research field. The first LC × GC setup and application were described in 2000 by Quigley et al. However, the system was used for headspace analysis of volatile organic compounds (VOCs) in water. A short primary LC column was operated using water as mobile phase, and the VOCs eluting from the first dimension ended up in drops of water formed at the tip of a capillary, housed in a "drop interface." The highly volatile compounds migrated to the surface of the drop and a substantial amount was carried away by a helium flow. The latter, enriched with VOCs, was directed to an injector for subsequent high-speed GC analysis. The approach described was certainly interesting but could not have been applied to the analysis of solutes, typically found in LC–GC applications. A great deal of LC–GC work has been devoted to the analysis of petrochemical products and food lipids (Grob, 1991) and when these heavier molecular weight compounds are considered, we can state that real LC × GC examples started to appear from 2003 on (de Koning et al., 2004a,b, 2006; Janssen et al., 2003; Sciarrone et al., 2008). An entirely automated LC × GC system, developed after an off-line optimization and a feasibility study (Janssen et al., 2003), was applied to the analysis of food triacylglycerols (TAGs) and fatty acid methyl esters (FAMEs) (de Koning et al., 2004a). The setup was used in combination with TOF MS and flame ionization detection (FID). LC × GC analyses were carried out using stop-flow conditions. The primary column flow was stopped after the transfer of a chromatography band from the LC to the GC instrument. The temperature-programmed GC analysis was then initiated, while the LC remained in the stop mode until the GC system reached a new ready status. The use of two types of interface was reported: namely, a six-port switching valve and a dual-side-port 100-μL syringe (Figure 11.1), both

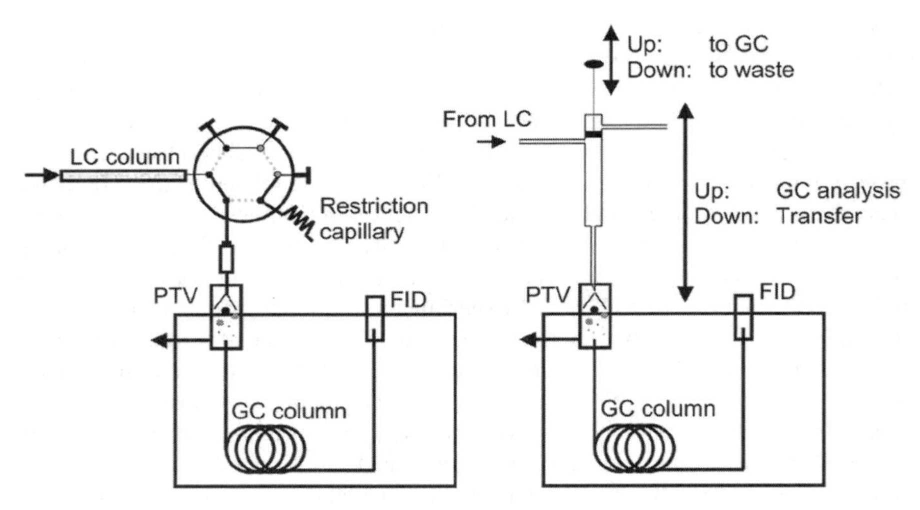

Figure 11.1 Valve-based (left) and the syringe-based (right) LC × GC interfaces. [From de Koning et al. (2004a), with permission. Copyright © 2004 by Wiley-VCH Verlag GmbH & Co. KGaA.]

of which provided satisfactory performance. In the valve system, a 100-μm i.d. fused-silica capillary (30 cm length) was employed to transfer the primary column effluent from the interface to the GC injector. At the beginning of each GC run, the liquid remaining inside the transfer capillary was eliminated through a restriction capillary, located in one of the valve ports. The syringe interface was characterized by a lower entrance that received effluent from the first dimension, while an upper exit was used to direct the effluent to waste. A plug was situated at the terminal part of the plunger, which was characterized by a lower outer diameter with respect to the internal diameter of the syringe barrel. In this way, the LC mobile phase flowed freely inside the syringe. In the waste mode, the plug was located below the LC effluent entrance and the mobile phase was directed to waste. In the transfer position, the plug was situated between the two lines and the effluent was directed to the GC. At the end of each transfer, the LC flow was stopped, the syringe withdrawn from the injector, and the GC analysis started. First-dimension effluent bands were subjected to splitting by using a high split flow, inside a hot PTV injector chamber. Consequently, only a minor percentage of each LC fraction was transferred to the secondary column, avoiding overloading. The authors emphasized that the initial GC temperature should enable both solvent elimination and analyte reconcentration.

For FAME analysis, a 250 mm × 2.0 mm i.d. Ag-loaded column was used as the primary column operated under step-gradient conditions (CH_2Cl_2 + CH_3OCH_3, then CH_3CN). A highly polar cyanopropylsiloxane capillary column (25 m × 0.25 mm i.d.) was employed for GC analyses. The LC flow rate was 200 μL/min, and 50-μL fractions were transferred to the second dimension and analyzed in about 7 min. The use of a column with an i.d. of 2 mm avoided the introduction of high amounts of solvent into the GC injector. The FAMEs were separated essentially on the basis of the double-bond (DB) number in the first dimension, and on the basis of carbon number (CN), DB number, and position of the double bond in the GC column.

Part of the TIC AgLC × GC–TOF MS chromatogram related to the saturated FAME zone is shown in Figure 11.2. As can be observed, the GC step enabled a detailed separation between DB0 FAMEs, with the generation of high-quality TOF MS spectra. The LC × GC FAME application reported is certainly worthwhile from a demonstrative viewpoint, but GC × GC is much more suitable for this specific application. In GC × GC using a nonpolar/polar column set, FAMEs are distributed in highly ordered structures on the basis of CN (C_{18}, C_{20}, etc.), DB number (0, 1, 2, etc.), and position (ω3, ω6, etc.). The formation of FAME group-type patterns enables reliable peak assignment even in the absence of MS detection and/or that of pure standard compounds (Tranchida et al., 2007).

One obvious concern related to use of the LC stop-flow mode was analyte diffusion and, consequently, band broadening. However, in an application lasting 10 h (!), it was found that after transfer of 100 fractions, no evident signs of band broadening were observed. An additional concern and disadvantage, still related to the stop-flow configuration, are the excessively long analysis times. The authors reported three options to circumvent the LC × GC run-time problem: (1) reduce the primary column sampling frequency (note that undersampling may be a risk because first-dimension resolution can be degraded, (2) use a faster GC method, and (3) carry out periodic and not continuous transfers (hence, extensive baseline zones could be directed to waste). It appeared that the combination of options 2 and 3 would be a wise choice.

Following the lipid study, a syringe-based 3D normal-phase (NP) LC × GC–TOF MS method was applied to an analysis of mineral oil (de Koning et al., 2004b). The first dimension, a 250 mm × 4.6 mm i.d. aminopropyl silica column, was operated using hexane as mobile phase (0.8 mL/min). Note that hexane is one of the easiest solvents to handle in LC × GC. The fractions transferred onto the secondary column (5% phenyl-methylsiloxane, 30 m × 0.25 mm i.d. × 0.25 μm d_f) were 6 s wide, corresponding to a volume of 80 μL, and were analyzed in about 15 min. The column combination was highly orthogonal. Analytes in the first dimension were resolved on the basis of polarity, while equal-polarity compounds were separated in the second dimension on the basis of boiling point (or molecule size). Only the LC fractions of interest were subjected to the second-dimension analysis, thus saving a considerable amount of time. The TOF MS was operated at a sampling frequency of 20 Hz, in the mass range 50 to 450 m/z. The bidimensional chromatogram of the NPLC × GC–TOF MS analysis of diesel oil is shown in Figure 11.3. In the LC step three chemical classes are isolated: (in order of polarity) saturated hydrocarbons, monoaromatics, and diaromatics. Compared to GC × GC, the LC × GC result was better in terms of interclass separation, but was considerabely inferior in terms of intraclass separation. An example of the insufficient degree of intraclass separation is illustrated

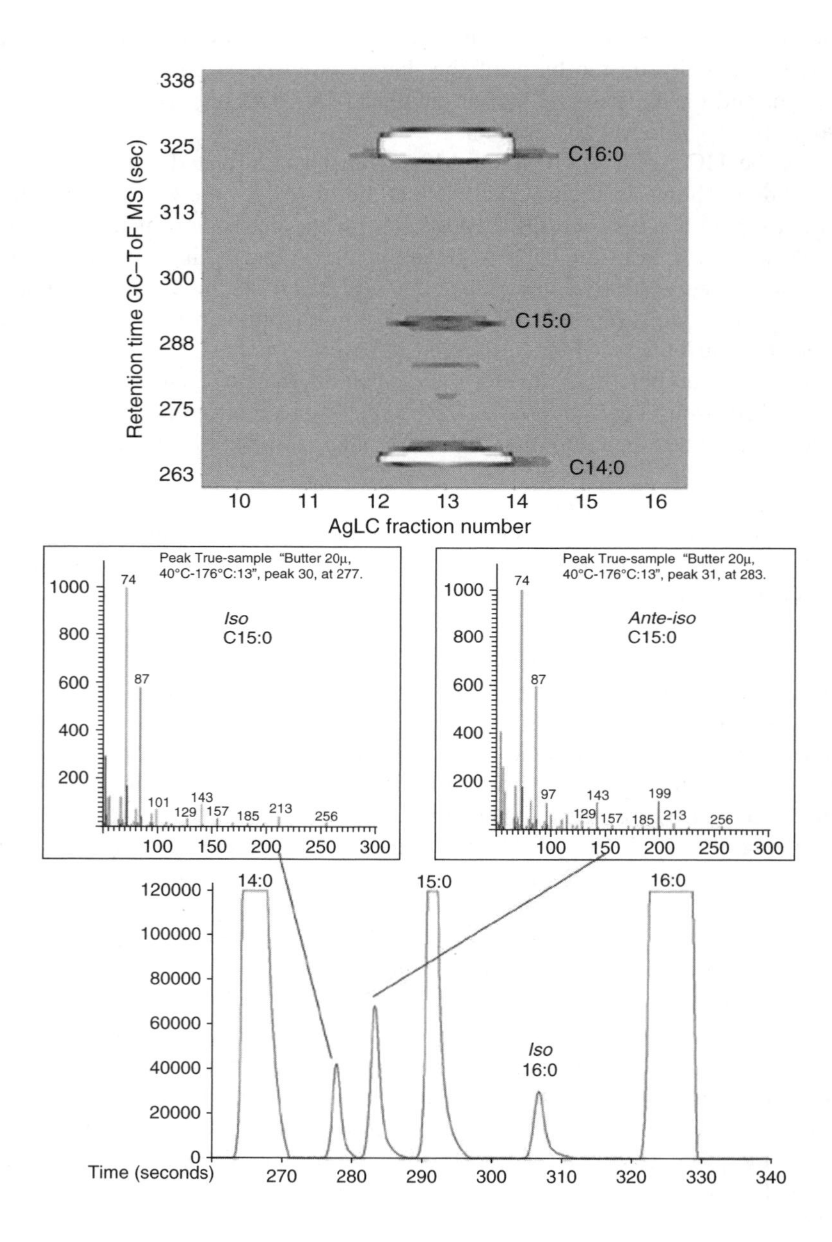

Figure 11.2 TIC AgLC × GC–TOF MS chromatogram expansion relative to the FAME DB0 region; unconverted GC–TOF MS chromatogram expansion relative to fraction 13; spectra relative to iso and anteiso-$C_{15:0}$. [From de Koning et al. (2004a), with permission. Copyright © 2004 by Wiley-VCH Verlag GmbH & Co. KGaA.]

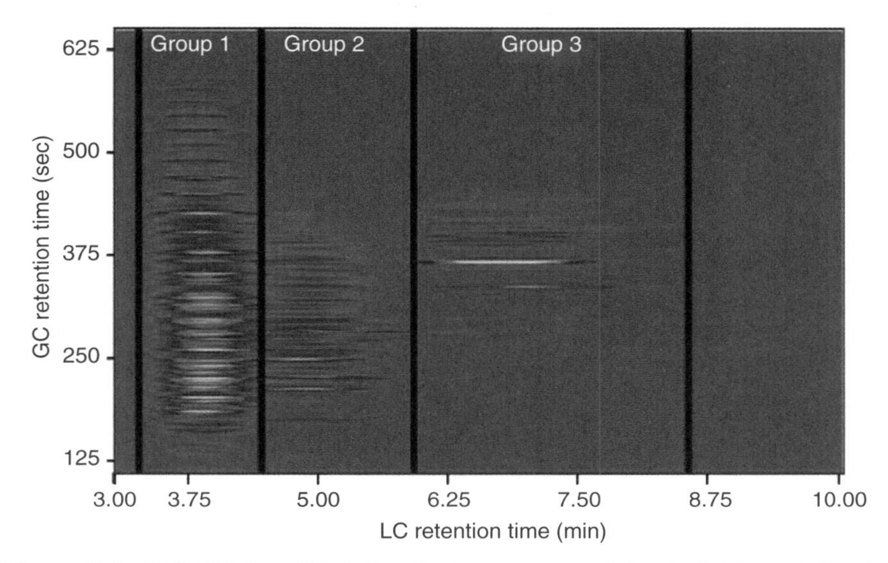

Figure 11.3 TIC NPLC × GC–TOF MS chromatogram of diesel oil. [From de Koning S et al. (2004b), with permission. Copyright © 2004 by Elsevier.]

in Figure 11.4, showing three GC chromatograms recorded at the start, in the middle, and at the end of the first LC peak. The crowded and dissimilar nature of the single-cut chromatograms clearly indicate that the resolving power of LC × GC alone is not sufficient for this type of application. As a consequence, exploitation of the MS dimension was necessary to further resolve compounds within each chemical class. Four extracted-ion NPLC × GC–TOF MS chromatogram expansions relative to four subclasses within the saturated hydrocarbon group are shown in Figure 11.5. Within each single saturated-hydrocarbon class, the extent of overlapping still present among the mono-, di-, and tricycloparaffins is evident. In the conclusion the authors stated that additional analytical benefits could be attained by combining an LC group-type pre-separation with the enhanced resolving power of GC × GC.

Off-line LC fractionation followed by GC × GC for similar applications has been described by several authors, but this is not the topic of this contribution and we refer the reader to the literature for details (e.g., Adam et al., 2007; Edam et al., 2005; Mao et al., 2008). At-line AgLC × GC × GC was also reported for the analysis of FAMEs (de Koning et al., 2006). Intact triglycerides (TAGs) were separated on the AgLC column on the basis of DB number. The fractions eluting from the AgLC column were transferred to autosampler vials for automated TAG–FAME transesterification prior to GC × GC separation. Again, the LC × GC × GC results (obtained in 72 h!) were interesting from a fundamental point of view, but similar results could have been achieved much faster through transesterification of the entire TAG mixture and careful optimization of a GC × GC method for FAME analysis.

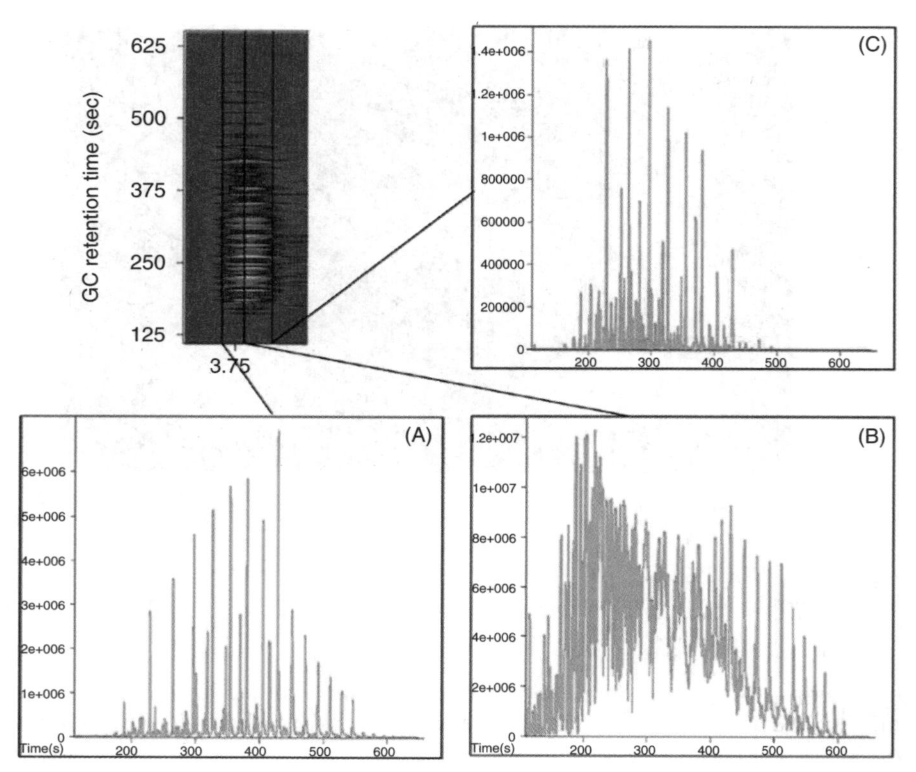

Figure 11.4 TIC NPLC × GC–TOF MS chromatogram expansion relative to the saturated hydrocarbon region, and related unconverted TIC GC–TOF MS chromatograms. [From de Koning et al. (2004b), with permission. Copyright © 2004 by Elsevier.]

The features of GC × GC using a rapid-scanning quadrupole mass spectrometer (qMS) as a detection system and LC × GC were combined in the analysis of a diesel sample. In the final off-line LC × GC × GC experiment (Sciarrone et al., 2008), all fractions were subjected to the various separation mechanisms. Initially, the diesel sample was subjected to apolarGC × polarGC–qMS analysis. The TIC chromatogram in Figure 11.6 shows a typical group-type separation: saturated hydrocarbons (SHs), monoaromatics (MCAHs), diaromatics (DCAHs), and tri- and tetracyclic aromatics (triCAHs and tetraCAHs). The latter are nicely separated but are characterized by a low signal-to-noise ratio. A stop-flow syringe-based LC × GC experiment was carried out on the diesel sample with the objective of defining the LC elution time windows of the aforementioned chemical classes and to evaluate the quality of the overall LC × GC separation. A dedicated LC × GC software enabled the control of both instruments through the respective native software. Sixteen liquid fractions of 250 μL eluting from a 20 cm × 4.6 mm i.d. aminopropyl silica column operated with hexane as mobile phase (0.5 mL/min) were transferred to a GC PTV injector. The internal syringe

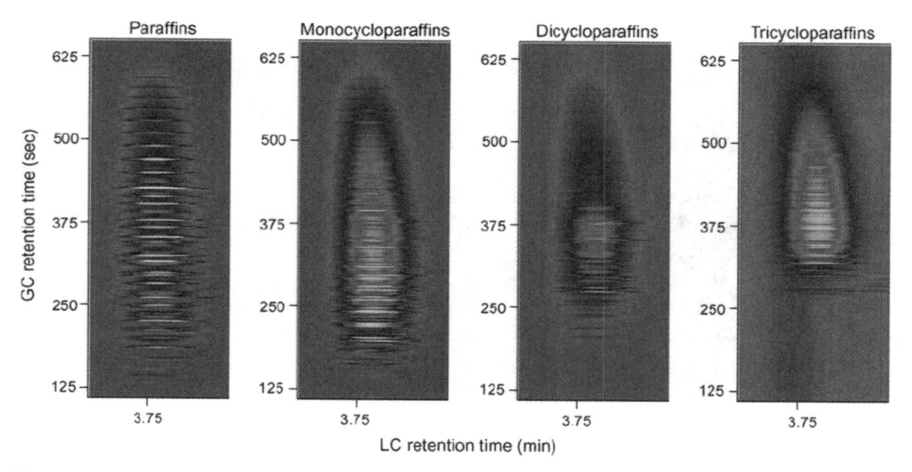

Figure 11.5 Extracted-ion NPLC × GC–TOF MS chromatogram of the characteristic mass fragments for the various hydrocarbon types in middle distillates as reported in ASTM method D2425. Expansions relative to the saturated hydrocarbon region. [From de Koning et al. (2004b), with permission. Copyright © 2004 by Elsevier.]

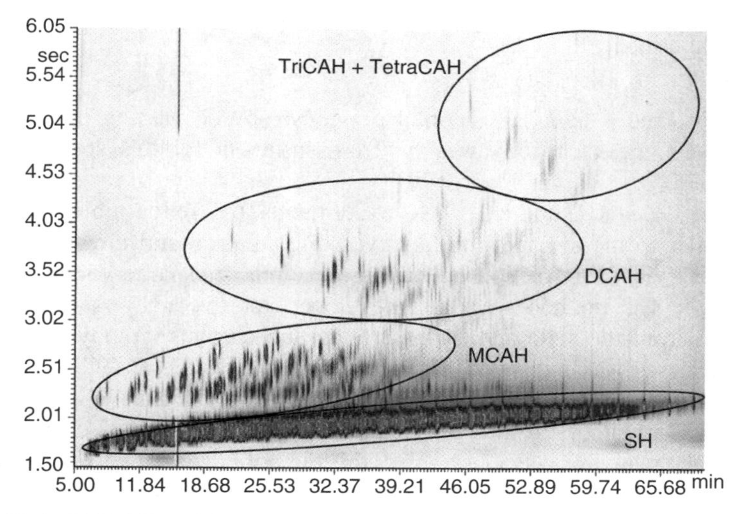

Figure 11.6 TIC GC × GC–qMS chromatogram of diesel. SH, saturated hydrocarbons; MCAH, monocyclic aromatics; DCAH, dicyclic aromatics; triCAH, tricyclic aromatics; tetraCAH, tetracyclic aromatics. For detailed experimental conditions, see Sciarrone et al. (2008). [From Sciarrone et al. (2008), with permission of the copyright owner.]

volume between the plug and the needle outlet was only 4 μL. This low volume enabled the almost immediate transfer of the LC effluent to the GC for PTV solvent venting. The diesel volatiles were separated rapidly on a nonpolar capillary column (GC analysis time: 13.25 min). The LC × GC chromatogram

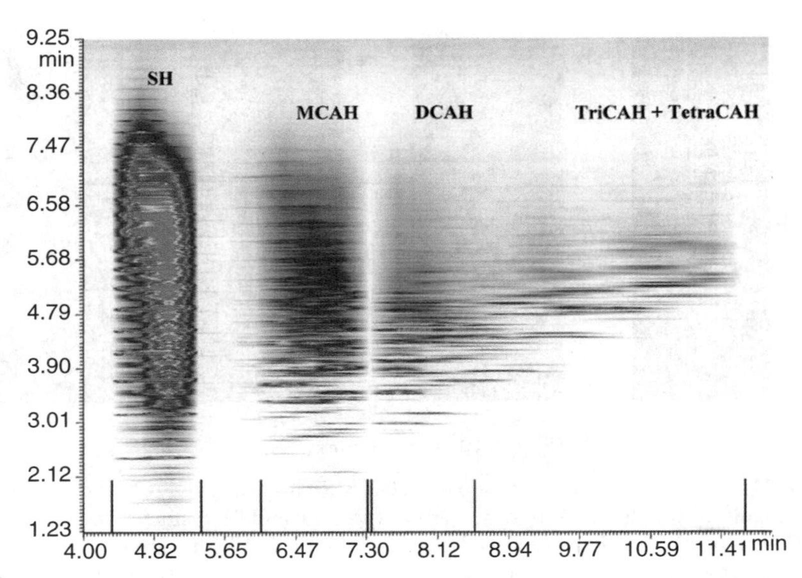

Figure 11.7 LC × GC–FID chromatogram of diesel oil. For detailed experimental conditions, see Sciarrone et al. (2008). [From Sciarrone et al. (2008), with permission of the copyright owner.]

is shown in Figure 11.7. As reported previously by de Koning et al. (2004b), the LC × GC peak capacity was far from sufficient for this specific type of application.

After the GC × GC and LC × GC experiments, the Teflon tubing connecting the LC outlet to the syringe interface was disconnected and directed manually to four glass vials for fraction collection. An optimized large-volume-injection (LVI) GC × GC method was then developed for the analysis of each fraction. Apart from the saturated hydrocarbons, the chromatography for all other chemical classes was improved over that of the direct diesel GC × GC application. The method developed can be defined as "comprehensive" because each LC peak was isolated and then subjected to a secondary analysis. Furthermore, although each peak was (under)sampled, this did not have a negative effect on the first-dimension resolution (only four peaks were present). The LC × LVI-GC × GC–qMS chromatogram relative to the fourth LC fraction, the tri- and tetracyclic aromatic group, is shown in Figure 11.8. The low signal-to-noise ratios of these diesel constituents in the direct GC × GC experiment were due to their limited amounts in the initial sample and to the high second-dimension retention. The first issue was resolved by LVI, while the second issue was contrasted, in part, through a $+10^{\circ}$C/s GC oven offset. The higher second-dimension elution temperatures had no negative effects on the resolution of the tri- and tetracyclic aromatics, while a beneficial effect on the sensitivity was noted as the second-dimension retention times decreased. The 20-Hz qMS spectra acquisition rate generated a sufficient number of data points for all tri- and tetracyclic aromatic compounds,

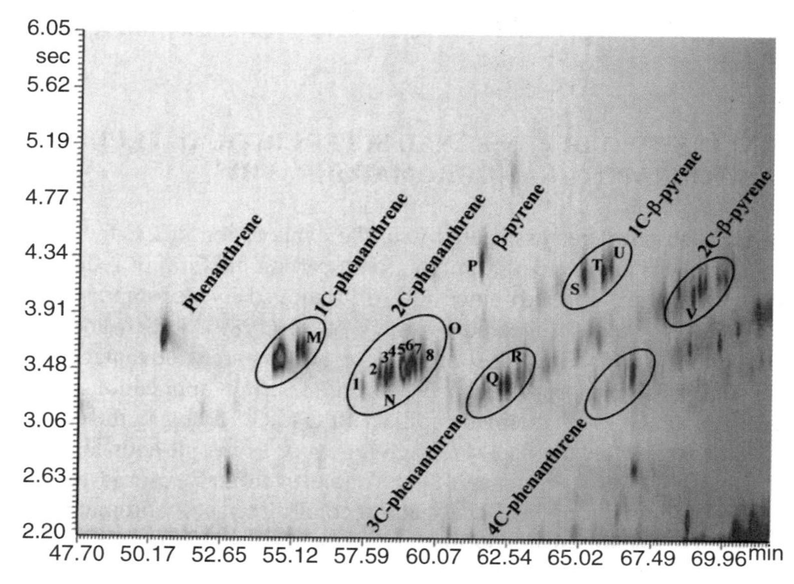

Figure 11.8 TIC LC × GC × GC–qMS chromatogram of tri- and tetra-aromatic compounds. For detailed experimental conditions, see Sciarrone et al. (2008). [From Sciarrone et al. (2008), with permission of the copyright owner.]

enabling adequate peak reconstruction and, hence, quantification. The authors concluded that the construction of an online LC × GC × GC–qMS instrument presents no substiantial technical difficulties. It was also added that the analytical potential of such a tetra-dimensional method described would be great, as well as a rather considerable economical cost per application (solvents and cryogenic gases).

Notwithstanding all features claimed for LC × GC, at the end of this section we should ask the question: Do we need LC × GC, and what is its future? The first observation is that the selectivity of LC through solid-phase extraction (SPE) is employed extensively in sample preparation (enrichment and/or fractionation) before GC analysis. SPE is now fully automated off- or online with the chromatographic system. When higher efficiency is needed, off-line LC fractionation is another route that should be considered. From a technological point of view, this is far simpler than LC × GC.

It is also striking that in recent years, developments in LC × GC have been scarce, and very little has been published. On the other hand, most LC × GC applications are overshadowed by the potential of GC × GC and LC × LC for similar applications. The syringe-based interface developed for LC–GC appears to be the most popular and powerful approach for LC × GC, but no (or little) instrumental progress has been made in recent years. Moreover, although stop-flow operation does work, it is also the greatest disadvantage of the method, due to the cost in time. In conclusion, LC × GC appears to be suited only for some niche applications, while LC–GC is promising for target-cutting applications.

11.2 ONLINE TWO-DIMENSIONAL SUPERCRITICAL FLUID CHROMATOGRAPHY–GAS CHROMATOGRAPHY

In SFC–GC, the problems associated with the evaporation of LC mobile phases before GC analysis are avoided (applying only carbon dioxide) or reduced (when organic modifiers are applied) since CO_2 is removed by decompressing into a gas. SFC–GC has been pioneered by Levy et al. (1989), Lynch and Heyward (1994), and Nam and King (1994). Despite the apparent advantages of SFC as the first dimension in comprehensive techniques, few applications have been described. Liu et al. (1993) coupled capillary SFC to GC by using the compressed carbon dioxide gas after the back-pressure regulator as the gaseous mobile phase for GC. Both columns were placed in the same oven and exposed to the same temperature programs, implying that each second-dimension chromatogram was acquired under isothermal conditions, as is the case in GC × GC. This was advantageous in terms of column re-equilibration time, but the first-dimension separation was negatively influenced by the increasing temperature. Moreover, the separation in the second dimension was largely compromised by using CO_2 as mobile phase, due to slow diffusion coefficients.

Venter and Rohwer (2004) and Venter et al. (2006) described SFC × GC for the group-type separation of petrochemical samples using packed-column SFC and for the characterization of oxygenates in petroleum products using capillary PLOT SFC. In the latter configuration, the first-dimension flow rate was stopped during the temperature gradient separation of every fraction in the second dimension. A high degree of orthogonality was achieved through the combination of polarity and volatility-based separations. Figure 11.9 shows the SFC × GC separation of a lead-free gasoline sample. The presence of the *tert*-amyl methyl ether (TAME) as well as small amounts of diisopropyl ether and diisoamyl ether could be confirmed. The x-axis separation is according to polarity on a 30 m × 0.32 mm i.d. CPSilica PLOT column with CO_2 as mobile phase at 150 bar and 28°C, while the y-axis represents a volatility separation on a 1 m × 0.25 mm i.d. stainless-steel column that was resistively heated at a rate of 450°C/min from—50 to 250°C.

11.3 ONLINE TWO-DIMENSIONAL SUPERCRITICAL FLUID CHROMATOGRAPHY–SUPERCRITICAL FLUID CHROMATOGRAPHY

Hirata et al. (2003) effectively coupled two packed-column SFC dimensions by means of an interface composed of a double three-port and a 10-port switching valve equipped with a trapping tube consisting of methyl silicone material on which the analytes were trapped after elution from the first dimension and

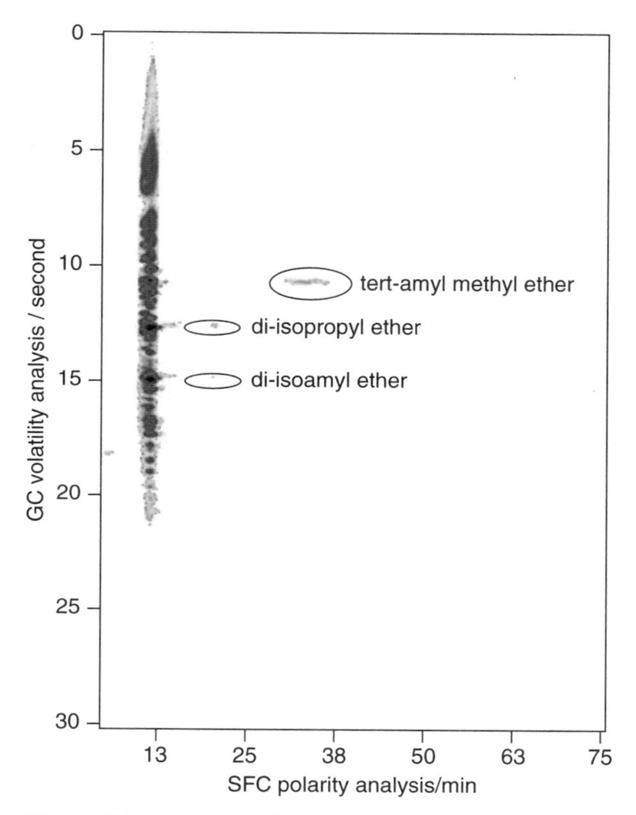

Figure 11.9 SFC × GC separation of lead-free gasoline. [From Venter et al. (2006), with permission. Copyright © 2006 by The American Chemical Society.]

decompression through the regulator. A schematic diagram of the configuration is shown in Figure 11.10. The first dimension was operated in the stop-flow mode. The selectivity in both dimensions was tuned through different temperatures (the first column under subcritical conditions and the second column under supercritical conditions) using the same octadecyl silica (ODS) stationary phase as illustrated for the analysis of TAGs (Hirata et al., 2003) or by using silica and ODS in the first and second dimension, respectively, for the separation of FAMEs (Hirata and Sogabe, 2004). Figure 11.11 shows the contour plot for the analysis of the TAGs in lard by SFC × SFC. SFC × SFC was later extended with SFE (Okamoto and Hirata, 2006). SFE–SFC × SFC was applied to the determination of styrene oligomers in polystyrene, and diastereoisomers could be resolved for linear tetramers and pentamers as well as for trimers containing the tetraline moiety. More recently, the same group described SFC × SFC with capillary columns in both dimensions (Hirata and Ozaki, 2006).

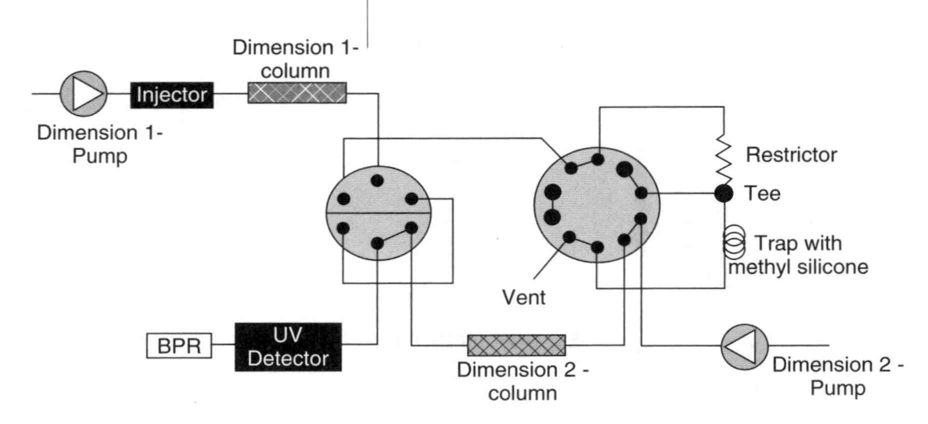

Figure 11.10 SFC × SFC configuration. [From Hirata et al. (2003).]

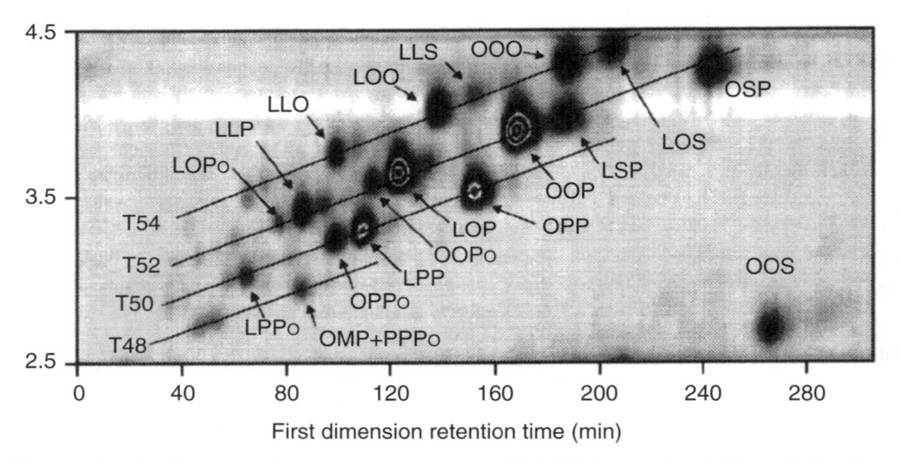

Figure 11.11 Contour plot of the separation of TAGS in lard by SFC × SFC. [From Hirata et al. (2003), with permission. Copyright © 2003 by Wiley-VCH Verlag GmbH & Co. KGaA.]

11.4 ONLINE TWO-DIMENSIONAL SUPERCRITICAL FLUID CHROMATOGRAPHY-LIQUID CHROMATOGRAPHY

SFC × LC was developed primarily in trying to improve the separation capabilities of normal-phase (NP) LC× reversed-phase (RP) LC, a combination highly susceptible to various problems, all related to the immiscibility of the mobile phases. By replacing NPLC by SFC, expansion of CO_2 when exposed to atmospheric pressure leads to fractions composed of solvents that are compatible with the secondary mobile phases. Additionally, the low viscosity of supercritical fluids, together with increased diffusivity, lead on the one hand, to shorter retention times, and on the other hand, to more efficient primary separations by coupling columns in series. Other advantages in comparison to NPLC are faster column re-equilibration and a wider range of experimental variables for separation optimization.

An interface design to perform comprehensive SFC × RPLC was recently described by François et al. (2008). The SFC × LC interface is shown in Figure 11.12. The interface is composed of a two-position 10-port switching valve equipped with two octadecyl silica (C_{18}) packed loops. To focus the analytes, a makeup flow of water is added to the SFC effluent prior to fraction collection in the packed loops. After SFC analysis, the effluent is depressurized after the back-pressure restrictor and mixed with water by means of a T-junction and transferred to the interface. The function of this interface is comparable to that of the conventional loop interface, widely used LC × LC. The two-position 10-port switching valve enables continuous sampling of first-dimension effluent alternatively in the two loops, after which their contents are redirected to the second dimension for fast separation. In contrast to the loops in LC × LC, which are for most applications empty storage loops, the analytes are trapped in loops packed with C_{18} stationary phase. The presence of packing material is of crucial importance, since the analytes enter the loops in a plug of expanded supercritical CO_2 together with small volumes of organic modifier. If no packing material is present, the strong flow of the expanded CO_2 gas mixed with modifier forces the analytes to elute the system through the waste line. Furthermore, the addition of water is necessary not only to reduce the interferences of residual CO_2 gas in the second-dimension separation after transfer (the *solvent displacement effect*), but additionally, to ensure effective focusing in the trap. When no makeup fluid is added, the presence of small volumes of modifier, which are usually "strong" RP solvents, results in breakthrough, and again, the analytes would disappear in the waste vessel. The effect of the addition of water is illustrated by off-line analyses of 5,7-dimethoxycoumarin. The compound was injected on a cyanopropyl silica

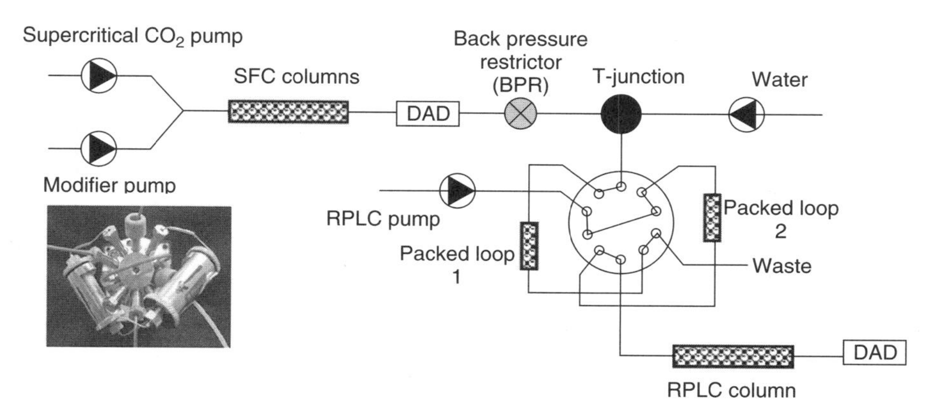

Figure 11.12 SFC × RPLC configuration with packed loops. [From François et al. (2008), with permission. Copyright © 2008 by Wiley-VCH Verlag GmbH & Co. KGaA.]

column (25 cm × 2 mm i.d. × 5 μm d_p) and analyzed by isocratic SFC with a mobile phase of supercritical CO_2 and 5% of EtOH at a flow rate of 0.5 mL/min. After effluent was mixed with water at a flow rate of 1.5 mL/min by the T-junction, the component was trapped in one of the C_{18} loops and the valve was turned manually to allow transfer of the loop content to the second-dimension column. In the case of the experiment where no makeup water was added, the entrance of the water line on the T-junction was replaced with a closed stainless-steel fitting. Figure 11.13 illustrates that the addition of water ensures effective focusing of 5,7-dimethoxycoumarin. The analysis of a lemon oil extract was performed using this configuration (François et al., 2008). The SFC column was 1 m long (four 25 cm × 2 mm i.d. × 5 μm d_p cyanopropyl silica columns) operated at a temperature of 40°C with CO_2 (A) and ethanol (B) as a modifier in gradient mode, and the second column was C_{18} 5 cm × 4.6 mm i.d. × 3.5 μm d_p (Agilent) operated at a flow rate of 4 mL/min with water (phase A) and acetonitrile (phase B) in a gradient as mobile phase.

The performance of the interface was improved by addition of an extra switching valve and T-junction. In this way, the second-dimension peak capacity could be increased, while the solvent displacement effect of water used as a rinsing medium for the loops prior to the transfer of the fractions to the second dimension, and as makeup fluid added to the SFC effluent to ensure analyte focusing, could be fully exploited (François and Sandra, 2009). Evaporative light scattering and ultraviolet (UV) detection with standard and high-pressure flow cells were evaluated in terms of data acquisition speed and suppression of signal interferences originating from the supercritical CO_2 expansion. Implementation of a high-pressure UV flow cell far outperformed the two other detection options. Figure 11.14 shows the SI-SFC × RPLC separations of the phenacyl esters of fish oil. For the SFC analysis, two silver-loaded columns of 25 cm × 4.6 mm i.d. × 5 μm d_p were coupled in series and operated at a temperature of 40°C. The mobile phase was operated at 1 mL/min and consisted of supercritical CO_2

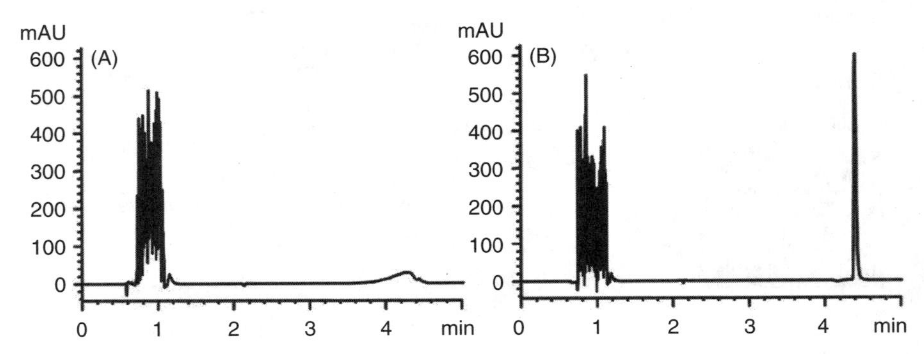

Figure 11.13 Influence of the addition of water to the SFC effluent: (A) no addition of water; (B) with addition of water at 1.5 mL/min. [From François et al. (2008), with permission. Copyright © 2008 by Wiley-VCH Verlag GmbH & Co. KGaA.]

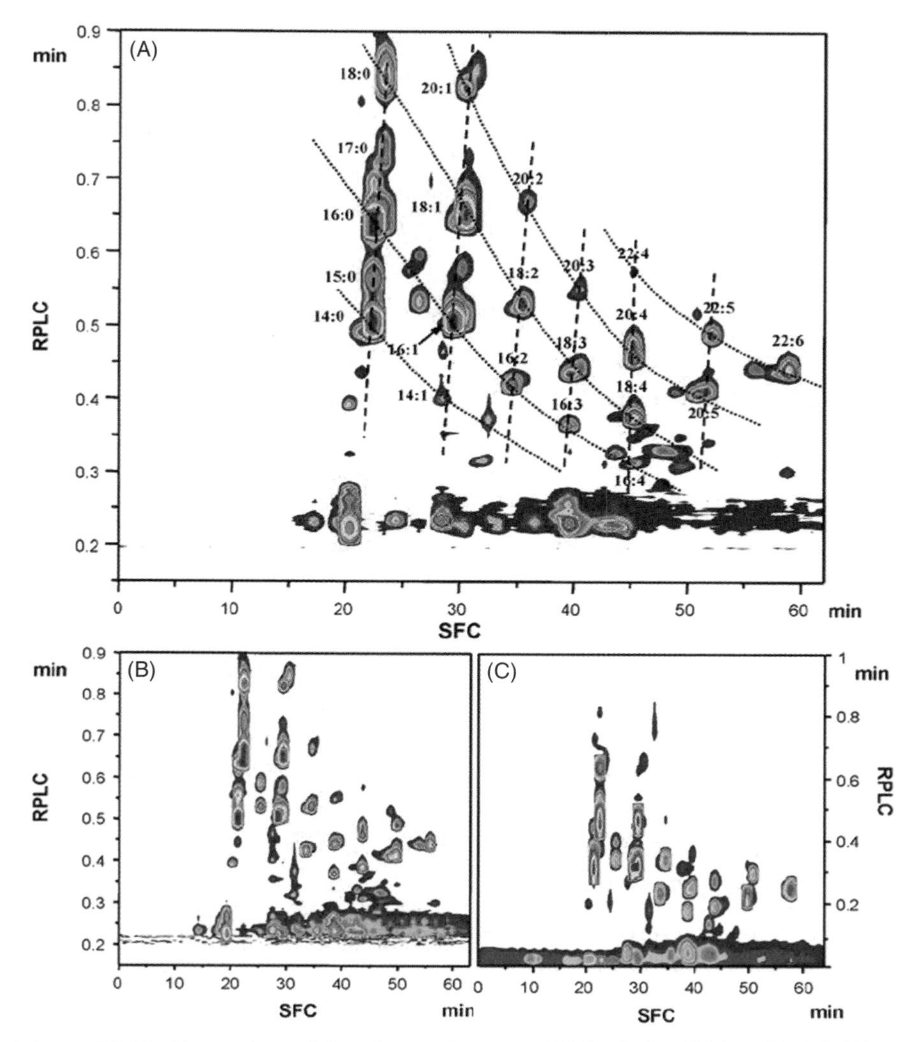

Figure 11.14 Separation of the phenacyl esters of fish oil by SFC × RPLC. [From François et al. (2009), with permission. Copyright © 2009 by Elsevier.]

(phase A) and ACN/IPA (6/4) (phase B). A pressure and modifier gradient was applied. The packed loops used for the trapping of the analytes from the SFC effluent were 7.5 mm × 4.6 mm i.d. × 5 μm d_p ODS stationary phase. The modulation time was 1 min. The second-dimension column was a C_{18} column 5 cm × 4.6 mm i.d. × 3.5 μm d_p operated at a flow rate of 5 mL/min at 30°C. The mobile phase consisted of water (phase A) and ACN (phase B), and a fast gradient was applied. Detection (80 Hz) was performed at 240 nm. The separation is characterized by a high degree of orthogonality, since the sample

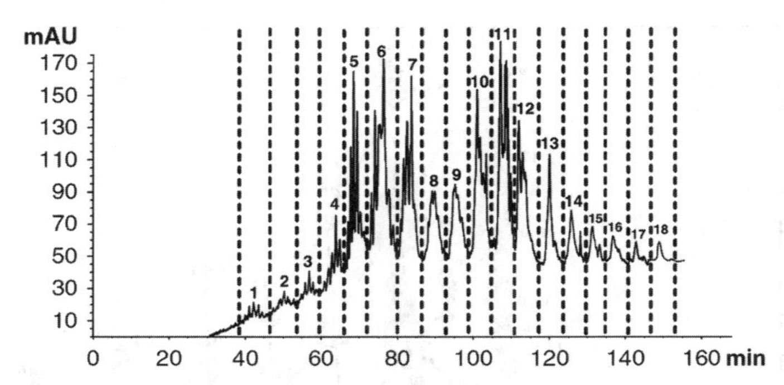

Figure 11.15 Separation of the triacylclycerols in fish oil by SFC. [From François et al. (2010), with permission. Copyright © 2010 by Wiley-VCH Verlag GmbH & Co. KGaA.]

compounds are spread randomly over the available retention plane, and almost the entire second-dimension retention time window is used (from 0.2 to 0.9 min). A correlation factor as low as $r^2 = 0.024$ was reported.

The same interface was applied for the separation of the intact triacylglycerols of fish oil (François et al., 2010). The first-dimension column was the same as described for phenacyl ester analysis, while the second column was a monolithic ODS column (10 cm × 4.6 mm i.d.) operated in nonaqueous RPLC mode with ACN (phase A) and IPA (phase B) in a gradient at a flow rate of 5 mL/min. The modulation time was 2 min. The UV detector was set at a wavelength of 210 nm. Figure 11.15 shows the first-dimension chromatogram of the fish oil on the two serially coupled silver-loaded columns. Separation occurs according to the number of double bonds, indicated by the bold numbers and dashed lines. The comprehensive SI-SFC × non-aqueous RPLC separation of the triacylglycerols in the fish oil sample is presented in Figure 11.16. The separation shows high orthogonality, and compounds with an identical number of double bonds, coeluting in the first dimension, are resolved according to hydrophobicity on the secondary column. The most striking observation in the contour plot (Figure 11.16A) and 3D representation (Figure 11.16B) is the high degree of complexity of the sample. Another observation is the limited time available for separation in the second dimension. Fast elution of the sample components is required within the modulation time with the highest possible efficiency, but the similarity of the compounds hampers the fulfillment of these requirements. This phenomenon also results in the need to comploy the relatively long modulation time of 2 min.

In the same report (François et al., 2010), the authors compared SI-SFC × non-aqueous RPLC and SI-SFC–nonaqueous RPLC for the fish oil sample. As expected, the off-line procedure clearly outperformed the comprehensive approach, which is caused mainly by the high peak capacity of the second dimension. The price that had to be paid was the very long analysis time.

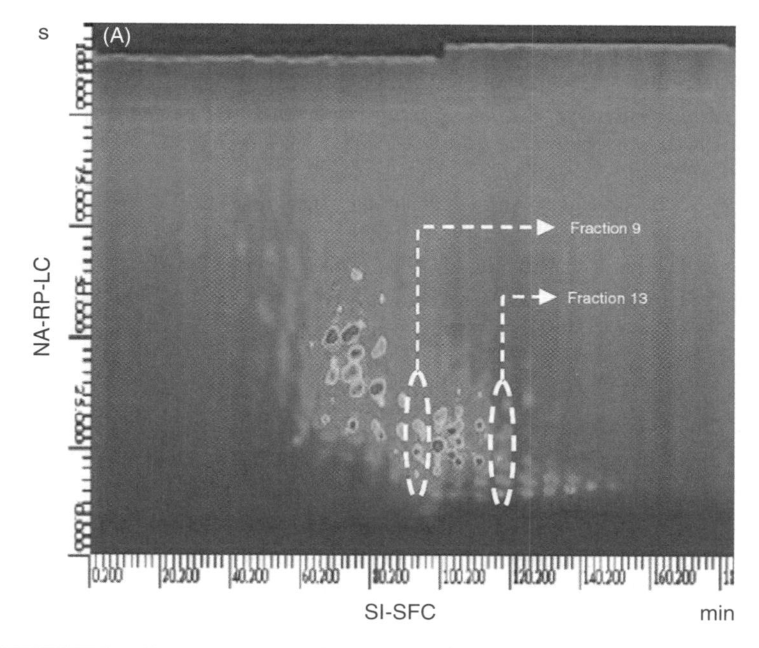

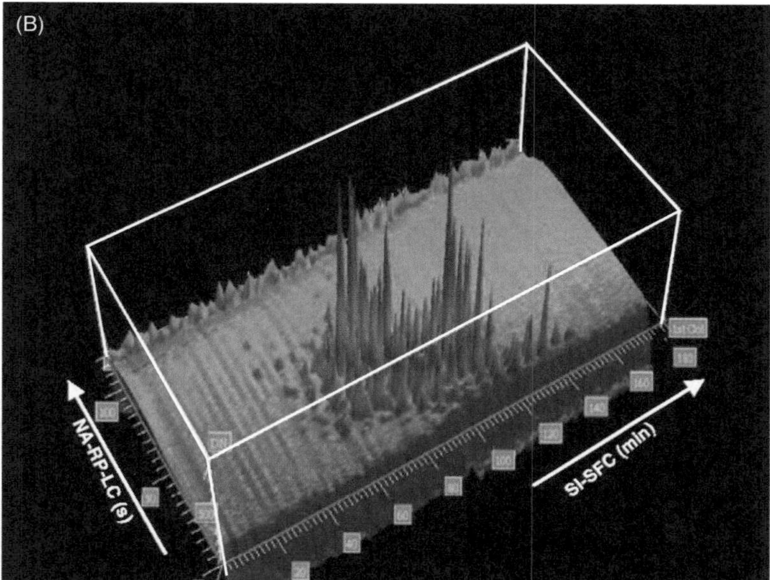

Figure 11.16 SI-SFC × non-aqueous RPLC separation of the TAGs in fish oil: (A) contour plot; (B) 3D representation. [From François et al. (2010), with permission. Copyright © 2010 by Wiley-VCH Verlag GmbH & Co. KGaA.]

REFERENCES

Adam F, Bertoncini F, Thiébaut D, Esnault S, Espinat D, Hennion MC. *J. Chromatogr. Sci*. 2007; 45:643–649.

Cortes HJ. *Multidimensional Chromatography Using On-Line Coupled High-Performance Liquid Chromatography and Capillary Gas Chromatography*. New York: Marcel Dekker, 1990, pp. 251–299.

de Koning S, Janssen H-J, van Deursen M, Brinkman UAT. *J. Sep. Sci*. 2004a; 27:397–409.

de Koning S, Janssen H-J, Brinkman UAT. *J. Chromatogr. A* 2004b; 1058:217–221.

de Koning S, Janssen H-J, Brinkman UAT. *LC–GC Eur*. 2006; 19:590–603.

Edam R, Blomberg J, Janssen H-G, Schoenmakers PJ. *J. Chromatogr. A* 2005; 1086:12–20.

François I, Sandra P. *J. Chromatogr. A* 2009; 1216:4005–4012.

François I, dos Santos-Pereira A, Lynen F, Sandra P. *J. Sep. Sci*. 2008; 31:3473–3478.

François I, dos Santos-Pereira A, Sandra P. *J. Sep. Sci*. 2010; 33:1504–1512.

Grob K. *On-Line Coupled LC–GC*. Heidelberg, Germany, Hüthig, 1991.

Hirata Y, Ozaki F. *Anal. Bioanal. Chem*. 2006; 384:1479–1484.

Hirata Y, Sogabe I. *Anal. Bioanal. Chem*. 2004; 378:1999–2003.

Hirata Y, Hashigushi T, Kawata E. *J. Sep. Sci*. 2003; 26:531–535.

Hyötyläinen T, Riekkola M-L. *J. Chromatogr. A* 2003; 1000:357–384.

Janssen H-J, Boers W, Steenbergen H, Horsten R, Flöter E. *J. Chromatogr. A* 2003; 1000:385–400.

Levy JM, Cavalier RA, Bosch TN, Rynaski AF, Huhak WE. *J. Chromatogr. Sci*. 1989; 27:341–346.

Liu Z, Ostrovsky I, Farnsworth PB, Lee ML. Chromatographia 1993; 35:567–574.

Lynch TP, Heyward, MP. *J. Chromatogr. Sci*. 1994; 32:534–540.

Mao D, Van De Weghe H, Diels L, De Brucker N, Lookman R, Vanermen G. *J. Chromatogr. A* 2008; 1179:33–40.

Mondello L, Dugo P, Dugo G, Lewis AC, Bartle KD. *J. Chromatogr. A* 1999; 842:373–390.

Nam K-S, King JW. *J. High Resolut. Chromatogr*. 1994; 17:577–582.

Okamoto D, Hirata Y. *Anal. Sci*. 2006; 22:1437–1440.

Quigley WWC, Fraga CG, Synovec RE. *J. Microcol. Sep*. 2000; 12:160–166.

Sciarrone D, Tranchida PQ, Costa R, Donato P, Ragonese C, Dugo P, Dugo G, Mondello L. *J. Sep. Sci*. 2008; 31:3329–3336.

Tranchida PQ, Donato P, Dugo P, Dugo G, Mondello L. *Trends Anal. Chem*. 2007; 26:191–205.

Venter A, Rohwer ER. *Anal. Chem*. 2004; 76:3699–3706.

Venter A, Makgwane PR, Rohwer ER. *Anal. Chem*. 2006; 78:2051–2054.

12

COMPREHENSIVE CHROMATOGRAPHY DATA INTERPRETATION TECHNOLOGIES

ELIZABETH M. HUMSTON AND ROBERT E. SYNOVEC
University of Washington, Seattle, Washington

The introduction of comprehensive multidimensional chromatography, in particular comprehensive two-dimensional (2D) gas chromatography (GC × GC) (Liu and Phillips, 1991), 2D liquid chromatography (LC × LC) (Bushey and Jorgenson, 1990), LC × GC, and so on, has paved the way for the analysis of complex and interesting sample types in a more complete way. Prior to multidimensional chromatography, these complex sample types were generally insufficiently separated, with only a single separation dimension. Without complete resolution, it is still possible to infer pattern-type information about a complex sample with one-dimensional (1D) separations. However, the ability to better isolate and identify individual analytes within a complex sample matrix with 2D separations (and mass spectral detection) and to further quantify peak areas or volumes makes it possible to address much more challenging questions about these samples. The powerful instrumentation provides a means for the acquisition of raw data for complex samples, but deciphering useful information from the complex data itself remains a challenge. Thus, sophisticated data interpretation technologies and subsequent development of user-friendly software are essential to take full advantage of relatively recent instrumentation developments in 2D comprehensive chromatography.

The scope of data interpretation technologies for comprehensive chromatography can be fairly broad, and data interpretation tasks are specific to each

Comprehensive Chromatography in Combination with Mass Spectrometry, First Edition.
Edited by Luigi Mondello.

study. Regardless, tasks such as data matrix visualization, peak identification, and quantification are generally considered some of the most basic tasks (Amador-Muñoz and Marriott, 2008). Frequently, data interpretation beyond these basic tasks is also desired. Because a complex sample type can be so well separated, instrumentation providing two dimensions of separation (e.g., GC × GC or LC × LC) is often applied to hypothesis-driven studies in which various classes of complex sample types are compared. For these instances, data interpretation technologies can allow for much more than simply finding, identifying, and quantifying peaks within the sample. Interpretation tools can also aid in classifying samples with pattern recognition, determining specific analytes that differ between sample classes with feature selection, and mathematically resolving, identifying, and quantifying overlapping analytes through deconvolution software. It is by applying these advanced data interpretation technologies that we can more fully take advantage of the benefits of multidimensional chromatography instrumentation.

In this chapter we discuss the fundamental differences in the data structure acquired with GC × GC, LC × LC, LC × GC, and so on, as compared to data from 1D instrumentation (i.e., GC or LC). In this chapter, the requirements and opportunities for data analysis associated with multidimensional data, in particular in the context of GC × GC, are explained and demonstrated. The shortcomings of extending 1D data analysis approaches to 2D (or 3D) data are described, and the benefits of fully utilizing the added information in the data structure with chemometrics are illuminated. Although the focus and examples presented herein are for GC × GC, the basic principles of the data analysis are readily applicable to other forms of comprehensive 2D separations.

12.1 HIGHER-ORDER DATA STRUCTURE

It may seem that the additional dimension (or dimensions, depending on detection type) of data could be dealt with simply through the extension of 1D data analysis techniques, but this is not the only or necessarily the best approach. It is true that data interpretation can be done by extension of 1D approaches, but this type of analysis neglects to fully utilize the additional benefits associated with the data structure from multidimensional chromatography. The fundamental differences in data structure lead to different approaches in the way that data interpretation should be addressed for 1D and 2D separations. The most basic structure difference is that instead of having a vector of first-order data (as with a 1D GC or LC separation), a 2D separation produces a matrix [e.g., GC × GC with univariate detection, such as flame ionization detection (FID), or LC × LC with single-wavelength absorbance detection] or cube (e.g., GC × GC with a multivariate detector such as mass spectrometry, or LC × LC with a full absorbance spectrum detection) of data with each sample that is analyzed. A graphical representation of a cube of data is provided in Figure 12.1 for GC × GC with mass spectral detection. Another specific aspect of this difference in data structure is that a 2D separation does not provide a single

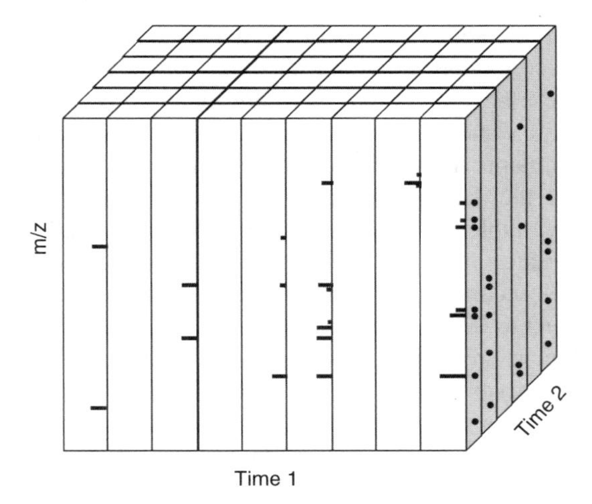

Figure 12.1 Representative data cube. A 1D separation produces the vector, time 1. A second dimension of separation results in a matrix with time 1 and time 2. The addition of a multivariate detector (i.e., MS) results in a cube with two dimensions of chromatographic separation and a third dimension of mass spectral information.

chromatographic peak per analyte. Instead, each analyte is distributed over a series of modulated peaks, all of which contain part of the peak signal "volume."

These data structure differences simultaneously complicate extensions of traditional 1D data analysis approaches and create opportunities for more sophisticated data interpretation technologies. The benefit of increased peak capacity for complex sample types with multidimensional chromatography is well known, but the increase in data interpretation options that are possible because of this specific data structure is also an important benefit that is often underemphasized and underutilized (Kowalski, 1975; Pierce et al., 2008). When, for example, 2D chromatography (i.e., GC $\times$ GC) is coupled to a flame ionization detector or any univariate detector, second-order 2D data matrices are generated. The addition of a multivariate detector, such as a time-of-flight mass spectrometer (TOF MS), further generates an additional dimension of data that results in third-order 3D data cubes. Because each dimension of data has, in principle, the opportunity to be complementary (as long as the separations are sufficiently orthogonal), these data structures allow for various mathematical approaches to data analysis and interpretation that cannot be accomplished with the first-order 1D data of one-dimensional chromatography (Kowalski, 1975; Pierce et al., 2008). The most useful information can be attained from these complex raw data by utilizing data interpretation technologies that take advantage of this higher-order data structure.

12.1.1 Benefits of Higher-Order Data

The higher-order data structure associated with multidimensional chromatography offers various benefits and advantages over first-order data. The primary advantages relate to the ability to generate better mathematical models to describe the data and to deconvolute individual analyte signals for better quantification. One reason for these benefits is that there are many correlated measurements in 2D chromatographic data from each series of modulated peaks. Because these correlated measurements are essentially redundancies, it is possible to generate models to describe the data that have better precision, and it is also possible to monitor multiple variables simultaneously (Synovec et al., 2003).

When 2D chromatography is coupled to a univariate detector, this 2D instrumentation produces second-order data that are, generally speaking, referred to as *bilinear* with respect to linear algebra constraints. Hence, the aspect of bilinearity (or trilinearity) that is frequently achieved with multidimensional chromatography also allows for advantages. Bilinearity (or trilinearity) is not always required to perform data analysis that is based on linear algebra approaches (Skov et al., 2009), but its presence does engender the straightforward chemometric analysis of multidimensional data. It should be noted that bilinear data are more than just data with two dimensions; the condition of bilinearity implies that each analyte has a well-behaved response in both separation dimensions (i.e., if data are bilinear, each analyte has a constant retention time in each dimension and a constant peak shape in that dimension). Thus, the data can be decomposed into peak profiles in each dimension for each analyte. Additionally, the sum of all of the components (analytes and noise) that are described by the product of each separation vector must represent the total response signal detected. These requirements are frequently met with GC × GC, LC × LC, and related 2D methods; thus, sufficient bilinearity is routinely achieved. A visual representation of chromatographic data with this bilinear data structure is shown in Figure 12.2. This second-order data structure, generated from 2D instrumentation, offers an advantage in terms of calibration (quantification) over first-order data that is referred to as the *second-order advantage*. This advantage allows for the quantification of an analyte in the presence of an interfering species with only a single calibration standard (Booksh and Kowalski, 1994; Fraga et al., 2000a).

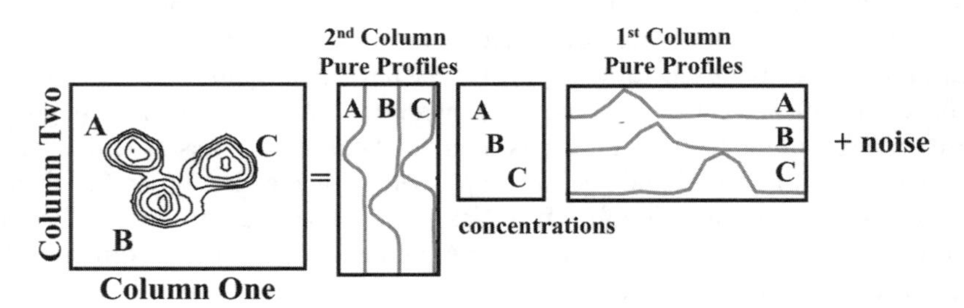

Figure 12.2 Bilinear data. The bilinear data structure and how it is decomposed into component vectors is demonstrated with chromatographic data. (From Synovee et al., 2003.)

A further advantage can be gained if a third unique dimension of data is added. This third-order data, termed *trilinear*, comes from either the combination of multiple bilinear separation runs acquired with 2D instrumentation or from 3D instrumentation, such as GC × GC coupled to a multivariate detector (e.g., MS). The requirements for a trilinear data structure are such as those for bilinear, but extended to a third dimension. Data with this structure possess the aptly named *third-order advantage*. This advantage allows for the detection and relative quantification of an analyte of interest in the presence of an unknown interfering analyte (or analytes) without the requirement of a standard (Booksh and Kowalski, 1994; Sinha et al., 2004a). Absolute quantification is still dependent on a calibration standard to relate the instrument response to a chemical concentration; however, the instrument response can readily be determined with only a single injection. These are very powerful advantages, as it means that neither chromatographic resolution nor numerous injections are required for accurate analyte quantification when GC × GC or LC × LC is employed with a multivariate detector.

Data analysis for multidimensional chromatography can be approached in a number of ways. Initially, it was primarily modifications of first-order data-handling approaches that were employed. As the field has developed, more analysis has been done with tools that utilize the higher-order data structure and its advantages. For the purpose of this chapter, it is instructive to start the discussion of data analysis with modifications of first-order data-handling approaches.

12.2 MODIFICATIONS OF FIRST-ORDER DATA-HANDLING APPROACHES

In the early days of GC × GC and LC × LC, commercially available instrumentation did not exist, let alone commercially available software for handling second (or third)-order data such as that generated by these instruments. For this reason, researchers were challenged to develop in-house software for data processing. Naturally, many of the early data interpretation technologies were modifications and extensions of those used with first-order data-from a 1D separation. This approach to data interpretation is sufficient for basic data-handling goals and is often able to provide reliable quantification information. However, these early approaches do not utilize the added benefits of the multidimensional data structure, but instead must find ways to correct for the issues related to the structure differences: specifically, the fact that each analyte is distributed over a series of modulated peaks in a 2D separation. Extensions of first-order approaches used with 1D data generally relate to data analysis tasks of *peak defining*, which includes peak finding, determination of retention times, identification, and peak quantification by integration.

12.2.1 Peak Finding

The task of locating peaks within a data matrix is accomplished by distinguishing between analyte signal and noise signal. This is the case with either 1D or 2D separation data. However, with 2D separations, peak finding must also address the fact that signal for each analyte is distributed over multiple modulations and one analyte will be found multiple times within the second-order data structure if a first-order peak-finding approach is applied. The two main approaches to peak finding are to use derivatives or to search for local maxima. Commonly, with a 1D separation, the first or second derivative is calculated across the data vector. This provides the location of the front, apex, and back of a peak while also assisting in identifying the presence of any shoulder peaks. These traditional 1D methods can be applied in the standard way to 2D data if the 2D data are first unfolded into a vector (i.e., treated as first-order data). As stated previously, this results in several entries (one for each second-dimension modulation) for each analyte. For ease of data interpretation, this is dealt with by combining the multiple hits from the 1D approach so that each peak has just one corresponding hit (Peters et al., 2007). The combination of multiple entries is not a trivial task and relies on the accurate grouping of each peak in each second-dimension modulation. This can be done through mass spectral matching (discussed herein) or through consistency of retention-time coordinates.

Alternatively, instead of using the derivative approach to peak finding, the local maxima that correspond to the peak apex can also be found. This can be done on data that have been unfolded into a 1D vector which will provide local maxima for each second-dimension modulation, as with the derivative approach, that need to be combined. Or, this can be accomplished while keeping the data in their 2D matrix form. By not unfolding the 2D data into a vector, the problem of combining multiple hits can be avoided. Algorithms exist to search for the local maxima (peaks) in both separation dimensions within the structured 2D data matrix (Mohler et al., 2008). This is an attractive option for peak finding, as in addition to eliminating the need for combining multiple hits, it better utilizes the added information in the second-order data structure. A watershed algorithm can also be used to find local maxima in the structured 2D form (Reichenbach et al., 2005) but may be plagued by retention-time shifting in the second dimension (Vivó-Truyols and Janssen, 2010).

The chromatographic location of a peak is typically recorded as a pair of retention time coordinates. The assignment of the retention time on the first column can be extracted from the modulated peaks. A good approximation can be made from the apex of the second-dimension chromatogram that contains the highest peak maximum for a specific analyte (Johnson et al., 2002), a normal distribution can be fit to the modulated peak profile from which the apex can be calculated (Marriott et al., 2000; Shellie et al., 2002), or the retention time can be determined by modeling the first-dimension peak using chromatographic parameters (Adcock et al., 2009). The retention time of the analytes on the second column is simply based on the data acquisition rate and the modulation period.

The combination of the two values provides the location in 2D space at which a peak was found.

12.2.2 Peak Quantification

Peak quantification for analytes within 1D separations is usually reported as peak height or peak area, as determined through integration, although chemometrics approaches are emerging (that are not based on height or area), which we discuss later. Peak quantification for 2D separations is generally not reported as peak height but is more commonly determined as a peak area or volume [e.g., for GC methods of the FID intensity or total ion current (TIC) for MS, and for LC methods of the ultraviolet–visible UV-Vis absorbance signal]. The peak height is not used because the intensity at the peak maximum varies with both the modulation period and the modulation phase with, for example, thermally modulated GC × GC. The more modulations across a first-dimension peak, the less peak intensity that each individual second-dimension modulation contains, and a peak that is modulated out of phase will also show a lower maximum intensity (Amador-Muñoz and Marriott, 2008; Ong and Marriott, 2002). However, the total area within all of the modulations to the second column is consistent. Because both peak signal area and peak signal volume are based on the area, variations in peak height are accounted for. The area can be determined by integration as an extension of the 1D approach, but again, instead of having a single value to represent each analyte as in a 1D separation, the peak area is distributed between several modulated peaks in a 2D separation for each analyte. The distribution of the peak area between the multiple modulation periods can also vary depending on a number of factors, such as peak symmetry, modulation phase, and signal/noise ratio (Amador-Muñoz and Marriott, 2008; Ong and Marriott, 2002). For these reasons, quantitative results tend to be inconsistent if not all of the peaks are used for quantification. Thus, if quantification is performed with integration approaches, it is necessary to determine the total modulated peak area or volume for the most accurate information.

By extending the first-order data analysis approach, typically all of the second column peaks are integrated, and then the individual second column peaks associated with the same analyte are grouped together and summed (Beens et al., 1998; Dallüge et al., 2002c; Korytár et al., 2002). As in peak finding, this step relies on the identification of each second-dimension peak so as to accurately group the modulated peak profiles that belong to the same analyte. This can be accomplished either by adding a separate summation program to traditional 1D integration software, or by developing software programs that combine the integration, identification, and summation. Several software programs developed in-house emerged in the literature shortly after the introduction of 2D chromatography, which acquired quantitative information by combining the two steps (Beens et al., 1998; Frysinger et al., 1999; Hyötyläinen et al., 2002). As an example, Tweedee, one of these programs for GC × GC, was effective at accomplishing the basic peak-defining tasks; peaks were found based on the first derivative calculation, each row of the data matrix (separation on the second column) was integrated separately, then the results were combined based on visual grouping through the interpretation of a contour plot into a peak table, providing peak locations and quantities (Beens et al., 1998).

In addition to the software developed in-house, several instrument manufacturers now provide commercial software for these basic data interpretations (Pierce et al., 2008). Chemstation (by Agilent), GC Solution (by Shimadzu), and XCalibur (by Thermo Fisher Scientific) are software packages for 1D instrumentation that have been used to collect and analyze 2D GC × GC data. When 1D software packages are used for data interpretation, the second step, combining the second-dimension modulations, is often quite tedious, which is a shortcoming of utilizing first-order data analysis approaches or extensions thereof for second-order data. For example, in 2000, after using GC–MS software for data processing, Dallüge et al. (2002b) exported the peak table results (containing thousands of peaks) to a spreadsheet program in which further screening and combining of second-dimension modulations was done manually. Commercial software modification has alleviated some of these tedious steps of data analysis, but the underlying math is generally still based on first-order approaches. Thermo Fisher Scientific has modified Hyperchrom software for use with 2D data that automatically identifies analytes based on retention time, integrates peak area, and generates tables of information. LECO has also modified its ChromaTOF software for use with GC × GC, which can be used to locate peaks, identify analytes based on mass spectral similarity, deconvolute, integrate, and create peak tables. Furthermore, most of these programs can also be used for visualization, which can be an important part of data interpretation as well.

12.3 VISUALIZATION

Visualization of raw chromatographic data is an important aspect of 2D chromatography and is often one of the first steps in data interpretation. Visualization is primarily a qualitative method for interpretation, but the ability to view the data as an image does allow for some preliminary comparisons of data that cannot be done easily with a matrix or cube of numbers. Furthermore, although there are more advanced data comparison tools (which we discuss further), a great deal of research is performed in which the primary comparison between samples is accomplished through visual inspection.

12.3.1 Qualitative Data

In most cases, detectors collect 2D data continuously as a 1D vector with each second-dimension separation stacked in series end to end, as defined by the modulation period selected, commonly referred to as *unfolded*. In this form it is possible to plot the data directly as a vector. Although these are 2D data (i.e., intrinsically second order), when they are stored in this structure as a 1D vector, they are treated as if they were first order. There are shortcomings to treating second-order data with first-order data analysis approaches, as described previously, and these shortcomings are also apparent in terms of visualization. Representing the data in this way, it is fairly easy to identify the peaks that comprise the separation on the first dimension, but it can often be quite difficult to interpret the second-dimension separation with this type of plot. For example, a subset of data (TIC is shown) generated with GC × GC-TOF MS from a matrix sample is provided in Figure 12.3A. This artificial matrix contains 12 analytes within the window that is provided. It is quite difficult to locate all 12 analytes with this visual representation, let alone to observe any relative separation information in the second dimension.

Although it is possible to zoom in on sections of the chromatogram to locate all 12 analytes more readily, as shown in Figure 12.3B–E with the subsections containing three, three, two, and four analytes, respectively, this is a tedious way to visualize and interpret the data set. It is better and more intuitive to rearrange the data in order to view them in a 2D representation, thus better demonstrating the second-order nature of the data structure. The data vector can be reshaped into a matrix based on the known modulation period and data acquisition rate of the detector (Kallio et al., 2009; Pierce et al., 2008). Each second-dimension chromatogram is moved from being initially stacked in series (i.e., unfolded raw data format) to being stacked in parallel, which allows for viewing the second-dimension separation much more clearly in the context of the first-dimension separation. There are many visualization options for viewing a 2D matrix. One of the most basic approaches to plotting 2D data is as an apex plot in which a dot or circle indicates each peak maximum. This is useful in that it shows all the data in 2D space and removes much of the complexity from the chromatogram, but in most cases all intensity information relating to individual analytes is lost (Dallüge et al., 2002a,b). For that reason, a contour plot is generally more useful because the separation in each dimension can still be seen clearly, and intensity information can be maintained in the color scale. The same subset of chromatographic data in the raw data format shown in Figure 12.3 is shown as a 2D contour plot in Figure 12.4. The 12 peaks that were difficult to view in the 1D plot (unless zoomed in upon) are all clearly visible in the 2D plot, and the relative second dimension separation can also be interpreted more easily. Three-dimensional (3D) views are also a useful option in which the intensity information is expanded out into the third dimension. In these surface plots, the peaks appear as "mountains" on an elevation map, with the mountain height corresponding to peak intensity. The same subset of chromatographic data provided in Figures 12.3 and 12.4 with a 1D and 2D representation, respectively, is shown as a 3D surface plot in Figure 12.5. Although this is a 3D representation of the data, the data are still second order, as the third dimension is not providing an additional dimension of information, just a visual representation of the signal intensity.

Several platforms are available for accomplishing these visualization goals, and most instrumentation software packages have built-in visualization tools. Although many of the instrumentation software packages offer the option to view the second-order data structure with 2D or 3D visualization tools, the underlying math for the data analysis does not fully utilize the additional data orders as part of analysis and thus these packages are still faced with the shortcomings of applying first-order data interpretation to second-order data. ChromaTOF (by LECO Corporation), which was used to generate Figures 12.3

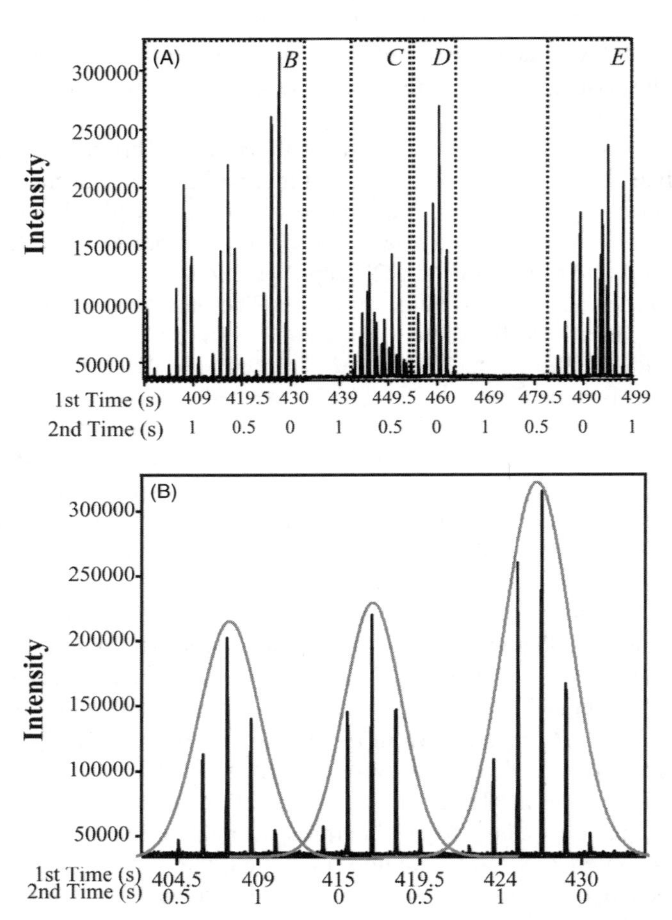

Figure 12.3 One-dimensional visualization. The TIC of a matrix sample, analyzed by GC × GC–TOF MS, containing 12 analytes is shown with a 1D visual representation. Visualization generated with LECO ChromaTOF software. Subsections of the 1D separation are shown in (B), (C), (D), and (E), containing three, three, two, and four analytes, respectively.

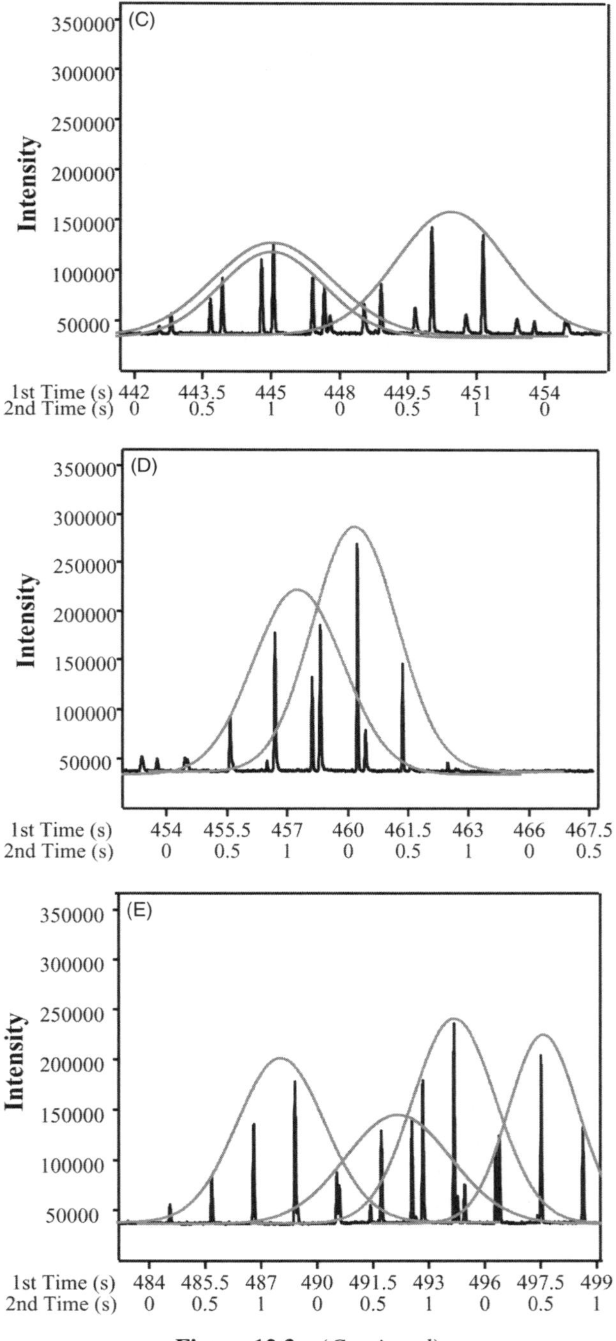

Figure 12.3 (*Continued*)

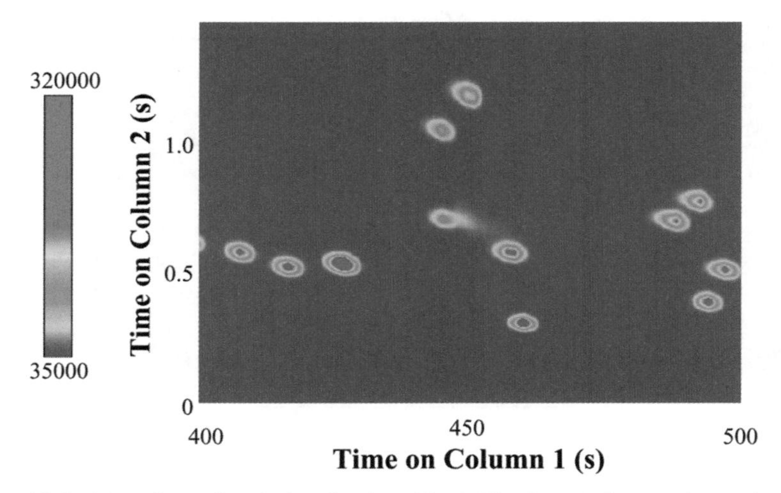

Figure 12.4 Two-dimensional visualization. The TIC of a matrix sample, analyzed by GC × GC–TOF MS, containing 12 analytes is shown as a 2D contour plot. Visualization generated with LECO ChromaTOF software.

to 12.5, has numerous data analysis capabilities and visualization options, including 2D or 3D plots, subtraction plots, and the ability to overlay multiple 1D unfolded chromatograms or to view individual mass channels. Transform (Noesys from Research Systems) provides a platform for viewing data in either 2D or 3D. HyperChrom Software (by Thermo Fisher Scientific) can be used for data analysis tasks and can also plot 1D, 2D, or 3D plots. Additionally, if raw data can be exported to Matlab, the numerous plotting options incorporated into the software can be utilized for visualization. GC Image (Reichenbach GC Image, LLC) offers 1D, 2D, and 3D plotting options and an opportunity for the acquisition of quantitative information (Reichenbach et al., 2003–2005).

12.3.2 Quantitative Data

Although visualization is primarily a qualitative technique, some quantitative data interpretation can be performed for GC × GC data by treating the chromatograms as images. The commercially available software program GC Image (Reichenbach GC Image, LLC) sums pixel intensity corresponding to a single analyte, thus providing quantitative information while also utilizing the second-order data structure (Reichenbach et al., 2003–2005). By representing the data matrix as an image, 3D imaging techniques can be applied to the data as they would be to a photograph. In practice, the first step is to subtract the background signal (i.e., base plane) from the 2D chromatographic data much as a baseline is subtracted from 1D data. The GC Image software contains an algorithm that estimates the

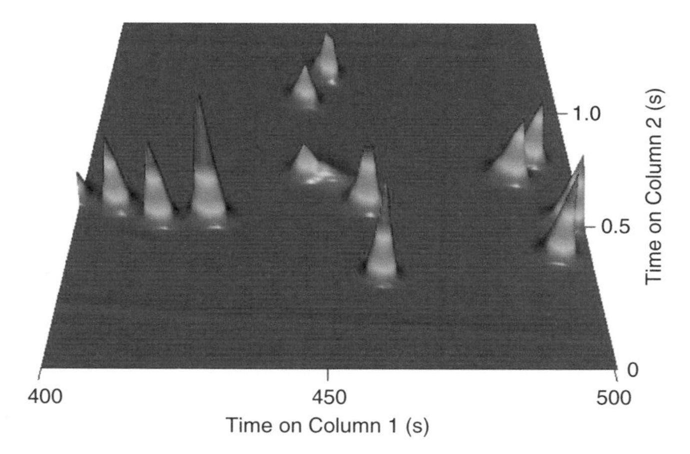

Figure 12.5 Three-dimensional visualization. The TIC of a matrix sample, analyzed by GC × GC–TOF MS, containing 12 analytes is shown with a 3D visual representation. Visualization generated with LECO ChromaTOF software.

background across the 2D space and then subtracts the local background from the entire chromatogram. This accomplishes a readjustment, after which the base plane is approximately zero (Reichenbach et al., 2003). Analytes can then be identified by classifying regions of the chromatogram as either part of the background or as part of a peak, referred to as "blobs" in this software package. The blobs are found by a "drain algorithm," which is the inverse of the more common watershed algorithm. By analogy, the chromatogram is first "filled with water" and is then slowly "drained." As the water drains, "islands" emerge out of the water, serving to locate peaks in order of intensity. The boundaries of each peak are also identified through this approach, and quantitative information can be acquired by combining the total pixel value within the boundary of each blob. This software is typically geared toward data matrices (acquired with univariate detectors) rather than the cubes that are produced with MS detection, but it does take advantage of the second-order data structure.

12.4 MASS SPECTRAL DETECTION

Up to this point, the difference between MS detection (in particular, TOF MS) and univariate detectors (such as FID for GC or single-wavelength absorbance detection for LC) has not been overly emphasized in terms of data interpretation (beyond the cube versus matrix aspect) because most of the data analysis approaches described previously could be implemented with either intensity information from FID or the TIC (or a single mass channel) from MS. However, in terms of identifying unknown analytes within a chromatogram, MS is generally required (Mondello et al., 2005).

Without MS detection, some analyte characteristics can be inferred from the structure of the chromatogram. In many 2D separations that utilize a large fraction of the 2D peak capacity available, achieved by pairing complementary separations, compounds of the same group type form peak structures within the chromatogram. These peak structures can be identified as continuous bands or clusters that stretch through the 2D separation space (Blomberg et al., 1997; de Geus et al., 2001; Giddings, 1995; Korytár et al., 2002). These structures occur in the separation of a sample that contains a large number of isomers, homologs, or congeners when the second-dimension stationary phase has been selected appropriately (Giddings, 1995). For example, analytes within the same group type tend to elute along a line, often with increased retention times on both columns through the 2D chromatographic space as the molecular mass increases. The actual angle of the line depends on the flow rate, temperature program, and various other instrumentation parameters, but is reproducible under a given set of conditions. If an unknown analyte falls within a structured chromatogram of which many of the other compounds are known, its identity could be determined more readily by which functional group peak structure line it falls within. This has been demonstrated successfully with vegetable oil samples in order to determine unknown fatty acid identities when retention coordinates of other fatty acids within the oil were known (Tranchida et al., 2008). Beyond identification such as this, MS is generally necessary.

MS detection provides a significant amount of additional information in the third dimension of 2D separation data that offers even more options in terms of data analysis and interpretation, as it results in third-order data. Preliminary identification can be achieved by matching the detector response (mass spectrum) associated with an eluting analyte to the mass spectra of standard compounds through mass spectral matching algorithms. The National Institute of Standards and Technology (NIST) has publicly available software, MS Search, that is capable of searching the collection of library spectra rapidly to identify the analyte with the highest match value (U.S. National Institute of Standards and Technology, 2002). There are numerous well-established libraries containing spectra for a wide range of compounds acquired with a range of MS detectors that can be scanned rapidly with the MS Search algorithm. For the most confident identification, it is generally well advised to generate in-house libraries whenever possible. This provides a mass spectrum acquired directly from the instrument used to analyze the unknown and also provides the researcher with retention-time information for analytes of interest. The most optimal analyte identification should be achieved with a combination of mass spectral matching and retention-time verification via the analysis of standards. One of the limitations of mass spectral matching alone is the occurrence of largely similar mass spectral patterns for multiple analytes. Even still, as these data are third order, the use of MS detection with 2D separations essentially provides an additional separation in the detector dimension (Mondello et al., 2005). At locations with overlapping compounds, the mass spectral signal is the combination of the mass spectra of each coeluting compound. If the analytes are known, unique mass channels can be used to distinguish between overlapping analytes. MS deconvolution algorithms can also separate the mass spectra. The ions associated with each of the overlapping analytes can be determined based on the constancy of ion ratios across the width of the peak, providing mass spectra for each analyte (Song et al., 2004).

Along these same lines, it is also possible to use MS matching algorithms to work in the other direction. Sometimes, instead of trying to determine the identity of an unknown analyte at a specific chromatographic location, the chromatographic location of a target analyte with a known mass spectrum is sought. This is possible through the use of data interpretation technologies with an algorithm termed DotMap (Sinha et al., 2004b). This algorithm uses the mass spectrum of a target analyte and calculates a match value rapidly at each point within the entire 2D separation space for GC × GC. The algorithm generates a contour plot, providing the chromatographic location with the highest match value to the target, thus the retention coordinates of the target analyte, and/or potential target analytes, to further explore via the injection of standards to acquire retention times.

12.5 CHEMOMETRICS

As described, the higher-order data structure possibilities (i.e., second or third order) associated with multidimensional data can create some challenges when trying to apply or modify 1D data-handling technologies (designed fundamentally for first-order data.) However, the higher-order data structure also provides opportunities for the use of data-handling approaches that are based on linear algebra and multivariate statistics. This area of data analysis, termed *chemometrics*, originated in the 1970s (Kowalski, 1975). The fundamental theme of chemometrics is that mathematics can be used to extract useful information from the data obtained from chemical analyses (Beebe et al., 1998; Kramer, 1998; Pierce et al., 2008; Sharaf et al., 1986). Multidimensional data can be quite complex, so it is often difficult (or impossible) to optimally extract the salient information from the data provided by 2D separations without chemometric techniques. It is critical that analysts meet the challenges to utilize more fully the rich data using chemometrics. Essentially, chemometrics is intrinsically tailored to more fully span the gap between raw complex data and useful information, much more so than using 1D data analysis approaches that have been adapted for multidimensional data.

The use of chemometrics is certainly not exclusive to multidimensional chromatography, as there are many examples of its application in other hyphenated techniques (e.g., GC–MS, LC–MS) and in 1D separations (e.g., stacked runs of GC–FID or LC–single wavelength absorbance detection). Chemometrics can readily be applied to many data analysis tasks and additionally, for instrument and experimental design optimization. Technically speaking, extracting useful information mathematically spans topics as diverse as pattern recognition, peak classification, target analysis, signal preprocessing, peak deconvolution, multivariate calibration, visualization, method optimization, retention calculation, prediction, and testing of separation orthogonality to highlight a few (Pierce et al., 2008; Sinha et al., 2004c; Synovec et al., 2003; van Mispelaar et al., 2003). Although all of these provide useful information, for the purposes of this chapter the topics relating to data interpretation (primarily pattern recognition, feature selection, deconvolution, and quantification) are covered in detail.

12.5.1 Pattern Recognition

One goal of data interpretation that chemometrics is well suited for is finding chromatographic patterns within a data set. In this sense, patterns are in reference to overall similarities and/or differences between samples based on the entire chromatographic data set rather than the peak structure patterns within a single chromatogram. However, initial pattern recognition between samples could eventually lead to the determination of peak structure information for a particular analyte within a chromatogram (i.e., functional group type), as the analytes contributing to pattern classification could be further processed and quantified, as discussed in detail herein. A key aspect of pattern recognition is to distinguish which samples are similar to one another and which are different from one another based on the entire chromatographic data set when class membership is not known, which allows for the elucidation of underlying sample-to-sample relationships in an unsupervised mode. The 2D chromatogram is considered a "fingerprint" for the samples in order to classify the samples qualitatively. Principal components analysis (PCA) is a pattern recognition tool that models the inherent variations within a data set and classifies samples into groups (or clusters) that are similar to each other based on their chemical fingerprints. For example, for many industrial and food samples, changes in production methodology produce subtle changes in the chemical fingerprint of the resulting product. PCA can be used to classify the samples based on their production method without manually locating the chromatographic differences, as products of the same method will generally group together based on their chemical similarities. PCA is frequently used with data from 1D chromatography and can also be employed successfully to data from 2D chromatography. PCA is applied to a 2D matrix of data, R. For 2D chromatography, R consists of the unfolded chromatograms, with the various samples compiled together as the second dimension. With MS detection, either a single mass channel or the TIC can be used. Briefly, the data matrix (R) is decomposed into two new matrices, scores (S) and loadings (L), plus an error matrix (E):

$$R = SL + E = \sum_{p=1}^{c} s_p \otimes l_p + E \qquad (12.1)$$

The scores matrix represents samples, and the loadings matrix represents variables, which are retention times in the chromatographic data. PCA breaks the data matrix down into several pairs of scores and loadings, termed *principal components* (PCs), to describe the variation between samples. The cross product of each score and loading pair describes a piece of the original data matrix, and combining all of these cross products (with the error matrix) would result in the original raw data matrix. The variable c is the number of PCs used in the decomposition of R, and the benefit of breaking the data into PCs is that the first PC represents and models the largest variations within R. Each additional PC then describes the next-largest variation within the data. Hence, the majority of the variance between the samples is modeled in the first few PCs (Beebe et al., 1998; Brereton, 2003).

The use of PCA facilitates pattern recognition because the variation between the samples is now described within the first few PCs rather than being hidden within the complex 2D data (i.e., the chemical fingerprint). The relative proximity of the scores on these PCs relates to the chemical similarities in the samples (while the loadings identify the variables that cause those similarities or differences). PCA has been applied effectively to (unfolded) 2D data to classify samples (and locate chemical differences) in a number of complex sample types, including oil (van Mispelaar et al., 2005), plants (Pierce et al., 2006a), and yeast (Mohler et al., 2006). Because no sample information need be known prior to analysis, PCA is an unsupervised method. This can be a benefit if the samples being compared are not part of a hypothesis-driven study and/or if the sample class assignment is not known. If additional information is available with regard to sample classification from the experimental design (e.g., if the researcher knew which samples were prepared via each production method), a supervised analysis can also be performed. In these cases the user inputs the additional sample or class information along with the data to determine differences in the samples.

12.5.2 Feature Selection

Another aspect of pattern recognition can be to determine the specific analytes or features (chemical differences) that are responsible for the patterns and sample classifications observed, referred to as *feature selection*. Feature selection differs from peak finding (discussed previously), as the goal is to find features that help distinguish sample classes that are ultimately identified as key analytes rather than simply to find all of the analytes within the matrix. This is often an important aspect of hypothesis-driven studies, as determining the differences between various sample class types is frequently a main objective. For example, GC × GC (or LC × LC) could be used to collect data from a set of healthy subjects and from a set of subjects with an illness. Feature selection can then be used to determine specific analytes that consistently differ between the two groups (i.e., sample classes), which can provide further insight into the illness and potentially even to identify biomarkers that could be applied later for diagnosis and/or prediction. The use of PCA can provide feature selection information, as the relative proximity of the loadings identifies the variables (features) that cause the similarities or differences between samples. The loading variables are an abstract result of the PCA calculations, but readily provide the chromatographic location of the sources of variation between these samples. The analyte at that particular location can be identified further as discussed previously, and quantified, as described below. As stated, PCA can be applied as either a supervised or an unsupervised approach to feature selection.

Feature selection is often accomplished more effectively in a supervised fashion if information on class assignment is available. One approach is to use an analysis of variance (ANOVA)-based algorithm termed the *Fisher ratio* (F-ratio). The F-ratio is calculated as

$$\text{F-ratio} = \frac{\sigma^2_{\text{between class}}}{\sigma^2_{\text{within class}}} \tag{12.2}$$

for each mass channel at each point in chromatographic space. Here $\sigma^2_{\text{between class}}$ is the variance between the classes, which in the example described would be the variance of the average of the healthy samples and the average of the illness samples, and $\sigma^2_{\text{within class}}$ is the variance within a given class, which would be the average of the variance of the healthy samples and the variance of the illness samples. This calculation can identify locations in the 1D or 2D chromatographic space with large class-to-class variation, compared to within-class variation when class information is available (Johnson and Synovec, 2002; Pierce et al., 2006b). These locations are the chemical differences between sample classes and ultimately relate to specific analytes. This is an especially useful approach to feature selection if large within-class variations prevent accurate clustering through PCA. This can be common with very similar samples and also with certain sample types, especially those of biological origin (i.e., human test subjects), which frequently have large within-class variations. Locations with large F-ratio values indicate retention times of features that help distinguish sample classes. The analytes at these retention times can eventually be identified with mass spectral matching and retention time verification, as described previously, as key analytes that contribute to the differences between samples. Incorporating within-class variation to the calculation helps filter out analytes that differ between sample classes but also within the sample class, as the corresponding chromatographic locations will have smaller F-ratios: for example, analytes that differ related to diet rather than illness in the example above. For the use of the F-ratio algorithm the sample class information must be known, which limits its applicability primarily to samples that are part of a hypothesis driven study. Additionally, the F-ratio algorithm tends to be most successful when only a small number of sample classes (i.e., two or three) are compared. The between-sample variation can be diminished by multiple sample classes when some of the sample classes are similar to each other.

Feature selection via ANOVA with sample class membership known a priori can often capture chemical differences when unsupervised PCA is unable to do so, because either the samples are too similar or there are numerous variations within a given sample class. Indeed, selecting features with F-ratio analysis to submit to PCA can result in sample clusters in cases where PCA on the entire chromatographic data set does not result in clustering (Johnson and Synovec, 2002). This has been demonstrated by Johnson et al. with highly similar jet fuel samples analyzed with GC $\times$ GC. Complete chromatographic data from jet fuels that differed by only 1% were subjected to PCA and failed to produce accurate clusters. It was then demonstrated that selecting features with ANOVA to submit to PCA could result in clustering and allowed for classifying the samples accurately based on the small differences between fuels (Johnson and Synovec, 2002).

If a large number of sample classes need to be compared, or in situations when the sample class information is unavailable, approaches other than the F-ratio are required for feature selection. One approach that is useful is the *signal ratio* (S-ratio) *algorithm* (Mohler et al., 2008). This algorithm computes the ratio between the maximum signal and the minimum signal at each point in the 2D separation space, regardless of which sample the maximum and minimum are found in. Because the maximum and minimum can be found in any sample, this approach makes it possible to detect differences between any of a number of sample classes even when only one sample differs from the rest. For example, a time-course study can be monitored with this approach, as any analyte that changes during the time of study can be determined, regardless of when the change occurs. An additional benefit of this algorithm is that it can be run in an unsupervised manner because the maximum and minimum are searched from all samples and no sample class information need be specified. However, it is also possible to calculate an S-ratio between specific samples in a supervised mode if sample class information is known. Although this algorithm can be confounded by within-class variation, as there is no compensation for this, its versatility makes it a useful tool for feature selection in many applications, such as metabolomics (Mohler et al., 2008; Shellie et al., 2005).

There are also simpler approaches to feature selection for situations in which only two sample classes need to be compared. In these instances it may be possible to calculate a difference or ratio between the two samples for feature selection (Hollingsworth et al., 2006; Nelson et al., 2006; Shellie et al., 2005; Tran and Marriott, 2007). However, these approaches may also be plagued by within-class variation, require class information to be known a priori, and are limited to small numbers of sample classes.

12.5.3 Quantification

Pattern recognition is primarily a qualitative approach to data analysis, but chemometrics can also provide quantitative information with multivariate calibration approaches (which are generalized forms of PCA). This is often done in conjunction with pattern recognition and feature selection, in that analytes found to have differences between sample classes can be targeted for further quantitative analysis. Chemometrics can be used to deconvolute pure peaks of interest from interfering analytes and noise to provide more reliable quantitative information than simply integrating the area under a peak. These techniques are based on decomposing the third-order data structure, the 3D data cube (R), into a number of factors (n) and an error matrix (E) as described by the equation

$$R = \sum_{n=1}^{N} x_n \otimes y_n \otimes z_n + E \tag{12.3}$$

Each individual factor (n) models a part of the data cube and can be expressed as the outer product of the component vectors (x, y, and z). The sum of all factors combined with the error matrix (E) describes the entire data cube. The model works to minimize the contribution of E so as to best describe the data by the n factors. For example, with GC × GC data, the vectors x and y describe the two separation dimensions, with the vector z differing depending on detection. With GC × GC–FID, the z dimension generally comprises stacked sample chromatograms, and with GC × GC–TOF MS the z dimension is mass spectral information. Each factor n represents a separate analyte component within the data cube. The decomposed vectors are shown graphically in Figure 12.6, with the z dimension left blank for clarity, as it could be either an additional dimension of data or stacked samples. There are two main algorithms that are based on this method of deconvolution: the generalized rank annihilation method (GRAM) and parallel factor analysis (PARAFAC).

GRAM is used primarily with bilinear data that have been acquired by 2D separations coupled to a univariate detector such as an FID with GC (or single-wavelength absorbance detection with LC). The algorithm works through a direct trilinear decomposition to resolve and quantify overlapped peaks within the 2D data. The third dimension, the z vector, comes from the combination of two bilinear samples. Combining a single calibration standard of a known concentration with an unknown sample creates trilinear data for GRAM (Bruckner et al., 1998; Fraga et al., 2000a,b; Sinha et al., 2004c). This is an example of the second-order advantage, in which an analyte of interest can be deconvoluted from interfering analytes with only a single calibration standard (Fraga et al., 2001a,b; Johnson

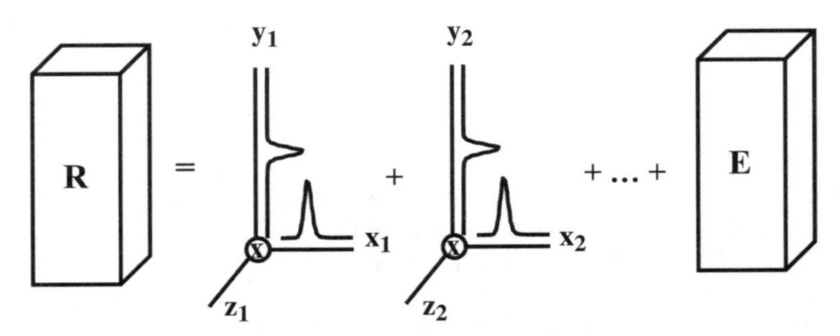

Figure 12.6 Matrix decomposition. PARAFAC and GRAM both work off the matrix decomposition shown graphically here.

et al., 2002). If the calibration standard and the unknown are both in the linear response range of the detector, the amount of unknown is readily determined. GRAM was one of the first examples of chemometrics used with GC × GC (Bruckner et al., 1998; Johnson et al., 2002; Sinha et al., 2003; Xie et al., 2003). GRAM has been shown to be more accurate and precise than quantification by traditional integration, because intensity from noise and overlapping peaks are removed, which enhances the signal/noise ratio for analytes of interest and lowers the overall limit of detection (Fraga et al., 2000a,b). Even without interfering peaks, GRAM provides an excellent approach to removing noise contributions and to quantify analytes more reliably. However, because samples have to be combined to create a data cube, GRAM can be quite sensitive to even minor shifts in retention time or peak shape (Bahowick and Synovec, 1995; Synovec et al., 2003). The standard addition method is frequently employed with GRAM to minimize run-to-run retention issues simultaneous with providing internal calibration for quantification.

Another solution to this issue, other than that of applying the standard addition method with rank alignment, is to acquire trilinear data without combining samples. Two-dimensional instrumentation coupled to a multivariate detector (such as MS) produces trilinear data with a single injection. This type of data can be analyzed with PARAFAC without requiring a reference standard. PARAFAC deconvolutes based on an iterative alternating least-squares approach (Bro, 1997). As in GRAM, the x and y vectors are the two separation dimensions, but the z vector is now the added third dimension of data from the instrument itself (Sinha et al., 2004b,d). It is also possible to use PARAFAC on the trilinear data acquired by combining multiple bilinear sample replicates (e.g., several GC × GC–FID runs), but this does not circumvent retention-time reproducibility issues. Utilizing the trilinearity produced by the additional dimension of data with PARAFAC is an example of the third-order advantage, which allows for the quantitative separation of an analyte in the presence of an interfering compound (i.e., deconvolution) with a single chromatogram and no standard (Booksh and Kowalski, 1994; Sinha et al., 2004a). PARAFAC has been applied to GC × GC–TOF MS data for many applications, providing reliable quantitative information of specific analytes within a complex sample matrix (Hoggard and Synovec, 2007; Hope et al., 2005; Humston et al., 2008, 2009a, 2010; Kramer, 1998; Mohler et al., 2006–2008; Sinha et al., 2004a,b; van Mispelaar et al., 2003). Like GRAM, PARAFAC provides noise reduction; thus, more reliable quantification can be achieved than with integration approaches. Early implementations of PARAFAC were done in a nonautomated fashion, as user input was required to determine the required number of factors, n. However, a PARAFAC graphical user interface (GUI) has been constructed by Hoggard et al. that can automatically select the appropriate n for a target analyte contained within a GC × GC–TOF MS data cube (Hoggard and Synovec, 2007). This has allowed for operation of the PARAFAC algorithm in a more automated fashion. The targeted approach to PARAFAC makes it possible to isolate a specific analyte of interest (often one that has been identified with feature selection) from the cube of data. For example, the targeted PARAFAC results for a single analyte isolated from overlapping analytes from the data shown in Figures 12.3 to 12.5 are shown in Figure 12.7. This report sheet provides information on peak signal volume, the match value to target mass spectrum, and on peak shape, to name a few. In Figures 12.3 to 12.5, the raw data in the middle section from 440 to 465 s, which was analyzed with PARAFAC, contained five analytes. In a targeted approach, only information for the target analyte [at retention time (455 s, 0.6 s)] is sought, so other analytes within the same window are not optimized, to save processing time. If information about all of the analytes is required, a nontarget PARAFAC approach can be taken in which all analytes within a set window are isolated from the matrix by deconvolution and optimized individually in a semiautomated fashion (Hoggard and Synovec, 2008). The same region of data was subjected to PARAFAC in a nontargeted approach, with the results shown in Figure 12.8. In this case, all five analytes within the window are isolated and identified via MS matching to library spectra, in a resolved form that can readily be quantified. Recently, the nontarget PARAFAC algorithm has been combined elegantly with classical least squares (CLS) to acquire absolute calibration information, with only a single injection in the presence of interfering analytes (Humston et al., 2009b). This approach is

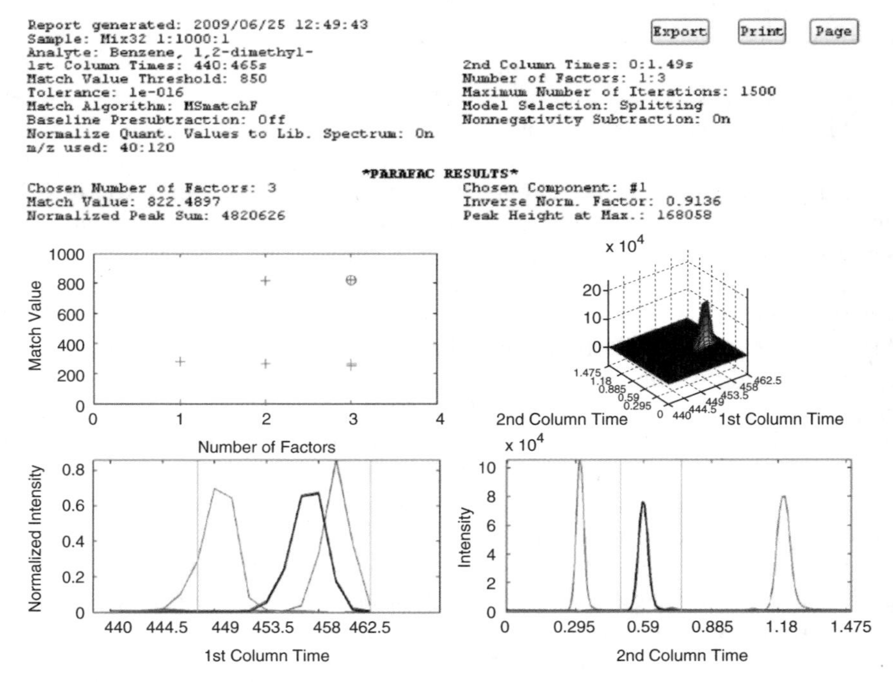

Figure 12.7 Target PARAFAC. The middle subset of the chromatographic data shown in Figures 12.3 to 12.5 is deconvoluted with PARAFAC. In the targeted approach, only the target analyte is optimized

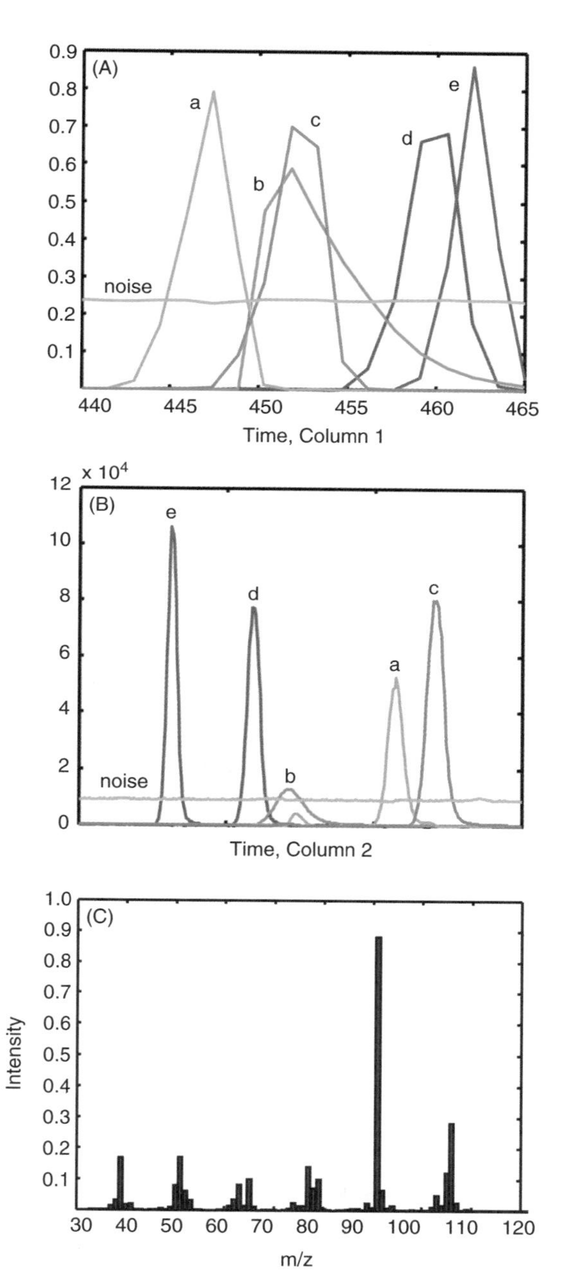

Figure 12.8 Nontarget PARAFAC. The middle subset of the chromatographic data shown in Figures 12.3 to 12.5 is deconvoluted with PARAFAC, here in a nontargeted approach. In this case, every analyte in the window is optimized for column 1 (A) and column 2 (B). The deconvoluted mass spectra for analyte b is shown in (C).

based on the addition of a known amount of ^{13}C-labeled standard that coelutes with the native ^{12}C version of the analyte. The combined peak profiles and mass spectra are isolated with nontarget PARAFAC. The resulting mass spectrum is a linear combination of the pure ^{12}C and ^{13}C reference spectra that can be separated mathematically to determine the relative amount of each. Since the ^{13}C concentration was a known amount, the initially unknown ^{12}C concentration is obtained. These quantification techniques allow for analysis without requiring high separation resolution while still acquiring excellent quantitative information, which can expand the scope of multidimensional chromatography to even more complex applications and/or speed up the separations by reducing the run time.

12.6 SUMMARY OF DATA INTERPRETATION TECHNOLOGIES

With the instrumentation advances made over the past few decades, it is now possible to turn a complex sample into complex raw multidimensional data. It is also possible to glean useful information about these complex samples from the raw data with recent data interpretation technology advances. Application of these data interpretation technologies (including chemometrics) allows for the analysis and interpretation of complex data sets. In the absence of these tools, many data analysis tasks are either tedious and prone to error or are not even possible. Although the field of chemometrics offers great promise for the future of multidimensional chromatography, its widespread applicability is still hindered by its limited availability and, in some cases, by the requirement of significant technical expertise by the user. Much of the algorithm development is still being done at an in-house level and is not always available commercially. Other areas for advancement potential are the further automation of these tools, as well as improving the speed of analysis. Still, the combination of these techniques has made possible the study of many interesting applications, and the future of separation science looks very bright as more chromatographers embrace the use of chemometrics.

REFERENCES

Adcock JL, Adams M, Mitrevski BS, Marriott PJ. *Anal. Chem*. 2009; 81:6797–6804.

Amador-Muñoz O, Marriott PJ. *J. Chromatogr. A* 2008; 1184:323–340.

Bahowick TJ, Synovec RE. *Anal. Chem*. 1995; 67:631–640.

Beebe KR, Pell RJ, Seasholtz MB. *Chemometrics: A Practical Guide*. New York: Wiley-Interscience, 1998.

Beens J, Boelens H, Tijssen R, Blomberg J. *J. High Resolut. Chromatogr*. 1998; 21:47–54.

Blomberg J, Schoenmakers PJ, Beens J, Tijssen R. *J. High Resolut. Chromatogr*. 1997; 20:539–544.

Booksh KS, Kowalski BR. *Anal. Chem*. 1994; 66: 782A–791A.

Brereton RG. *Chemometrics: Data Analysis for the Laboratory and Chemical Plant*. New York: Wiley, 2003.

Bro R. *Chemometr. Intell. Lab. Syst*. 1997; 38:149–171.

Bruckner CA, Prazen BJ, Synovec RE. *Anal. Chem*. 1998; 70:2796–2804.

Bushey MM, Jorgenson JW. *Anal. Chem*. 1990; 62:161–167.

Dallüge J, van Rijn M, Beens J, Vreuls RJ, Brinkman UATh. *J. Chromatogr. A* 2002a; 965:207–217.

Dallüge J, van Stee LLP, Xu X, Williams J, Beens J, Vreuls RJJ, Brinkman UATh. *J. Chromatogr. A* 2002b; 974:169–184.

Dallüge J, Vreuls RJ, Beens J, Brinkman UATh. *J. Sep. Sci*. 2002c; 25:201–214.

de Geus HJ, Aidos I, de Boer J, Luten JB, Brinkman UATh. *J. Chromatogr. A* 2001; 910:95–103.

Fraga CG, Prazen BJ, Synovec RE. *J. High Resolut. Chromatogr*. 2000a; 23:215–224.

Fraga CG, Prazen BJ, Synovec RE. *Anal. Chem*. 2000b; 72:4154–4162.

Fraga CG, Bruckner CA, Synovec RE. *Anal. Chem*. 2001a; 73:675–683.

Fraga CG, Prazen BJ, Synovec RE. *Anal. Chem*. 2001b; 73:5833–5840.

Frysinger GS, Gaines RB, Ledford EBJ. *J. High Resolut. Chromatogr*. 1999; 22:195–200.

Giddings JC. *J. Chromatogr. A* 1995; 703:3–15.

Hoggard JC. Synovec RE. *Anal. Chem*. 2007; 79:1611–1619.

Hoggard JC. Synovec RE. *Anal. Chem*. 2008; 80:6677–6688.

Hollingsworth BV, Reichenbach SE, Tao Q, Visvanathan A. *J. Chromatogr. A* 2006; 1105:51–58.

Hope JL, Prazen BJ, Nilsson EJ, Lidstrom ME, Synovec RE. *Talanta* 2005; 65:380–388.

Humston EM, Dombek KM, Hoggard JC, Young ET, Synovec RE. *Anal. Chem*. 2008; 80:8002–8011.

Humston EM, Zhang Y, Brabeck GF, McShea A, Synovec RE. *J. Sep. Sci*. 2009a; 32:2289–2295.

Humston EM, Hoggard JC, Synovec RE. *Anal. Chem*. 2009b; 82:41–43.

Humston EM, Knowles JK, McShea A, Synovec RE. *J. Chromatogr. A* 2010; 1217:1963–1970.

Hyötyläinen T, Kallio M, Hartonen K, Jussila M, Palonen S, Riekkola ML. *Anal. Chem*. 2002; 74:4441–4446.

Johnson KJ, Synovec RE. *Chemometr. Intell. Lab. Syst*. 2002; 60:225–237.

Johnson KJ, Prazen BJ, Olund RK, Synovec RE. *J. Sep. Sci*. 2002; 25:297–303.

Kallio M, Kivilompolo M, Varjo S, Jussila M, Hyötyläinen T. *J. Chromatogr. A* 2009; 1216:2923–2927.

Korytár P, Leonards PEG, de Boer J, Brinkman UATh. *J. Chromatogr. A* 2002; 958:203–218.

Kowalski BR. *J. Chem. Inf. Comput. Sci*. 1975; 15:201–203.

Kramer R. *Chemometric Techniques for Quantitative Analysis*. New York: Marcel Dekker, 1998.

Liu Z, Phillips JB. *J. Chromatogr. Sci*. 1991; 29: 227.

Marriott PJ, Kinghorn RM, Ong RCY, Morrison P, Haglund P, Harju M. *J. High Resolut. Chromatogr*. 2000; 23:253–258.

Mohler RE, Dombek KM, Hoggard JC, Young ET, Synovec RE. *Anal. Chem*. 2006; 78:2700–2709.

Mohler RE, Dombek KM, Hoggard JC, Pierce KM, Young ET, Synovec RE. *Analyst* 2007; 132:756–767.

Mohler RE, Tu BP, Dombek KM, Hoggard JC, Young ET, Synovec RE. *J. Chromatogr. A* 2008; 1186:401–411.

Mondello L, Casilli A, Tranchida PQ, Dugo G, Dugo P. *J. Chromatogr. A* 2005; 1067:235–243.

Nelson RK, Kile BM, Plata DL, Sylva SP, Xu L, Reddy CM, Gaines RB, Frysinger GS, Reichenbach SE. *Environ. Forens*. 2006; 7:33–44.

Ong RCY, Marriott P. *J. Chromatogr. Sci*. 2002; 40:276–291.

Peters S, Vivó-Truyols G, Marriott PJ, Schoenmakers PJ. *J. Chromatogr. A* 2007; 1156:14–24.

Pierce KM, Hope JL, Hoggard JC, Synovec RE. *Talanta* 2006a; 70:797–804.

Pierce KM, Hoggard JC, Hope JL, Rainey PM, Hoofnagle AN, Jack RM, Wright BW, Synovec RE. *Anal. Chem*. 2006b; 78:5068–5075.

Pierce KM, Hoggard JC, Mohler RE, Synovec RE. *J. Chromatogr. A* 2008; 1184:341–352.

Reichenbach SE, Ni M, Zhang D, Ledford EB. *J. Chromatogr. A* 2003; 985:47–56.

Reichenbach SE, Ni M, Kottapalli V, Visvanathan A. *Chemometr. Intell. Lab. Syst*. 2004; 71:107–120.

Reichenbach SE, Kottapalli V, Ni M, Visvanathan A. *J. Chromatogr. A* 2005; 1071:263–269.

Sharaf MA, Illman DL, Kowalski BR. *Chemometrics*. New York: Wiley, 1986.

Shellie RA, Xie LL, Marriott PJ. *J. Chromatogr. A* 2002; 968:161–170.

Shellie R, Welthagen W, Zrostlíková J, Spranger J, Ristow M, Fiehn O, Zimmermann R. *J. Chromatogr. A* 2005; 1086:83–90.

Sinha AE, Johnson KJ, Prazen BJ, Lucas SV, Fraga CG, Synovec RE. *J. Chromatogr. A* 2003; 983:195–204.

Sinha AE, Fraga CG, Prazen BJ, Synovec RE. *J. Chromatogr. A* 2004a; 1027:269–277.

Sinha AE, Hope JL, Prazen BJ, Nilsson EJ, Jack RM, Synovec RE. *J. Chromatogr. A* 2004b; 1058:209–215.

Sinha AE, Prazen BJ, Synovec RE. *Anal. Bioanal. Chem*. 2004c; 378:1948–1951.

Sinha AE, Hope JL, Prazen BJ, Fraga CG, Nilsson EJ, Synovec RE. *J. Chromatogr. A* 2004d; 1056:145–154.

Skov T, Hoggard JC, Bro R, Synovec RE. *J. Chromatogr. A* 2009; 1216: 4020–4029.

Song SM, Marriott P, Wynne P. *J. Chromatogr. A* 2004; 1058:223–232.

Synovec RE, Prazen BJ, Johnson KJ, Fraga CG, Bruckner CA. In: Brown PR, Grushka E, Eds., *Advances in Chromatography*. New York: Marcel Dekker, 2003, pp. 1–42.

Tran TC, Marriott PJ. *Atmos. Environ*. 2007; 41:5756–5768.

Tranchida P, Giannino A, Mondello M, Sciarrone D, Dugo P, Dugo G, Mondello L. *J. Sep. Sci*. 2008; 31:1797–1802.

U.S. National Institute of Standards and Technology, MS Search Program, Gaithersburg, MD, 2002.

van Mispelaar VG, Tas AC, Smilde AK, Schoenmakers PJ, van Asten AC. *J. Chromatogr. A* 2003; 1019:15–29.

van Mispelaar VG, Smilde AK, de Noord OE, Blomberg J, Schoenmakers PJ. *J. Chromatogr. A* 2005; 1096:156–164.

Vivó-Truyols G, Janssen HG. *J. Chromatogr. A* 2010; 1217:1375–1385.

Xie L, Marriott PJ, Adams M. *Anal. Chim. Acta* 2003; 500:211–222.

INDEX

Comprehensive Chromatography in Combination with Mass Spectrometry, First Edition.
Edited by Luigi Mondello.

V

W

X

Z

WILEY-INTERSCIENCE SERIES IN MASS SPECTROMETRY

Series Editors

Dominic M. Desiderio
Departments of Neurology and Biochemistry
University of Tennessee Health Science Center

Nico M. M. Nibbering
Vrije Universiteit Amsterdam, The Netherlands

John R. de Laeter • *Applications of Inorganic Mass Spectrometry*
Michael Kinter and Nicholas E. Sherman • *Protein Sequencing and Identification Using Tandem Mass Spectrometry*
Chhabil Dass • *Principles and Practice of Biological Mass Spectrometry*
Mike S. Lee • *LC/MS Applications in Drug Development*
Jerzy Silberring and Rolf Eckman • *Mass Spectrometry and Hyphenated Techniques in Neuropeptide Research*
J. Wayne Rabalais • *Principles and Applications of Ion Scattering Spectrometry: Surface Chemical and Structural Analysis*
Mahmoud Hamdan and Pier Giorgio Righetti • *Proteomics Today: Protein Assessment and Biomarkers Using Mass Spectrometry, 2D Electrophoresis, and Microarray Technology*
Igor A. Kaltashov and Stephen J. Eyles • *Mass Spectrometry in Biophysics: Confirmation and Dynamics of Biomolecules*
Isabella Dalle-Donne, Andrea Scaloni, and D. Allan Butterfield • *Redox Proteomics: From Protein Modifications to Cellular Dysfunction and Diseases*
Silas G. Villas-Boas, Ute Roessner, Michael A.E. Hansen, Jorn Smedsgaard, and Jens Nielsen • *Metabolome Analysis: An Introduction*
Mahmoud H. Hamdan • *Cancer Biomarkers: Analytical Techniques for Discovery*
Chabbil Dass • *Fundamentals of Contemporary Mass Spectrometry*
Kevin M. Downard (Editor) • *Mass Spectrometry of Protein Interactions*
Nobuhiro Takahashi and Toshiaki Isobe • *Proteomic Biology Using LC-MS: Large Scale Analysis of Cellular Dynamics and Function*
Agnieszka Kraj and Jerzy Silberring (Editors) • *Proteomics: Introduction to Methods and Applications*
Ganesh Kumar Agrawal and Randeep Rakwal (Editors) • *Plant Proteomics: Technologies, Strategies, and Applications*
Rolf Ekman, Jerzy Silberring, Ann M. Westman-Brinkmalm, and Agnieszka Kraj (Editors) • *Mass Spectrometry: Instrumentation, Interpretation, and Applications*
Christoph A. Schalley and Andreas Springer • *Mass Spectrometry and Gas-Phase Chemistry of Non-Covalent Complexes*
Riccardo Flamini and Pietro Traldi • *Mass Spectrometry in Grape and Wine Chemistry*
Mario Thevis • *Mass Spectrometry in Sports Drug Testing: Characterization of Prohibited Substances and Doping Control Analytical Assays*
Sara Castiglioni, Ettore Zuccato, and Roberto Fanelli • *Illicit Drugs in the Environment: Occurrence, Analysis, and Fate Using Mass Spectrometry*
Ángel Garciá and Yotis A. Senis (Editors) • *Platelet Proteomics: Principles, Analysis, and Applications*
Luigi Mondello • *Comprehensive Chromatography in Combination with Mass Spectrometry*